Essentials of
USING AND UNDERSTANDING
MATHEMATICS

Essentials of USING AND UNDERSTANDING MATHEMATICS
A QUANTITATIVE REASONING APPROACH

INSTRUCTOR'S EDITION

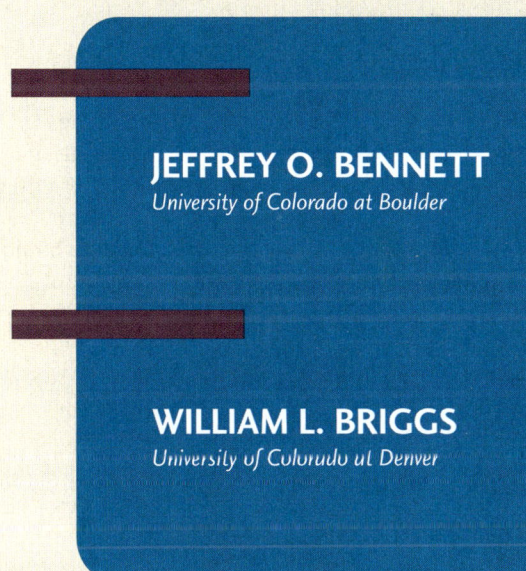

JEFFREY O. BENNETT
University of Colorado at Boulder

WILLIAM L. BRIGGS
University of Colorado at Denver

Boston San Francisco New York
London Toronto Sydney Tokyo Singapore Madrid
Mexico City Munich Paris Cape Town Hong Kong Montreal

Publisher: Greg Tobin
Sponsoring Editor: Anne Kelly
Project Editor: Rachel S. Reeve
Editorial Assistant: Jaime Bailey
Senior Production Supervisor: Karen Wernholm
Production Services: Lifland et al., Bookmakers
Marketing Manager: Becky Anderson
Marketing Coordinator: Julia Coen
Senior Software Project Manager/InterAct Math: Alicia Anderson
Senior Software Project Supervisor/TestGen-EQ: David Malone
Manufacturing Buyer: Evelyn Beaton
Senior Designer: Barbara T. Atkinson
Text Design: Geri Davis, The Davis Group Inc.
Cover Design: Leslie Haimes
Compositor: WestWords, Inc.
Illustrations: Precision Graphics
Cover Fractal: Anne Burns

For permission to use copyrighted material, grateful acknowledgment is made to the copyright holders on pages 455 and 456, which are hereby made part of this copyright page.

Library of Congress Cataloging-in-Publication Data

Bennett, Jeffrey O.
 Essentials of using and understanding mathematics : a quantitative reasoning approach / Jeffrey O. Bennett, William L. Briggs.
 p. cm.
 Includes bibliographical references and index.
 ISBN 0-201-79387-3
 1. Mathematics. I. Briggs, William L. II. Title.

QA39.2 .B47655 2002
510—dc21

2001056073

ISBN (Instructor's Edition): 0-201-79388-1

Copyright © 2003 Pearson Education, Inc. All rights reserved. No part of this publication may be reproduced, stored in a retrieval system, or transmitted, in any form or by any means, electronic, mechanical, photocopying, recording, or otherwise, without the prior written permission of the publisher. Printed in the United States of America.

1 2 3 4 5 6 7 8 9 10—VH—05 04 03 02

This book is dedicated to everyone who wants a better understanding of our world, and especially to those who have struggled with mathematics in the past. We hope this book will help you achieve your goals.

And it is dedicated to those who make our own lives brighter, especially Lisa, Julie, Katie, Grant, and Brooke.

CONTENTS

Preface xi

Instructor's Material xxi

Prologue: Literacy for the Modern World 1

Part 1 Logic and Problem Solving

1 THINKING CRITICALLY 13

1A Recognizing Fallacies 14
1B Propositions and Truth Values 21
1C Sets and Venn Diagrams 35
 A Brief Review: Sets of Numbers 36
1D Critical Thinking in Everyday Life 54

2 APPROACHES TO PROBLEM SOLVING 69

2A The Problem-Solving Power of Units 70
 A Brief Review: Working with Fractions 72
2B Standardized Units: More Problem-Solving Power 87
 A Brief Review: Powers of 10 91
2C Problem-Solving Guidelines and Hints 101

Part 2 Quantitative Information in Everyday Life

3 NUMBERS IN THE REAL WORLD 117

3A Uses and Abuses of Percentages 118
 A Brief Review: Percentages 119
 A Brief Review: What Is a Ratio? 125
3B Putting Numbers in Perspective 136
 A Brief Review: Working with Scientific Notation 137

CONTENTS

 3C Dealing with Uncertainty 151
 A Brief Review: Rounding 153
 3D How Numbers Deceive: Polygraphs, Mammograms, and More 164

4 FINANCIAL MANAGEMENT 177

 4A The Power of Compounding 178
 A Brief Review: Three Basic Rules of Algebra 189
 4B Savings Plans 194
 A Brief Review: Algebra with Powers and Roots 200
 4C Loan Payments, Credit Cards, and Mortgages 205

Part 3 *Probability and Statistics*

5 STATISTICAL REASONING 223

 5A Fundamentals of Statistics 224
 5B Should You Believe a Statistical Study? 239
 5C Statistical Tables and Graphs 249
 5D Graphics in the Media 261
 5E Correlation and Causality 276
 5F Characterizing a Data Distribution 290

6 PROBABILITY: LIVING WITH THE ODDS 303

 6A Fundamentals of Probability 304
 A Brief Review: The Multiplication Principle 307
 6B Combining Probabilities 316
 6C The Law of Averages 326
 6D Counting and Probability 337
 A Brief Review: Factorials 340

Part 4 *Modeling*

7 EXPONENTIAL ASTONISHMENT 351

 7A Growth: Linear Versus Exponential 352
 7B Doubling Time and Half-Life 360
 A Brief Review: Logarithms 367
 7C Exponential Modeling 371
 A Brief Review: Algebra with Logarithms 373

Part 5 Further Applications

8 MATHEMATICS AND THE ARTS 387
- 8A Mathematics and Music 388
- 8B Perspective and Symmetry 395
- 8C Proportion and the Golden Ratio 409

9 MATHEMATICS AND POLITICS 419
- 9A Voting: Does the Majority Always Rule? 420
- 9B Apportionment: The House of Representatives and Beyond 434

SUGGESTED READINGS 453
CREDITS 455
ANSWERS A-1
INDEX I-1

PREFACE

> *Human history becomes more and more a race between education and catastrophe.*
>
> —H. G. Wells, *The Outline of History,* 1920

TO THE STUDENT

There is no escaping the importance of mathematics in the modern world. However, for most people, the importance of mathematics lies not in its abstract ideas, but in its application to personal and social issues. This book is designed with such practical considerations in mind. In particular, we've designed this book with three specific purposes:

- To prepare you for the mathematics you will encounter in other **college** courses, particularly core courses in social and natural sciences.
- To develop your ability to reason with quantitative information in a way that will help you achieve success in your **career.**
- To provide you with the critical thinking and quantitative reasoning skills needed to understand major issues in **life.**

We hope this book will be useful to everyone, but it is designed primarily for those of you who are not planning to major in a field that requires advanced mathematical skills. In particular, if you've ever felt any fear or anxiety about mathematics, we've written this book with you in mind. Perhaps you will discover that mathematics is much more important and relevant to your life than you guessed.

Wherever your interests lie—social sciences, environmental issues, politics, business and economics, art and music, or any of many other topics—you will find relevant examples in this book. But the most important idea to take away from this book is that mathematics can help you understand a variety of topics and issues. Once you have completed your study of this book, you should be prepared to understand most quantitative issues that you will encounter.

TO THE INSTRUCTOR

This book represents a new approach to mathematics for general education students—that is, students who are planning to major in nonmathematical fields, such as liberal arts, fine arts, business, and education. Most students have never been taught how to combine critical thinking and mathematical skills in a way

that applies to their lives. In the jargon adopted by the MAA, AMATYC, and other mathematical organizations, students need to learn **quantitative reasoning.** This book is designed to help you succeed in teaching quantitative reasoning to your students.

THE CHALLENGE: WINNING OVER YOUR STUDENTS

Perhaps the greatest challenge in teaching mathematics to general education students lies in winning them over—that is, convincing them that you have something useful to teach them. This challenge arises because, by the time they reach college, many students have come to dislike or fear mathematics. Indeed, the vast majority of students in general education mathematics courses are there not by choice, but because such courses are required for graduation. Reaching your students therefore requires not only that you teach them useful material, but that you first get them to listen to you long enough and carefully enough so that they'll be able to see its usefulness.

We've built this book around two important strategies that are designed to help you win students over:

1. Confront negative attitudes about mathematics head on, showing students that their fear or loathing is ungrounded and that mathematics actually is relevant to their lives. This strategy is embodied in the Prologue of this book (pages 1–12), which we urge you to emphasize in class. It continues implicitly throughout the rest of the text.
2. Focus on goals that are meaningful to students—namely, on the goals of learning mathematics for *college, career,* and *life.* Your students will then learn mathematics because they will see how it affects their lives.

THE KEY TO SUCCESS: A CONTEXT-DRIVEN APPROACH

Broadly speaking, approaches to teaching mathematics can be divided into two categories:

- A *content-driven* approach is organized by mathematical ideas. After each mathematical topic is presented, examples of its applications are shown.
- A *context-driven* approach is organized by practical contexts. Applications are the drivers, and mathematical ideas are presented as needed for the applications.

The same content can be covered through either approach, but the context-driven approach has an enormous advantage for general education students: It shows them directly how mathematics is important to college, career, and life. In contrast, the content-driven approach tends to come across as "learn this content because it's good for you," causing many students to tune out before you ever reach

the practical applications. For more details, see our article "General Education Mathematics: New Approaches for a New Millennium" (*AMATYC Review,* Fall 1999).

ALTERNATIVE VERSIONS OF THE BOOK

This book has been under development since 1987, when we first wrote and used a set of class notes for a new quantitative reasoning course at the University of Colorado, Boulder. The notes evolved from semester to semester. Early on, we also developed and tested many new ideas with the help of a grant from the National Science Foundation's program in Undergraduate Course and Curriculum Development (NSF Grant USE-915060). Once we decided to publish our work, we were faced with the challenge of turning our class notes into a book that would be useful to a wide range of instructors. Based on feedback from many instructors and reviewers over the years, we refined the book through numerous editions, including a Class Test Edition (1995), a Preliminary Edition (1996), and the First Edition (1999). Through this process, we have learned that quantitative reasoning courses are quite diverse, and some instructors require more material than others. As a result, we have decided to offer the Second Edition in two alternative versions:

- *Using and Understanding Mathematics: A Quantitative Reasoning Approach, Second Edition* is the full version of the book. It has enough material to fit most courses in liberal arts mathematics, for either one or two semesters.
- *Essentials of Using and Understanding Mathematics: A Quantitative Reasoning Approach* is a condensed version of the full book. It is designed for those instructors who seek only the "core" material for their courses.

By comparing the tables of contents of the two versions, you should be able to determine which version best meets your needs as an instructor.

MODULAR STRUCTURE OF THE BOOK

We'd all love to have a year or more to teach mathematics to our general education students. Unfortunately, most schools have only a one-quarter or one-semester mathematics requirement, so we can cover only a fraction of the material we'd cover in an ideal course. This book is therefore organized with a modular structure that allows you to create a course to meet your particular concerns and constraints. The chapters are organized broadly by contextual areas. Each chapter, in turn, is divided into a set of self-contained *units* that focus on particular concepts or applications. In most cases, you can cover chapters in any order; and while the units within each chapter build sequentially in terms of sophistication, in many cases you can skip certain units, particularly those toward the end of the chapter.

PREREQUISITE MATHEMATICAL BACKGROUND

Because of its modular structure, this book can be used by students with a wide range of mathematical backgrounds. Many of the units require nothing more than arithmetic and a willingness to think about quantitative issues in new ways. Other

units use techniques of algebra or geometry, which are reviewed as they arise. Thus, this book should be accessible to any student who has completed two or more years of high school mathematics. Note, however, that this book is *not* remedial: While much of the book relies on mathematical techniques from secondary school, the techniques always arise in the context of issues and applications that require students to demonstrate a high level of critical thinking.

PEDAGOGICAL FEATURES

Besides the main narrative of the text, this book includes the following features, each designed with a specific pedagogical purpose in mind.

Chapter Overview. Appearing at the beginning of each chapter, this feature offers a unit-by-unit overview of the chapter. It is designed both to show students how the chapter is organized and to help instructors decide which units to cover in class.

Time Out to Think. Appearing throughout the book, the *Time Out to Think* features pose short conceptual questions designed to help students reflect on important new ideas. They also serve as excellent starting points for classroom discussions.

Summary Boxes. Flowing right along with the narrative, the book contains boxes that summarize key ideas, definitions, or formulas.

Examples and Case Studies. Numbered examples are designed to build understanding and to offer practice with the types of questions that appear in the exercises. Occasional case studies go into somewhat more depth than the numbered examples.

Margin Features. This book includes a wide margin so that students have room to make notes while studying. The margin also contains material that should spur student interest but is peripheral to the main discussion. This material comes in three basic forms:

- **By the Way** features contain interesting notes and asides relevant to the topic at hand.
- **Historical Note** remarks give historical context to the ideas presented in the chapter.
- **Technical Notes** contain details that are important mathematically, but generally do not affect students' understanding of the material.

A Brief Review. These boxed features review key mathematical skills that students should have learned previously, but in which many students still need review and practice. They appear in the book wherever a particular skill is first needed, and exercises based on the review boxes can be found at the end of the unit.

Practical Matters. These boxed features focus on topics of particularly high practical importance, such as how to watch out for bad deals on telephone service, how to avoid credit card trouble, and how to save money and energy at the same time.

Thinking About. These boxed features build upon ideas in the main narrative but go somewhat beyond the level of other material in the book. The box on Zeno's paradox is an example.

Chapter Summary. Appearing at the end of each chapter, this feature offers a detailed outline of the chapter that students can use as a study guide.

ASSESSMENT TOOLS

Each unit ends with questions designed for student assessment, organized into the following sections:

Review Questions. These questions ask students to summarize the important ideas covered in the unit. They are designed primarily for self-study and generally can be answered simply by reviewing the text.

In the News. These questions ask students to find examples from the news or from daily life of the ideas presented in the unit. These questions may be assigned as homework or used for class discussions.

Exercises. These questions offer practice with the concepts covered in the unit. The exercises can be used for homework assignments or for self-study (answers to most odd-numbered exercises appear in the back of the book).

Web Projects. The Web Projects require searching for data or other information on the Web. They can be used for extended projects, discussion, group activities, or essays.

SUPPLEMENTS

SUPPLEMENTS FOR THE INSTRUCTOR

Instructor's Edition
ISBN 0-201-79388-1
 This version of the text includes answers to all of the problems and a chapter-by-chapter guide, as well as sample course outlines.

Instructor's Guide and Solutions Manual
ISBN 0-321-10905-8
 This manual contains model curricula, unit-by-unit teaching hints, answers or discussion points for all the *Time Out to Think* questions, and detailed, worked out solutions to all of the exercises in the text.

Instructor's Testing Manual
ISBN 0-321-10906-6
 The Testing Manual includes three alternative tests per chapter. These items may be used as actual tests or as references for creating tests.

TestGen-EQ with QuizMaster-EQ
ISBN 0-201-79390-3
 TestGen-EQ is a computerized test generator with problems organized specifically for this textbook. Its user-friendly graphical interface enables instructors to

select, view, edit, and add test items, then print tests in a variety of fonts and forms. A built-in question editor gives the user the power to create graphs, import graphics, insert mathematical symbols and templates, and insert variable numbers or text. An "Export to HTML" feature lets instructors create practice tests that can be posted to a Web site. Tests created with TestGen-EQ can also be displayed on Web pages when the free TestGen-EQ plug-in has been installed with Internet Explorer or Netscape Navigator. Tests created with TestGen-EQ can be used with QuizMaster-EQ, which enables students to take exams on a computer network. QuizMaster-EQ automatically grades the exams, stores results on disk, and allows the instructor to view or print a variety of reports for individual students, classes, or courses. Contact your Addison-Wesley Sales Consultant.

Videotapes
ISBN 0-321-10908-2

Videotapes correlated to each important topic are available to departments. Contact your Addison-Wesley Sales Consultant.

MyMathLab.com
www.mymathlab.com

MyMathLab.com is a complete, on-line course for Addison-Wesley mathematics textbooks that integrates interactive, multimedia instruction correlated to the textbook content. MyMathLab can be easily customized to suit the needs of students and instructors and provides a comprehensive and efficient on-line course-management system that allows for diagnosis, assessment, and tracking of students' progress.

SUPPLEMENTS FOR THE STUDENT

Student's Study Guide and Solutions Manual
ISBN 0-321-10907-4

This manual contains hints for learning and studying as well as detailed, worked-out solutions to all the odd-numbered unit exercises. Students will find this manual very helpful.

AW Math Tutor Center
http://www.aw.com/tutorcenter

The AW Math Tutor Center is staffed by qualified mathematics instructors who provide students with tutoring on text examples, exercises, and problems. Tutoring is provided by telephone, fax, and email. For qualified adopters only.

Digital Video Tutor
ISBN 0-321-11621-6

The Digital Video Tutor provides our videos on CD-ROM, making it easy and convenient for students to watch video segments from a computer at home or on campus. This complete video set is ideal for distance learning or extra instruction.

InterAct Math® Tutorials
www.InterActMath.com
 This interactive tutorial Web site provides algorithmically generated practice exercises. A detailed worked-out example and guided solution accompany each practice exercise.

WEB SITE
http://www.aw.com/bennett-briggs
 The Web site contains additional resources for instructors and students, such as PowerPoint slides, links for the Web Projects, quantitative issues from In the News, and More Brief Review using InterAct Math online.

ACKNOWLEDGMENTS

A textbook may carry author names, but it is the result of hard work by a long list of committed individuals. This book has been under development for more than 15 years, and even its beginnings were a group effort, as one of the authors was a member of a committee at the University of Colorado that worked to establish one of the nation's first courses in "quantitative reasoning." Since that beginning, the book has benefited from input and feedback from many dozens of faculty members and hundreds of students.

 There are too many people to thank for us to name them all, but first and foremost we extend our thanks to Bill Poole, whose faith in this project allowed it to grow from typewritten class notes into a true textbook. We'd also like to thank the other members of our outstanding publishing team, including Greg Tobin, Anne Kelly, Rachel Reeve, Becky Anderson, Carter Fenton, Michael Boezi, Karen Wernholm, and Peggy McMahon. Special thanks to Elka Block and Roberta Lewis, for their superb work as development editors for the first and second editions, respectively. Special thanks also to Sally Lifland, for her outstanding efforts in producing this book. We also thank Cherilynn Morrow, who helped frame the concept for the book and worked as a co-author during its early development.

 We'd like to thank the following people for their help:

 Lou Barnes, *Boulder West Financial Services*
 Carol Bellisio, *Monmouth University*
 Bob Bernhardt, *East Carolina University*
 Terence R. Blows, *Northern Arizona University*
 W. Wayne Bosché, Jr., *Dalton College*
 Michael Bradshaw, *Caldwell Community College and Technical Institute*
 W. E. Briggs, *University of Colorado, Boulder*
 Margaret Cibes, *Trinity College*
 Walter Czarnec, *Framingham State College*
 Adrian Daigle, *University of Colorado, Boulder*
 Andrew J. Dane, *Angelo State University*
 Greg Dietrich, *Florida Community College at Jacksonville*

Marsha J. Driskill, *Aims Community College*
John Emert, *Ball State University*
Kathy Eppler, *Salt Lake Community College*
Kellie Evans, *York College of Pennsylvania*
Fred Feldon, *Coastline Community College*
David E. Flesner, *Gettysburg College*
Pat Foard, *South Plains College*
Shane Goodwin, *Brigham Young University–Idaho*
Barbara Grover, *Salt Lake Community College*
Louise Hainline, *Brooklyn College*
Ward Heilman, *Bridgewater State University*
Peg Hovde, *Grossmont College*
Andrew Hugine, *South Carolina State University*
Lynn R. Hun, *Dixie College*
Hal Huntsman, *University of Colorado, Boulder*
Joel Irish, *University of Southern Maine*
David Jabon, *DePaul University*
Melvin F. Janowitz, *University of Massachusetts, Amherst*
Craig Johnson, *Brigham Young University–Idaho*
Vijay S. Joshi, *Virginia Intermont College*
Anton Kaul, *University of South Florida*
Bonnie Kelly, *University of South Carolina*
William Kiley, *George Mason University*
Jim Koehler, *University of Colorado, Denver*
Robert Kuenzi, *University of Wisconsin, Oshkosh*
Deann Leoni, *Edmonds Community College*
Linda Lester, *Wright State University*
Paul Lorczak, *MathSoft, Inc.*
Jay Malmstrom, *Oklahoma City Community College*
Erich McAlister, *University of Colorado, Boulder*
Patricia McNicholas, *Robert Morris College*
Elaine Spendlove Merrill, *Brigham Young University–Hawaii*
Colm Mulcahy, *Spelman College*
Paul O'Heron, *Broome Community College*
Mary K. Patton, *University of Illinois at Springfield*
Frank Pecchioni, *Jefferson Community College*
Evelyn Pupplo-Cody, *Marshall University*
Scott Reed, *College of Lake County*

Anne Roberts, *University of Utah*
Lori Rosenthal, *Austin Community College*
Hugo Rossi, *University of Utah*
Doris Schraeder, *McLennan Community College*
Dee Dee Shaulis, *University of Colorado, Boulder*
Judith Silver, *Marshall University*
Alu Srinivasan, *Temple University*
John Supra, *University of Colorado, Boulder*
Scott Surgent, *Arizona State University*
Timothy C. Swyter, *Frederick Community College*
David Theobald, *University of Colorado, Boulder*
Terry Tolle, *Southwestern Community College*
Christina Vertullo, *Marist College*
Pam Wahl, *Middlesex Community College*
Thomas Wangler, *Benedictine University*
Richard Watkins, *Tidewater Community College*
Emily Whaley, *DeKalb College*
David Wilson, *University of Colorado, Boulder*
Robert Woods, *Broome Community College*
Fred Worth, *Henderson State University*
Margaret Yoder, *Eastern Kentucky University*
Fredric Zerla, *University of Southern Florida*
Donald J. Zielke, *Concordia Lutheran College*

Jeffrey O. Bennett
William L. Briggs

INSTRUCTOR'S MATERIAL

You are reading the Instructor's Edition (IE) of *Essentials of Using and Understanding Mathematics*. This IE differs from the student edition only in that it contains this extra section of resources for instructors and, in the back of the book, all of the answers to the exercises (as opposed to answers to the odd-numbered exercises only). The Instructor's Material contained here is an abridged version of what is found in the *Instructor's Guide and Solutions Manual* (IGSM, 0-321-10905-8):

Using This Book—a brief overview of the book's purpose and structure …………………………………………………………………………… p. xxii

A Handout on Homework Quality—a set of guidelines on how homework should be presented when it is turned in (which you may freely photocopy for your students) ……………………………………… IGSM

Activity—instructions and a handout (which you may freely photocopy) for a suggested activity for the first day of class …………………… IGSM

Sample Course Outlines—suggestions on which units to cover for courses of various emphases and durations …………………………… p. xxv

Suggestions on Teaching Strategies—a few suggestions on teaching strategies useful in a general education mathematics course …………… IGSM

Chapter-by-Chapter Guide—an overview of teaching issues related to each chapter, including prerequisites and teaching hints for each unit …………………………………………………………………… IGSM

Time Out to Think—answers or discussion points for the *Time Out to Think* questions in the book ………………………………………… IGSM

USING THIS BOOK

This book is designed to be useful in a variety of general education mathematics courses, including but not limited to the following:

- Core mathematics courses satisfying requirements for liberal arts or fine arts students at two- and four-year colleges and universities
- General mathematics courses for business students at two- and four-year colleges and universities
- General mathematics courses for pre-service and in-service teachers at two- and four-year colleges and universities
- Senior-level mathematics courses for high school students

A QUANTITATIVE REASONING APPROACH

The subtitle of this book advertises "A Quantitative Reasoning Approach." But what exactly is this?

The idea of "quantitative reasoning" goes back at least to the early 1980s (possibly much earlier), when it was raised as a catch-all for the kinds of mathematical skills needed by all Americans. It has since been defined in greater detail by several different groups, including

- the Mathematical Association of America (MAA), in its report *Quantitative Reasoning for College Graduates: A Complement to the Standards* (1995)
- the American Mathematical Association for Two-Year Colleges (AMATYC), in its report *Crossroads in Mathematics: Standards for Introductory College Mathematics before Calculus* (D. Cohen, ed., 1995)
- the College Board, in its book *Why Numbers Count: Quantitative Literacy for Tomorrow's America* (Lynn Arthur Steen, ed., 1997)
- the National Council on Education in the Disciplines, in its book *Mathematics and Democracy: The Case for Quantitative Literacy* (Lynn Arthur Steen, ed., 2001).

Our own definition of quantitative reasoning largely overlaps these definitions, but originally grew out of discussions during 1987 by an interdisciplinary faculty committee at the University of Colorado, Boulder (author Jeffrey Bennett was an advisory member of the committee). This committee recommended the creation of a new type of college mathematics course, which was instituted at Boulder in 1988 under the direction of author Bennett. Our current definition of quantitative reasoning emphasizes the following four broad content areas:

1. **Logic, Critical Thinking, and Problem Solving.** Students should learn skills that will enable them to construct a logical argument based on rules of inference and to develop strategies for solving quantitative problems.
2. **Number Sense and Estimation.** Students should become "numerate," or able to make sense of the numbers that confront them in the modern world. For example, students should be able to give meaning to the term *a billion dollars* and distinguish it from *a million dollars* or *a trillion dol-*

lars. Developing number sense involves being able to make simple calculations and estimates to put numbers in perspective. Other crucial aspects of number sense include being able to interpret the many different uses and abuses of percentages in the media, understanding the role of uncertainty in most real problems, and comprehending issues of personal finance.

3. **Statistical Interpretation and Basic Probability.** Reports about statistical research are ubiquitous in the news. Students must have the tools needed to *interpret* this research. In particular, they should understand the basic ideas behind statistical research, how to interpret statistical graphics, and the difference between correlation and causality. At a slightly higher level, they should know the difference among various measures of central tendency (e.g., mean, median, mode), the meaning of variation, the importance of the normal distribution, and basic ideas of statistical inference. Probability is important both because of its use in statistical inference and because of its involvement in so many topics of relevance to most Americans, including lotteries, casino gambling, and risk assessment.

4. **Interpreting Mathematical Models.** Many major issues today are studied through mathematical models; for example, economic models underlie much of the political wrangling over budgetary and tax policies, while climate models form the basis for controversies over potential impacts of global warming. While the creation of such models generally requires much more advanced mathematics than we can cover in general education courses, we can easily teach students how to interpret what they read or hear about models. The skills required for interpreting models include knowing the difference between linear and exponential growth, understanding the concept of a function and its domain of applicability, and knowing how to put together a simple linear or exponential function.

Core Quantitative Reasoning Content: Chapters 1–7

This textbook is designed to help instructors teach the four content areas described above. The first seven chapters of this book parallel these content areas quite closely:

- Chapters 1 and 2 emphasize content area (1). Chapter 1 focuses on logic and critical thinking, while Chapter 2 builds problem-solving skills.
- Chapters 3 and 4 emphasize content area (2). Chapter 3 covers basic ideas of number sense and estimation, while Chapter 4 focuses on finance.
- Chapters 5 and 6 emphasize content area (3). Chapter 5 covers basic concepts needed for interpreting statistical research, and Chapter 6 discusses probability.
- Chapter 7 emphasizes content area (4), focusing on the difference between linear and exponential growth and the use of exponential models.

Further Applications: Chapters 8 and 9

The final two chapters of the book go somewhat beyond the core topics of quantitative reasoning to give students a greater

appreciation for the role that these topics play in modern society. In particular,

- Chapter 8 emphasizes the connections between mathematics and the arts, which should be of special interest to fine arts majors.
- Chapter 9 emphasizes the connections between mathematics and politics, with units on voting methods and apportionment.

PREREQUISITE MATHEMATICAL BACKGROUND

This book is designed primarily as a college-level text, so we generally assume that students have taken at least one course in high school algebra. Note, however, that the book includes *Brief Review* boxes to remind students of key mathematical skills as they arise, so it should be possible for students to cover this text even if they enter with very weak mathematical skills. In addition, because of the book's modular structure (see below), it is possible for instructors to pick and choose among topics in accordance with the skill level of students.

MODULAR STRUCTURE OF THE BOOK

This book has a modular structure, designed to allow maximum flexibility in creating a course that meets your particular needs. The chapters are organized broadly by general topics. Each chapter, in turn, is divided into units that focus on particular issues or applications. Because the units are largely self-contained, you can pick and choose among them in designing a course. The chapter-by-chapter guide in the IGSM lists the prerequisites, if any, for each unit.

PEDAGOGICAL FEATURES

Besides the main narrative of the text, this book contains several pedagogical features designed to enhance student learning. You can find a complete list of the features, with descriptions, in the book's Preface (page xiv). Here, we describe a few ways in which you might use the features as you teach:

- Use the *Chapter Overview* and *Chapter Summary* to help you decide which units to cover in class.
- Use the *Summary Boxes* and the many *Examples* and *Case Studies* as the basis for class presentations; you may also wish to cover material in the *Brief Review* boxes if your students need the review.
- Use the *Time Out to Think* questions as starting points for classroom discussion.
- Ask your students to use the end-of-unit *Review Questions* as a study aid.
- Draw from the end-of-unit questions under *In the News* and *Exercises* in assigning homework. Because answers to the odd-numbered exercises appear in the back of the book, you may wish to assign only even-numbered exercises while suggesting that students use the odd-numbered ones for additional practice.
- If you wish to have students work on extended projects, choose from among the *Web Projects* that appear at the end of each unit.

SAMPLE COURSE OUTLINES

Roughly speaking, you'll need to consider three main dimensions when designing your course: the prerequisite background of your students, the length of your course, and any special areas you intend to emphasize. For simplicity in developing sample course outlines, we've identified the following choices for the three dimensions:

Dimension 1: Student Level	Dimension 2: Course Length	Dimension 3: Course Emphasis
Level 1: No algebra prerequisite	**Q:** one-quarter course	**G:** General—a taste of many topics
Level 2: Elementary algebra prerequisite	**S:** one-semester course	
Level 3: Intermediate algebra prerequisite		

If you have a two-semester course or wish to have a specialized course emphasis (i.e., anything besides the "general" emphasis under Dimension 3 above), we urge you to consider the full version of *Using and Understanding Mathematics*; you may want to cover some of the material in the full book that is not in *Essentials of Using and Understanding Mathematics*. Here we offer sample course outlines for the following combinations of these dimensions:

1. (Level 1, Q, G): general one-quarter course for students with only one or two years of high school mathematics
2. (Level 1, S, G): similar to (1) above, but for one semester rather than one quarter
3. (Level 2, S, G): similar to (2) above, but at a slightly faster pace for students with an elementary algebra prerequisite
4. (Level 3, S, G): similar to (2) above, but at a faster pace for students with an intermediate algebra prerequisite

SUGGESTED OUTLINE FOR LEVEL 1, ONE QUARTER, GENERAL EMPHASIS

General Notes:

- This outline assumes a pace of approximately 2 units per week, which should be realistic for Level 1 students.
- Note that we have chosen to focus on parts of Chapters 1–5 and 7. If you wish to cover other topics, you might consider reducing the coverage of logic in Chapter 1 or the coverage of finance in Chapter 4.
- We recommend assigning 5 to 10 exercises for each unit. You may also want to assign one of the *In the News* questions for each unit.

Week	Suggested Topics	Reading
1	Introduction: welcome, including discussion of course content and expectations regarding work load; you may wish to conduct the activity described in the IGSM. Cover the Prologue and Units 1A and 1B.	Prologue, Units 1A, 1B
2	Cover sets and Venn diagrams in Unit 1C; to keep the time devoted to this topic reasonable, you may wish to stop before the Venn diagrams for three sets in Unit 1C. Then cover basic ideas of problem solving through unit analysis in Unit 2A.	Units 1C, 2A
3	Continue problem solving with Unit 2B; discuss the four-step problem-solving process in Unit 2C, but skip the rest of Unit 2C and assign homework only from Units 2A and 2B.	Units 2B, 2C (4-step process only)
4	Spend this week on uses and abuses of percentages (Unit 3A) and putting numbers in perspective (Unit 3B).	Units 3A, 3B
5	Cover uncertainty in Unit 3C and begin financial management with Unit 4A.	Units 3C, 4A
6	Cover savings plans in Unit 4B and loan payments in Unit 4C.	Units 4B, 4C
7	Introduce fundamental ideas of statistics (Unit 5A), but devote most of your time to the question of whether to believe a statistical result (discussed in Unit 5B).	Units 5A, 5B
8	Cover statistical tables and graphs in Units 5C and 5D.	Units 5C, 5D
9	Conclude coverage of statistics by discussing the difference between correlation and causality in Unit 5E. Then begin discussion of linear versus exponential growth (Unit 7A).	Units 5E, 7A
10	Discuss doubling time and half-life (Unit 7B). Use any extra time for review.	Unit 7B

SUGGESTED OUTLINE FOR LEVEL 1, ONE SEMESTER, GENERAL EMPHASIS

General Notes:

- This outline is quite similar to the previous outline, except that it assumes a semester of 15 weeks rather than a quarter of 10 weeks.
- We have included topics only for the first 14 weeks, on the assumption that the last week may be devoted to review.
- For a few of the weeks, we've slowed the pace to a single unit, to allow more time for review and consolidation.

Week	Suggested Topics	Reading
1	Introduction: welcome, including discussion of course content and expectations regarding work load; you may wish to conduct the activity described in the IGSM. Cover the Prologue and Units 1A and 1B.	Prologue, Units 1A, 1B
2	Cover sets and Venn diagrams in Unit 1C; to keep the time devoted to this topic reasonable, you may wish to stop before the Venn diagrams for three sets in Unit 1C. Then cover basic ideas of problem solving through unit analysis in Unit 2A.	Units 1C, 2A
3	Continue problem solving with Unit 2B; discuss the four-step problem-solving process in Unit 2C, but skip the rest of Unit 2C and assign homework only from Units 2A and 2B.	Units 2B, 2C (4-step process only)
4	Spend this week on uses and abuses of percentages (Unit 3A) and putting numbers in perspective (Unit 3B).	Units 3A, 3B
5	Conclude coverage of Chapter 3 with a discussion of uncertainty in Unit 3C.	Unit 3C
6	Begin full coverage of the financial management chapter with compound interest (Unit 4A) and savings plans (Unit 4B).	Units 4A, 4B
7	Cover loan payments (Unit 4C).	Unit 4C
8	Begin discussing basic ideas of statistics (Unit 5A).	Unit 5A
9	Continue coverage of statistics with Units 5B and 5C.	Units 5B, 5C
10	Conclude coverage of statistics with Units 5D and 5E.	Units 5D, 5E
11	Give a brief introduction to probability with Units 6A and 6B.	Units 6A, 6B
12	Discuss the law of large numbers (Unit 6C).	Unit 6C
13	Cover basics of exponential growth (Unit 7A) and doubling time and half-life (Unit 7B).	Units 7A, 7B
14	Cover basic voting ideas (Unit 9A).	Unit 9A
15	Class wrap-up.	—

SUGGESTED OUTLINE FOR LEVEL 2, ONE SEMESTER, GENERAL EMPHASIS

General Notes:

- This outline assumes a more rapid pace through the early units in the book, which should be reasonable for Level 2 students. The pace slows as you move into the more advanced material.

Week	Suggested Topics	Reading
1	Introduction: welcome, including discussion of course content and expectations regarding work load; you may wish to conduct the activity described in the IGSM. Cover the prologue and units 1A and 1B.	Prologue, Units 1A, 1B
2	Cover sets and Venn diagrams in Unit 1C. Begin coverage of critical thinking in Unit 1D.	Units 1C, 1D
3	Conclude Unit 1D. Then cover basic ideas of problem solving through unit analysis in Unit 2A.	Unit 2A
4	Continue problem solving with Units 2B and 2C, emphasizing the four-step problem-solving process in Unit 2C.	Units 2B, 2C
5	Spend this week on uses and abuses of percentages (Unit 3A) and putting numbers in perspective (Unit 3B).	Units 3A, 3B
6	Conclude coverage of Chapter 3 with a discussion of uncertainty in Unit 3C and deceptions in Unit 3D.	Units 3C, 3D
7	Begin full coverage of the financial management chapter with compound interest (Unit 4A) and savings plans (Unit 4B).	Units 4A, 4B
8	Cover loan payments (Unit 4C). Then begin discussing basic ideas of statistics (Unit 5A).	Units 4C, 5A
9	Continue coverage of statistics with Units 5B and 5C.	Units 5B, 5C
10	Continue coverage of statistics with Units 5D and 5E.	Units 5D, 5E
11	Conclude coverage of statistics and begin probability.	Units 5F, 6A
12	Continue coverage of probability (Unit 6B) and discuss the law of large numbers (Unit 6C).	Units 6B, 6C
13	Begin discussion of linear versus exponential growth (Unit 7A) and cover doubling time and half-life (Unit 7B).	Units 7A, 7B
14	Discuss issues of voting (Unit 9A) and apportionment (Unit 9B).	Units 9A, 9B
15	Class wrap-up.	—

SUGGESTED OUTLINE FOR LEVEL 3, ONE SEMESTER, GENERAL EMPHASIS

General Notes:

- This outline assumes a more rapid pace and covers the entire book, which should be reasonable for Level 3 students.

Week	Suggested Topics	Reading
1	Introduction: welcome, including discussion of course content and expectations regarding work load; you may wish to conduct the activity described in the IGSM. Cover the prologue and units 1A and 1B.	Prologue, Units 1A, 1B
2	Cover sets and Venn diagrams in Unit 1C and critical thinking in Unit 1D.	Units 1C, 1D
3	Cover the first two units of Chapter 2 on problem solving.	Units 2A, 2B
4	Complete Chapter 2 and cover uses and abuses of percentages (Unit 3A).	Units 2C, 3A
5	Cover Unit 3B on putting numbers in perspective and Unit 3C on uncertainty.	Units 3B, 3C
6	Complete Chapter 3 and begin coverage of financial management with compound interest (Unit 4A).	Units 3D, 4A
7	Cover savings plans (Unit 4B) and loan payments (Unit 4C).	Units 4B, 4C
8	Begin coverage of statistics; note that Unit 5C should go very quickly for Level 3 students.	Units 5A–5C
9	Conclude coverage of statistics; note that Unit 5F should go very quickly for Level 3 students.	Units 5D–5F
10	Give a brief introduction to probability with Units 6A and 6B.	Units 6A, 6B
11	Discuss the law of large numbers (Unit 6C) and counting techniques (Unit 6D).	Units 6C, 6D
12	Cover all of Chapter 7 on exponential growth. The first unit should go very rapidly for Level 3 students.	Units 7A–7C
13	Cover mathematics and the arts in Chapter 8.	Units 8A–8C
14	Discuss issues of voting (Unit 9A) and apportionment (Unit 9B).	Units 9A, 9B
15	Class wrap-up.	—

Essentials of USING AND UNDERSTANDING MATHEMATICS

PROLOGUE

LITERACY FOR THE MODERN WORLD

Mathematics is the key to opportunity. No longer just the language of science, mathematics now contributes in direct and fundamental ways to business, finance, health, and defense. For students, it opens doors to careers. For citizens, it enables informed decisions. For nations, it provides knowledge to compete in a technological economy. To participate fully in the world of the future, America must tap the power of mathematics.

—from *Everybody Counts*

WHAT IS QUANTITATIVE LITERACY?

Literacy is the ability to read and write, and it comes in varying degrees. Some people can recognize only a few words and write only their names; others read and write in many languages. A primary goal of our educational system is to provide citizens with a level of literacy sufficient to read and write about the important issues of our time.

Today, the abilities to interpret and reason with **quantitative information**—information that involves mathematical ideas or numbers—are crucial aspects of literacy. This so-called **quantitative literacy** is essential to understanding issues that appear in the news every day. The process of interpreting and reasoning with quantitative information is called **quantitative reasoning.** The purpose of this book is to help you gain skills in quantitative reasoning as it applies to issues you will encounter in

- your subsequent course work
- your career, and
- your daily life.

QUANTITATIVE REASONING AND CULTURE

Quantitative reasoning enriches the appreciation of both ancient and modern culture. The historical record shows that all great cultures devoted substantial energy to mathematics and to science (or to observational studies that predated modern science). Without a sense of how quantitative concepts are used in art, architecture, and science, you cannot fully appreciate the incredible achievements of the Mayans in Central America, the builders of the great city of Zimbabwe in Africa, the ancient Egyptians and Greeks, the early Polynesian sailors, and countless others.

A ROCK, A RIVER, A TREE
Hosts to species long since departed
Marked the mastodon,
The dinosaur, who left dry tokens
Of their sojourn here
On our planet floor,
Any broad alarm of their hastening doom
Is lost in the gloom of dust and ages.

—MAYA ANGELOU, *On the Pulse of the Morning*

Similarly, quantitative concepts can help you understand and appreciate the great works of modern artists. You can more fully appreciate the excerpt here from a Maya Angelou poem if you know that dinosaurs have been extinct for 65 million years and are able to put this amount of time into perspective. The ties between mathematics and music can be found in modern and classical music, as well as in the digital production of music. Indeed, it is hard to find popular works of art, film, or literature that do not make at least some indirect reference to mathematics.

QUANTITATIVE REASONING IN THE WORK FORCE

Quantitative reasoning is important in the work force. A lack of quantitative skills puts out of reach many of the most challenging and highest paying jobs. Table 1 defines skill levels in language and mathematics on a scale of 1 to 6, and Table 2 shows the typical levels needed in many jobs.

TABLE 1 Skill Levels

LEVEL	LANGUAGE SKILLS	MATH SKILLS
1	Recognizes 2500 two- or three-syllable words. Reads at a rate of 95–120 words per minute. Writes and speaks simple sentences.	Adds and subtracts two-digit numbers. Does simple calculations with money, volume, length, and weight.
2	Recognizes 5000–6000 words. Reads 190–215 words per minute. Reads adventure stories and comic books, as well as instructions for assembling model cars. Writes compound and complex sentences with proper grammar and punctuation.	Adds, subtracts, multiplies, and divides all units of measure. Computes ratio, rate, and percentage. Draws and interprets bar graphs.
3	Reads novels and magazines, as well as safety rules and equipment instructions. Writes reports with proper format and punctuation. Speaks well before an audience.	Understands basic geometry and algebra. Calculates discount, interest, profit and loss, markup, and commissions.
4	Reads novels, poems, newspapers, and manuals. Prepares business letters, summaries, and reports. Participates in panel discussions and debates. Speaks extemporaneously on a variety of subjects.	Has true quantitative reasoning abilities. Understands logic, problem solving, ideas of statistics and probability, and modeling.
5	Reads literature, book and play reviews, scientific and technical journals, financial reports, and legal documents. Can write editorials, speeches, and critiques.	Knows calculus and statistics. Is able to deal with econometrics.
6	Same types of skills as level 5, but more advanced.	Works with advanced calculus, modern algebra, and statistics.

TABLE 2 Skill-Level Requirements

OCCUPATION	LANGUAGE LEVEL	MATH LEVEL	OCCUPATION	LANGUAGE LEVEL	MATH LEVEL
Biochemist	6	6	Web page designer	5	4
Computer engineer	6	6	Corporate executive	5	5
Mathematician	6	6	Computer sales agent	4	4
Cardiologist	6	5	Athlete's agent	4	4
Social psychologist	6	5	Management trainee	4	4
Lawyer	6	4	Insurance sales agent	4	4
Tax attorney	6	4	Retail store manager	4	4
Newspaper editor	6	4	Cement mason	3	3
Accountant	5	5	Poultry farmer	3	3
Personnel manager	5	4	Tile setter	3	3
Corporate president	5	5	Travel agent	3	3
Weather forecaster	5	5	Janitor	3	2
Secondary teacher	5	5	Short-order cook	3	2
Elementary teacher	5	4	Assembly-line worker	2	2
Financial analyst	5	5	Toll collector	2	2
Journalist	5	4	Laundry worker	1	1

SOURCE: Both tables adapted from *Education: The Knowledge Gap*, supplement to *The Wall Street Journal*, February 9, 1990.

[Calvin and Hobbes comic strip by Bill Watterson]

CALVIN and HOBBES © 1990 Watterson. Reprinted with permission of UNIVERSAL PRESS SYNDICATE. All rights reserved.

Note that the occupations requiring high skill levels are generally the most prestigious and highest paying. Note also that most of these occupations call for high skill levels in *both* language and math, refuting the myth that if you're good at language you don't have to be good at mathematics, and vice versa.

MISCONCEPTIONS ABOUT MATHEMATICS

Do you consider yourself to have "math phobia" (fear of mathematics) or "math loathing" (dislike of mathematics)? We hope not; but if you do, you aren't alone. Many adults harbor fear or loathing of mathematics, and, unfortunately, these attitudes are often reinforced by classes that present mathematics as an obscure and sterile subject.

In reality, mathematics is not nearly so dry as it sometimes seems in school. Indeed, attitudes toward mathematics often are directed not at what mathematics really is but at some common misconceptions about mathematics. Let's investigate a few of these misconceptions and the reality behind them.

In a sense, we are all mathematicians—and superb ones. It makes no difference what you do. Your real forte lies in navigating the complexities of social networks, weighing passions against histories, calculating reactions, and generally managing a system of information that, when all laid out, would boggle a computer.

—A. K. DEWDNEY,
200% of Nothing

MISCONCEPTION ONE: MATH REQUIRES A SPECIAL BRAIN

One of the most pervasive misconceptions is that some people just aren't good at mathematics because learning mathematics requires special and rare abilities. The reality is that nearly everyone can do mathematics. All it takes is self-confidence and hard work—

thinking about...
PEOPLE WHO STUDIED MATHEMATICS

The critical thinking skills developed through the study of mathematics are valuable in many careers. The following is only a small sample of people who studied mathematics but became famous for work in other fields. (Most of the names on this list are from an article by Steven G. Buyske, "Famous Non-mathematicians," *American Mathematical Monthly*, November 1993.)

Corazon Aquino, former president of the Philippines, a mathematics minor; **Harry Blackmun,** former Supreme Court justice, summa cum laude in mathematics, Harvard University; **Lewis Carroll** (Charles Dodgson), author of *Alice in Wonderland*, a mathematician; **David Dinkins,** former mayor of New York City, BA in mathematics, Howard University; **Alberto Fujimori,** former president of Peru, MS in mathematics, University of Wisconsin; **Art Garfunkel,** musician, MA in mathematics, Columbia University; **Grace Hopper,** computer pioneer and first woman Rear Admiral in the U.S. Navy, Ph.D. in mathematics, Yale; **Mae Jemison,** first African-American woman in space, studied mathematics as part of her degree in chemical engineering from Stanford University; **John Maynard Keynes,** economist, MA in mathematics, Cambridge University; **Lee Hsien Loong,** politician in Singapore, BA in mathematics, Cambridge University; **Edwin Moses,** three-time Olympic Champion in the 400-meter hurdles, studied mathematics as part of his degree in physics from Morehouse College; **Florence Nightingale,** pioneer in nursing, studied mathematics and applied it to her work; **William Perry,** former secretary of defense, Ph.D. in mathematics, Pennsylvania State University; **Sally Ride,** first American woman in space, studied mathematics as part of her Ph.D. in physics from Stanford University; **David Robinson,** basketball star, bachelor's degree in mathematics, U.S. Naval Academy; **Alexander Solzhenitsyn,** Nobel prize winning Russian author, degrees in mathematics and physics from the University of Rostov; **Bram Stoker,** author of *Dracula*, studied mathematics at Trinity University, Dublin; **Laurence Tribe,** Harvard law professor, summa cum laude in mathematics, Harvard University; **Virginia Wade,** Wimbledon champion, bachelor's degree in mathematics, Sussex University.

the same qualities needed to learn to read, to master a musical instrument, or to become skilled at a sport. Indeed, the belief that mathematics requires special talent found in a few elite people is peculiar to the United States. In other countries, particularly in Europe and Asia, *all* students are expected to become proficient in mathematics.

Of course, different people learn mathematics at different rates and in different ways. For example, some people learn by concentrating on concrete problems, others by thinking visually, and still others by thinking abstractly. No matter what type of thinking style you prefer, you can succeed in mathematics.

MISCONCEPTION TWO: THE MATH IN MODERN ISSUES IS TOO COMPLEX

Some people claim that the advanced mathematical concepts underlying many modern issues are too complex for the average person to understand. While it is true that only a few people receive the training needed to work with or discover

Equity for all requires excellence for all; both thrive when expectations are high.
—from *Everybody Counts*

The Far Side ® by Gary Larson © 1990 FarWorks, Inc. All Rights Reserved. Used with permission.

advanced mathematical concepts, most people are capable of understanding enough about the mathematical basis of important issues to develop informed and reasoned opinions.

The situation is similar in other fields. For example, years of study and practice are required to become a proficient professional writer, but most people can read a book. It takes hard work and law school to become a lawyer, but most people can understand how the law affects them. And though few have the musical talent of Mozart, anyone can learn to appreciate his music. Mathematics is no different. If you've made it this far in school, you can understand enough mathematics to succeed as an individual and a concerned citizen. Don't let anyone tell you otherwise!

MISCONCEPTION THREE: MATH MAKES YOU LESS SENSITIVE

Some people believe that learning mathematics will somehow make them less sensitive to the romantic and aesthetic aspects of life. In fact, understanding the mathematics that explains the colors of a sunset or the geometric beauty in a work of art can only enhance aesthetic appreciation. Furthermore, many people find beauty and elegance in mathematics itself. It's no accident that people trained in mathematics have made important contributions to art, music, and many other fields.

MISCONCEPTION FOUR: MATH MAKES NO ALLOWANCE FOR CREATIVITY

The "turn the crank" nature of the problems in many textbooks may give the impression that mathematics stifles creativity. Some of the facts, formalisms, and skills required for mathematical proficiency are fairly cut and dried, but *using* these mathematical tools takes creativity. Consider designing and building a home. The task demands specific skills to lay the foundation, frame in the structure, install plumbing and wiring, and paint walls. But building the home involves much more: Creativity is needed to develop the architectural design, respond to on-the-spot problems during construction, and factor in constraints based on budgets and building codes. The mathematical skills you've learned in school are like the skills of carpentry or plumbing. Applying mathematics is like the creative process of building a home.

Probably the most harmful misconception is that mathematics is essentially a matter of computation. Believing this is roughly equivalent to believing that writing essays is the same as typing them.

—John Allen Paulos

MISCONCEPTION FIVE: MATH PROVIDES EXACT ANSWERS

A mathematical formula will yield a specific result, and in school that result may be marked right or wrong. But when you use mathematics in real-life situations, answers are never so clear-cut. For example:

> A bank offers simple interest of 5%, paid at the end of one year (that is, after one year the bank pays you 5% of your account balance). If you deposit $1000 today and make no further deposits or withdrawals, how much will you have in your account after one year?

A straight mathematical calculation seems simple enough: 5% of $1000 is $50; so you should have $1050 at the end of a year. But will you? How will your balance be affected by service charges or taxes on interest earned? What if the bank fails? What if the bank is located in a country in which the currency collapses during the year? Choosing a bank in which to invest your money is a *real* mathematics problem that doesn't necessarily have a simple or definitive solution.

Skills are to mathematics what scales are to music or spelling is to writing. The objective of learning is to write, to play music, or to solve problems—not just to master skills.

—from *Everybody Counts*

MISCONCEPTION SIX: MATH IS IRRELEVANT TO MY LIFE

No matter what your path in college, career, and life, you will find mathematics involved in many ways. A major goal of this text is to show you hundreds of examples in which mathematics applies to everyone's life. We hope you will find that mathematics is not only relevant but also interesting and enjoyable.

WHAT *IS* MATHEMATICS?

In discussing misconceptions, we identified what mathematics is *not*; now let's look at what mathematics *is*. The word *mathematics* is derived from the Greek word *mathematikos*, which means "inclined to learn." Thus, literally speaking, to be mathematical is to be curious, open-minded, and interested in always learning more! Today, we tend to look at mathematics in three different ways: as the sum of its branches, as a way to model the world, and as a language.

MATHEMATICS AS THE SUM OF ITS BRANCHES

As you progressed through school, you undoubtedly learned to associate mathematics with some of its branches. Among the better known branches of mathematics are these:

- **logic**—the study of principles of reasoning;
- **arithmetic**—methods for operating on numbers;
- **algebra**—methods for working with unknown quantities;
- **geometry**—the study of size and shape;
- **trigonometry**—the study of triangles and their uses;
- **probability**—the study of chance;
- **statistics**—methods for analyzing data; and
- **calculus**—the study of quantities that change.

One way to view mathematics is as the sum of its branches. Indeed, most "traditional" mathematics books focus on one branch of mathematics at a time. In this book, we emphasize the connections between the branches rather than their differences. Nevertheless, familiarity with the various branches of mathematics is helpful, even if only to know when you are using them.

MATHEMATICS AS A WAY TO MODEL THE WORLD

Mathematics also may be viewed as a tool for creating **models**, or representations that allow us to study real phenomena. Modeling is not unique to mathematics. For example, a road map is a model that represents the roads in some region.

FIGURE 1

Mathematical models can be as simple as a single equation that predicts how the money in your bank account will grow or as complex as a set of thousands of interrelated equations and parameters used to represent the global climate. By studying models, we gain insight into otherwise unmanageable problems. A global climate model, for example, can help us understand weather systems and ask "what if" questions about how human activity may affect the climate. When a model is used to make a prediction that does *not* come true, it points out areas where further research is needed. Today, mathematical modeling is used in nearly every field of study. Figure 1 indicates some of the many disciplines that use mathematical modeling to solve problems.

MATHEMATICS AS A LANGUAGE

A third way to look at mathematics is as a language with its own vocabulary and grammar. Indeed, mathematics often is called "the language of nature" because it is so useful for modeling the natural world. As in any language, different degrees of fluency are possible. From this point of view, quantitative literacy is the level of fluency required for success in today's world.

The idea of mathematics as a language also is useful in thinking about how to *learn* mathematics. Table 3 compares learning mathematics to learning a language and learning art.

Mathematics is a way of thinking that can help make muddy relationships clear. It is a language that allows us to translate the complexity of the world into manageable patterns.

—K. C. COLE,
The Universe and the Teacup

TABLE 3 Learning Mathematics: An Analogy to Language and Art

LEARNING A LANGUAGE	LEARNING THE LANGUAGE OF ART	LEARNING THE LANGUAGE OF MATHEMATICS
Become familiar with many styles of speaking and writing, such as essays, poetry, and drama.	Become familiar with many styles of art, such as classical, renaissance, impressionist, and modern.	Become familiar with techniques from many branches of mathematics, such as arithmetic, algebra, and geometry.
Place literature in context by studying its history and the social conditions under which it was created.	Place art in context by studying its history and the social conditions under which it was created.	Place mathematics in context by studying its history, purposes, and applications.
Learn the elements of language—such as words, parts of speech (nouns, verbs, etc.), and rules of grammar—and practice their proper use.	Learn the elements of visual form—such as lines, shapes, colors, and textures—and practice using them in your own art work.	Learn the elements of mathematics—such as numbers, variables, and operations—and practice using them to solve simple problems.
Develop the ability to analyze the language in complex forms critically—novels, short stories, essays, poems, speeches, debates, and similar works.	Develop the ability to analyze works of art critically—painting, sculpture, architecture, photography, and similar works.	Develop the ability to analyze quantitative information critically—mathematical models, statistical studies, economic forecasts, investment strategies, and similar works.
Use language creatively for your own purposes, such as writing a term paper or story or engaging in debate.	Use your sense of art creatively, such as in designing your house, taking a photograph, or making a sculpture.	Use mathematics creatively to solve problems you encounter and to help you understand important issues in the world around you.

HOW TO SUCCEED IN MATHEMATICS

If you are reading this book, you probably are enrolled in a mathematics course. The keys to success in your course include approaching the material with an open and optimistic frame of mind, paying close attention to how useful and enjoyable mathematics can be in *your* life, and studying effectively and efficiently. The following sections offer a few specific hints that may be of use as you study.

USING THIS BOOK

Before we get into more general strategies for studying, here are a few guidelines that will help you use *this* book most effectively.

- Before doing any assigned exercises, read assigned material *twice*:
 - On the first pass, read quickly to gain a "feel" for the material and concepts presented.
 - On the second pass, read the material in more depth and work through the examples carefully.
- During the second reading, take notes that will help you when you go back to study later. In particular:
 - *Use the margins!* The wide margins in this textbook are designed to give you plenty of room for making notes as you study.
 - Don't highlight—underline! Using a pen or pencil to underline material requires greater care than highlighting and therefore helps to keep you alert as you study.
- After you complete the reading, and again when studying for exams, make sure you can answer the review questions at the end of each unit.
- You'll learn best by *doing*, so do plenty of the end-of-unit exercises. In particular, try some of the exercises that have answers in the back of the book, in addition to those assigned by your instructor.

BUDGETING YOUR TIME

A general rule of thumb for college classes is that you should expect to study about 2 to 3 hours per week *outside* class for each unit of credit. For example, a student taking 15 credit hours should spend 30 to 45 hours each week studying outside of class. Combined with time in class, this works out to a total of 45 to 60 hours per week—not much more than the time required of a typical job, and you get to choose your own hours. Of course, if you are working or taking care of a family while you attend school, you will need to budget your time carefully.

The following table gives some rough guidelines for how you might divide your studying time. If you are spending fewer hours than these guidelines suggest, you could probably improve your grade by studying more. If you are spending more hours than these guidelines suggest, you may be studying inefficiently; in that case, you should talk to your instructor about how to study more effectively for a mathematics class.

If Your Course Is	Time for Reading the Assigned Text (per Week)	Time for Homework Assignments (per Week)	Time for Review and Test Preparation (Average per Week)	Total Study Time (per Week)
3 credits	1 to 2 hours	3 to 5 hours	2 hours	6 to 9 hours
4 credits	2 to 3 hours	3 to 6 hours	3 hours	8 to 12 hours
5 credits	2 to 4 hours	4 to 7 hours	4 hours	10 to 15 hours

GENERAL STRATEGIES FOR STUDYING

- Don't miss class. Listening to lectures and participating in discussions is much more effective than reading someone else's notes. Active participation will help you retain what you are learning.
- Budget your time effectively. An hour or two each day is more effective, and far less painful, than studying all night before homework is due or before exams. It's also good to study shortly after class, while the lessons are still fresh in your mind.
- If a concept gives you trouble, do additional reading or problem solving beyond what has been assigned. If you still have trouble, ask for help; you surely can find friends, colleagues, or teachers who will be glad to help you learn.
- Working together with friends can be valuable, as you will solidify and enhance your own understanding when discussing concepts with others. However, be sure that you learn *with* your friends and do not become dependent on them.

PREPARING FOR EXAMS

- Rework exercises and other assignments; try additional exercises to be sure you understand the concepts. Study your performance on assignments, quizzes, or exams from earlier in the semester.
- Study your notes from lectures and discussions. Pay attention to what your instructor expects you to know for an exam.
- Reread the relevant sections in the textbook, paying special attention to notes you have made in the margins.
- Study individually before joining a study group with friends. Study groups are effective only if *every* individual comes prepared to contribute.
- Don't stay up too late before an exam. Don't eat a big meal within an hour of the exam (thinking is more difficult when blood is being diverted to the digestive system).
- Try to relax before and during the exam. If you have studied effectively, you are capable of doing well. Staying relaxed will help you think clearly.

Discussion Questions

1. **Mathematics in Modern Issues.** Describe at least one way that mathematics is involved in each issue below.

 Example: The spread of AIDS: Mathematics is used to study the probability of contracting AIDS.

 a. The long-term viability of the Social Security system
 b. The appropriate level for the federal gasoline tax
 c. The fairness of medical insurance rates
 d. Job discrimination against women or ethnic minorities
 e. Effects of population growth (or decline) on your community
 f. Possible bias in standardized tests (e.g., the SAT)
 g. The degree of risk posed by carbon dioxide emissions
 h. Immigration policy of the United States
 i. Violence in public schools
 j. An issue of your choice from today's newspaper

2. **Quantitative Concepts in the News.** Identify the major unresolved issue discussed on the front page of today's newspaper. List at least three areas in which quantitative concepts play a role in the policy considerations of this issue.

3. **Mathematics and the Arts.** Choose a well-known historical figure in a field of art in which you have a personal interest (e.g., a painter, sculptor, musician, or architect). Briefly describe how mathematics played a role in or influenced that person's work.

4. **Quantitative Literature.** Choose a favorite work of literature (poem, play, short story, or novel). Describe one or more instances in which quantitative reasoning is helpful in understanding the subtleties intended by the author.

5. **Your Quantitative Major.** Identify ways in which quantitative reasoning is important within your major field of study. (If you haven't yet chosen a major, pick a field that you are considering for your major.)

6. **Career Preparation.** Realizing that most Americans change careers several times during their lives, identify at least three occupations in Table 2 that interest you. Do you have the necessary skills for them at this time? If not, how can you acquire these skills?

7. **Attitudes Toward Mathematics.** Most children have a natural affinity for mathematics; they take pride in their counting skills and enjoy puzzles, building blocks, and computers. Unfortunately, this natural interest seems to be snuffed out in most people by the time they reach adulthood. What is your attitude toward mathematics? If you have a negative attitude, can you identify when in your childhood that attitude developed? If you have a positive attitude, can you explain why? How might you encourage someone with a negative attitude to become more positive?

1

THINKING CRITICALLY

> *Civilized life depends upon the success of reason in social intercourse, the prevalence of logic over violence in interpersonal conflict.*
> —Juliana Geran Pilon

Critical thinking is an approach to problem solving and decision making that involves careful reading (or listening), sharp thinking, logical analysis, good visualization, and healthy skepticism. You probably already apply critical thinking to many problems that affect you personally, but you may never have studied it in a course. In this first chapter, you will study a variety of skills—many of them skills of logic—that will be useful to you not only in the rest of this book, but also in the thinking you do in daily life.

UNIT 1A Recognizing Fallacies: We will begin our study of critical thinking by looking at deceptive arguments, or fallacies.

UNIT 1B Propositions and Truth Values: We will define and study basic components of logic including propositions, truth values, and truth tables. We will also consider the logical connectors *and, or,* and *if . . . then.*

UNIT 1C Sets and Venn Diagrams: We will explore how many propositions express relationships between sets, or categories, and how Venn diagrams can help us visualize these relationships.

UNIT 1D Critical Thinking in Everyday Life: The skills built in the first three units are useful in and of themselves, but our primary goal is to become proficient at analyzing information carefully. We will put the experience gained in earlier units to work in studying critical thinking situations encountered in everyday life.

UNIT 1A

RECOGNIZING FALLACIES

We often try to persuade other people of our views. These discussions take many different forms, some of which are more useful than others. Consider the following "argument" between two classmates.

BOB: The death penalty is immoral.
SUSAN: No it isn't.
BOB: Yes it is! Judges who give the death penalty should be impeached.
SUSAN: You don't even know how the death penalty is decided.
BOB: I know a lot more than you know!
SUSAN: I can't talk to you; you're an idiot!

This type of heated conversation is fairly common, but it accomplishes little. Rather than helping Bob or Susan to sway the other, it is likely to leave both of them upset and angry. Fortunately, there is a better way to argue—with skills of **logic,** the study of the methods and principles of reasoning. Arguing logically may not change Bob's or Susan's position, but it can help them understand each other.

In this book, the term **argument** refers to reasoned or thoughtful processes. Specifically, an argument is a set of facts or assumptions, called **premises,** that are used to support a **conclusion.** Some arguments provide strong support for their conclusions; others do not. An argument that fails to make a compelling case for its conclusion may contain some error in reasoning, or **fallacy** (from the Latin for "deceit" or "trick"). In other words, a fallacious argument tries to persuade in a way that doesn't really make sense when analyzed carefully.

People generally quarrel because they cannot argue.

—G. K. CHESTERTON (1874–1936), English author

> **DEFINITIONS**
>
> **Logic** is the study of the methods and principles of reasoning.
>
> An **argument** is a set of facts or assumptions, called **premises,** used to support a **conclusion.**
>
> A **fallacy** is a deceptive argument; its conclusion is not well supported by its premises.

by the way...
More than $100 billion per year is spent on advertising in the United States, and the best advertising firms do extensive research before embarking on new campaigns. It is estimated that the average American has seen over 1 million advertisements by age 18.

You might expect fallacies to be rare, but they are not. If you study advertisements, political campaigns, or editorial commentaries, you'll find that many of them are filled with fallacies. Moreover, fallacies are not always easy to recognize, in part because public relations specialists have spent billions of dollars researching how subtle fallacies can be used to persuade us to buy products, vote for candidates, or support particular policies.

Because fallacies are so common, it is important to be able to recognize them. Thus, we begin our study of critical thinking with examples of a few of the most common fallacies. The fallacy in each example has a fancy name, but learning the names

is far less important than learning to recognize the faulty reasoning. The experience you gain by analyzing fallacies will provide a foundation upon which to build additional critical thinking skills.

EXAMPLE 1 *Appeal to Popularity*

"Ford makes the best automobile in America; after all, more people drive Fords than any other American car."

Analysis The first step in dealing with any argument is recognizing which statements are premises and which are conclusions. This argument tries to make the case that *Ford makes the best automobiles in America,* so this statement is its conclusion. The only evidence it offers to support this conclusion is the statement *more people drive Fords than any other American car;* this is the argument's one and only premise. Overall, this argument has the form

> Premise: More people drive Fords than any other American car.
> Conclusion: Ford makes the best automobiles in America.

Note that the original written argument states the conclusion before the premise; such "backwards" structures are common in everyday speech and are perfectly legitimate as long as the argument is well reasoned. In this case, however, the reasoning is faulty: The fact that more people drive Fords does not, in itself, mean that they are the best cars. This argument suffers from the fallacy of *appeal to popularity* (or *appeal to majority*), in which the fact that large numbers of people believe or act some way is used inappropriately as evidence that the belief or action is correct.

We can represent the general form of this fallacy with a diagram in which the letter *p* stands for a particular statement (Figure 1.1). In this case, *p* stands for the statement *Ford makes the best automobiles in America.*

> Many people believe
> *p* is true;
> therefore ...
>
> *p* is true.

FIGURE 1.1
The fallacy of appeal to popularity. The letters *p* and *q* (used in later diagrams) represent statements.

EXAMPLE 2 *False Cause*

"I placed the quartz crystal on my forehead, and in five minutes my headache was gone. The crystal made my headache go away."

Analysis We identify the premises and conclusion of this argument as follows:

> Premise: I placed the quartz crystal on my forehead.
> Premise: Five minutes later my headache was gone.
> Conclusion: The crystal made my headache go away.

The premises tell us that one thing (crystal on forehead) happened before another (headache went away), but they don't prove any connection between them. That is, we cannot conclude that the crystal *caused* the headache to go away. This argument suffers from the fallacy of *false cause,* in which the fact that one event came before another is incorrectly taken as evidence that the first event *caused* the second event. We can represent this fallacy with a diagram in which *A* and *B* represent two different events (Figure 1.2); in this case, *A* is the event of putting the crystal on the forehead and *B* is the event of the headache going away. (We'll discuss how cause *can* be established in Chapter 5.)

> *A* came before *B*;
> therefore ...
>
> *A* caused *B*.

FIGURE 1.2
The fallacy of false cause. The letters *A* and *B* represent events.

EXAMPLE 3 Appeal to Ignorance

"Scientists have not proved that global warming will have any dire consequences for the human race. Therefore, the global warming catastrophes that environmentalists shout about are bunk."

Analysis If we strip the argument to its core, it says this:

Premise: There's no proof that global warming will have any dire consequences.

Conclusion: Global warming will not have any dire consequences.

The fallacy should be clear: A *lack* of proof for dire consequences does not mean that they won't occur. This fallacy is called *appeal to ignorance* because it uses ignorance (lack of knowledge) about the truth of a proposition to conclude the opposite (Figure 1.3). Another form of this fallacy arises when something (such as UFOs) has not been proved *non*existent; appeal to ignorance would lead one to conclude wrongly that it *does* exist.

p has not been proven false or nonexistent; therefore ...

↓

p is true or exists.

FIGURE 1.3
The fallacy of appeal to ignorance.

Time out to think

Suppose a person is tried for a crime and found *not* guilty. Can you conclude that the person is innocent? Why or why not? Why do you think our legal system demands that prosecutors prove guilt, rather than demanding that defendants (suspects) prove innocence? How is this idea related to the fallacy of appeal to ignorance?

EXAMPLE 4 Hasty Generalization

"Two cases of childhood leukemia have occurred along the street where the high-voltage power lines run. The power lines must be the cause of these illnesses."

Analysis The premise of this argument cites two cases in which leukemia occurred on the street with the power lines. But two cases are not enough to establish a pattern, let alone to conclude that the power lines are the cause of the illnesses. This argument illustrates the fallacy of *hasty generalization,* in which a conclusion is drawn from an inadequate number of cases or cases that have not been sufficiently analyzed. If any connection between power lines and leukemia exists, it will have to be established with far more evidence than is provided in this argument (see Unit 5E). We can represent this fallacy with a diagram in which *A* and *B* represent two linked events (Figure 1.4).

A and *B* are linked one or a few times; therefore ...

↓

A causes *B* (or vice versa).

FIGURE 1.4
The fallacy of hasty generalization.

EXAMPLE 5 Limited Choice

"You don't support the President, so you are not a patriotic American."

Analysis This argument has the form

Premise: You don't support the President.

Conclusion: You are not a patriotic American.

The argument suggests that there are only two types of Americans: patriotic ones who support the President and unpatriotic ones who don't. But there are many other possibilities, such as being patriotic while disliking a particular president. The fallacy is called *limited choice* (or *false dilemma*) because it artificially precludes choices that ought to be considered; Figure 1.5 shows one common form of this fallacy. Limited choice also arises with questions such as "Have you stopped smoking?" Both *yes* and *no* answers imply that you smoked in the past. Thus, the question precludes the possibility that you never smoked. (In legal proceedings, questions of this type are disallowed because they attempt to "lead the witness.") Another simple and common form of this fallacy is "You're wrong, so I must be right." ■

p is false;
therefore ...
↓
only q can be true.

FIGURE 1.5
The fallacy of limited choice.

EXAMPLE 6 *Appeal to Emotion*

In a commercial for Michelin tires, a picture of a baby is shown with the words "because so much is riding on your tires."

Analysis If we can consider this an argument at all, it has the form

> Premise: You love your baby.
> Conclusion: You should buy Michelin tires.

The argument makes no appeal to logic; the advertisers just hope that the positive emotions you feel for a baby will make you want to buy their tires. This attempt to evoke an emotional response as a tool of persuasion represents the fallacy of *appeal to emotion*; Figure 1.6 shows its form when the emotional response is positive. Sometimes the appeal is to negative emotions. For example, the statement *if my opponent is elected, your tax burden will rise* tries to convince you that electing the opponent will lead to consequences you won't like. (In this negative form, the fallacy is sometimes called *appeal to force*.) ■

p is associated with a positive emotional response;
therefore ...
↓
p is true.

FIGURE 1.6
The fallacy of appeal to emotion.

EXAMPLE 7 *Personal Attack*

> GWEN: You should stop drinking because it is hurting your grades, endangering people when you drink and drive, and destroying your relationship with your family.
> MERLE: I've seen you drink a few too many on occasion yourself!

Analysis Gwen's argument is well reasoned, with premises offering strong support for her conclusion that Merle should stop drinking. Merle rejects this argument by noting that Gwen sometimes drinks too much herself. Even if Merle's claim is true, it is irrelevant to Gwen's point. Merle has resorted to attacking Gwen personally rather than arguing logically, so we call this fallacy *personal attack* (Figure 1.7). (It is also called *ad hominem,* Latin for "to the person.")

The fallacy of personal attack can also apply to groups. For example, you might hear someone say, "This new bill will be an environmental disaster because its sponsors received large campaign contributions from heavy industrial polluters." This argument is fallacious because it doesn't challenge the bill; instead, it questions the motives of the sponsors. ■

I have a problem with the person or group claiming p.
↓
p is *not* true.

FIGURE 1.7
The fallacy of personal attack.

Time out to think

A person's (or group's) character, circumstances, and motives occasionally *are* logically relevant to an argument. That is why, for example, witnesses in criminal cases often are asked questions about their personal lives. If you were a judge, how would you decide when to allow such questions?

EXAMPLE 8 Circular Reasoning

"Society has an obligation to shelter the homeless because the needy have a right to the resources of the community."

Analysis This argument states the conclusion (*society has an obligation to shelter the homeless*) before the premise (*the needy have a right to the resources of the community*). But if you read them carefully, you'll see that the premise and the conclusion both say essentially the same thing; after all, society has an obligation to provide shelter only if the needy have a right to societal resources. Thus, this argument suffers from *circular reasoning* (Figure 1.8).

FIGURE 1.8 The fallacy of circular reasoning.

EXAMPLE 9 Diversion (Red Herring)

"We should not continue to fund genetic research because there are so many ethical issues involved. Ethics is at the heart of our society, and we cannot afford to have too many ethical loose ends."

Analysis The argument begins with its conclusion—*we should not continue to fund genetic research.* However, the discussion is all about ethics. This argument represents the fallacy of *diversion* (Figure 1.9) because it attempts to divert attention from the real issue (funding for genetics) by focusing on another issue (ethics). The issue to which attention is diverted is sometimes called a *red herring*. (A herring is a fish that turns red when rotten. Use of the term *red herring* to mean a diversion can be traced back to the 19th century, when British fugitives discovered that they could divert bloodhounds from their pursuit by rubbing a red herring across their trail.) Note that personal attacks (see Example 7) are often used as diversions.

FIGURE 1.9 The fallacy of diversion.

EXAMPLE 10 Straw Man

Columnist William F. Buckley has written many articles arguing in favor of legalizing drugs. A speaker who opposes legalization begins by stating, "Mr. Buckley doesn't think there's anything wrong with drug use, but I do."

Analysis William F. Buckley has never said that drug use is acceptable; his arguments for legalization are based on considerations such as effects on the crime rate, cost to society, and difficulties of overcoming addiction. Thus, the speaker has distorted Buckley's views. Any argument based on a distortion of someone's ideas or beliefs is called a *straw man* (Figure 1.10); the term comes from the idea that the speaker has used a poor representation of a person's beliefs in the same way that a straw man is a poor representation of a real man. A straw man is similar to a diversion; the primary difference is that a diversion argues against an unrelated issue, while the straw man argues against a distorted version of the real issue.

FIGURE 1.10 The straw man fallacy.

Review Questions

1. What is logic? Briefly explain how logic can be useful.
2. How do we define *argument*? What is the basic structure of an argument?
3. What is a fallacy? Why is it important to be able to recognize fallacies?
4. In your own words, describe how each argument in the examples is deceptive. Make up your own similar examples for each type of fallacy.

In the News

1. **Editorial Fallacies.** Examine editorials and letters to the editor in your local newspaper. Find at least three examples of fallacies. In each case, describe how the argument is deceptive. If the fallacy represents one or more of the common types described in this unit, describe the types.
2. **Fallacies in Advertising.** Pick a single night and a single commercial television channel and analyze the advertisements shown over a 1-hour period. Describe how each advertisement tries to persuade the viewer, and discuss whether the argument is logical or fallacious. What fraction of the advertisements involve fallacies? Are any fallacy types more common than others?
3. **Fallacies in Politics.** Discuss the tactics used by both sides in a current or recent political campaign. How much of the campaign is/was based on logic and how much on fallacies? Describe some of the fallacies used in the campaign. Overall, do you believe that fallacies influenced (or will influence) the outcome of the vote?
4. **Personal Fallacies.** Describe an instance in which you were persuaded of something that you later decided was untrue. Explain how you were persuaded and why you later changed your mind. Did you fall victim to any fallacies? If so, how might you prevent the same thing from happening in the future?

Exercises

Recognizing Fallacies. Each of the arguments given or described in Exercises 1–16 involves some kind of fallacy. Identify the premises and the conclusion of the argument, and then explain why the argument is deceptive. If the fallacy happens to fit one or more of the types described in the examples of this unit, name the types and explain how they apply.

1. Following President Reagan's defense buildup, the Soviet Union began the process of democratization that ultimately led to its breakup. Therefore, Reagan is responsible for the changes that led to the demise of the Soviet Union.
2. The Golden Rule is a sound ethical principle because it is basic to every system of ethics in every culture.
3. At a hearing to decide a child custody case, a lawyer asks the mother: "Have you stopped using drugs?"
4. Everyone I know is voting for the Governor, so you should vote for her too.
5. My mom says that I should never smoke, but I'm not going to pay any attention because she smoked when she was my age.
6. Most of the great mathematicians in history have been men, so men must be better at mathematics than women.
7. I believe in telepathy, because no one has ever proven that it doesn't exist.
8. My two best teachers were both women, proving that women make better teachers than men.
9. There's no way that Senator Smith's bill can help the cause of gun control because he is one of the biggest recipients of campaign contributions from the National Rifle Association.
10. Violent crime by youth has risen in virtual lockstep with increased violence on television, thus proving that television violence leads to real violence.
11. Since 1960, the percentage of the population over 18 who smoke has decreased from 40% to about 20%. During the same period, the percentage of overweight people has increased from 25% to 35%. Clearly, quitting smoking leads to overeating.

12. On average, boys score higher than girls on standardized mathematics tests (such as the SAT), which proves that boys have a greater aptitude for mathematics than girls. Similarly, girls score higher on verbal tests, proving that girls are better at verbal skills.

13. My little boy loves dolls and my little girl loves trucks, so there's no truth to the claim that boys are more interested in mechanical toys while girls prefer maternal toys.

14. In the 1964 presidential campaign, President Johnson ran a television ad against Barry Goldwater (the Republican candidate who ultimately lost the election) that showed a little girl picking daisy petals and counting them, "One, two, three " In the background, a gruff male voice began a countdown, "Ten, nine, eight . . . ," slowly drowning out the little girl's voice. As the countdown ended, the screen lit up with an atomic explosion. President Johnson's voice then said, "These are the stakes: to make a world in which all of God's children can live or go into the dark."

15. During the 2000 presidential campaign, George W. Bush favored an end to the estate tax. Because estate taxes fall most heavily on the wealthy, the wealthy would realize the greatest benefits from this change. The Gore campaign often argued against Bush's position with statements that began, "George Bush thinks that rich people aren't rich enough "

16. During the 2000 presidential campaign, Al Gore proposed regulatory changes that he claimed would reduce the danger of global warming. The Bush campaign countered with statements that began, "Al Gore believes that government is the solution to all our problems "

Web Projects

Find useful links for Web Projects on the text Web site: www.aw.com/bennett-briggs

1. **Campaign Site.** Find the Web site for a candidate or initiative (for or against) in an upcoming or recent election. Study the arguments given on the site, looking for examples of fallacies. Write a brief summary of the fallacies. Overall, does the site make a strong or weak case for its candidate or position? Explain.

2. **Gun Control Debate.** One of the most contentious debates in American politics concerns the question of whether the United States should enact stricter gun control laws. Visit one or more of the many Web sites for each side in the gun control debate. Identify at least two fallacies given by each side. Without stating which side you support, give your opinion of the logical quality of the arguments on the Web sites for the two sides; that is, do the arguments on one involve more fallacies than those on the other side? Explain.

3. **Death Penalty Debate.** The death penalty is another contentious issue in American politics. Visit one or more of the many Web sites for each side in the death penalty debate. Identify at least two fallacies given by each side. Without stating which side you support, give your opinion of the logical quality of the arguments on the Web sites for the two sides; that is, do the arguments on one involve more fallacies than those on the other side? Explain.

4. **Health Care Debate.** Choose some aspect of the health care debate in the United States, such as whether there should be coverage for prescription drugs or national insurance for all citizens. Find a Web site for each side of the debate. Without stating which side you support, give your opinion of the logical quality of the arguments on the Web sites for the two sides; that is, do the arguments on one involve more fallacies than those on the other side? Explain.

5. **Fallacy Web Sites.** There are many Web sites devoted to the study of fallacies. Visit one, and choose a fallacy of a type not covered in this unit. Explain the fallacy, and give an example of it.

UNIT 1B

PROPOSITIONS AND TRUTH VALUES

Having discussed fallacies in Unit 1A, we now move to studying proper arguments. The building blocks of arguments are called **propositions**—statements that make (propose) a claim that may be either true or false. A proposition must have the structure of a complete sentence and must make a distinct assertion or denial. For example:

- *Joan is sitting in the chair* is a proposition because it is a complete sentence that makes an assertion.
- *I did not take the pen* is a proposition because it is a complete sentence that makes a denial.
- *Are you going to the store?* is not a proposition because it is a question; it does not assert or deny anything.
- *three miles south of here* is not a proposition both because it does not make any claim and because it does not have the structure of a complete sentence.
- *7 + 9 = 2* is a proposition, although it is false, because it can be read as a complete sentence and makes a distinct claim.

> **DEFINITION**
>
> A **proposition** makes a claim (either an assertion or a denial) that may be either true or false; it must have the structure of a complete sentence.

"Contrariwise," continued Tweedledee, "if it was so, it might be; and if it were so, it would be; but as it isn't, it ain't. That's logic."
—LEWIS CARROLL, *Through the Looking Glass*

NEGATION (OPPOSITES)

The opposite of a proposition is called its **negation**. For example, the negation of *Joan is sitting in the chair* is *Joan is not sitting in the chair*, and the negation of $7 + 9 = 2$ is $7 + 9 \neq 2$. If we represent a proposition with a letter such as p, then its negation is *not p*. Note that negations are also propositions, since they have the structure of a complete sentence and may be either true or false.

A proposition has a **truth value** of either true (T) or false (F). If a proposition is true, its negation must be false, and vice versa; thus, a proposition and its negation cannot both be true or both be false. We can represent these facts with a simple **truth table**—a table that has a row for each possible set of truth values. The following truth table shows the possible truth values for a proposition p and its negation *not p*. It has two rows because there are only two possibilities: The first row shows that if p is true (T), *not p* is false (F). The second row shows that if p is false (F), *not p* is true (T).

p	not p
T	F
F	T

by the way...
Although we make only limited use of symbols for logic in this book, an entire branch of mathematics, sometimes called *symbolic logic*, deals with analyzing logical structure through symbols rather than words. In symbolic logic, negation is denoted by a $\sim$ (tilde). Thus, the negation of proposition p (not p) is written $\sim p$.

> **DEFINITIONS**
>
> Any proposition has two possible **truth values:** T = true or F = false.
>
> The **negation** of a proposition *p* is another proposition that makes the opposite claim of *p*. It is written *not p* and has the opposite truth value of *p*.
>
> A **truth table** is a table with a row for each possible set of truth values for the propositions being considered.

EXAMPLE 1 *Negation*

Consider the proposition *Betsy is the fastest runner on the team*. Write its negation. If the negation is false, is Betsy really the fastest runner on the team?

Solution The negation of the given proposition is *Betsy is not the fastest runner on the team*. If the negation is false, the original statement must be true, meaning that Betsy *is* the fastest runner on the team.

Double Negation

> *I cannot say that I do not disagree with you.*
> —GROUCHO MARX

The Groucho Marx quotation in the margin may be an extreme example, but many everyday statements contain double (or multiple) negatives—that is, they involve more than one negation.

We've already seen that a negation, *not p*, must have the opposite truth value of the original proposition *p*. The double negative, *not not p*, must therefore have the same truth value as the original proposition *p*. We can show this fact with a truth table. The first column contains the two possible truth values for *p*; then two additional columns show the corresponding truth values for *not p* and *not not p*.

p	not p	not not p
T	F	T
F	T	F

The only subtlety arises from the fact that, in ordinary language, double negatives rarely involve phrases like "not not." As the following two examples show, we must analyze wording carefully to recognize double negatives.

EXAMPLE 2 *Radiation and Health*

After reviewing data showing an association between low-level radiation and cancer among older workers at the Oak Ridge National Laboratory, Dr. Steve Wing (a health scientist from the University of North Carolina, Chapel Hill) was asked about the possibility of a similar association among younger workers at another national laboratory. He was quoted as saying (*Boulder Daily Camera*, 7/2/00):

> *My opinion is that it's unlikely that there is no association.*

Does Dr. Wing think there is an association between low-level radiation and cancer?

Solution Because of the words "unlikely" and "no association," Dr. Wing's statement contains a double negation. To see the effects of these words clearly, let's start with a simpler proposition:

$$p = \textit{it's likely that there is an association}$$

The word "unlikely" gives us the statement *it's unlikely that there is an association*, which we identify as *not p*. The words "no association" transform this last statement into Dr. Wing's statement, *it's unlikely that there is no association*, which we recognize as *not not p*. Because the double negation has the same truth value as the original proposition, we conclude that Dr. Wing believes it likely that there *is* an association between low-level radiation and cancer.

EXAMPLE 3 *The 2000 Miranda Ruling*

In a June 2000 decision (*Dickerson v. United States*), the U.S. Supreme Court voted 7–2 to uphold the basic requirements of the 1966 Miranda decision (*Miranda v. State of Arizona*). That decision required that suspects taken into custody be informed of their constitutional rights, such as the right to remain silent and the right to legal counsel. In his majority opinion, Chief Justice William Rehnquist wrote:

> . . . [legal] principles weigh heavily against overruling [Miranda] now.

According to this statement, did he feel that legal principles support or oppose the original Miranda decision?

Solution Analyzed carefully, the Chief Justice's statement contains a double negation. The first negation comes from the term "overruling," which alone would imply opposition to the original decision. But the statement argued *against* overruling. Thus, the statement concludes that legal principles support the original Miranda decision.

Negations with *All, Some,* or *None*

Negating propositions that contain words such as *all, some, none,* or *no* can be a bit subtle. For example, what is the negation of the proposition *all birds can fly?* It's tempting to write the negation as *all birds cannot fly.* But a bit of thought shows this is not correct.

The original proposition, *all birds can fly,* is false because there are a few species of birds (such as penguins and ostriches) that do not fly. The second statement, *all birds cannot fly,* is also false; it cannot be the negation of the first statement, because a statement and its negation must have opposite truth values. In fact, all it takes to negate the first statement is to claim that at least one bird cannot fly. Therefore, the simplest way to negate *all birds can fly* is to write *some birds cannot fly,* which is a true statement.

Similarly, the statement *no vegetables contain protein* is false because some vegetables do contain protein. But the negation cannot be *all vegetables contain protein,* because this statement is also false. The negation of *no vegetables contain protein* is the true statement *some vegetables contain protein.*

historical note

Based largely on his confession, Ernesto Miranda was convicted of a 1963 rape and kidnapping. His lawyer argued that the confession should not have been admitted as evidence because Miranda had not been told of his right to remain silent. The Supreme Court agreed and overturned his conviction. However, he was again convicted when retried (on the basis of evidence besides the confession). After his release from prison, Miranda was stabbed to death during a bar fight in 1976. A suspect in his killing chose to remain silent upon arrest, and police never filed charges.

> *If people don't want to come out to the baseball park, then nobody's going to stop them.*
>
> —YOGI BERRA

EXAMPLE 4 *All and Some Negations*

Give the truth value of each statement and write a negation of the statement.

a. All state capitals are the state's largest city.
b. Some states have no personal income tax.

Solution

a. The original statement is false because many state capitals are not their state's largest city; for example, California's capital of Sacramento is far smaller than Los Angeles. The negation is the true statement *some state capitals are not the state's largest city.*

b. The original statement is true; for example, Wyoming has no personal income tax. The negation is *no states have personal income taxes,* which is false. (Note: In this particular example, the statement *all states have personal income taxes* also has the opposite value of the original statement. However, in general, the negation of a *some states* statement must be a *no states* statement; see p. 44.)

Logical Connectors

Propositions are often joined together with *logical connectors*—words such as *and, or,* and *if... then*. For example, consider the following two propositions:

$$p = \text{the test was hard}$$
$$q = \text{I got an A}$$

If we join the two propositions with *and,* we get a new proposition (sometimes called a *compound proposition*) that reads *the test was hard and I got an A.* If we join them with *or,* we get the statement *the test was hard or I got an A.* Although such statements are familiar from everyday speech, in logic we must analyze them carefully.

And Statements (Conjunctions)

In formal logic, an *and* statement is called a **conjunction.** According to the rules of logic, the conjunction *p and q* is true only if *p* and *q* are *both* true. For example, the statement *the test was hard and I got an A* is true only if it *was* a hard test and you *did* get an A.

Each proposition of the conjunction *p and q* can be either true or false. Thus, there are four possible cases: (1) *p* and *q* are both true; (2) *p* is true and *q* is false; (3) *p* is false and *q* is true; and (4) *p* and *q* are both false. The first two columns in the following truth table show the four cases, with one case in each row. The third column shows the resulting truth values for the conjunction *p and q*. Note that the *and* statement is true only in the first case, where both individual propositions are true.

> *by the way...*
>
> In symbolic logic, the connector *and* is denoted by the symbol $\wedge$. Thus, the conjunction *p and q* is written symbolically as $p \wedge q$.

p	q	p and q
T	T	T
T	F	F
F	T	F
F	F	F

> **The Logic of *And***
> Given two propositions *p* and *q*, the statement *p and q* is called their **conjunction**. It is true only if *p* and *q* are *both* true.

EXAMPLE 5 And Statements

Evaluate the truth value of the following two statements.

 a. The capital of France is Paris and Antarctica is cold.
 b. The capital of France is Paris and the capital of America is Madrid.

Solution

 a. The statement contains two distinct propositions: *The capital of France is Paris* and *Antarctica is cold*. Because both propositions are true, their conjunction is also true.
 b. The statement contains two distinct propositions: *The capital of France is Paris* and *the capital of America is Madrid*. Although the first proposition is true, the second is false. Thus, their conjunction is false. ■

EXAMPLE 6 Triple Conjunction

Suppose you are given three individual propositions *p*, *q*, and *r*. Make a truth table for the conjunction *p and q and r*. Under what circumstances is the conjunction true?

Solution When we considered the statement *p and q*, we saw that there were four possible cases for truth values. For each of these four cases, proposition *r* can be either true or false. Thus, the statement *p and q and r* has eight possible cases for truth values. The following truth table contains a row for each of the eight cases; note that the four cases for *p and q* each appear twice, once with *r* true and once with *r* false.

p	q	r	p and q and r
T	T	T	T
T	T	F	F
T	F	T	F
T	F	F	F
F	T	T	F
F	T	F	F
F	F	T	F
F	F	F	F

historical note

The Greek philosopher Aristotle (384–322 B.C.) made the first known attempt to put logic on a rigorous foundation. He believed that truth could be established from three basic laws: (1) A thing is itself. (2) A statement is either true or false. (3) No statement is both true and false. Aristotle's laws were used by Euclid (c. 325–270 B.C.) to establish the foundations of geometry and still lie at the heart of much of modern science, industry, and business.

The conjuction *p and q and r* is true only if all three statements are true; this case is shown in the first row. You can see why if you remember that the conjunction of two statements is true only if both statements are true. Because we can think of *p and q* as a single statement (which is true only when *p* and *q* are both true), the conjunction of this statement with *r* can be true only if *r* is also true.

Time out to think

Given four propositions *p*, *q*, *r*, and *s*, how many rows are required for a truth table of the conjunction *p and q and r and s*? Under what circumstances is the conjunction true?

Or Statements (Disjunctions)

The connector *or* can have two different meanings. If a health insurance policy says that it covers hospitalization in cases of illness *or* injury, it probably means that it covers either illness or injury *or both*. This is an example of the **inclusive *or*** that means "either or both." In contrast, when a restaurant offers a choice of soup *or* salad, you probably are not supposed to choose both. This is an example of the **exclusive *or*** that means "one or the other."

> **Two Types of *Or***
>
> The word *or* can be interpreted in two distinct ways:
>
> - An **inclusive *or*** means "either or both."
> - An **exclusive *or*** means "one or the other, but not both."

Time out to think

Kevin's insurance policy states that his house is insured for earthquakes, fire, *or* robbery. Suppose a major earthquake levels much of his house, and the rest burns in a fire caused by earthquake-damaged gas lines; then his remaining valuables are looted in the aftermath. Explain why Kevin hopes that the *or* in his insurance policy is *inclusive* rather than exclusive.

In general, the only way to determine whether an *or* statement is inclusive or exclusive is by its context. However, in logic we always assume that *or* is *inclusive* unless told otherwise.

A compound statement made with *or* is called a **disjunction.** Because we assume that the *or* is inclusive, the disjunction *p or q* is true if either or both of the propositions are true. Thus, as shown in the following truth table, a disjunction *p or q* is false only if both individual propositions are false.

> **by the way ...**
> In symbolic logic, the connector *or* is denoted by the symbol $\vee$. Thus, the disjunction *p or q* is written symbolically as $p \vee q$.

p	q	p or q
T	T	T
T	F	T
F	T	T
F	F	F

> **The Logic of *Or***
>
> Given two propositions *p* and *q*, the statement *p or q* is called their **disjunction.** In logic, we always assume that *or* is *inclusive* unless told otherwise. Thus, the disjunction is true if either or both of the propositions are true, which means that it is false only if *both* propositions are false.

by the way...
Digital computers use electronic circuitry to represent only two numbers: 0 and 1. A 1 means a circuit carries current and a 0 means a circuit carries no current. Computer scientists can think of 1 as true and 0 as false. They use the logical connectors *and, or,* and *not* to design circuits that accomplish particular tasks. For example, an *and* circuit allows current to pass (its value is 1 = true) only if both incoming circuits carry current.

EXAMPLE 7 Smart Cows?

Consider the statement *airplanes can fly or cows can read*. Is it true?

Solution The statement is a disjunction of two propositions: *airplanes can fly* and *cows can read*. The first proposition is clearly true, while the second is clearly false. As shown in the second row of the preceding truth table, the disjunction *p or q* is true in this case. Thus, the statement *airplanes can fly or cows can read* is true. ■

EXAMPLE 8 Key Word Search

You are doing a research report for which you need to find articles about the federal debt that appeared in either *Time* or *Newsweek* between 1999 and 2001. You have access to a database that allows you to search on date, magazine, and title. How should you structure the search?

Solution In general, using OR widens the search whereas using AND narrows the search. For the date, the key word search

1999 OR *2000* OR *2001*

will generate a list with articles published in any of the three years (because *or* is assumed to be inclusive). Similarly, for the magazine, you would search on

Time OR *Newsweek*

to ensure that both magazines are included on the list.

The title requires a bit more thought. Searching on *federal* OR *debt* would lead to a list with many articles unrelated to the federal debt, such as articles on federal crimes or corporate debt. Searching on *federal* AND *debt* would limit the search to articles containing both words in the title, but might miss relevant articles with different key words in their titles; for example, it would miss an article with the title "the United States budget debt." One way to solve this problem is to use a more complex set of search criteria, such as

(*federal* OR *United States* OR *government*) AND (*debt* OR *budget*)

This search would find articles with any of the following word combinations in their titles: *federal debt, United States debt, government debt, federal budget, United States budget,* or *government budget.*

If . . . Then Statements (Conditionals)

Another common way to connect propositions is with the words "if . . . then," as in the statement "If all politicians are liars, then Representative Smith is a liar." Statements of this type are called **conditional propositions** (or *implications*) because they propose something to be true (the *then* part of the statement) on the *condition* that something else is true (the *if* part of the statement).

We can represent a conditional proposition as *if p, then q*. Proposition *p* is called the **hypothesis** (or *antecedent*) and proposition *q* is called the **conclusion** (or *consequent*).

Let's use an example to examine the truth table for a conditional proposition. Suppose that while running for Congress, candidate Jones claimed:

> *If I am elected, then the minimum wage will increase.*

This proposition has the standard form *if p, then q*, where *p = I am elected* and *q = the minimum wage will increase*. Because each individual proposition can be either true or false, we must consider four possible cases for the truth value of *if p, then q*:

1. ***p* and *q* both true.** In this case, Jones *was* elected (*p* true) and the minimum wage increased (*q* true). Jones kept her campaign promise; her claim, *if I am elected, then the minimum wage will increase,* was true.
2. ***p* true and *q* false.** In this case, Jones was elected, but the minimum wage did not increase. Because things did not turn out as she promised, her claim was false.
3. ***p* false and *q* true.** This is the case in which Jones was *not* elected, yet the minimum wage still increased. The conditional statement makes a claim about what should happen only in the event that Jones is elected. Because she was not elected, she surely did not break any campaign promise, regardless of whether or not the minimum wage increased. According to the rules of logic, Jones's original statement was true.
4. ***p* and *q* both false.** Now we have the case in which Jones was not elected and the minimum wage did not increase. Again, because she was not elected, she surely did not violate her campaign promise, even though the minimum wage did not increase. The rules of logic dictate that the original statement was true.

> **by the way ...**
> In symbolic logic, a right arrow (⇒) is used to represent a conditional; $p \Rightarrow q$ is read "*p* implies *q*."

Time out to think

Suppose Jones had made the following campaign promise: *If I am elected, I will personally eliminate all poverty on Earth.* According to the rules of logic for cases (3) and (4) above, we consider this statement to be true in the event that Jones is *not* elected. Is this logical definition of truth the same one that you would use if you heard her make this promise? Explain.

In summary, the statement *if p, then q* is true in all cases except when the hypothesis *p* is true and the conclusion *q* is false. Here is the truth table for a conditional proposition:

p	q	if p, then q
T	T	T
T	F	F
F	T	T
F	F	T

The Logic of *If . . . Then*

A statement of the form *if p, then q* is called a **conditional proposition** (or *implication*). Proposition *p* is called the **hypothesis** and proposition *q* is called the **conclusion.** The conditional *if p, then q* is true in all cases except the case in which *p* is true and *q* is false.

EXAMPLE 9 *Conditional Truths*

Evaluate the truth of the statement *if* $2 + 2 = 5$, *then* $3 + 3 = 4$.

Solution The statement has the form *if p, then q*, where *p* is $2 + 2 = 5$ and *q* is $3 + 3 = 4$. Both *p* and *q* are clearly false. However, according to the rules of logic, the conditional *if p, then q* is true any time *p* is false, regardless of what *q* says. Thus, the statement *if* $2 + 2 = 5$, *then* $3 + 3 = 4$ is true. ∎

Alternative Phrasings of Conditionals

In ordinary language, conditional statements don't always appear in the standard form *if p, then q*. In such cases, it can be useful to rephrase the statements into the standard form. For example, the claim *I'm not coming back if I leave* has the same meaning as *if I leave, then I'm not coming back*. Similarly, the statement *more rain will lead to a flood* can be recast as *if there is more rain, then there will be a flood*.

Two particularly important ways of phrasing conditionals use the words *necessary* and *sufficient*, respectively. The statement *if p, then q* means that *p* cannot be true unless *q* is true. Therefore, the statement *if p, then q* is equivalent to the statement *q is necessary for p*. For example, the statement *breathing is necessary for living* can be rephrased as *if you are living, then you are breathing*. This does not mean that breathing is the only necessity for living; it may be one of many necessities (such as eating, breathing, and having a heartbeat).

Similarly, the statement *if p, then q* means that *q* will be true if *p* is true. Therefore, the statement *if p, then q* is equivalent to the statement *p is sufficient for q*. For example, the statement *being in Denver is sufficient for being in Colorado* can be rephrased as *if you are in Denver, then you are in Colorado*. This does not mean that being in Denver is necessary for being in Colorado, because many other places are also in Colorado.

historical note

The system of logic presented in this chapter was first developed by the ancient Greeks and is called *binary logic* because a proposition is either true or false, but not both and certainly nothing in between. Recently, mathematicians have developed systems of logic in which other truth values are possible. One form, called *fuzzy logic*, allows for a continuous range of values between absolutely true and absolutely false. Fuzzy logic is likely to be increasingly important in new technologies.

> **Alternative Phrasings of Conditionals**
>
> The following are common alternative ways of stating *if p, then q*:
>
> > *p is sufficient for q* *p will lead to q* *p implies q*
> > *q is necessary for p* *q if p*

EXAMPLE 10 *Rephrasing Conditional Propositions*

Recast each of the following statements in the form *if p, then q*. Discuss whether the statement is true or false.

a. A rise in sea level will be devastating to Florida.
b. A red tag on an item is sufficient to mean it's on sale.
c. Being male is necessary for being president of the United States.

Solution

a. This statement is equivalent to *if sea level rises, then Florida will suffer devastation*. The truth of the statement depends on how much sea level rises. A few inches might have only limited effects, while a few feet could inundate most of the state.
b. This statement can be written as *if an item is marked with a red tag, then it is on sale*. This statement is true only if all red tag items really are on sale.
c. This statement can be expressed as *if a person is president, then the person is male*. While it is true that all American presidents to date have been male, many other nations have had female presidents and it is reasonable to think that the United States may have a female president in the future. Thus, the statement is false. ■

Converse, Inverse, and Contrapositive

The order of the propositions does not matter in a conjunction or disjunction. For example, *p and q* has the same meaning as *q and p*, and *p or q* has the same meaning as *q or p*. However, when we switch the order of the propositions in a conditional, we create a completely different proposition called the **converse**. The following box summarizes definitions for the converse and two other variations on a conditional proposition *if p, then q*.

> **VARIATIONS ON THE CONDITIONAL**
>
Name	Form	Example
> | Conditional | *if p, then q* | If you are sleeping, then you are breathing. |
> | Converse | *if q, then p* | If you are breathing, then you are sleeping. |
> | Inverse | *if not p, then not q* | If you are not sleeping, then you are not breathing. |
> | Contrapositive | *if not q, then not p* | If you are not breathing, then you are not sleeping. |

We can determine the truth values for the converse, inverse, and contrapositive with a truth table. Because all the statements consist of the same two propositions (p, q), the table has four rows. The first two columns show the truth values for p and q, respectively. The next two columns show the truth values for the negations *not p* and *not q*, which are needed for the inverse and contrapositive. The fifth column shows the truth values found previously for the conditional proposition *if p, then q*. We then find the truth values for the converse, inverse, and contrapositive by applying the rule that a conditional is false only when the hypothesis is true and the conclusion is false.

p	q	not p	not q	if p, then q	if q, then p (converse)	if not p, then not q (inverse)	if not q, then not p (contrapositive)
T	T	F	F	T	T	T	T
T	F	F	T	F	T	T	F
F	T	T	F	T	F	F	T
F	F	T	T	T	T	T	T

Note that the truth values for the conditional *if p, then q* are the same as those for its contrapositive. We therefore say that a conditional and its contrapositive are **logically equivalent:** If one is true, so is the other, and vice versa. Similarly, the table shows that the converse and inverse are logically equivalent.

> **DEFINITION**
>
> Two statements are **logically equivalent** if they share the same truth values: If one is true, so is the other, and if one is false, so is the other.

EXAMPLE 11 *Logical Equivalence*

Consider the statement *if a creature is a whale, then it is a mammal*. Write its converse, inverse, and contrapositive. Evaluate the truth of each statement. Which statements are logically equivalent?

Solution The statement has the form *if p, then q*, where p = *a creature is a whale* and q = *a creature is a mammal*. Thus, we find

> converse (*if q, then p*): *if a creature is a mammal, then it is a whale*. This statement is false, because most mammals are not whales.
>
> inverse (*if not p, then not q*): *if a creature is not a whale, then it is not a mammal*. This statement is also false; for example, dogs are not whales, but they are mammals.
>
> contrapositive (*if not q, then not p*): *if a creature is not a mammal, then it is not a whale*. This statement is true, because all whales are mammals.

Note that the original proposition and its contrapositive have the same meaning and are logically equivalent. Similarly, the converse and inverse have the same meaning and are logically equivalent.

TECHNICAL NOTE
The contrapositive is sometimes said to be the "inverse of the converse." You can see why by starting with the converse of *if p, then q*, which is *if q, then p*. Now, making the inverse of this converse gives us the statement *if not q, then not p*, which is the contrapositive of the original conditional *if p, then q*.

Review Questions

1. What is a proposition? Give a few examples, and explain why each is a proposition.
2. What do we mean by the negation of a proposition? Make a simple truth table for a proposition and its negation.
3. Give several examples of propositions and their negations, including some that use words such as *all*, *some*, or *none*.
4. Define *conjunction*, *disjunction*, and *conditional*, and give an example of each in words.
5. What is the difference between an inclusive *or* and an exclusive *or*? Give an example of each.
6. Make a truth table for each of the following: *p and q*; *p or q*; *if p, then q*. Explain all the truth values in the tables.
7. Describe how to make the converse, inverse, and contrapositive of a conditional proposition. Make a truth table for each. Which statement is logically equivalent to the original conditional?

In the News

1. **Logical *Or*.** Find a news article or advertisement in which the connector *or* is used. Is it used in an inclusive or exclusive sense? Explain its meaning in the given context.
2. **Multiple Negation.** Find a news article or advertisement in which a double (or multiple) negation is used. Explain the meaning of the sentence in which it occurs.
3. **Conditional News.** Find a news article or editorial in which a conditional statement is used. Does it appear in *if . . . then* form or can it be rephrased in this form? From the context of the article, can you comment on the truth of the propositions involved? Is the entire conditional true or false? Explain.

Exercises

Is It a Proposition? For Exercises 1–6, determine whether each statement is a proposition and give your reasons.

1. High taxes penalize the wealthy.
2. Over the hill and far away.
3. All cabbages are vegetables.
4. Some of her whiskers are gray.
5. Have you seen her latest movie?
6. My dog is a Rottweiler.

Negation. For Exercises 7–14, write the negation of each proposition. Then state the truth value of the original proposition and of its negation.

7. London is the capital of England.
8. The Gettysburg Address was given by George Washington.
9. Caesar was a Roman.
10. Mark Twain wrote *Tom Sawyer*.
11. All cowboys wear boots.
12. No person is above the law.
13. Some dogs are obedient.
14. Some folks don't know how to listen.

Multiple Negations. Each of the statements in Exercises 15–20 contains a double or multiple negation. Analyze each statement, explaining the negation and what the statement means.

15. The city council did not approve the no-confidence vote for the police chief.
16. The Senator opposes the ban on anti-war demonstrations.
17. Congress voted against overriding the veto of the tax increase bill.

18. The Senate voted to override the President's veto on the disaster relief bill.
19. The overturning of the anti-discrimination policy violates the Constitution.
20. The ban on anti-nuclear demonstrations was lifted.

And Statements. Each of the statements in Exercises 21–26 connects two individual propositions with the word *and*. State the two individual propositions and, if possible, whether each is true or false. Then state whether the entire statement is true or false, and explain why.

21. Quebec is the capital of Canada and Moscow is the capital of Russia.
22. Emily Dickinson was an American poet and Boris Yeltsin was a Russian president.
23. Ben is married and Ben is a bachelor.
24. $3 \times 7 = 21$ and $6 + 3 = 9$.
25. Grass is green and skies are blue.
26. Some people are nice and some people are mean.

Interpreting Or. For each of the statements in Exercises 27–32, state whether *or* is being used in the exclusive or inclusive sense. Explain.

27. The menu offers a choice of appetizer or dessert.
28. The car warranty covers parts for 3 years or 36,000 miles.
29. If I win the lottery, I will go to Brazil or Nepal.
30. I will wear either boots or shoes.
31. The road will be made of asphalt or concrete.
32. The insurance policy covers fire or theft.

Or Statements. Each of the statements in Exercises 33–38 connects two individual propositions with the word *or*. State the two individual propositions and, if possible, whether each is true or false. Then state whether the entire statement is true or false, and explain why.

33. Seattle is the capital of the United States or Paris is the capital of France.
34. Julia Roberts was an 18th-century monk or Mark Twain was a writer.
35. Bill Clinton or Martin Luther King was a U.S. president.
36. Ben is married or Ben is a bachelor.
37. $8 \times 4 = 32$ or $7 \times 3 = 22$.
38. $3 \times 7 = 21$ or $6 + 3 = 9$.

Key Word Searches. Describe how you would structure a key word search to find the information described in each of Exercises 39–42.

39. Articles about earthquakes in California
40. Articles that mention either earthquakes or volcanic eruptions (or both) that occurred in Japan
41. Articles about Martin Luther King's assassination
42. Articles about how the Cold War affected relations between the United States and Argentina

If . . . Then Statements. For Exercises 43–50, identify the hypothesis and conclusion in each statement, and state their truth values. Then determine whether the entire statement is true or false, and explain why.

43. If San Francisco is in California, then San Francisco is in the Pacific time zone.
44. If London is the capital of England, then Paris is the capital of China.
45. If Paris is the capital of China, then London is the capital of England.
46. If the Pope is unmarried, then the Pope is a bachelor.
47. If sparrows can fly, then sparrows are birds.
48. If butterflies can fly, then butterflies are birds.
49. If $2 + 2 = 5$, then primates are not mammals.
50. If $5 \times 7 = 0$, then fish can fly.

Rephrasing Conditionals. Rephrase each statement in Exercises 51–56 as a conditional proposition with the form *if p, then q*. Identify *p* and *q*. Then state whether the conditional proposition is true or false, and explain why.

51. A resident of Miami is a resident of the United States.
52. She's a lawyer if she's a member of Congress.
53. Playing the saxophone is necessary for being considered a musician.
54. Taking a biology course is sufficient to meet a one-course science requirement.
55. Being a male is necessary for having prostate cancer.
56. Living in Santa Fe is sufficient to be living in New Mexico.

Converse, Inverse, and Contrapositive. For Exercises 57–62, write the converse, inverse, and contrapositive of the given proposition and state whether each is true or false. Which statements are logically equivalent?

57. If Marco lives in Chicago, then he lives in the United States.

58. If the patient is alive, then the patient is sleeping.
59. If I go swimming, then I will get wet.
60. If it's a penguin, then it can fly.
61. If the car's engine is running, there is gas in the tank.
62. If you live near the ocean, then you live in Hawaii.

Famous Quotes. Rephrase each of the quotations in Exercises 63–66 in conditional (*if p, then q*) form. Then state whether you think the proposition is true or false, and defend your opinion.

63. The young man who has not wept is a savage, and the old man who will not laugh is a fool. —GEORGE SANTAYANA
64. Laugh and the world laughs with you. —ANTHONY BURGESS
65. It is my experience that folks who have no vices have very few virtues. —ABRAHAM LINCOLN
66. The man who reads nothing at all is better educated than the man who reads nothing but newspapers. —THOMAS JEFFERSON

Writing Conditional Propositions. In Exercises 67–70, write your own example of a proposition that has the indicated characteristics. If it is not possible to write such a proposition, explain why.

67. A true conditional proposition whose converse is false
68. A true conditional proposition whose converse is true
69. A false conditional proposition whose converse is true
70. A false conditional proposition whose converse is false
71. **Library Search.** Suppose that you wanted to find the call number of the book *Finnegan's Wake* by James Joyce. For each of the following key word search combinations, state whether the book would be on the search list. Explain.
 a. Finnegan
 b. Finnegan and Joyce
 c. Joyce or Yeats
 d. Joyce and James
 e. Joyce and Yeats
 f. Finnegan and (Joyce or Yeats)

72. **Alimony Tax Laws.** The federal tax policy on alimony payments is as follows:
 (1) Alimony you receive after you remarry is taxable if the payer did not know you had remarried.
 (2) Alimony you receive after you remarry is not taxable if the payer did know you had remarried.
 (3) Alimony you pay is never deductible by the payer.
 Rephrase these three statements as conditional propositions.

73. **Artificial Intelligence.** Consider the following quote from *TIME Magazine* (Ray Kurzweil, May 22, 2000):
 Let's consider the requirements for a computer to exhibit human-level intelligence.... A necessary (but not sufficient) condition is the requisite processing power, which I estimate at about 20 million billion calculations per second.
 Rephrase this statement as an *if... then* proposition and explain what it means.

Truth Tables. Exercises 74–79 each list two statements in which *p*, *q*, and *r* represent propositions. Make a truth table for each of the two given statements. Are the two statements logically equivalent? Explain. (*Hint:* Evaluate expressions within the parentheses first.)

74. *not* (*p* and *q*); *not p* or *not q*
75. *not* (*p* or *q*); *not p* and *not q*
76. *not* (*p* and *q*); *not p* and *not q*
77. *not* (*p* or *q*); *not p* or *not q*
78. (*p* and *q*) or *r*; (*p* or *r*) and (*p* or *q*)
79. (*p* or *q*) and *r*; (*p* and *r*) or (*q* and *r*)

Web Project

Find useful links for Web Projects on the text Web site: www.aw.com/bennett-briggs

1. **Internet Searches.** Investigate rules for key word searches on several Web sites (such as Yahoo or Britannica) or search engines (such as Google or Lycos). Are the rules consistent with the laws of logic described in this chapter? Of the sites you checked, which offers the most "user-friendly" search? Explain why you like it.

UNIT 1C

SETS AND VENN DIAGRAMS

We've seen that propositions come in many forms. The only general requirement is that a proposition must make a clear claim (assertion or denial). In this unit, we focus on propositions that claim a relationship between two categories of things. For example, the proposition *all whales are mammals* makes the claim that the category *whales* is entirely contained within the category *mammals*.

Propositions of this type are most easily studied with the aid of two key ideas. The first is the idea of a *set*, which is really just another way of describing a collection or category. The second is the idea of a *Venn diagram*, which is a simple visual way of illustrating relationships between sets. Both ideas are very useful for organizing information and hence are important tools of critical thinking.

RELATIONSHIPS AMONG SETS

A **set** is a collection of objects, living or nonliving. The **members** of a set are the specific objects within it. For example:

- The members of the set *days of the week* are the individual days Sunday, Monday, Tuesday, Wednesday, Thursday, Friday, and Saturday.
- The members of the set *American military services* are the Army, Navy, Air Force, and Marine Corps.
- The members of the set *Academy Award winning actresses* are the individual actresses who have won Academy Awards.

Set Notation

Sets are often described by listing all their individual members within a pair of braces, { }. For example, the set *American military services* can be written as

{Army, Navy, Air Force, Marine Corps}

Notice that every member is listed within the braces, with each member separated from the next by a comma.

Some sets have so many members that it would be difficult or impossible to list all of them. In that case, we can use three dots, . . . , to indicate that the list continues in the same basic manner. (The three dots are called an *ellipsis,* but most people just say "dot-dot-dot" when reading them.) If the dots come at the end of a list, they indicate that the list continues indefinitely. For example, we can write the set of *dog breeds* as

{Rottweiler, German Shepherd, Poodle, . . . }

The three dots indicate that the list continues as you would expect—in this case, with the names of all other dog breeds. Note that it does not matter how many members you list, as long as the pattern is clear; listing three members explicitly, as in the dog breed list, is usually enough.

by the way . . .

The four military services are organized under the United States Department of Defense. The United States Coast Guard is often grouped with these services, but it is administered by the Department of Transportation.

The whole is more than the sum of its parts.

— ARISTOTLE, Metaphysica

A brief review...

SETS OF NUMBERS

In mathematics, we commonly work with the following important sets of numbers.

The set of **natural numbers** (or *counting numbers*) is

$$\{1, 2, 3, \ldots\}$$

We can represent the natural numbers on a *number line* with equally spaced dots beginning at 1 and continuing to the right forever.

The set of **whole numbers** is the same as the set of natural numbers except it includes zero. Thus, its members are

$$\{0, 1, 2, 3, \ldots\}$$

We can represent the whole numbers on a number line with equally spaced dots beginning at 0 and continuing to the right forever.

The set of **integers** includes the whole numbers and their negatives. Thus, its members are

$$\{\ldots, -3, -2, -1, 0, 1, 2, 3, \ldots\}$$

On the number line, the integers extend forever both to the left and to the right.

The set of **rational numbers** includes the integers and the fractions that can be made by dividing one integer by another, as long as we don't divide by zero. (The word *rational* refers to a *ratio* of integers.) In other words, rational numbers can be expressed in the form

$$\frac{x}{y}, \text{ where } x \text{ and } y \text{ are integers and } y \neq 0$$

(Recall that the symbol $\neq$ means "is not equal to.")

When expressed in decimal form, rational numbers are either terminating decimals with a finite number of digits (such as 0.25, which is $\frac{1}{4}$) or repeating decimals in which a pattern repeats over and over (such as 0.333..., which is $\frac{1}{3}$).

Irrational numbers are numbers that *cannot* be expressed in the form $\frac{x}{y}$. When written as decimals, irrational numbers neither terminate nor have a repeating pattern. For example, the number $\sqrt{2}$ is irrational because it cannot be expressed exactly in a form x/y; as a decimal, we can write it as 1.414213562..., where the dots mean that the digits continue forever with no pattern. The number π is also an irrational number, which as a decimal is written 3.14159265....

The set of **real numbers** consists of both rational and irrational numbers; hence it is represented by the entire number line. Each point on the number line has a corresponding real number, and each real number has a corresponding point on the number line. In other words, the real numbers are the integers and "everything in between." A few selected real numbers are shown on the number line below.

Examples:

- The number 25 is a natural number, which means it is also a whole number, an integer, a rational number, and a real number.
- The number −6 is an integer, which means it is also a rational number and a real number.
- The number $\frac{2}{3}$ is a rational number, which means it is also a real number.
- The number 7.98418... is an irrational number; the dots indicate that the digits continue forever with no particular pattern. It is also a real number.

In other cases, the dots may come in the middle of the list, representing members that were not explicitly written. For example, the set

$$\{A, B, C, \ldots, Z\}$$

represents the entire alphabet in capital letters; the dots are simply a convenience, saving us from having to write every letter.

Although it is less common, the three dots can also come at the beginning of a list. For example, the set

$$\{\ldots, -3, -2, -1\}$$

represents all negative integers; here, the dots indicate that the list continues to the left, to ever smaller (more negative) numbers.

> **DEFINITIONS**
>
> A **set** is a collection of objects; the individual objects are the **members** of the set. We often describe sets by listing their members within a pair of braces, { }. If there are too many members to list, we can use three dots, . . . , to indicate a continuing pattern.

Time out to think

How would you use the braces notation to describe the set of students in your mathematics class? How about the set of countries you have visited? Choose one more example of a set that affects you personally, and state how you would describe it with the braces notation.

EXAMPLE 1 *Set Notation*

Use braces to write the contents of each of the following sets:

a. the set of *planets in our solar system*
b. the set of *years of the Cold War*, generally taken to have started in 1945 and ended in 1991
c. the set of *natural numbers greater than 5*

Solution

a. The set of *planets in our solar system* is {Mercury, Venus, Earth, Mars, Jupiter, Saturn, Uranus, Neptune, Pluto}.

b. The set of *years of the Cold War* is {1945, 1946, 1947, . . . , 1991}; the dots indicate that the list includes all the years between 1947 and 1991, even though they are not listed explicitly.

c. The set of *natural numbers greater than 5* is {6, 7, 8, . . . }; the dots indicate that the list continues to ever-higher numbers.

Illustrating Relationships with Venn Diagrams

The English logician John Venn (1834–1923) invented a simple visual way of describing relationships between sets. His diagrams, now called **Venn diagrams,** use circles to represent sets. Venn diagrams are fairly intuitive and best learned through examples.

Consider the sets *whales* and *mammals*. Because every member of the set *whales* is also a member of the set *mammals*, we say that the set *whales* is a **subset** of the set *mammals*. As shown in Figure 1.11, we represent this relationship in a Venn diagram by drawing the circle for whales *inside* the circle for mammals.

FIGURE 1.11 The set *whales* is a subset of the set *mammals*.

historical note

John Venn was an ordained priest who wrote books on logic, statistics, and probability. He also wrote a history of Cambridge University, where he was both a student and a teacher. Though well respected for many of his contributions to logic, he is best known for the diagrams that bear his name.

TECHNICAL NOTE
Some logicians distinguish between Venn diagrams and Euler (pronounced "oiler," after Leonhard Euler) diagrams, depending on the precise way the circles are used. In this book, we use the term *Venn diagrams* for all circle diagrams.

The circles are enclosed by a rectangle, so this diagram has three regions:

- The inside of the *whales* circle represents all whales.
- The region outside the *whales* circle but inside the *mammals* circle represents mammals that are not whales (such as cows, bears, and people).
- The region outside the *mammals* circle represents non-mammals; from the context, we can interpret this region to represent animals (or living things) that are not mammals, such as birds, fish, and insects.

Notice that the diagram illustrates only the relationship between the sets; the sizes of the circles do not matter.

Next, consider the sets *dogs* and *cats*. A pet can be either a dog or a cat, but not both. We draw the Venn diagram with separated circles that do not touch, and we say that *dogs* and *cats* are **disjoint sets** (Figure 1.12). Again, we enclose the circles in a rectangle. This time, the context suggests that the region outside both circles represents pets that are neither dogs nor cats, such as birds and hamsters.

For our last general case, consider the sets *nurses* and *women*. As shown in Figure 1.13, these are **overlapping sets** because it is possible for a person to be both a woman and a nurse. Because of the overlapping region, this diagram has four regions:

- The overlapping region represents people who are both women and nurses—that is, female nurses.
- The non-overlapping region of the *nurses* circle represents nurses who are not women—that is, male nurses.
- The non-overlapping region of the *women* circle represents women who are not nurses.
- From the context, we interpret the region outside both circles to represent people who are neither nurses nor women—that is, men who are not nurses.

FIGURE 1.12 The set *dogs* is disjoint from the set *cats*.

FIGURE 1.13 The sets *nurses* and *women* are overlapping.

Note that the sizes of the regions are not important; for example, the small size of the overlapping region does not imply that female nurses are less common than male nurses. In fact, it is possible that some of the regions might have *no* members. For example, imagine that Figure 1.13 represents a single clinic where all four nurses are male. In that case, the overlapping region, which represents female nurses, would have no members. Speaking more generally, we use overlapping circles whenever two sets *might* have members in common.

> ### Set Relationships and Venn Diagrams
> Two sets *A* and *B* may be related in three basic ways:
>
> - *A* may be a **subset** of *B*, meaning that all members of *A* are also members of *B*. The Venn diagram for this case shows the circle for *A* inside the circle for *B*.
> - *A* may be **disjoint** from *B*, meaning that the two sets have no members in common. The Venn diagram for this case consists of separated circles that do not touch.
> - *A* and *B* may be **overlapping** sets, meaning that the two sets share some of the same members. The Venn diagram for this case consists of two overlapping circles. We also use overlapping circles for cases in which we are not sure if the two sets share common members.

TECHNICAL NOTE Mathematically, there is a fourth way that two sets can be related: They can be *equal*, meaning that they have precisely the same members.

EXAMPLE 2 Venn Diagrams

Describe the relationship between the given pairs of sets and draw a Venn diagram showing this relationship. Interpret all the regions of the Venn diagram.

a. *Democrats* and *Republicans* (party affiliations)
b. *Nobel Prize winners* and *Pulitzer Prize winners*

Solution

a. A person can be registered for only one political party, so the sets *Democrats* and *Republicans* are disjoint. Figure 1.14 shows the Venn diagram. The region outside both circles represents people who are neither Democrats nor Republicans—that is, people who are registered for other political parties, who are independent, or who are not registered.

FIGURE 1.14

b. It is possible for a person to win both a Nobel Prize and a Pulitzer Prize, so these are overlapping sets. Figure 1.15 shows the Venn diagram. The region outside both circles represents people who have not won either prize, which is most of humanity.

FIGURE 1.15

by the way...
Among the people who have won both a Nobel Prize (in literature) and a Pulitzer Prize are Toni Morrison, Ernest Hemingway, William Faulkner, John Steinbeck, and Saul Bellow.

EXAMPLE 3 Sets of Numbers

Draw a Venn diagram showing the relationships among the sets of *natural numbers*, *whole numbers*, *rational numbers*, and *real numbers*. Where are irrational numbers found in this diagram? (If you've forgotten the meanings of these number sets, see *A Brief Review* on p. 36.)

Solution Natural numbers are also whole numbers; thus, the natural numbers are a subset of the whole numbers. Similarly, all whole numbers are integers, so the whole numbers are a subset of the integers. The integers, in turn, are a subset of the rational numbers, and the rational numbers are a subset of the real numbers. Figure 1.16 shows these relationships with a set of nested circles. Because irrational numbers are real numbers that are not rational, they are represented by the region outside the *rational numbers* circle but inside the *real numbers* circle.

FIGURE 1.16

CATEGORICAL PROPOSITIONS

Now that we have discussed general relationships between sets, we are ready to study propositions that make claims about sets. For example, the proposition *all whales are mammals* makes a very specific claim—namely, that the set *whales* is a subset of the set *mammals*. Propositions of this type are called **categorical propositions** because they claim a particular relationship between two categories or sets.

Like all propositions, categorical propositions must have the structure of a complete sentence. Categorical propositions also have another important general feature: Of the two sets in a categorical proposition, one set appears in the subject of the sentence and the other appears in the predicate. For example, in the proposition *all whales are mammals*, the set *whales* is the **subject set** and the set *mammals* is the **predicate set.** We usually use the letter S to represent the subject set and P for the predicate set, so we can rewrite *all whales are mammals* as

all S are P, where S = *whales* and P = *mammals*

As shown in the following box, categorical propositions take one of four basic forms.

THE FOUR STANDARD CATEGORICAL PROPOSITIONS

Form	Example	Subject Set (S)	Predicate Set (P)
All S are P	All whales are mammals	whales	mammals
No S are P	No fish are mammals	fish	mammals
Some S are P	Some doctors are women	doctors	women
Some S are not P	Some teachers are not men	teachers	men

Because categorical propositions express a relationship between two sets, we can use what we've already learned about Venn diagrams to make visual representations of categorical propositions. However, in the case of overlapping sets, we must be specific in stating which parts of the overlap are addressed by the proposition. The following example should make this idea clear.

EXAMPLE 4 *Venn Diagrams for Categorical Propositions*

Draw a Venn diagram to represent each of the four sample propositions in the box above.

Solution The proposition *all whales are mammals* claims that the set $S =$ *whales* is a subset of $P =$ *mammals*. Figure 1.17 shows the Venn diagram. Note that all propositions of the form *all S are P* would have the same basic Venn diagram.

FIGURE 1.17 The Venn diagram for *all S are P*.

The proposition *no fish are mammals* tells us that the set $S =$ *fish* is disjoint from the set $P =$ *mammals*; that is, they have no common members. The Venn diagram therefore shows two separated circles (Figure 1.18).

FIGURE 1.18 The Venn diagram for *no S are P*.

The Venn diagram for the proposition *some doctors are women* requires overlapping circles for S = *doctors* and P = *women*. However, we must be specific about the claim made in the proposition. In this case, the proposition asserts that there *are* some people (at least one) who are both doctors and women; we indicate this by putting an X in the overlapping region of the diagram (Figure 1.19). Note that the proposition does not tell us whether there are also doctors who are not women, nor does it tell us whether there are women who are not doctors. Without further information, we do not know whether the non-overlapping regions contain any members.

FIGURE 1.19 The Venn diagram for *some S are P*. The X indicates that the overlapping region has at least one member; no claim is made about whether other regions also have members.

The proposition *some teachers are not men* also requires overlapping circles (Figure 1.20). Specifically, this proposition asserts that there are some people (at least one) who are in the S = *teachers* circle but not in the P = *men* circle. Therefore, we put an X in the non-overlapping region of the *teachers* circle. No claim is made about whether there are teachers who are also men or men who are not teachers.

FIGURE 1.20 The Venn diagram for *some S are not P*. The X indicates the region claimed to have at least one member; no claim is made about other regions.

Time out to think

The principal of an elementary school states that at her school some teachers are not men. Can you conclude that some of the teachers *are* men? Why or why not? Can you conclude that *none* of the teachers are men? Why or why not?

Putting Categorical Propositions in Standard Form

Many statements in everyday speech make claims about relationships between two categories, but don't look precisely like one of the four standard forms for categorical propositions. It's often useful to rephrase such statements in one of the standard forms. For example, the statement *all diamonds are valuable* can be rephrased to read *all diamonds are things of value*. Now the statement has the form *all S are P*, where S = *diamonds* and P = *things of value*.

EXAMPLE 5 *Rephrasing in Standard Form*

Rephrase each of the following statements in one of the four standard forms for categorical propositions. Then draw the Venn diagram.

 a. Some birds can fly. **b.** Elephants never forget.

Solution

 a. *Some birds can fly* can be rephrased as *some birds are animals that can fly*. It now has the form *some S are P*, where S = *birds* and P = *animals that can fly*. Figure 1.21a shows the Venn diagram, with an X in the overlapping region to indicate the claim that there *are* some birds that can fly.

 b. *Elephants never forget* can be rephrased as *no elephants are creatures that forget*. This proposition has the form *no S are P*, shown in Figure 1.21b, where S = *elephants* and P = *creatures that forget*.

FIGURE 1.21

Negations of Categorical Propositions

In Unit 1B, we discussed how to negate propositions that contain the word *all*, *some*, *none*, or *no*. We now see that these propositions were categorical propositions. The following table summarizes what we found earlier, but now using standard forms.

NEGATIONS OF CATEGORICAL PROPOSITIONS

Proposition	Negation
All S are P	Some S are not P
No S are P	Some S are P
Some S are P	No S are P
Some S are not P	All S are P

EXAMPLE 6 Negations

State the negation for each of the propositions in Example 5. Confirm that the proposition and its negation have opposite truth values.

Solution
 a. The statement *some birds are animals that can fly* is true. From the box above, the negation of this *some S are P* statement is *no S are P*, which in this case means *no birds are animals that can fly*. This statement is false because there *are* birds that fly, so the original proposition and its negation have opposite truth values.
 b. The proposition *no elephants are creatures that forget* has the form *no S are P*, where *S = elephants* and *P = creatures that forget*. Its negation has the form *some S are P*, which means *some elephants are creatures that forget*. Although it is hard to say whether either statement is really true, it is clear that the proposition and its negation have opposite truth values. ■

VENN DIAGRAMS WITH THREE SETS

Venn diagrams are particularly useful for dealing with three sets that may overlap one another. For example, suppose you are conducting a study to learn how teenage employment rates differ between boys and girls and between honor students and others. For each teenager in your study, you need to record the answers to these three questions:

- Is the teenager a boy or a girl?
- Is the teenager an honor student or not?
- Is the teenager employed or not?

Figure 1.22 shows a Venn diagram that can help you organize this information. Notice that it has three circles, representing the sets *boys, honor students,* and

FIGURE 1.22 A Venn diagram for three overlapping sets.

employed. Because the circles overlap one another, they form a total of eight regions (including the region outside all three circles). Each region is labeled in the figure; be sure that you understand how the labels correspond to the overlaps shown.

Time out to think

Could the same information be recorded in a diagram in which the three circles represented the sets *girls* (rather than boys), *honor students*, and *employed*? Could it be recorded in a diagram with the three circles representing the sets *girls*, *boys*, and *unemployed*? Explain.

EXAMPLE 7 Makeup of the Study

You hire an assistant to help you with the study of teenage employment described above. He focuses on a small group of teenagers who are all enrolled in the same school. He reports the following facts about this group:

- Some of the honor-student boys are unemployed.
- Some of the non-honor-student girls are employed.

Put X's in the appropriate places in Figure 1.22 to indicate the regions that you can be sure have members. Based on this report, do you know whether any of the school's honor-student girls are unemployed? Why or why not?

Solution Figure 1.23 shows the X's in the correct regions of the diagram (compare with the labels in Figure 1.22). The region corresponding to unemployed honor-student girls lies outside all three circles. There is no X in this region because the given information does not tell us whether this region has members. Thus, among the teenagers in the school, we do not know whether any honor-student girls are unemployed.

FIGURE 1.23 The X's mark regions known to have members.

EXAMPLE 8 Color Monitors

Color television and computer monitors make all the colors you see by combining *pixels* (short for *picture elements*) that display just three colors: red, green, and blue. In pairs, these combinations of colors give the following results (with colors at equal strength):

Combination	Result
Red-green	Yellow
Red-blue	Purple (or magenta)
Blue-green	Light blue (or cyan)

White is made by combining all three colors, and black is made by using none of the colors. Draw a Venn diagram to represent all this color information.

Solution The Venn diagram in Figure 1.24 represents the color combinations with three overlapping circles. The basic colors (red, green, and blue) appear in regions where there is no overlap. The red-green, red-blue, and blue-green combinations

appear in regions where two circles overlap. White appears in the central region, where all three colors are present, while black appears outside all three circles, where no colors are present. A real monitor can make a much wider range of colors by varying the relative strengths of the colors.

FIGURE 1.24 The color combinations possible from red, green, and blue (in equal strength).

Time out to think

If you have a magnifying glass, hold it close to your television to see the individual red, blue, and green dots. If you don't have a magnifying glass, try splashing a few drops of water on the screen, and observe the colors that you see. How do you think a high-resolution monitor (such as a monitor for HDTV) differs from a lower-resolution one?

MORE USES OF VENN DIAGRAMS

So far we have used Venn diagrams only to describe relationships, such as whether two sets overlap or whether overlapping sets share common members. Venn diagrams can be even more useful when we add specific information, such as the number of members in each set or overlapping region.

Consider the study summarized in Table 1.1. This study was designed to learn whether a pregnant mother's status as a smoker or nonsmoker affects whether she delivers a low or normal birth weight baby. Notice that the table shows four numbers, which correspond to the four possible combinations of the baby's birth weight status and the mother's smoking status. To be sure that you are reading the table correctly, make sure you see how it shows the following facts:

- 18 babies were born with low birth weight to smoking mothers.
- 132 babies were born with normal birth weight to smoking mothers.
- 14 babies were born with low birth weight to nonsmoking mothers.
- 186 babies were born with normal birth weight to nonsmoking mothers.

TABLE 1.1 Distribution of 350 Births by Birth Weight Status and Mother's Smoking Status

		Baby's birth weight status	
		LOW BIRTH WEIGHT	NORMAL BIRTH WEIGHT
Mother's smoking status	SMOKER	18	132
	NONSMOKER	14	186

U.S. National Center for Health Statistics.

Although a table is a good way to summarize data, sometimes a visual representation is easier to understand. Because there are four combinations of birth weight and smoking status, we can use a Venn diagram with two overlapping circles (creating four regions on the diagram). Figure 1.25 shows one way of making the diagram. The circles represent the sets *smoking mothers* and *low birth weight babies*. The labels show how each region corresponds to one of the combinations in Table 1.1. This Venn diagram differs from those we have studied earlier in that it also shows the number of members in each region, taken from Table 1.1.

FIGURE 1.25 Venn diagram for the data in Table 1.1.

by the way...

Over 400,000 premature or low birth weight babies are born each year in the United States. These babies are much more likely than others to suffer health problems including brain hemorrhages, chronic lung disease, respiratory distress, visual impairment, cerebral palsy, and a variety of developmental problems. They often require weeks or months of high-technology intensive and prolonged out-patient care.

It is now easy to see how smoking affected babies in the study. First, notice that normal birth weight babies were much more common than low birth weight babies among both smokers and nonsmokers. However, notice also that the smoking mothers had a lower proportion of normal birth weight babies and a higher proportion of low birth weight babies than the nonsmokers. This suggests that smoking increases the risk of having a low birth weight baby, a fact that has been borne out by careful statistical analysis of this and other studies.

Time out to think

Explain why the Venn diagram also could be drawn with circles for the sets *nonsmoking mothers* and *normal birth weight babies*. Draw the diagram for this case, and put the four numbers in the right places.

EXAMPLE 9 Blood Types

Human blood is often classified according to whether three antigens, A, B, and Rh, are present or absent. Blood type is stated first in terms of the antigens A and B: Blood containing only A is called type A, blood containing only B is called type B, blood containing both A and B is called type AB, and blood containing neither A nor B is called type O. The presence or absence of Rh is indicated by adding the word *positive* (present) or *negative* (absent) or its symbol. Table 1.2 shows the eight blood types that result and the percentage of people with each type in the U.S. population. Draw a Venn diagram to illustrate these data.

TABLE 1.2 Blood Types in the U.S. Population

BLOOD TYPE	PERCENTAGE OF POPULATION
A positive	34%
B positive	8%
AB positive	3%
O positive	35%
A negative	8%
B negative	2%
AB negative	1%
O negative	9%

Solution We can think of the three antigens as three sets *A*, *B*, and *Rh* (positive) and draw a Venn diagram with three overlapping circles. Figure 1.26 shows the eight regions, each labeled with its type and percentage of the population. For example, the central region corresponds to the presence of all three antigens (AB positive), so it is labeled with 3%. You should check that all eight regions are labeled according to the data from Table 1.2.

FIGURE 1.26 Venn diagram for blood types.

historical note

Human blood groups were discovered in 1901 by Austrian biochemist Karl Landsteiner. In 1909, he classified the groups he had discovered as A, B, AB, and O. He also showed that blood transfusions could be done successfully if the donor and recipient were of the same blood type. For this work, he was awarded the 1930 Nobel Prize in Medicine.

Review Questions

1. What is a set? Describe the use of braces notation for listing the members of a set.
2. What is a Venn diagram? How do we show that one set is a subset of another in a Venn diagram? How do we show disjoint sets? How do we show overlapping sets?
3. List the four standard categorical propositions. Give an example of each type, and draw a Venn diagram for each of your examples.
4. Briefly discuss how you can put a categorical proposition into one of the standard forms if it is not in such a form already.
5. State the negation of each of the four standard categorical propositions. Give an example of each showing that the proposition and its negation have opposite truth values.
6. Explain how to draw a Venn diagram for three overlapping sets. Discuss the types of information that can be shown in such diagrams.
7. Explain how to read a table such as Table 1.1 and how to show the information in a Venn diagram.
8. Briefly show why a Venn diagram with two overlapping circles consists of four distinct regions, while a Venn diagram with three overlapping circles consists of eight distinct regions.

In the News

1. **Categorical Propositions.** Find at least three examples of categorical propositions in news articles or advertisements. State the sets involved in each proposition and draw a Venn diagram for each proposition.
2. **Venn Diagrams in Your Life.** Describe a situation in your own life that could be described or organized using a Venn diagram.
3. **Quantitative Diagram.** Find a news article or research report with a table similar to Table 1.1, 1.2, or 1.3. Draw a Venn diagram to represent the data in the table.

Exercises

Classifying Numbers. In Exercises 1 and 2, consider the sets *natural numbers, whole numbers, integers, rational numbers,* and *real numbers*. Identify from the list the *simplest* set that describes each number given. Explain.

1. a. 2.3 b. $-\frac{3}{2}$ c. 3
 d. -5 e. 100.1 f. -6.1
2. a. 23.8 b. $\sqrt{3}$ c. 15
 d. -25 e. $\frac{4}{5}$ f. -14.1

3. **Set Notation.** Use braces to write the members of each of the following sets, or state that the set has no members.
 a. The months of the year
 b. The natural numbers from 1 to 100 (inclusive)
 c. The first three U.S. presidents
 d. The states whose names start with N

4. **Set Notation.** Use braces to write the members of each of the following sets, or state that the set has no members.
 a. The states whose names start with C
 b. The letters in the second half of the alphabet
 c. The natural numbers from 100 to 200 (inclusive)
 d. The last three U.S. presidents

Venn Diagrams for Two Sets. For Exercises 5–12, draw a Venn diagram for the given sets. In words, explain why you drew one set as a subset of the other, disjoint sets, or overlapping sets.

5. doctors and women
6. poetry and literature
7. negative integers and natural numbers
8. whole numbers and positive integers
9. words and verbs
10. pianists and musicians
11. painters and artists
12. genetics courses and science courses

Categorical Propositions. Each of Exercises 13–20 gives a categorical proposition. If it is not already in standard form, rephrase in standard form. State the subject and predicate sets, and draw a Venn diagram for the proposition. Label all regions of the diagram clearly.

13. All bachelors are men.
14. Some pilots are women.

15. All U.S. presidents have been men.
16. Some mammals swim.
17. Fish don't fly.
18. Dogs can't swim.
19. Every nurse knows CPR.
20. Lawyers are highly educated.

Categorical Negations. Each of Exercises 21–26 gives a categorical proposition. If it is not already in standard form, rephrase in standard form. State the subject and predicate sets, and state the negation of the proposition. Give the truth values for both the original proposition and its negation.

21. Some U.S. Senators are Republican.
22. All snakes are mammals.
23. All doctors are left-handed.
24. Some countries are not countries with coastlines.
25. Some people are not sociable.
26. Some days are rainy.

Venn Diagrams for Three Sets. For Exercises 27–32, draw a Venn diagram with three overlapping circles for the three given sets. On your diagram, label the contents of every region; if a region has no members, state that fact clearly.

27. women, dentists, and airline pilots
28. basketball players, professional athletes, and men
29. songs, poetry, and published works
30. oceans, bodies of salt water, and bodies of fresh water
31. clothes, hats, and shirts
32. states of the United States, places in North America, and places with coastlines on the Pacific Ocean.

A Venn Diagram with Numbers. In Exercises 33 and 34, consider the following Venn diagram that represents people in a room. The number in each region is the number of people matching the characteristics of that region.

people in a room
Republicans: 7
Overlap: 6
women: 9
Outside: 6

33. a. How many people are in the room?
 b. How many women are in the room?
 c. How many men are in the room?
 d. How many Republicans are in the room?

34. a. How many men are Republicans?
 b. How many men are not Republicans?
 c. How many women are Republicans?
 d. How many women are not Republicans?

Venn Diagrams for Sets. In Exercises 35–38, draw a Venn diagram to represent the given information.

35. Of the children on a school bus, 12 play soccer only, 6 play softball only, 2 play soccer and softball, and 4 play neither softball nor soccer.

36. A movie critic rated 15 feature movies and 10 documentaries. He gave 7 feature movies a "thumbs up" and the other feature movies a "thumbs down." He gave 9 of the documentaries a "thumbs up."

37. A doctor testing a new antibiotic puts 15 patients on a daily dose of the drug and 15 patients on a daily dose of a placebo (a substance that looks like the actual drug, but has no active ingredients). Twelve of the patients on the actual drug show improvement after a week, while 6 of the patients on the placebo show improvement.

38. A survey of 100 Toyota and Nissan owners revealed that all owners drove one vehicle and it was either an SUV or a pick-up. Half of the 60 Toyota owners drove SUVs, while one-fourth of the Nissan owners drove SUVs.

Venn Diagram Answers. In Exercises 39 and 40, draw a two circle Venn diagram that contains all the listed information. Label each region clearly. Then use the diagram to answer the questions.

39. Of the 43 students in the Cosmos Club, 16 are taking at least mathematics, 23 are taking at least biology, and 9 are taking mathematics only. How many students are taking biology only? How many are taking neither biology nor mathematics?

40. Of the 17 students in the Cycling Club, 12 own at least a road bike, 8 own at least a mountain bike, and 0 own no bike. How many students own both types of bike?

41. **Majors and Gender.** A scan of all biology and business majors at a college shows the following breakdown by gender. Draw a Venn diagram for the data.

	Biology	Business
Women	32	110
Men	21	87

42. **Tomatoes and Cancer.** A 1999 study by the Harvard Medical School (*Journal of the National Cancer Institute*, February 17, 1999) reviewed 72 previous studies of the effect of tomatoes on cancer. The data showed convincingly that "high consumers of tomatoes and tomato products are at substantially decreased risk of numerous cancers, although probably not all cancers." Consider the following table that shows the incidence of oral cancer for a group of people who ate an average of one tomato a day and another group of people who ate fewer than three tomatoes per week. Draw a Venn diagram for the data.

	No oral cancer	Oral cancer
One tomato per day	191	9
Fewer than three tomatoes per week	164	16

43. **Coffee and Gall Stones.** A study on the effect of coffee on gall stones (*Journal of the American Medical Association*, June 9, 1999) resulted (in small part) in the data shown below. The category *Coffee* means more than four cups of caffeinated coffee per day. The category *No coffee* means no caffeinated coffee. Draw a Venn diagram for the data.

	Gall stone disease	No disease
No coffee	385	14,068
Coffee	91	4,806

44. **Interpreting a Survey.** A survey asked newspaper readers which of the following three newspapers they read daily: *New York Times (NYT)*, *Washington Post (WP)*, and *Wall Street Journal (WSJ)*. The results are shown below.

Paper(s)	Readers	Paper(s)	Readers
NYT only	24	NYT and WSJ only	14
WSJ only	27	NYT and WP only	16
WP only	26	WP and WSJ only	13
None	15	All three	8

Answer the following questions, interpreting the connecting words carefully.

a. Draw a Venn diagram for the table.
b. How many people read *WP* and *WSJ*?
c. How many people read *NYT* and *WSJ* and *WP*?
d. How many people read *NYT* or *WSJ* or *WP*?
e. How many people read *NYT* but not *WP*?

45. **Interpreting Sales.** A review of sales at a large automobile dealership for the past month shows that the following numbers of buyers chose air conditioning (AC), four-wheel-drive (4WD), and cassette players (CP) in their cars.

Option(s)	Buyers	Option(s)	Buyers
AC only	45	AC and 4WD only	21
4WD only	8	AC and CP only	16
CP only	12	CP and 4WD only	8
None	3	All three	10

a. Draw a Venn diagram for the table.
b. How many people chose AC and 4WD?
c. How many people chose AC but not 4WD?
d. How many people chose AC or 4WD or CP?
e. How many people chose exactly two options?

46. **Analyzing Diets.** The following table shows ten common foods and whether they are significant sources of protein, fat, and carbohydrate. Organize these data in a Venn diagram. (Hint: You'll need circles for protein, fat, and carbohydrate; then put the name of each food in the appropriate region of the diagram.)

Food	Protein	Fat	Carbohydrate
Meat	X	X	X
Rice	X		X
Nuts	X	X	X
Potatoes			X
Apples			X
Cheese	X	X	
Eggs	X	X	
Lettuce			
Coffee			
Water			

More Than Three Sets. In Exercises 47–50, use your knowledge of the listed sets to draw a Venn diagram illustrating the relationships among them.

47. animals, house pets, dogs, cats, canaries
48. things that fly, birds, eagles, airplanes, jets, hang gliders
49. food, ice cream, dessert, fruit, snack, citrus fruit, strawberries
50. people, men, women, surgeons, NFL football players, quarterbacks, lawyers

Organizing Propositions. Each of Exercises 51–53 lists a set of propositions. Draw a Venn diagram that represents all the information in the propositions and use it (and no other assumptions) to answer the questions that follow. Explain your reasoning.

51. All hairy animals are mammals. No mammals are fish. Some mammals can swim. No fish can walk on land.

 Questions: Could there be hairy fish? Could there be hairy animals that swim? Could there be walking mammals? Could there be hairy animals that walk on land?

52. All meat has protein. All dairy products have protein. Some beans have protein. All beans, but no meat or dairy products, are plants.

 Questions: Could there be beans that are dairy products? Could there be meat that is a dairy product? Could there be dairy products that are plants? Could there be plants with protein?

53. No Republicans are Democrats. No Republicans are Green Party members. All Republicans are conservative. Some liberals are Democrats. No liberals are conservatives.

 Questions: Could there be conservative Democrats? Could there be liberal Green Party members? Could there be liberal Republicans?

54. **Organizing Politics.** You are at a conference attended by men and women of various political parties. The conference organizer tells you that none of the women are Republicans and some (not all) of the Democrats are women.

 a. Draw a Venn diagram to organize the given information.
 b. Based on the given information, is it possible to meet a woman who is neither a Republican nor a Democrat?
 c. Based on the given information, is it possible that there are any male Republicans?

55. **Is It Possible?** Shoppers are given a choice of one of three brands of vanilla ice cream and then asked if they approve or disapprove of the flavor. Is it possible to draw a Venn diagram with two circles that shows how many shoppers approved or disapproved of each brand? Explain.

56. *N* **Sets.** A computer store is offering a basic computer with four options A, B, C, and D, each of which buyers can select.

 a. How many different sets of options can buyers choose? (For example, one choice is no options, another choice is only options A and C, and another choice is all options.)
 b. Is it possible to make a Venn diagram using circles to show the number of buyers selecting each of the choices?
 c. Suppose the store offered five options A, B, C, D, and E. How many different sets of options would be available?
 d. Can you generalize parts a–c? If the store offers n options, each of which can be either selected or not selected, how many different sets of options are available?

Web Projects

Find useful links for Web Projects on the text Web site: www.aw.com/bennett-briggs

1. **State Politics.** Determine from the Web how many states have a Republican majority in the State House and how many states have a Republican majority in the State Senate. Draw a Venn diagram to illustrate the situation.

2. **U.S. Presidents.** Use the Web to learn the following about each past American president:
 • Bachelor or married (classify as married if married for part of the term)
 • Inaugurated before or after age 50
 • Served one term (or less) or more than one term
 Make a three-circle Venn diagram to represent your results.

3. **Uses of Venn Diagrams.** Visit the Web site for VennDiagram.com and read about the top 10 uses of Venn diagrams. Choose at least one of these uses and apply it to a specific problem in your own life.

UNIT 1D

CRITICAL THINKING IN EVERYDAY LIFE

The skills discussed in the preceding units are all useful in their own right. But critical thinking involves much more than isolated skills. It also involves knowing when to use particular skills and when to invent entirely new strategies to analyze an argument or make a decision. Indeed, as we stated at the very beginning of this chapter, critical thinking involves careful reading (or listening), sharp thinking, logical analysis, good visualization, and healthy skepticism.

Because it is so wide-ranging, critical thinking cannot be described by any simple step-by-step procedure. Instead, it is developed through experience and by questioning and analyzing every argument or decision you face. Nevertheless, a few general guidelines or hints can be useful. In this unit, we offer a few such hints.

HINT 1: READ (OR LISTEN) CAREFULLY

Language can be used in complex ways that require careful effort to understand. Always read (or listen) carefully to make sure you've grasped precisely what was said. Also be sure that you've distinguished what was actually said, what was assumed, and what must be determined.

EXAMPLE 1 *Colorado's Amendment 2*

In 1992, Colorado voters were asked to vote *yes* or *no* on the following ballot question.

> *Shall there be an amendment to the Colorado constitution to prohibit the state of Colorado and any of its political subdivisions from adopting or enforcing any law or policy which provides that homosexual, lesbian, or bisexual orientation, conduct, or relationships constitutes or entitles a person to claim any minority or protected status, quota preferences, or discrimination?*

Explain the meaning of *yes* and *no* votes.

Solution The question contains a lot of obscure wording, making it difficult to see the key point. But, if we read carefully, we find that the question essentially asks: *Should the state prohibit laws that grant gay people minority status or protection from discrimination?* Even in this simplified form, the question is still somewhat confusing because it is asked in negative terms. That is, a *yes* vote is a vote *against* gay rights, while a *no* vote is a vote *for* gay rights. At least some evidence suggests that the wording confused many voters. Pre-election polls found that most voters supported gay rights, implying that the proposed amendment would be defeated. However, the amendment passed (53% to 47%), which means either that the polls were quite inaccurate or that many voters were confused and voted opposite to their beliefs.

Time out to think

If voters really did misunderstand the wording of the Colorado ballot initiative, would the confusion have affected both sides equally, or would it have favored one side more than the other? Explain.

by the way...

Despite passing, the Colorado amendment never took effect. A legal challenge led to an immediate injunction against the law after the 1992 election. In 1996, the law was permanently struck down by the U.S. Supreme Court. The justices said the amendment denied gays a political right enjoyed by everyone else—the chance to seek protection from discrimination in employment, housing, and public accommodations—and therefore violated the U.S. Constitution.

HINT 2: LOOK FOR HIDDEN ASSUMPTIONS

The arguments we study in logic usually consist of clear premises leading to the conclusion. But many real arguments lack such clarity, relying instead on ambiguous terms or hidden assumptions. Often, the speaker (or writer) may think these things "obvious," but listeners (or readers) may not agree. Indeed, an argument that seems very convincing to the speaker may actually be quite weak to a listener who is not aware of the hidden assumptions.

EXAMPLE 2 *Building More Prisons*

Analyze the following argument:

> *We should build more prisons because incarcerating more criminals will reduce the crime rate.*

Solution This argument looks deceptively simple because it is so short. It concludes that we should build more prisons, based on a single premise—that incarcerating more criminals will reduce the crime rate. If we recognize the premise as a conditional statement, the argument has the form

> Premise: If we incarcerate more criminals, then the crime rate will be reduced.
>
> Conclusion: We should build more prisons.

Viewed this way, the argument hardly makes any sense at all, because the premise says nothing about building prisons. Clearly, the speaker must have intended some hidden assumptions to be obvious to the listeners.

We can begin to make sense of the argument by identifying possible hidden assumptions. For example, a plausible argument might begin like this:

> Hidden Assumption 1: If we build more prisons, then more criminals can be incarcerated.
>
> Stated Premise: If we incarcerate more criminals, then the crime rate will be reduced.
>
> Hidden Assumption 2: If the crime rate is reduced, then we will have a more desirable society.

With the two hidden assumptions, this part of the argument becomes a *chain of conditionals*. That is, it has the structure

> If p, then q.
> If q, then r.
> If r, then s.

Clearly, such an argument leads to this conclusion: If p, then s.

In this case, this conclusion represents the statement *If we build more prisons, then we will have a more desirable society.* If we also assume that it is a good idea to enact policies that lead to a more desirable policy (which some people might debate, since we have not considered the costs of the policy), then we finally have the original conclusion that *We should build more prisons.*

Note that we've been able to make a case for the original conclusion only by adding hidden assumptions, which means that the original argument was not well

structured. Moreover, both the hidden and stated assumptions are highly debatable. For example, many people would argue with the first hidden assumption, because incarcerating more criminals requires not only more prison space, but also a more efficient court system. Similarly, the stated premise is open to debate, because not all social researchers agree that a higher incarceration rate automatically reduces the crime rate. In summary, given its poor structure and debatable assumptions, the original argument has *not* made a good case for its conclusion. ∎

HINT 3: IDENTIFY THE REAL ISSUE

People often argue about one question when there is a different deeper issue involved. Identifying the real issue in a debate is often difficult, because people can be experts at hiding their true intentions. Fortunately, by analyzing arguments carefully, we can often determine whether the real issue is hidden, even if we don't know exactly what it is.

EXAMPLE 3 *Banning Rock Concerts*

Analyze the following segment of an editorial from a local newspaper.

> With last Saturday's sellout crowd at the Moonlight Amphitheater, it is clear that the parking problem has become hopeless. Concert goers parked along residential streets up to a mile away from the amphitheater, badly overcrowding sidewalks, blocking driveways, and disrupting traffic. In light of this parking problem, future rock concerts should be canceled.

Solution The argument makes several claims about parking problems, but stripped of the details the argument boils down to this:

Premise: There is a parking problem for rock concerts at the theater.

Conclusion: Future rock concerts at the theater must be canceled.

The rest of the argument simply lists reasons why a parking problem might be considered serious. But is a parking problem really a good enough reason to cancel future rock concerts? After all, a parking problem could have many solutions. For example, it might be possible to create new parking lots, to use shuttle buses, to step up enforcement of parking violations, or to encourage carpooling. The weakness of the argument should make us wonder whether the editorial writers are *really* concerned about parking or whether they are using this issue to hide their real concern. Perhaps they oppose the rock concerts themselves for some reason. Or perhaps they are responding to a few vocal residents who complained about the parking problem. It's hard to know for sure, but it seems unlikely that parking is the real issue. ∎

HINT 4: USE VISUAL AIDS

When arguments, instructions, or options become complex, it can be useful to organize information visually. Useful tools include tables, Venn diagrams, flow charts, and other types of pictures.

EXAMPLE 4 *Reading IRS Guidelines*

According to Internal Revenue Service (IRS) guidelines on who must file a federal income tax return, a single person who can be claimed as a dependent on another person's tax return and is *either* 65 or older *or* blind must file a return if

historical note

Can logic settle all arguments? The German mathematician Gottfried Wilhelm von Leibniz (1646–1716) thought so and took the first steps in a two-century search for a "calculus of reasoning." An outgrowth of this search called *symbolic logic* is now used extensively in mathematics and computing. But Leibniz's dream of resolving arguments through logic was not to be. In 1931, the Austrian mathematician Kurt Gödel discovered that no system of logic can solve all mathematical problems, let alone problems of ethics or morality.

(i) earned income was more than $5350 ($6400 if 65 or older and blind); or
(ii) unearned income was more than $1750 ($2800 if 65 or older and blind); or
(iii) gross income was more than the total of earned income (up to $4300) or $700, whichever is greater, plus $1050 ($2100 if 65 or over and blind).

Decide whether these guidelines apply to each of the following individuals; if the guidelines do apply, use them to decide whether the individual must file a return.

- Chase is a 15-year-old boy who is claimed as a dependent on his parent's tax return. He has an earned income of $6500. Does this policy apply to Chase? If so, must he file an income tax return?
- Carmen is a single 24-year-old blind woman who is claimed as a dependent on her parent's tax return. Her earned income is $6500, and her unearned income is $750. Does this policy apply to Carmen? If so, must she file an income tax return?
- Serena is a single 68-year-old blind woman who is claimed as a dependent on her son's tax return. Her gross income is $2500, and her earned income is $550. Does this policy apply to Serena? If so, must she file an income tax return?

FIGURE 1.27 Flow chart to help interpret a set of IRS guidelines.

58 CHAPTER 1 THINKING CRITICALLY

Solution As this example demonstrates, IRS policies can be notoriously difficult to interpret. A diagram or a flow chart, as shown in Figure 1.27, can simplify matters. Notice that conditions (i)–(iii) are separated by *or*, so if any one condition is satisfied, then a tax return must be filed. We can use the flow chart to see how the guidelines affect the three individuals.

- Chase is neither over 65 nor blind, so the guidelines do not apply to him.
- Because Carmen is blind, single, and claimed as a dependent on her parent's tax return, the guidelines apply to her. Because her earned income exceeds $5350, she must file a tax return by condition (i).
- The guidelines apply to Serena because she is single, blind, over 65, and claimed as a dependent on her son's tax return. Condition (iii) applies to her case. The greater of Serena's earned income ($550) and $700 is $700. Thus, because Serena is blind *and* over 65, she must file a tax return if her gross income exceeds $700 + $2100 = $2800. But her gross income is only $2500, so she does not need to file a return.

HINT 5: UNDERSTAND ALL THE OPTIONS

We regularly make decisions in situations where we have several options. For example, we face decisions about which insurance policy to choose, which auto loan to take, or which model of new computer to buy. The key to such decisions is making sure that you understand how each option would affect you.

EXAMPLE 5 Which Airline Ticket to Buy?

Airlines typically offer many different prices for the same trip. Suppose you are planning a trip six months in advance and discover that you have two choices in purchasing an airline ticket:

(A) The lowest fare is $1100, but 25% of the fare is nonrefundable if you change or cancel the ticket.

(B) A fully refundable ticket is available for $1900.

Analyze the situation.

Solution We can think of each of the two options as a pair of conditional propositions. Under option (A), you will lose 25% of $1100, or $275, if you cancel your trip. Thus, option A represents the following pair of conditional propositions:

(1A) If you purchase ticket A and *go* on the trip, then you will pay $1100.
(2A) If you purchase ticket A and *cancel* the trip, then you will pay $275.

Similarly, option B represents the following pair of conditional propositions:

(1B) If you purchase ticket B and *go* on the trip, then you will pay $1900.
(2B) If you purchase ticket B and *cancel* the trip, then you will pay 0.

Figure 1.28 represents the four possibilities. Clearly, option A is the better buy if you go on the trip, and option B is the better buy if you end up canceling your trip. However, because you are planning six months in advance, it's impossible to foresee all the circumstances that might lead you to cancel your trip. Therefore, you might want to

FIGURE 1.28

analyze the *difference* between the two tickets under the two possibilities (going on the trip or canceling).

If you go on the trip: ticket B costs $800 more than ticket A.
If you cancel the trip: ticket A costs $275 more than ticket B.

In effect, you must decide which ticket to purchase by balancing the risk of spending an extra $800 if you go on the trip against spending an extra $275 if you cancel. How would *you* decide?

Time out to think

Consider the two ticket options in Example 5 from the point of view of the airline. How does offering the two options help the airline maximize its revenue?

HINT 6: WATCH FOR FINE PRINT AND MISSING INFORMATION

There's an old saying that "the devil is in the details," and it is probably nowhere more applicable than in the fine print of offers, deals, and other documents. What looks like a great deal without the fine print may be a poor one with it. Worse yet, sometimes even the fine print neglects to state important information. Use your powers of critical thinking to decide what information you need; if anything is missing, be sure to ask before you act.

EXAMPLE 6 *Shipping and Handling*

Katie received a mail offer for a CD club. In large print, it offered "12 CDs for 1¢." The reply card asked Katie to sign her agreement to the following Club terms:

> I've enclosed $1.50 (that's 1¢ for 12 CDs, plus $1.49 for enrollment charge). I agree to buy 6 CDs at regular Club prices within the next two years.
>
> A shipping/handling charge is added to each shipment. Applicable sales tax added to each order.

She signed up and sent in $1.50, but when the CDs arrived they came with an invoice for $36.50. What happened?

Solution The fine print warned of a shipping/handling charge, but did not state how much it would be. It turned out to be quite significant. To fulfill her agreement, Katie now needs to purchase 6 more CDs at "regular Club prices." Based on this first experience, Katie should be concerned that these prices may turn out to be much higher than she would pay elsewhere. Furthermore, each of these CDs will have a significant shipping/handling charge, and perhaps sales tax as well. It's still possible that the end result will be a decent deal, but certainly not 12 CDs for 1¢.

HINT 7: ARE OTHER CONCLUSIONS POSSIBLE?

You should never accept the first argument or choice you hear. Even when an argument proves its conclusion, it's still possible that there are other conclusions that have been left unstated. And remember that many real arguments

involve matters of opinion, which cannot be absolutely proved no matter how strong the argument may seem.

EXAMPLE 7 *Nuclear Deterrence*

Evaluate the following argument and discuss its conclusion.

> *The development of nuclear weapons changed the way world leaders think about potential conflicts. A single nuclear weapon can kill millions of people, and the arsenals of the United States and the Soviet Union contained enough power to kill everyone on Earth many times over. This potential for catastrophic damage led to the idea of nuclear deterrence, which held that the United States and the Soviet Union would be deterred from direct warfare by the fear of nuclear war. For the more than 45 years of the Cold War, the United States and the Soviet Union never did fight directly. This was one of the longest periods in human history during which two major enemies avoided direct war. We can only conclude that nuclear deterrence prevented war between the United States and the Soviet Union.*

Solution The conclusion of this argument is the claim that nuclear deterrence prevented war between the United States and the Soviet Union. The facts cited in support of this conclusion are all true and all appear relevant to the conclusion. Thus, the argument appears quite strong. Nevertheless, it does not *prove* that nuclear deterrence worked, and we can imagine other possible explanations for why the United States and the Soviet Union avoided war. For example, it might have been because of economic factors, or the emergence of a united Western Europe after World War II, or changes in non-nuclear weaponry. Indeed, because the argument concerns why history unfolded in a particular way, it is probably impossible for us to ever know for certain whether the conclusion is true or false. ■

practical matters

BEWARE OF "UP TO" DEALS

You've probably heard offers for cheap long distance calling, such as 99¢ for calls of "up to 20 minutes." At first, it might sound like a great deal, because most people focus on 20 minutes and 99¢. Indeed, if you make a 20-minute phone call, the rate works out to about 5¢ per minute, which is quite low. But the words "up to" hold hidden danger. Suppose you make a call and reach only your friend's answering machine. Even though it may take you only 30 seconds (or less) to leave your message and hang up, you'll still pay 99¢ for the call under this plan—which works out to about $2 per minute in this case. If your calls are longer than 20 minutes, there's usually a higher rate that kicks in after 20 minutes.

More generally, you should beware of any deal that contains the words "up to." For example, you may receive offers promising that you can earn "up to $5000 a month" while working from home or that you can get "up to five free oil changes" if you pay an up-front charge that is much more expensive than a single oil change. Such deals are usually structured to be great if you just touch the "up to" limit, but are often quite poor in all other circumstances. Unless you've carefully thought out how you will take advantage of the good parts of the deal, it's best to assume that "up to" deals are up to no good.

Time out to think

Can you think of any facts that would tend to undercut the argument that nuclear deterrence prevented war between the United States and the Soviet Union? Overall, in your judgment, is the argument in Example 7 convincing? Why or why not?

PUTTING IT ALL TOGETHER

The examples we've studied show only a few of the many ways in which critical thinking affects our daily lives. We live in a complex world, and unless you are careful in all your thinking and decision making, you are liable to be led astray. Learn to pick apart every argument or choice you are given. With time, you will gain the experience needed to make wise decisions in school, in your career, and in your life.

Review Questions

1. Describe critical thinking and why it is important to everyone.
2. Summarize the hints given in this unit, and explain how each is important to critical thinking.
3. Give a few examples of situations in which you used critical thinking in your own life.

In the News

1. **Argument Analysis.** Find a news article or editorial in which a clear position is taken on a controversial issue. Analyze the argument using the hints and strategies given in this unit. Whether or not you agree with the final conclusion, state whether the argument is effective.
2. **Personal Decisions.** Discuss a situation in your own life in which you needed to think carefully before making a critical decision. Did you use critical thinking strategies? In retrospect, would you have made a different decision using critical thinking strategies?
3. **Ambiguous Terms.** Many arguments and promotions are difficult to analyze because of vague, ambiguous, or overused terms (for example, "natural food" or "human rights"). In news articles or advertising, find three terms that you believe are vague or ambiguous. Explain how the terms could be more clearly defined.

Exercises

Ambiguity in the News. Exercises 1–4 are direct quotes from news sources. In each case, describe how the quote is ambiguous. What further information would you need to remove the ambiguity?

1. LOS ANGELES—Two jumbo jets with more than 350 people aboard nearly collided over the city during a landing attempt . . .
2. There has been an alarming 25% rise in [child abuse] cases between 1988 and 1993, a year when 2.9 million incidents were reported to child welfare agencies.
3. France must decrease its public deficit from 6% of its gross domestic product, or $83 billion, to 3%.
4. The United States was found to regularly engage in human rights abuses.
5. **Reading a Ballot Initiative.** Consider the following ballot initiative that appeared in a statewide election:

 Shall there be an amendment to the state constitution stating that the state and any city, town, or county may control the promotion of obscenity to the full extent permitted by the First Amendment to the U.S. Constitution, and thereby preventing the state courts from interpreting the right of free expression more broadly under the state constitution than under the First Amendment to the U.S. Constitution in the area of obscenity?

 a. According to the text of this initiative, is the state constitution more or less strict about obscenity than the First Amendment to the U.S. Constitution? Explain.
 b. Explain the effect of a *yes* vote on this initiative.
 c. True or false: If the initiative passed, then the state would have a stricter law on obscenity than specified in the U.S. Constitution. Explain.

6. **Reading a Ballot Initiative.** Consider the following ballot initiative that appeared in a statewide election:

 Shall there be an amendment to the state constitution concerning term limits declarations that may be voluntarily submitted by candidates for the U.S. Congress, and in connection therewith, specifying when such declarations must be submitted to the

secretary of state; providing that a candidate shall not be refused placement on the ballot if the candidate does not submit a declaration; providing that candidates may voluntarily declare that the candidate will not serve more than three terms as a U.S. Representative or more than two terms as a U.S. Senator or may voluntarily declare that the candidate has chosen not to accept term limits?

a. Does the state have mandatory term limits for Representatives and Senators?

b. If the initiative is passed, must a candidate accept term limits if elected? Explain.

c. If the initiative is passed, will there be a penalty if a candidate refuses to accept term limits? Explain.

7. **Choosing Airline Tickets.** You are planning a trip to Japan six months in advance and discover that you have two choices in purchasing an airline ticket:

(A) The lowest fare is $1600, but 25% of the fare is nonrefundable if you change or cancel the ticket.

(B) A fully refundable ticket is available for $2900.

Describe all of your options, and discuss how you would decide which ticket to buy.

8. **Choosing Long Distance Carriers.** Suppose you currently use a long distance carrier that offers a uniform rate of $0.10 per minute on all calls (any time of day or night). You receive an offer to switch to a long distance carrier that offers the following rates:

- $0.25 per minute for daytime calls on a weekday
- $0.10 per minute for nighttime calls on a weekday
- $0.05 per minute for all weekend calls

What additional information would you like to have in order to make a decision about switching phone companies? For your own personal situation, which option is best? Describe the factors that led to your decision.

9. **IRS Guidelines.** IRS guidelines state that a single person under 65 years of age who can be claimed as a dependent on another person's tax return must file a return if

(i) unearned income was over $700; or

(ii) earned income was over $4300; or

(iii) total of earned and unearned income was more than the greater of $700 or earned income (up to $4050) plus $250.

For each situation below, state whether the person must file a return.

a. Trip had earned income of $600 and unearned income of $200. Must he file a federal tax return?

b. Sally had earned income of $680 and no unearned income. Must she file a federal tax return?

c. Monica had earned income of $3900 and unearned income of $300. Must she file a federal tax return?

d. Horace had earned income of $4000 and unearned income of $200. Must he file a federal tax return?

e. Delila had earned income of $4200 and unearned income of $200. Must she file a federal tax return?

10. **IRS Guidelines.** Consider the following IRS instructions:

You must make estimated tax payments if you expect to owe at least $500 in tax and you expect your withholding and credits to be less than the smaller of:

1. *90% of the tax shown on next year's tax return, or*
2. *The tax shown on last year's tax return (110% of that amount if you are not a farmer or fisherman and the adjusted gross income shown on that return is more than $150,000 or, if filing separately next year, more than $75,000).*

For each situation below, state whether you must make estimated tax payments during the course of a year.

a. You expect to pay $450 in taxes next year.

b. You are a teacher and expect to pay more than $500 in taxes. Your adjusted gross income is $35,000. You expect your withholding and credits to be 85% of your tax for next year, and you expect next year's taxes to equal last year's taxes.

c. You are a teacher and expect to pay more than $500 in taxes. Your adjusted gross income is $35,000. You expect your withholding and credits to be 95% of your tax for next year and 120% of your tax for last year.

d. You are a lawyer and expect to pay more than $500 in taxes. Your adjusted gross income is $185,000. You expect your withholding and credits to be 80% of your tax for next year and 115% of your tax for last year.

e. You are a farmer and expect to pay more than $500 in taxes. Your adjusted gross income is $40,000. You expect your withholding and credits to be 95% of your tax for next year and 95% of your tax for last year.

11. **What Is a Farmer?** The Internal Revenue Service (IRS) provides the following test for who is considered a farmer:

A landowner is materially participating in farming when any one of the following four tests are met: (1) the owner does any three of the following four activities: (a) advances, pays, or assumes liability for at least half of the direct costs of crop production,

(b) furnishes at least half of the tools, equipment, and livestock used in crop production, (c) periodically advises and consults with the tenant, and (d) periodically inspects the production activities; (2) regularly and frequently makes or takes an important part in making management decisions, substantially contributing to or affecting the success of the enterprise; (3) works 100 hours or more in activities connected with crop production over a period of five weeks or more; or (4) otherwise engages in activities that demonstrate that the landowner is materially participating in farming.

a. Make a flow chart that illustrates the various branches of this test.

b. Jack works 150 hours every 4 weeks in activities connected with crop production. Is the test clear on Jack's situation? Does Jack qualify as participating in farming?

c. Jill works 150 hours every 8 weeks in activities connected with crop production. Is the test clear on Jill's situation? Does Jill qualify as participating in farming?

d. Ivan assumes liability for 75% of the direct costs of crop production and furnishes 60% of the tools used in crop production. Does Ivan qualify as participating in farming?

e. Joan assumes liability for 65% of the direct costs of crop production, meets weekly with the tenant, and inspects production activities monthly. Does Joan qualify as participating in farming?

12. **Reading a Lease.** Consider the following excerpt from the contract for the lease of an apartment:

 Landlord shall return the security deposit to resident within one month after termination of this lease or surrender and acceptance of the premises, whichever occurs last.

 a. Suppose your lease terminates on June 30, you move out of the apartment on June 5, and you receive your security deposit on June 28. Has the landlord complied with the terms of the lease? Explain.

 b. Suppose your lease terminates on June 30, you move out of the apartment on June 5, and you receive your security deposit on July 2. Has the landlord complied with the terms of the lease? Explain.

 c. Suppose your lease terminates on June 30, you move out of the apartment on June 5, and you receive your security deposit on July 7. Has the landlord complied with the terms of the lease? Explain.

13. **Comparing Candidates.** Two assistant district attorneys, Alice and Zack, are running for the position of district attorney. The candidates make the following statements:

 ZACK: In the last five years, Alice has prosecuted 126 defendants charged with sexual assault on children. Of those, only nine cases even went to trial, and of those, only four cases resulted in conviction and a prison sentence.

 ALICE: I have prosecuted 253 cases involving sexual assault on children. More than 40% of the 253 cases resulted in prison or jail sentences as part of the original plea agreement.

 a. Comment on whether the two candidates' numbers for cases prosecuted by Alice are consistent.

 b. Comment on whether the two candidates' numbers for convictions obtained by Alice are consistent.

 c. Would you use these statements as grounds for making a decision on whom to vote for in the district attorney race? If not, what would you do to obtain more complete information?

14. **Drugs and Athletes.** An official of the U.S. Olympic Committee (U.S.O.C) stated in a July 2000 press conference that recent track and field performances have been so unbelievable that athletes must be using drugs. Evaluate this argument.

15. **A Financial Decision.** Consider the following statement:

 I need a special computer for a project I will be working on for the next three months; after that, I will no longer need this computer. I can lease the computer for $350 per month, or I can buy it for $2100. If I resell

it after the three months, I can expect to get $1200. And don't forget the 5% sales tax.

Identify all the essential facts for both options; distinguish between "hard facts" and estimates that could affect the outcome of the argument. What is the most economical choice for acquiring a computer? What would you recommend in this situation? Why?

16. **Reading Your Water Bill.** The following explanation is given on a municipal water bill:

 Water charges are composed of two parts: a fixed service charge, based on meter size, and a quantity charge. The quantity charge consists of three different blocks of rates. As the amount of water use increases and moves into the next rate block, the cost per thousand gallons of water increases. Each account's Average Winter Consumption (AWC) is used to determine the point at which water use moves into the next higher block. AWC is an average based on actual water use reflected on a user's water bill from December through March. With the exception of irrigation-only accounts, which will begin billing at Block 2 rates, all consumption up to and including an account's AWC is billed at the Block 1 rate. All consumption greater than and including the account's AWC but less than 350% of the account's AWC is billed at the Block 2 rate. All consumption greater than 350% of an account's AWC will be billed at the Block 3 rate.

 a. Suppose you have a residential account and in June you use 250% of your AWC. Explain how your total water bill for the month will be determined.

 b. Suppose you have a residential account and in August you use 400% of your AWC. Explain how your total water bill for the month will be determined.

 c. Suppose you have an irrigation-only account and in May you use 450% of your AWC. Explain, if possible, how your total water bill for the month will be determined.

 d. For the same AWC and the same amount of water used, would your water bill be higher with a residential account or an irrigation-only account?

Decision Making. In Exercises 17–21, analyze each situation and explain how you would make a decision.

17. You need three sticks of butter for baking. Your local store sells individual sticks for 40¢ each and packages of four sticks for $1.25.

18. You and your spouse are expecting a baby. Your current health insurance costs $115 per month, but doesn't cover prenatal care or delivery. You can upgrade to a policy that covers prenatal care and delivery, but your new premium will be $275 per month. The cost of prenatal care and delivery is approximately $4000.

19. You are planning a trip to visit your family two months in advance. The airline offers you two choices: (1) You may purchase a ticket for $350, and the ticket price can be refunded for a $75 penalty; or (2) you may purchase a fully refundable ticket for $600.

20. You own a small business and need to be in another city on Monday for a meeting. You could fly to the meeting on Monday morning and return Monday evening on a ticket priced at $750. However, if you included a Saturday night stay in your trip, the ticket price would be only $335. A hotel will cost $105 per night, and you estimate that meals away from home will cost you $55 per day.

21. You fly frequently between two cities 1,500 miles apart. Average round-trip cost on Airline A is $350. Airline B offers the same trip for only $325. However, Airline A has a frequent flyer program in which you earn a free round-trip ticket after you fly 25,000 miles. Airline B does not have a frequent flyer program.

Critical Thinking. Exercises 22–31 give short arguments in support of a specific conclusion. Use all the critical thinking tools at your disposal to analyze the argument and determine if the conclusion is reached in a convincing way. Write out your analysis carefully and cite all aspects of the analysis.

22. **Athletic Priority.** The athletic program is given more money than any academic department, so this university must value athletics over academics. Also, the football coach is the highest paid university employee.

23. **Oil Crisis.** Dependency on foreign oil has put the U.S. economy at great risk, yet the nation has at least a decade's worth of oil untouched in wilderness areas and offshore. The federal government should immediately open all potential oil fields to drilling, regardless of the environmental consequences.

24. **Teacher Supply.** During the next decade, the number of Americans aged 5–15 will increase. Furthermore, the number of Americans aged 25–35 will decline. Therefore, the demand for teachers will increase during the next decade.

25. **Health Care.** That we have a health care crisis is clear: Large numbers of Americans are uninsured, and the cost of health care to our nation is skyrocketing. I believe that

26. **The Fed.** The Federal Reserve Board should lower interest rates, because the economy is slowing down and when the economy slows down there is less investment.

27. **Global Warming.** The populations of China and India dwarf that of the United States, yet they use far less energy. In the future, the people of China and India no doubt will demand a standard of living comparable to that in the United States. To achieve it, the governments of China and India are likely to exploit their enormous reserves of coal as an energy source. Their burning of so much fossil fuel undoubtedly will cause a worldwide environmental catastrophe.

28. **Human Population.** Overpopulation is not a real problem. Modern technology, especially in bioengineering, will enable scientists to develop far more efficient agriculture. In addition, advances in irrigation technology, along with the development of crops that can grow in salt water, will enable the conversion of much of the world's desert wastelands into productive farms. As a result, agribusiness will be able to produce enough food for at least 50 billion people, or some eight times the current world population. Some people claim that such large-scale agriculture would cause environmental damage, but this result is unlikely: The human ingenuity of a much larger population undoubtedly will prevent such damage.

29. **Classroom Calculators.** Fourth-grade students should use pencil and paper to learn to add fractions and strengthen their mental skills. Drill problems also induce good discipline. Therefore, fourth-grade students should not be allowed to use calculators. Anyway, their teachers and parents never used calculators when they were learning.

30. **Classroom Computers.** Word processors with spell checkers allow students to grow up without any spelling skills. Now, word processors have grammar checkers, so students will not learn grammar. What's next? Word processors should not be allowed in the schools.

31. **College Enrollment.** American public universities are approaching a crisis. State funding is declining as priorities shift to public school systems and prisons. At the same time, population projections indicate a 30% increase in high school graduates in the next 15 years. Many universities are at capacity and could not expand even if funds were available. Even if the number of university graduates remains level, the United States will have one of the highest university graduation rates of any country in the world. The solution is to raise both admission standards and tuition at four-year colleges and universities, thereby limiting the number of students admitted to these institutions.

32. **Poetry and Mathematics.** Consider the following poem by the English poet A. E. Housman (1859–1936).

Loveliest of Trees

Loveliest of trees, the cherry now
 Is hung with bloom along the bough
And stands about the woodland ride
 Wearing white for Eastertide.
Now, of my threescore years and ten,
 Twenty will not come again,
And take from seventy springs a score,
 It only leaves me fifty more.
And since to look at things in bloom
 Fifty springs are little room,
About the woodlands I will go
 To see the cherry hung with snow.

How old was the poet at the time he wrote this poem? (Hint: A score is 20.) Based on your reading of this poem, how much longer does the poet expect to live? Explain.

Web Projects

Find useful links for Web Projects on the text Web site: www.aw.com/bennett-briggs

1. **Interpreting the Second Amendment.** Much of the debate over gun control revolves around interpretations of the Second Amendment of the U.S. Constitution. It reads:

 A well-regulated militia being necessary to the security of a free State, the right of the people to keep and bear arms shall not be infringed.

 Gun rights advocates tend to focus on the "right of the people to keep and bear arms." Gun control advocates tend to focus on "a well-regulated militia." Visit a few of the many Web sites on each side of this issue to find interpretations of the Second Amendment that both support and oppose gun control. Based on what you learn, do you believe that the Second Amendment allows for gun control laws? Defend your opinion.

2. **Ballot Initiatives.** Investigate a particular ballot initiative in your state, county, or city (or choose one in another area from a Web site). Explain the important arguments on each side of the issue. If possible, find a statement of the initiative as it appeared on the ballot. Is the statement of the initiative clear? Explain.

3. **Fuzzy Logic.** Use the Web to learn about *fuzzy logic*, which allows for a continuous range of values between absolutely true and absolutely false (as opposed to the traditional logic discussed in this chapter, in which a proposition can be only true or false, and nothing in between). Discuss situations in which fuzzy logic would be useful and more realistic than traditional logic. Then investigate and report on at least one use of fuzzy logic in modern technology.

CHAPTER 1 SUMMARY

UNIT	KEY TERMS	KEY IDEAS AND SKILLS
1A	logic argument premises, conclusions fallacy	**Identify** an argument's premises and conclusions. **Recognize** fallacious arguments, which contain logical flaws so that their conclusions are not well supported by their premises.
1B	proposition truth values, truth tables negation: *not p* conjunction: *p and q* disjunction: *p or q* conditional: *if p, then q* logical equivalence	**Understand** truth tables for negation, conjunction, disjunction, conditional. **Inclusive versus exclusive *or*** Inclusive *or* means *either or both*. Exclusive *or* means *one or the other, but not both*. **Variations on the conditional** *if p, then q* **converse:** *if q, then p*. **inverse:** *if not p, then not q*. **contrapositive:** *if not q, then not p*.
1C	set categorical propositions Venn diagrams	**Set relationships** subsets, disjoint sets, overlapping sets **Four categorical propositions** All *S* are *P*; No *S* are *P*; Some *S* are *P*; Some *S* are not *P* **Uses of Venn diagrams** illustrating set relationships organizing information
1D	critical thinking	1. Read (or listen) carefully. 2. Look for hidden assumptions. 3. Identify the real issue. 4. Use visual aids. 5. Understand the options. 6. Watch for fine print and missing information. 7. Are other conclusions possible?

2

APPROACHES TO PROBLEM SOLVING

*Nothing in life is to be feared.
It is only to be understood.*

—Marie Curie

In your past mathematics classes, it might have seemed that mathematics problems involved only numbers and symbols. But the mathematical problems that you encounter in other classes, jobs, and daily life are almost always posed in words. That's why Chapter 1 focused on logic and critical thinking: They help you find the key ideas buried in problems that are posed in words. In this chapter, we will begin to consider *quantitative* problem solving, in which the problems involve words *and* numbers or equations.

UNIT 2A The Problem-Solving Power of Units: We will consider the powerful technique of working with units, both to solve problems and to check answers.

UNIT 2B Standardized Units: More Problem-Solving Power: We will review standardized units of both the U.S. customary and the metric system. We will also explore units for problems involving energy.

UNIT 2C Problem-Solving Guidelines and Hints: We will extend the techniques of the previous units to form a general set of guidelines and hints for effective problem solving.

UNIT 2A

THE PROBLEM-SOLVING POWER OF UNITS

Just as critical thinking can be approached in many different ways, there's no single best approach to solving quantitative problems. However, people have spent lifetimes coming up with strategies and guidelines that can help. The overall goal of this chapter is to help you learn to apply some of these strategies and guidelines.

We will begin by looking at how the units of a problem can be useful in figuring out what the problem is about, as well as for solving the problem and checking its answer. This work will also provide valuable experience with problem solving—experience that we will build upon when we consider general guidelines for problem solving in Unit 2C.

YOU CAN'T ADD APPLES AND ORANGES

It may seem like a cliché, but mathematically it is true that you can't add apples and oranges. The reason is that we can add or subtract only quantities of the same type. For example, we can add 5 apples and 3 apples to get 8 apples. But apples and oranges are different things, so if we try to add 5 apples and 3 oranges, all we can say is that we have 5 apples and 3 oranges.

More generally, numbers in real problems usually represent a quantity of *something*, which we can describe by its **units**. If we count the number of apples in a basket, the result has units of *apples*. If we measure the length of a room, the result may have units of *feet*. If we ask how fast a car is traveling, the answer may have units of *miles per hour*.

TECHNICAL NOTE
When abbreviating units, we do not distinguish between singular and plural. For example, we use *mi* for both mile and miles.

> **DEFINITION**
>
> The **units** of a quantity describe what is being measured or counted.

We can add or subtract numbers only when they have the same units. However, we can multiply and divide quantities with the same or different units, much as we multiply and divide numbers. A few examples should clarify the idea:

- If you drove 100 miles in two hours, your average speed was the distance traveled (100 mi) divided by the time it took (2 hr):

$$\frac{100 \text{ mi}}{2 \text{ hr}} = \frac{50 \text{ mi}}{1 \text{ hr}} = 50 \, \frac{\text{mi}}{\text{hr}}$$

We read this answer as "50 miles per hour"; note that the word *per* means "divided by" (or "for every").

- To find the *area* of a room, we multiply its length by its width (Figure 2.1). If the room is 12 feet long and 10 feet wide, its area is

$$12 \text{ ft} \times 10 \text{ ft} = 120 \text{ ft} \times \text{ft} = 120 \text{ ft}^2$$

We read this area as "120 square feet," remembering that the word *square* implies something raised to the second power. Note that we multiply the

FIGURE 2.1 Each small square is one foot on a side and therefore has an area of 1 square foot. The room fits a total of 120 of these squares, so its total area is 120 square feet.

numbers (12 × 10 = 120) separately from the units (ft × ft = ft²) but keep track of both.

- To find the *volume* of a box, we multiply its width, depth, and height (Figure 2.2). If the box is 6 inches wide, 4 inches deep, and 10 inches long, its volume is

$$6 \text{ in.} \times 4 \text{ in.} \times 10 \text{ in.} = 240 \text{ in.} \times \text{in.} \times \text{in.} = 240 \text{ in.}^3$$

We read this volume as "240 cubic inches," remembering that the word *cubic* implies something raised to the third power. (The abbreviation "in." includes a period.)

- To find the energy used by a light bulb, we multiply its power rating (in watts or kilowatts) by the length of time it is turned on. If a movie set uses a 0.5-kilowatt light bulb for 6 hours, its energy usage is

$$0.5 \text{ kilowatt} \times 6 \text{ hr} = 3 \text{ kilowatt} \times \text{hr}$$

We read this answer as "3 kilowatt-hours," where the hyphen between the units implies multiplication. (See Unit 2B for further discussion of energy units.)

FIGURE 2.2 Each small cube is one inch on a side and therefore has a volume of 1 cubic inch. The box fits a total of 240 of these cubes, so its total volume is 240 cubic inches.

READING UNITS

Operation	Key word or symbol	Example
Division	*per*	Read miles ÷ hours as "miles per hour."
Raising to second power	*square*	Read ft × ft, or ft², as "square feet" or "feet squared."
Raising to third power	*cube* or *cubic*	Read ft × ft × ft, or ft³, as "cubic feet" or "feet cubed."
Multiplication	*hyphen*	Read kilowatts × hours as "kilowatt-hours."

EXAMPLE 1 *Identifying Units*

Identify the units of the answer for each of the following cases.

a. The price you paid for gasoline, found by dividing its total cost in dollars by the number of gallons of gas that you bought.
b. The area of a circle, found with the formula πr^2 (where r is the radius of the circle), in which the radius is measured in centimeters. (Note that π is a pure number and has no units.)
c. A volume found by multiplying an area measured in acres by a depth measured in feet.

Solution

a. The price of the gasoline has units of dollars divided by gallons, which we write as $/gal and read as "dollars per gallon."
b. The area of the circle has units of centimeters to the second power, which we write as cm² and read as "square centimeters."
c. In this case, the volume has units of acres × feet, which we read as "acre-feet." This unit of volume is commonly used by hydrologists (water engineers) in the United States.

Time out to think

Pick up a newspaper or magazine and look for numbers. For at least five cases, identify the units of the numbers.

a brief review . . .

WORKING WITH FRACTIONS

We can express a fraction in three basic ways: as a *common fraction* such as $\frac{1}{2}$; as a fraction in *decimal form* such as 0.5; and as a *percentage* such as 50%. Here, we review the first two forms (see p. 119 for a review of percentages).

Common Fractions
Common fractions have the form a/b, where a and b can be any numbers as long as b is not zero; the number on top is called the **numerator** and the number on the bottom is called the **denominator**. A fraction represents division:

$$\text{numerator} \rightarrow \frac{a}{b} \quad \text{means} \quad a \div b$$
$$\text{denominator} \rightarrow$$

We can write an integer as a fraction with a denominator of 1. For example:

$$3 = \frac{3}{1} \quad \text{or} \quad -4 = \frac{-4}{1}$$

Adding and Subtracting Fractions If two fractions have a common (same) denominator, we can add or subtract them by adding or subtracting their numerators. For example:

$$\frac{1}{5} + \frac{2}{5} = \frac{1+2}{5} = \frac{3}{5} \quad \text{or} \quad \frac{7}{9} - \frac{2}{9} = \frac{7-2}{9} = \frac{5}{9}$$

Otherwise, we must write the fractions with a common denominator before adding or subtracting. For example, we can add $\frac{1}{2} + \frac{1}{3}$ by writing them with a common denominator of 6 as $\frac{3}{6}$ and $\frac{2}{6}$, respectively:

$$\frac{1}{2} + \frac{1}{3} = \frac{3}{6} + \frac{2}{6} = \frac{3+2}{6} = \frac{5}{6}$$

Multiplying Fractions To multiply fractions, we multiply the numerators and denominators separately. For example:

$$\frac{1}{3} \times \frac{2}{5} = \frac{1 \times 2}{3 \times 5} = \frac{2}{15}$$

Sometimes we can simplify fractions at the same time we multiply them by *canceling* terms that occur in both the numerator and the denominator. For example:

$$\frac{3}{4} \times \frac{5}{3} = \frac{\cancel{3} \times 5}{4 \times \cancel{3}} = \frac{5}{4}$$

Reciprocals and Division Two numbers are **reciprocals** if their product is 1. For example:

2 and $\frac{1}{2}$ are reciprocals because $2 \times \frac{1}{2} = 1$

$\frac{4}{3}$ and $\frac{3}{4}$ are reciprocals because $\frac{4}{3} \times \frac{3}{4} = 1$

We find the reciprocal of any fraction by inverting it (interchanging the numerator and the denominator); for

UNIT CONVERSIONS

Many everyday problems involve nothing more than converting one set of units to another. For example, you might want to convert a distance from kilometers to miles or a recipe measure from quarts to cups.

The basic "trick" of unit conversions is to devise an appropriate way of multiplying by 1. For example, the following are all different ways of writing 1.

$$1 = \frac{1}{1} = \frac{8}{8} = \frac{\frac{1}{4}}{\frac{1}{4}} = \frac{1 \text{ kilogram}}{1 \text{ kilogram}} = \frac{1 \text{ week}}{7 \text{ days}} = \frac{12 \text{ inches}}{1 \text{ foot}}$$

an integer like 2, we think of it as $\frac{2}{1}$ to find that its reciprocal is $\frac{1}{2}$. In general:

- The reciprocal of a is $\dfrac{1}{a}$ ($a \neq 0$).
- The reciprocal $\dfrac{a}{b}$ is $\dfrac{b}{a}$ ($a \neq 0, b \neq 0$).

To divide fractions, we replace the division with multiplication by the reciprocal. This process is sometimes called *invert* and *multiply*, as shown in the following examples:

$$10 \div \frac{1}{2} = 10 \times \underbrace{\frac{2}{1}}_{\text{invert and multiply}} = 20$$

$$\frac{3}{4} \div \frac{2}{5} = \frac{3}{4} \times \underbrace{\frac{5}{2}}_{\text{invert and multiply}} = \frac{15}{8}$$

Rules Summary
Addition/subtraction (common denominator required):

$$\frac{a}{c} + \frac{b}{c} = \frac{a+b}{c} \quad \text{or} \quad \frac{a}{c} - \frac{b}{c} = \frac{a-b}{c}$$

Multiplication:

$$\frac{a}{b} \times \frac{c}{d} = \frac{a \times c}{b \times d}$$

Division (invert and multiply):

$$\frac{a}{b} \div \frac{c}{d} = \frac{a}{b} \times \frac{d}{c}$$

Decimal Form
For a fraction in decimal form, each digit corresponds to a certain place value, which is always a power of 10 (such as 10, 100, 1000, . . .). The following example shows values for the decimal places in the number 3.141.

$$\underset{\underset{(1)}{\text{ones}}}{3} . \underset{\underset{(0.1 = \frac{1}{10})}{\text{tenths}}}{1} \underset{\underset{(0.01 = \frac{1}{100})}{\text{hundredths}}}{4} \underset{\underset{(0.001 = \frac{1}{1000})}{\text{thousandths}}}{1}$$

Converting to Common Form Converting a fraction from decimal to common form requires only recognizing the value of the last digit in the decimal. For example:

$$0.4 = \frac{4}{10} \quad 3.15 = \frac{315}{100} \quad 0.097 = \frac{97}{1000}$$

Converting to Decimal Form To convert a common fraction into decimal form, we carry out the division implied by the fraction. For example:

$$\frac{1}{4} = 1 \div 4 = 0.25$$

Many common fractions cannot be written *exactly* in decimal form, but instead have an endless repeating pattern of digits. For example, the decimal form of $\frac{1}{3}$ contains an endless string of 3s:

$$\frac{1}{3} = 0.3333333 \ldots$$

Mathematically, we represent the repeating digits by putting a bar over them; for example, the bar in $0.\overline{3}$ means that the 3 repeats indefinitely. In practice, we usually round repeating decimals to one or a few digits; for example, we might round the decimal form of $\frac{1}{3}$ to 0.33.

The last expression shows the necessity of stating units: 12 ÷ 1 is *not* 1, but 12 in. ÷ 1 ft *is* 1 because 12 inches and 1 foot are equal.

Conversion Factors

The statement 12 in. = 1 ft is an example of a **conversion factor.** We can write this conversion factor in three equivalent ways (the latter two forms come from dividing on both sides of the first equation):

$$12 \text{ in.} = 1 \text{ ft} \quad \text{or} \quad \frac{12 \text{ in.}}{1 \text{ ft}} = 1 \quad \text{or} \quad \frac{1 \text{ ft}}{12 \text{ in.}} = 1$$

The key to unit conversions is to identify the needed conversion factor and use it in the form that helps you change from one set of units to another.

EXAMPLE 2 Feet to Inches

Convert a distance of 7 feet into inches.

Solution We begin with *feet* and want to end up with *inches*. We therefore need the conversion factor between *inches* and *feet* in a form that has *feet* in the denominator, so that *feet* will cancel:

$$7 \text{ ft} \times \frac{12 \text{ in}}{1 \text{ ft}} = 84 \text{ in.}$$

Note that we really just multiplied by 1 in a form that allowed us to change the units. Seven feet is the same as 84 inches. ■

EXAMPLE 3 Inches to Feet

Convert a length of 102 inches to feet.

Solution This time we use the conversion factor in the form with *inches* in the denominator, so that *inches* will cancel:

$$102 \text{ in.} \times \frac{1 \text{ ft}}{12 \text{ in.}} = 8.5 \text{ ft}$$

Thus, 102 inches is equivalent to 8.5 feet. ■

EXAMPLE 4 Using a Chain of Conversions

How many seconds are there in one day?

Solution Most of us don't immediately know the answer to this question, but we do know that 1 day = 24 hr, 1 hr = 60 min, and 1 min = 60 s. We can answer the question by setting up a *chain* of unit conversions in which we start with *day* and end up with *seconds*.

$$1 \text{ day} \times \frac{24 \text{ hr}}{1 \text{ day}} \times \frac{60 \text{ min}}{1 \text{ hr}} \times \frac{60 \text{ s}}{1 \text{ min}} = 86{,}400 \text{ s}$$

By using the conversion factors needed to cancel the appropriate units, we are left with the answer in *seconds*. There are 86,400 seconds in one day. ■

FOXTROT © 1995 by Bill Amend. Reprinted with permission of UNIVERSAL PRESS SYNDICATE. All rights reserved.

Conversions with Units Raised to Powers

We must take special care when converting units raised to powers. For example, suppose we want to know the number of square feet in a square yard. We may not know the conversion factor between square yards (yd^2) and square feet (ft^2), but we know that 1 yd = 3 ft. Therefore, we can replace 1 yard by 3 feet when we write out 1 square yard:

$$1 \text{ yd}^2 = 1 \text{ yd} \times 1 \text{ yd} = 3 \text{ ft} \times 3 \text{ ft} = 9 \text{ ft}^2$$

That is, 1 square yard is the same as 9 square feet. We can also find this conversion factor by squaring both sides of the yards-to-feet conversion:

$$1 \text{ yd} = 3 \text{ ft} \xrightarrow{\text{square both sides}} (1 \text{ yd})^2 = (3 \text{ ft})^2 \xrightarrow{\text{square each term}} 1 \text{ yd}^2 = 9 \text{ ft}^2$$

Figure 2.3 confirms that, indeed, 9 square feet fit exactly into 1 square yard. As usual, we can write the conversion factor in three equivalent forms:

$$1 \text{ yd}^2 = 9 \text{ ft}^2 \quad \text{or} \quad \frac{1 \text{ yd}^2}{9 \text{ ft}^2} = 1 \quad \text{or} \quad \frac{9 \text{ ft}^2}{1 \text{ yd}^2} = 1$$

EXAMPLE 5 *Carpeting a Room*

You want to carpet a room that measures 10 feet by 12 feet, making an area of 120 square feet. But carpet is usually sold by the square yard rather than by the square foot. How many square yards of carpet do you need?

Solution We need to convert the room's area from units of *square feet* to *square yards*, so we use the conversion factor in the form that has square feet (ft^2) in the denominator:

$$120 \text{ ft}^2 \times \frac{1 \text{ yd}^2}{9 \text{ ft}^2} = \frac{120}{9} \text{ yd}^2 \approx 13.3 \text{ yd}^2$$

(The symbol ≈ means "approximately equal to.") Note that we rounded the answer to the nearest 0.1 square yard. Most stores will not sell fractions of a square yard, so you will need to buy 14 square yards of carpet. ∎

FIGURE 2.3 This diagram shows that 1 square yard is the same as 9 square feet.

EXAMPLE 6 Cubic Units: Purchasing Garden Soil

You are preparing a vegetable garden that is 40 feet long and 16 feet wide, and you need enough soil to fill it to a depth of 1 foot. The landscape supply store sells soil by the cubic yard. How much soil should you order?

Solution To find the volume of soil that you need, we multiply the garden's length (40 feet), width (16 feet), and depth (1 foot):

$$40 \text{ ft} \times 16 \text{ ft} \times 1 \text{ ft} = 640 \text{ ft}^3$$

Because soil is sold by the cubic yard, we need to convert this volume from units of *cubic feet* to *cubic yards*. We know that 1 yd = 3 ft, so we find the cubic conversion factor by cubing both sides of this equation:

$$1 \text{ yd} = 3 \text{ ft} \xrightarrow{\text{cube both sides}} (1 \text{ yd})^3 = (3 \text{ ft})^3 \xrightarrow{\text{cube each term}} 1 \text{ yd}^3 = 27 \text{ ft}^3$$

In the last step, we recognized that $3^3 = 3 \times 3 \times 3 = 27$. As usual, we can write this conversion factor in three equivalent forms:

$$1 \text{ yd}^3 = 27 \text{ ft}^3 \quad \text{or} \quad \frac{1 \text{ yd}^3}{27 \text{ ft}^3} = 1 \quad \text{or} \quad \frac{27 \text{ ft}^3}{1 \text{ yd}^3} = 1$$

To convert the soil volume from cubic feet to cubic yards, we use the conversion factor in the form that has cubic feet (ft^3) in the denominator:

$$640 \text{ ft}^3 \times \frac{1 \text{ yd}^3}{27 \text{ ft}^3} = \frac{640}{27} \text{ yd}^3 \approx 23.7 \text{ yd}^3$$

You will need to order about 24 cubic yards of soil for your garden.

Currency Conversions

Different countries use different money, or **currency.** If you travel internationally, you'll need to change money from one currency to another. Converting between currencies is a unit conversion problem in which the conversion factors are the exchange rates.

Table 2.1 shows a typical table of currency exchange rates. The *Dollars per Foreign* column gives the conversion factor in the form useful for converting from foreign currency into U.S. dollars. For example, the row for Mexico shows 0.1073 in this column, which means

$$1 \text{ peso} = \$0.1073$$

The *Foreign per Dollar* column gives the conversion factor in the form useful for converting U.S. dollars into foreign currency; in this case,

$$\$1 = 9.3197 \text{ pesos}$$

Time out to think

Find the reciprocals of the numbers in the *Dollars per Foreign* column of Table 2.1. Are your results the numbers in the *Foreign per Dollar* column? Explain.

TABLE 2.1	Sample Currency Exchange Rates (July 2000)	
CURRENCY	DOLLARS PER FOREIGN	FOREIGN PER DOLLAR
British pound	1.5144	0.6603
Canadian dollar	0.6764	1.4784
European euro	0.9544	1.0478
Japanese yen	0.0093	107.53
Mexican peso	0.1073	9.3197

by the way...
Currency exchange rates change nearly every day. Sometimes, especially when markets are volatile, currency rates change hour by hour or even minute by minute. You can find exchange rates listed daily in many newspapers, and up-to-the-minute rates are posted on the Web.

EXAMPLE 7 Price Conversion

At a French department store, the price for a pair of Levi jeans is 45 euros. What is the price in U.S. dollars? Use the exchange rates in Table 2.1.

Solution From the *Dollars per Foreign* column in Table 2.1, we see that

$$1 \text{ euro} = \$0.9544$$

As usual, we can write this conversion factor in two other equivalent forms:

$$\frac{1 \text{ euro}}{\$0.9544} = 1$$

or

$$\frac{\$0.9544}{1 \text{ euro}} = 1$$

We use the latter form to convert the price from euros to dollars:

$$45 \text{ euro} \times \frac{\$0.9544}{1 \text{ euro}} \approx \$42.95$$

The price of 45 euros is equivalent to $42.95.

EXAMPLE 8 Buying Currency

You are on holiday in Mexico and need cash. How many pesos can you buy with $100? Use the exchange rates in Table 2.1 and assume the transaction does not involve any fees.

Solution From the *Foreign per Dollar* column in Table 2.1, we see that

$$\$1 = 9.3197 \text{ pesos}$$

historical note
The euro came into being as a new currency for 11 countries of the European Union in January 1999. These countries immediately tied their currency values to the euro but, to ease the transition, decided to keep their national currencies in circulation for three additional years. As of January 2002, the national currencies were replaced by euro bills and coins.

Two other equivalent forms of this conversion are

$$\frac{9.3197 \text{ peso}}{\$1} = 1$$

or

$$\frac{\$1}{9.3197 \text{ peso}} = 1$$

The first form allows us to convert dollars to pesos:

$$\cancel{\$}100 \times \frac{9.3197 \text{ peso}}{\cancel{\$1}} = 931.97 \text{ peso}$$

Your $100 will buy 931.97 pesos. ∎

practical matters

CHANGING MONEY IN FOREIGN COUNTRIES

It can cost money to change money from one currency to another, so you should always look for the best deal. Two factors affect the cost of changing currency: (1) the exchange rate and (2) fees for the exchange.

Published exchange rates are usually "wholesale" rates available only to banks. Most money changers—including airport change stations, hotels, and exchange booths on streets—make money by giving you a rate that is not as good as the wholesale rate. For example, if the wholesale rate is 9.3 pesos to the dollar, a money changer might give you only 9.1 or 8.95 pesos per dollar.

Once you find a good exchange rate, you must still watch out for additional fees. For example, many money changers (especially in hotels, stores, and street booths) charge a fee every time you make an exchange. Here are a few general hints for changing currency:

• Before you travel, get a small amount of your destination's currency from your local bank. That way, if you need cash upon arrival, you won't be forced to accept a poor exchange rate or high fees.

• Once you get to your destination, banks usually offer better exchange rates than other money changers. The best deal may be your ATM card, which generally gets bank exchange rates with the added convenience of ATM machines (at least in countries where ATM machines are common). But be sure you're aware of any fees: Before your trip, find out whether your bank charges a fee for foreign ATM transactions; any fee charged by the ATM owner will usually (but not always) be stated during your transaction.

• Consider using a credit card for purchases and hotel and restaurant bills. Credit cards generally offer good exchange rates and, unlike cash, can be replaced if lost or stolen. But be sure to check whether your credit card company adds any fees for foreign purchases. And avoid using your credit card to get cash unless it is an emergency, because most credit cards charge high fees and interest for cash advances.

• Travelers' checks also offer safety in case of loss or theft, but watch out for fees. Some banks charge a fee when you purchase travelers' checks at home, and you're likely to pay another fee when you change the travelers' checks to local currency at your destination.

Time out to think

Find the current value of the Mexican peso in dollars. How does it compare to the value listed in Table 2.1? If it is significantly different, suggest some possible economic reasons for the change in the peso's value.

PROBLEM SOLVING WITH UNITS

Now that we have covered basic ideas of units and conversions, we are ready to apply them to problem solving. In some cases, units can help us decide how to solve a problem. In all cases, units offer a useful way of checking answers. The following box summarizes how we work with units; as you study, notice how these ideas are applied in the examples throughout this chapter and the rest of this book.

Working with Units

Step 1: Identify the units involved in the problem. Use the units to help you decide how to approach the problem and what units to expect in the answer.

Step 2: Perform any operations (such as multiplication or division) on both the numbers and their associated units. Remember:

- You cannot add or subtract numbers with different units (such as apples and oranges), but you can combine units through multiplication, division, or raising to powers.
- To make your work with units easier, replace division with multiplication by the reciprocal. For example,

$$\text{instead of dividing by } 60 \, \frac{\text{s}}{\text{min}}, \text{ multiply by } \frac{1 \, \text{min}}{60 \, \text{s}}.$$

Step 3: When you complete your calculations, make sure that your answer has the units you expected. If it doesn't, then you've done something wrong.

EXAMPLE 9 Distance, Time, and Speed

A car is traveling 25 miles every half-hour. How fast is it going?

Solution The "how fast" suggests that the final answer should be a speed. In this case, since the problem involves units of miles and hours, we should expect a speed in *miles per hour*. Knowing that these units mean miles *divided by* hours helps us realize that we should divide the distance traveled by the time it takes:

$$25 \, \text{mi} \div \frac{1}{2} \, \text{hr} = 25 \, \text{mi} \times \frac{2}{1 \, \text{hr}} = 50 \, \frac{\text{mi}}{\text{hr}}$$

Note that we replaced the division with multiplication by the reciprocal; be sure you understand why the reciprocal of $\frac{1}{2}$ hr is $\frac{2}{1 \, \text{hr}}$ hr. The car is traveling at a speed of 50 miles per hour.

by the way...

You can use your familiarity with *miles per hour* to find the important general relationship among distance, time, and speed. Simply remember that miles are a distance and hours are a time, so a speed like miles per hour must be a distance (such as miles) divided by a time (such as hours):

$$\text{speed} = \frac{\text{distance}}{\text{time}}$$

> **historical note**
>
> An *acre* is a unit of area originally defined as the amount of land a pair of oxen could plow in a day. Today it is defined as 43,560 square feet. One square mile is 640 acres.

EXAMPLE 10 Buying Farm Land

You are buying 30 acres of farm land at a cost of $12,000 per acre. What is the total cost?

Solution The question asks about total cost, so the answer should be in *dollars*. We multiply the acreage by the cost per acre:

$$30 \text{ acres} \times \frac{\$12{,}000}{1 \text{ acre}} = \$360{,}000$$

Note how the unit *acre* "cancels" to leave the answer in units of dollars, as we expected. The land will cost $360,000. ∎

Using Units to Check Answers

Checking answers is one of the most important reasons for clearly showing all units. If the answer does not come out with the expected units, then something has gone wrong. The following two examples illustrate this idea. Of course, you should always check your method and your calculations as well as your units; an answer with the correct units could still contain other errors.

EXAMPLE 11 Flight Time

You overhear someone claiming that you can figure out how long it would take to fly to Beijing by dividing an airplane's speed by the distance to Beijing. Can he be right?

Solution Look at the units involved in the claim. Let's assume that the airplane's speed is measured in *miles per hour* and the distance to Beijing is measured in *miles*. In that case, dividing the airplane's speed by the distance to Beijing gives an answer that has units of

$$\frac{\text{mi}}{\text{hr}} \div \text{mi} = \frac{\text{mi}}{\text{hr}} \times \frac{1}{\text{mi}} = \frac{1}{\text{hr}}$$

As usual, we replaced the division (by mi) with multiplication by the reciprocal (1/mi). The claim concerns the flying time, which should have units of hours. But the result above has units of 1/hr, which is *not* the same as hours, so the claim must be wrong. The correct way to find the travel time is by dividing the distance by the airplane's speed; you should confirm for yourself that dividing a distance in miles by a speed in miles per hour gives a result with units of hours. ∎

EXAMPLE 12 Exam Check

You are a grader for a math course. An exam question reads: "Eli purchased 5 pounds of apples at a price of 50 cents per pound. How much did he pay for the apples?" On the paper you are grading, a student has written: "50 ÷ 5 = 10. He paid 10 cents." Write a note to the student explaining what went wrong.

Solution Dear student—First, notice that your answer does not make sense; if one pound of apples costs 50¢, how could five pounds cost only 10¢? You could have prevented your error by keeping track of the units. In the exam question, the number 50

> **by the way...**
>
> The abbreviation for pound, *lb*, comes from the Latin *libra*, meaning "scales."

has units of *cents per pound* and the number 5 has units of *pounds*. Thus, your calculation of "50 ÷ 5 = 10" means

$$50 \frac{¢}{lb} \div 5 \text{ lb} = 50 \frac{¢}{lb} \times \frac{1}{5 \text{ lb}} = 10 \frac{¢}{lb^2}$$

(As usual, we replaced the division with multiplication by the reciprocal.) Your calculation gives units of "cents per square pound," so it cannot be correct for a question that asks for a simple price. The correct calculation multiplies the price per pound by the weight in pounds:

$$50 \frac{¢}{\cancel{lb}} \times 5 \cancel{\text{ lb}} = 250¢ = \$2.50$$

The units now work out as they should: The 5 pounds of apples cost $2.50.

Using Units to Help You Solve Problems

As we've seen, knowing the units to expect in the answer can help us determine how to solve a problem. The following two examples illustrate how powerful this idea can be. The first shows how units can help us get "unstuck" on an everyday sort of problem, while the second shows how units can help us come up with a strategy for a more complex problem—in this case, a problem relevant to the issue of global warming.

EXAMPLE 13 *Gas Mileage*

After a long day of driving, your destination is only 90 miles away. You know that your car gets 25 miles per gallon. How much gas do you need in your tank if you are to reach your destination without stopping? If you have a 12-gallon tank and the fuel gauge shows it is one-quarter full, will you make it? Show how units can help answer this question.

Solution The question asks "how much gas," so we are looking for an answer with units of *gallons*. We have two other pieces of information available: the distance of 90 *miles* and the car's gas mileage of 25 *miles per gallon*. Thus, we are looking for a way to combine these two pieces of information to get an answer in gallons. We can do it by dividing the distance by the mileage:

$$90 \text{ mi} \div \frac{25 \text{ mi}}{1 \text{ gal}} = 90 \cancel{\text{ mi}} \times \frac{1 \text{ gal}}{25 \cancel{\text{ mi}}} = 3.6 \text{ gal}$$

You'll need 3.6 gallons of gas to drive the 90 miles. One-quarter of a 12-gallon tank is only 3 gallons, so you'd better stop for gas.

EXAMPLE 14 *Melting of the Ice Caps*

We usually think of "ice ages" as times when glaciers covered much of North America and Europe. But in some sense the Earth is still in an ice age today, because geological evidence suggests that the Earth had no polar ice at all during much of its history. Some time in the future, it is likely that the Earth will again have no polar ice.

Measurements of polar ice show that, if all the ice melts, about 25 million cubic kilometers of water will be added to the oceans—most of it coming from Antarctica (Figure 2.4). How much will sea level rise as a result? Useful data: The total surface area of the Earth's oceans is about 340 million square kilometers.

Solution Because the question asks how much sea level will rise, we are looking for an answer with units of height, such as kilometers or meters. We are given two pieces of information that seem relevant: a volume of water in *cubic kilometers* and a surface area for the oceans in *square kilometers*. If we divide these two numbers, we get a result with units of kilometers:

$$\frac{25 \text{ million km}^3}{340 \text{ million km}^2} \approx 0.074 \text{ km} = 74 \text{ m}$$

Of course, the fact that this result has the expected units doesn't make it right; we must also ask whether it makes sense. In this case, it does: Dividing a volume by an area does indeed give a depth (because units of km^3 divided by units of km^2 give units of km), and the depth we've found represents the rise in sea level as long as the melting does not appreciably change the surface area of the oceans. (Because most continents have coastal hills or mountains that are higher than the expected sea level rise, ocean water would not spread very far inland; thus, the ocean surface area would increase only slightly.) Perhaps the greatest surprise is the size of the answer: Complete polar melting would cause sea level to rise by some 74 meters, which is more than 240 feet. Thus, if this ever occurs, future archaeologists will have to be extreme deep sea divers to visit the ruins of modern coastal cities.

FIGURE 2.4

by the way...
The Antarctic ice sheet contains more than 90% of the Earth's ice. This ice sheet covers an area of 13 million square kilometers, with an average depth of 2 kilometers (1.2 mi); in some places, the ice is more than 4.5 kilometers (2.8 mi) thick.

Time out to think

Although it would probably take thousands of years for the ice caps to melt fully, some melting may already be occurring due to global warming. Suppose that 1% of the ice melts during the next 100 years, raising sea level by about 2.4 feet. How would such a sea level rise affect human civilization? Explain.

Review Questions

1. What are units? Describe how to read units that involve division, multiplication, squares, and cubes.

2. Explain why a unit conversion really just involves multiplying by 1.

3. Describe the three forms in which we can write any conversion factor. If given the conversion in one form, such as 1 lb = 16 oz, how do you find the other two forms?

4. Explain in words or with a picture why there are 9 square feet in 1 square yard and 27 cubic feet in 1 cubic yard. Then describe generally how to find conversion factors involving squares or cubes.

5. Describe how to read and use the currency data in Table 2.1.

6. Briefly describe how units can help you check your answers and solve problems. Give examples.

In the News

1. **Are the Units Clear?** Find a news story that involves numerical data. Are all the numbers in the story given with meaningful units, or is the meaning of some of the units unclear? Briefly summarize how well (or not well) the article uses units.

2. **Units on the Highway.** Next time you are on the highway, look for three signs that use numbers (such as speed limits or distances to nearest exits). Are the units of the numbers given? If not, how are you expected to know the units? In cases where the units are not given, do you think the units would be obvious to everyone? Why or why not?

3. **False Advertising?** A Goodyear tire commercial began by stating that the Goodyear Aquatread tire can "channel away" 1 gallon of water per second. The announcer then said: "One gallon per second—that's 396 gallons per mile." What's wrong with this statement? What point do you think the advertisement was trying to make? Can you find other examples of advertisements that misstate units?

Exercises

Working with Fractions. Exercises 1–6 use skills covered in the Brief Review on pp. 72 and 73.

1. Evaluate each of the following.
 a. $\frac{4}{3} \times \frac{1}{2}$ b. $\frac{4}{3} + \frac{1}{2}$ c. $\frac{4}{3} \div \frac{1}{2}$ d. $\frac{4}{3} - \frac{1}{2}$
 e. $\frac{7}{20} + \frac{3}{5}$ f. $\frac{12}{13} - \frac{1}{4}$ g. $\frac{7}{17} \times \frac{2}{7}$ h. $\frac{1}{3} + \frac{1}{5}$

2. Evaluate each of the following.
 a. $\frac{14}{5} - \frac{3}{10}$ b. $\frac{5}{9} \div \frac{4}{9}$ c. $\frac{2}{3} + \frac{1}{4}$ d. $\frac{1}{2} + \frac{1}{3} + \frac{1}{4}$
 e. $\frac{1}{6} \times \frac{6}{11}$ f. $\frac{1}{8} - \frac{1}{16}$ g. $\frac{3}{5} \div \frac{5}{3}$ h. $\frac{1}{2} \times \frac{1}{3} \times \frac{1}{4}$

3. Write each of the following as a common fraction.
 a. 0.3 b. 0.124 c. 0.78 d. 0.005
 e. 1.84 f. 3.009 g. 0.0001 h. 0.1001

4. Write each of the following as a common fraction.
 a. 0.7 b. 1.4 c. 0.45 d. 0.988
 e. 0.401 f. 1.001 g. 3.07 h. 0.00002

5. Convert each of the following common fractions into decimal form. If necessary, round to the nearest thousandth.
 a. $\frac{3}{2}$ b. $\frac{4}{5}$ c. $\frac{1}{6}$ d. $\frac{1}{7}$
 e. $\frac{2}{9}$ f. $\frac{7}{11}$ g. $\frac{88}{91}$ h. $\frac{122}{30}$

6. Convert each of the following common fractions into decimal form. If necessary, round to the nearest thousandth.
 a. $\frac{7}{12}$ b. $\frac{27}{70}$ c. $\frac{24}{45}$ d. $\frac{18}{25}$
 e. $\frac{537}{912}$ f. $\frac{243}{81}$ g. $\frac{5}{6}$ h. $\frac{16}{3}$

7. **Identifying Units.** Identify the units you would expect for each of the following. State the units both in words (e.g., dollars per gallon) and mathematically (e.g., $/gal).
 a. The price of apples, found by dividing their total weight in pounds by their total cost in dollars.
 b. A speed found by dividing a distance measured in kilometers by a time measured in seconds.
 c. The installation cost of floor tile, found by dividing the total cost of the installation in dollars by the area of the room in square feet.
 d. The flow rate of a river in which 5000 cubic feet of water flow past a particular location every second.
 e. The atmospheric pressure, measured by finding the weight of atmosphere (in pounds) above each square inch of surface on the ground.

8. **Identifying Units.** Identify the units you would expect for each of the following. State the units both in words (e.g., dollars per gallon) and mathematically (e.g., $/gal).
 a. The price of fabric, found by dividing its cost in dollars by its area in square feet.

b. The gas mileage of a car, found by dividing the distance in miles that it travels by the amount of gas in gallons that it uses.

c. The cost for grass seed when you buy enough to cover 80 square yards at a total price of $160.

d. The density of a rock, found by dividing its weight in grams by its volume in cubic centimeters.

e. A car engine torque calculated by multiplying a force in pounds by a distance in feet.

9. **Area and Volume Calculations.**

 a. A swimming pool 3 meters deep, 10 meters long, and 5 meters wide is filled with water. What is the area of the water's surface? What volume of water does the pool contain?

 b. A box measures 22 inches by 15 inches by 12 inches. What is the area of its largest sides? What is the volume of the box?

 c. A skyscraper is 1000 feet high with a 25,000-square-foot base. What is its total volume?

10. **Area and Volume Calculations.**

 a. A warehouse is 40 yards long and 25 yards wide and piled with cartons to a height of 3 yards. What is the area of the warehouse floor? What is the total volume of the cartons? (Assume there is no space between cartons.)

 b. The bed of a pickup truck is 3.5 feet deep, 12 feet long, and 5 feet wide. What is the area of the bed's floor? What is the volume of the bed?

 c. A can has a circular base with an area of 6 square inches and is 4 inches tall. What is its total volume?

11. **Unit Conversions.**

 a. Convert a distance of 18 yards into feet.

 b. Convert a distance of 18 yards into inches.

 c. Convert a weight of 142 ounces into pounds; there are 16 ounces in 1 pound.

 d. Convert a lot size of $\frac{1}{5}$ acre to square feet (1 acre = 43,560 ft^2).

 e. There are 8 ounces in a cup, 4 cups in a quart, and 4 quarts in a gallon. Using a chain with these conversions, convert 6 gallons into ounces.

12. **Unit Conversions.**

 a. Convert a distance of 7 miles into yards; there are 1760 yards in a mile.

 b. Using the fact that there are 1760 yards in a mile and 3 feet in a yard, convert a distance of 3 miles into feet.

 c. Use a chain of conversions with familiar measures of time to convert 4 weeks into minutes.

 d. Convert a park size of 3.5 square miles to acres (1 acre = $\frac{1}{640}$ mi^2).

 e. A car is driving at 100 kilometers per hour. What is its speed in kilometers per second?

13. **Squared Units.**

 a. Find a conversion factor between square inches and square feet. Write it in three forms.

 b. How many square inches are in 5 square feet?

 c. A football field is 100 yards long and 60 yards wide. Find its area in square yards and square feet.

14. **Cubic Units.**

 a. Find a conversion factor between cubic inches and cubic feet. Write it in three forms.

 b. How many cubic inches are in 3 cubic yards?

 c. A cargo container is 50 feet long, 10 feet wide, and 8 feet tall. Find its volume in cubic feet and in cubic yards.

15. **Currency Conversions.** Answer the following questions using the exchange rates in Table 2.1.

 a. Which is worth more, 1 British pound or 1 dollar? Explain.

 b. How many Japanese yen can you buy for $100?

 c. You return from a trip with 2500 Mexican pesos. How much are your pesos worth in dollars?

 d. Gasoline in Germany sells for about 5.50 euros per gallon. What is this price in dollars per gallon?

16. **Currency Conversions.** Answer the following questions using the exchange rates in Table 2.1.

 a. Which is worth more, 1 Mexican peso or 1 Japanese yen? Explain.

 b. How many Canadian dollars can you buy for $100?

 c. You return from a trip with 75 British pounds. How much are your pounds worth in dollars?

 d. Apples in Japan sell for about 250 yen each. If you buy 4 apples, how much have you spent in dollars?

Working with Units. Use units to help you answer the questions in Exercises 17–30.

17. You are buying 4.7 pounds of apples priced at $1.29 per pound. How much will you pay?

18. You are buying 2.8 kilograms of cherries priced at $3.50 per kilogram. How much will you pay?

19. A car travels 14 miles in 15 minutes. How fast is it going in miles per hour?

20. An airplane travels 95 miles in 10 minutes. How fast is it going in miles per hour?
21. You purchased 40 acres of farm land for $200,000. How much did you pay per acre?
22. A 40-acre orchard produces 12,000 apples. What is its yield in apples per acre?
23. You need 10 square yards of cloth priced at $2 per square yard. How much will the cloth cost?
24. You are buying floor tile to cover a room that measures 20 feet by 25 feet. The tile is priced at $7.50 per square foot. How much will the tile cost?
25. You are buying carpet to cover a floor area of 1520 square feet. The carpet costs $18 per square yard. How much will the carpet cost?
26. You are buying artificial turf to cover a game field that is 150 feet long and 100 feet wide. The turf costs $7.50 per square yard. How much will the turf cost?
27. You take a trip in which you drive 1200 miles in 20 hours. What is your average speed for the entire trip?
28. You work 40 hours per week and are paid $13.50 per hour. If you work all 52 weeks in a year, how much will you earn?
29. If you sleep an average of 7.5 hours each night, how many hours do you sleep in a year?
30. An average human heart beats 60 times per minute. If an average human being lives to the age of 75, how many times does the average heart beat in a lifetime?

What Went Wrong? Exercises 31–34 include an exam question and a solution given by a student. State whether each solution is right or wrong. If it is wrong, write a note to the student explaining why the answer is wrong and how to solve the problem correctly.

31. *Exam Question:* A candy store sells chocolate for $7.70 per pound. The piece you want to buy weighs 0.11 pound. How much will it cost, to the nearest cent? (Neglect sales tax.)
 Student Solution: 0.11 ÷ 7.70 = 0.014. It will cost 1.4¢.
32. *Exam Question:* You ride your bike up a steep mountain road at 5 miles per hour. How far do you go in 3 hours?
 Student Solution: 5 ÷ 3 = 1.7. I rode 1.7 miles.
33. *Exam Question:* You can buy a 50-pound bag of flour for $11 or you can buy a 1-pound bag for $0.39. Compare the per pound cost for the large and small bags.
 Student Solution: The large bag price is 50 ÷ $11 = $4.50 per pound, which is much more than the 39¢ per pound price of the small bag.
34. *Exam Question:* The average person needs 1500 Calories a day. A can of Coke contains 140 Calories. How many Cokes would you need to drink to fill your daily caloric needs? (Note: This diet may not meet other nutritional needs!)
 Student Solution: 1500 × 140 = 210,000. You would need to drink 210,000 Cokes to meet your daily caloric needs.
35. **Filling a Pool.** A swimming pool is 75 feet long and 54 feet wide. It is only partially filled with water, so the water surface is 6 inches below where it is supposed to be. How much water would it take, in cubic feet, to raise the water level by 6 inches?
36. **House Footprint.** A local zoning ordinance says that a house's "footprint" (area of its ground floor) cannot occupy more than $\frac{1}{4}$ of the lot it is built on. Suppose you own a $\frac{1}{4}$-acre lot. What is the maximum allowed footprint for your house, in square feet (1 acre = 43,560 ft^2)?

37. **Full of Hot Air.** The average person breathes 6 times per minute (at rest), inhaling and exhaling half a liter of air each time. How much "hot air" (the air is warmed by the body), in liters, does the average person exhale each day?

38. **E-Books.** Computer memory is measured in units of bytes, where one byte is enough memory to store one character (a letter in the alphabet or a number). How many typical pages of text can be stored on a 500-megabyte hard drive (a megabyte is one million bytes)? Assume 2000 bytes per page.

39. **Dog Years.** Sometimes the age of dogs is described in a unit called "dog years." A commonly used conversion is that 1 real year equals 7 dog years.

 a. If your dog is 15 real years old, what is her age in dog years?

 b. People often refer to the third year in the life of a human child as the "terrible twos" stage. If dogs have a terrible twos stage in their third dog year, how old are they, in real time, during this stage?

 c. Based on what you know about dogs, do you think the common conversion of 1 real year to 7 dog years seems reasonable? Explain.

40. **Glen Canyon Flood.** In April 1996, the Department of the Interior released a "spike flood" from the Glen Canyon Dam on the Colorado River. Its purpose was to restore the river and the habitats along its banks, particularly in the Grand Canyon. The reservoir behind the dam contains about 1.2 trillion (1,200,000,000,000) cubic feet of water. The release from the dam lasted a week at a rate of 25,800 cubic feet of water per second. About how much water was released during the 1-week flood? What fraction of the total water in the reservoir was released during the flood?

41. **Home Project.** Assume you are building a simple shed and doing some landscaping. The shed will be the shape of a box that is 10 feet long, 10 feet wide, and 8 feet tall. The area you are landscaping is 75 feet long and 40 feet wide. For each of the following questions, find the price of the needed item at a local store.

 a. How much would it cost to plant the entire landscaping area with grass seed? (Hint: Grass seed is usually rated by how many square feet can be covered by each pound of seed.)

 b. Suppose you decide to cover the landscaping area with sod (i.e., rolls of pre-grown grass). How much will the sod cost?

 c. You decide that before any planting, you need to add 8 inches of high-quality top soil to the entire landscape area. How much will the soil cost?

 d. How much will it cost to paint the exterior walls of the shed? (Hint: Paint cans generally tell you how much area can be covered with the paint they contain.)

Web Projects

Find useful links for Web Projects on the text Web site: www.aw.com/bennett-briggs

1. **South American Adventure.** Suppose you are planning an extended trip through many countries in South America. Use one of the many currency exchange sites on the Web to get all the exchange rates you'll need. Make a brief table showing each currency you'll need and the value of each currency in dollars.

2. **Polar Ice Melting.** Starting with a search on "glaciers" or "glaciology" (the study of glaciers), use the Web to learn more about polar ice melting. Focus on one aspect of the issue, such as whether global warming is causing melting, or the environmental impacts of melting, or the geological history of ice ages. Write a one-page summary of what you learn.

UNIT 2B

STANDARDIZED UNITS: MORE PROBLEM-SOLVING POWER

We've seen that working with units can be a powerful problem-solving technique—as long as the units are clear and meaningful. When units (such as feet, miles, or kilograms) are *standardized*, everyone agrees on their meaning. We will now review the two systems of standardized units in common use—the U.S. system and the international metric system—and how to convert between them. We'll also explore standardized units for energy, which will allow us to study a broader range of problems.

STANDARDIZED UNIT SYSTEMS: U.S. AND METRIC

Today, only one system of standardized units enjoys wide international use: the international metric system, also known as SI (from the French *Système International d'Unités*). However, the United States still uses the vestiges of an ancient measurement system sometimes called the English system but more formally known as the *U.S. customary system* (USCS). Let's briefly review each system.

The U.S. Customary System

Units of length in the U.S. customary system come from ancient units based on body parts. For example, the *foot* was once the length of the foot of whoever was doing the measuring. The ancient Romans discovered that most adult feet are about as long as 12 thumb-widths; our word *inch* comes from the Latin *uncia*, meaning "thumb-width." The Romans paced out longer distances; our word *mile* comes from the Latin *milia passum*, meaning "one thousand paces."

Time out to think

How many of *your* thumb-widths fit along your bare foot? Were the Romans correct in believing there are 12 thumb-widths in a foot?

As you might imagine, lengths that vary from person to person can lead to difficulties. For example, if you are buying 10 feet of rope, is it the length of your foot or the seller's foot that should be used to measure the rope? Many cultures therefore standardized units according to body measures of one particular person, such as a king. But such measures still changed every time one king was replaced by another. The first permanent standardization began when English King Henry I (1100–1135) set the *yard* as the measurement from the tip of *his* nose to the tip of *his* thumb on *his* outstretched arm. Today, all lengths in the U.S. customary system are based on the inch, which is defined by law to be exactly 2.54 centimeters. Table 2.2 summarizes units of length in the U.S. customary system.

For measuring weight, the U.S. customary system actually has three distinct sets of units. Traditionally, jewelers used *troy* measures of weight, pharmacists used *apothecary* measures of weight, and most other commerce was conducted with *avoirdupois* measures of weight. Table 2.3 summarizes all three systems, but only the

Thumb-width (inch)
foot (foot)
pace (1000 paces = mile)

historical note

The roots of U.S. customary units reach back to ancient Middle Eastern civilizations, including the Egyptians, Sumerians, Babylonians, and Hebrews. The units were first standardized a few hundred years ago by the English government, which introduced them to the American colonies.

TABLE 2.2 USCS Lengths (common abbreviations in parentheses)

1 inch (in.) = 2.54 centimeters		1 furlong	= 40 rods = $\frac{1}{8}$ mile
1 foot (ft) = 12 inches		1 mile (mi)	= 1760 yards = 5280 feet
1 yard (yd) = 3 feet		1 nautical mile	= 1.852 km ≈ 6076.1 feet
1 rod = 5.5 yards		1 land league	= 3 miles
1 fathom = 6 feet		1 marine league	= 3 nautical miles

TABLE 2.3 USCS Weights (common abbreviations in parentheses)

AVOIRDUPOIS MEASURES	TROY MEASURES	APOTHECARY MEASURES
1 grain = 0.0648 gram	1 grain = 0.0648 gram	1 grain = 0.0648 gram
1 ounce (oz) = 437.5 grains	1 carat ≈ 0.2 gram = 3.086 grains	1 scruple = 20 grains
1 pound (lb) = 16 ounces	1 pennyweight = 24 grains	1 dram = 3 scruples
1 ton = 2000 pounds	1 troy ounce = 480 grains	1 apoth. ounce = 8 drams
1 long ton = 2240 pounds	1 troy pound = 12 troy ounces	1 apoth. pound = 12 ounces

TABLE 2.4 USCS Volumes (common abbreviations in parentheses)

LIQUID MEASURES		DRY MEASURES	
1 tablespoon (tbsp or T)	= 3 teaspoons (tsp or t)	1 in.3	≈ 16.387 cm^3
1 fluid ounce (fl oz)	= 2 tablespoons = 1.805 in.3	1 ft^3	= (12 in.)3 = 1728 in.3
1 cup (c)	= 8 fluid ounces	1 yd^3	= (3 ft)3 = 27 ft^3
1 pint (pt)	= 16 fluid ounces = 28.88 in.3	1 dry pint (pt)	= 33.60 in.3
1 quart (qt)	= 2 pints = 57.75 in.3	1 dry quart (qt)	= 2 dry pints = 67.2 in.3
1 gallon (gal)	= 4 quarts	1 peck	= 8 dry quarts
1 barrel of petroleum	= 42 gallons	1 bushel	= 4 pecks
1 barrel of liquid	= 31 gallons	1 cord	= 128 ft^3

historical note

The *avoirdupois* measures were standardized by London merchants around the year 1300. The term *avoirdupois* comes from the French for "goods of weight." The *troy* system was probably named for the town of Troyes, France, and was established in the 15th century. The apothecary system was developed by medieval pharmacists (*apothecary* is a synonym for *pharmacy*).

Flour Water

Volume of dry pint = 33.60 in.3 Volume of liquid pint = 28.88 in.3

FIGURE 2.5 In the U.S. customary system, a dry pint (such as a pint of flour) contains a greater volume than a liquid pint (such as a pint of water).

avoirdupois measures are commonly used today. The basic unit of weight in all three sets is the *grain*, an ancient unit originally based on the weight of a typical grain of wheat.

The basic U.S. customary unit of volume is the cubic inch (in.3), but from here the system becomes complicated, because units with the same name represent different volumes depending on whether we're working with dry or liquid materials. For example, a *dry* pint is 33.60 cubic inches, while a *liquid* pint is only 28.88 cubic inches (Figure 2.5). Thus, a container that holds one pint of water is too small for one pint of flour. (If you travel, it can be even more complicated, because British customary measures differ from U.S. measures. For example, a British pint of beer is about 20% larger than a U.S. pint of beer.) Table 2.4 summarizes USCS measures of volume.

EXAMPLE 1 *The Kentucky Derby*

In the Kentucky Derby, horses race a distance of 10.2 furlongs. How many miles is the race?

Solution From Table 2.2, 1 furlong = $\frac{1}{8}$ mi, which is the same as 0.125 mile. We can write this conversion factor in two other equivalent forms:

$$\frac{1 \text{ furlong}}{0.125 \text{ mi}} = 1$$

or

$$\frac{1 \text{ furlong}}{0.125 \text{ mi}} = 1$$

The second form allows us to convert furlongs to miles:

$$10.2 \text{ furlong} \times \frac{0.125 \text{ mi}}{1 \text{ furlong}} = 1.275 \text{ mi}$$

The Kentucky Derby is a race of 1.275 miles.

practical matters

GEMS AND GOLD JEWELRY

If you've ever bought jewelry, you've probably seen labels stating *karats* or *carats*. But you may be surprised to know that these are not just alternative spellings of the same word. A *carat* is a unit of weight defined to be exactly 0.2 gram (see Table 2.3). A *karat* is a measure of the purity of gold: 24-karat gold is 100% pure, 18-karat gold is 75% pure, 14-karat gold is 50% pure (because $\frac{14}{24} \approx 0.58$), and so on.

If you are buying gold jewelry, look for a label stating both its purity in karats and its weight (usually given in grams); the amount of gold you are actually buying depends on both. For example:

- Ten grams of 18-karat gold actually contain 7.5 grams of pure gold, since 18-karat gold is 75% pure.
- Ten grams of 14-karat gold actually contain 5.8 grams of pure gold, since 14-karat gold is 58% pure.

Although a higher karat value is purer, it comes with an important practical tradeoff: Gold is a soft metal that deforms easily, and the metals (usually silver and copper) mixed into lower-karat gold give it added strength. Thus, for example, a 14-karat gold ring is stronger and more durable than an 18-karat gold ring.

You'll deal with carats if you are buying gems such as diamonds or emeralds. Because 1 carat is 0.2 gram, the number of carats tells you the precise weight of the gem. However, several factors besides weight are important to the price, such as the gem's shape and color.

If you are gem shopping, you'll need to make tradeoffs among weight, color, and clarity (and sometimes other factors as well) to stay within your budget. For example, if you are looking for a round diamond, $10,000 might buy you either a 1-carat diamond with good color and clarity or a 2-carat diamond with poorer color and clarity. Given the cost of most gems, you should spend some time doing price comparisons before buying.

> **by the way...**
> The deepest point in the oceans lies in the Marianas Trench in the western Pacific Ocean. Its depth is 36,198 feet (11,033 m), or almost 7 miles.

EXAMPLE 2 *20,000 Leagues Under the Sea*

In Jules Verne's novel *20,000 Leagues Under the Sea,* does the title refer to an ocean depth? How do you know?

Solution From Table 2.2, 1 marine league = 3 nautical miles. Thus, 20,000 leagues is

$$20{,}000 \text{ leagues} \times \frac{3 \text{ naut. mi}}{1 \text{ league}} = 60{,}000 \text{ naut. mi}$$

Because the oceans are at most a few miles deep, 20,000 leagues cannot possibly refer to an ocean depth. The book's title refers to the distance traveled by submarine under the sea.

The International Metric System

The international metric system was invented in France late in the 18th century and is now used throughout the world. The metric system was designed for two primary reasons: (1) to replace many customary units with just a few basic units and (2) to simplify conversions through use of a decimal (base 10) system. Similar ideas had already been applied to currency; for example, in 1785, the United States replaced the customary currencies of the individual states with a national currency, the dollar (devised by Thomas Jefferson). Like the metric system, the dollar is based on a decimal relationship (because 100¢ = $1).

The basic units of length, mass, time, and volume in the metric system are

- the **meter** for length, abbreviated m
- the **kilogram** for mass, abbreviated kg
- the **second** for time, abbreviated s
- the **liter** for volume, abbreviated L

These basic units can be combined with a prefix that indicates multiplication by a power of 10. For example, *kilo* means 1000 so a kilometer is 1000 meters, and *micro* means *one-millionth* so a microgram is a millionth of a gram. Table 2.5 lists common metric prefixes.

> **historical note**
> In 1790, Thomas Jefferson (then Secretary of State) proposed adoption of the metric system to Congress. Had Congress agreed, the United States would have been the *first* country to adopt the metric system—ahead of France, which adopted it in 1795. In 1875, the United States was one of 17 original signatories on the *Treaty of the Meter,* which promoted and standardized the metric system. Congress replaced customary units with metric units as the legal standards in 1893. Today, through the National Institute of Standards and Technology, the United States helps to refine and extend the metric system.

Time out to think

People often say that expensive things cost "megabucks." What does this statements mean literally? Can you think of other cases where metric prefixes have entered popular language?

TABLE 2.5 Metric Prefixes

\- Small Values \-			\- Large Values \-		
PREFIX	ABBREV.	VALUE	PREFIX	ABBREV.	VALUE
deci	d	10^{-1} (one-tenth)	deca	da	10^{1} (ten)
centi	c	10^{-2} (one-hundredth)	hecto	h	10^{2} (hundred)
milli	m	10^{-3} (one-thousandth)	kilo	k	10^{3} (thousand)
micro	μ	10^{-6} (one-millionth)	mega	M	10^{6} (million)
nano	n	10^{-9} (one-billionth)	giga	G	10^{9} (billion)
pico	p	10^{-12} (one-trillionth)	tera	T	10^{12} (trillion)

NOTE: The abbreviation for micro, μ, is the Greek letter mu.

a brief review...

POWERS OF 10

Powers of 10 indicate how many times to multiply 10 by itself. For example:

$$10^2 = 10 \times 10 = 100$$

$$10^6 = 10 \times 10 \times 10 \times 10 \times 10 \times 10 = 1{,}000{,}000$$

Negative powers indicate reciprocals of corresponding positive powers. For example:

$$10^{-2} = \frac{1}{10^2} = \frac{1}{100} = 0.01$$

$$10^{-6} = \frac{1}{10^6} = \frac{1}{1{,}000{,}000} = 0.000001$$

Thus, powers of 10 follow two basic rules:

1. A positive exponent tells how many 0s follow the 1. For example, 10^0 is a 1 followed by no 0s, 10^8 is a 1 followed by eight 0s.
2. A negative exponent tells how many places are to the right of the decimal point, including the 1. For example, $10^{-1} = 0.1$ has one place to the right of the decimal point; $10^{-6} = 0.000001$ has six places to the right of the decimal point.

Multiplying and Dividing Powers of 10

Multiplying powers of 10 simply requires adding exponents. For example:

$$10^4 \times 10^7 = \underbrace{10{,}000}_{10^4} \times \underbrace{10{,}000{,}000}_{10^7} = \underbrace{100{,}000{,}000{,}000}_{=10^{4+7}=10^{11}}$$

$$10^5 \times 10^{-3} = \underbrace{100{,}000}_{10^5} \times \underbrace{0.001}_{10^{-3}} = \underbrace{100}_{=10^{5+(-3)}=10^2}$$

$$10^{-8} \times 10^{-5} = \underbrace{0.00000001}_{10^{-8}} \times \underbrace{0.00001}_{10^{-5}} = \underbrace{0.0000000000001}_{=10^{-8+(-5)}=10^{-13}}$$

Dividing powers of 10 requires subtracting exponents. For example:

$$\frac{10^5}{10^3} = \underbrace{100{,}000}_{10^5} \div \underbrace{1000}_{10^3} = \underbrace{100}_{=10^{5-3}=10^2}$$

$$\frac{10^3}{10^7} = \underbrace{1000}_{10^3} \div \underbrace{10{,}000{,}000}_{10^7} = \underbrace{0.0001}_{=10^{3-7}=10^{-4}}$$

$$\frac{10^{-4}}{10^{-6}} = \underbrace{0.0001}_{10^{-4}} \div \underbrace{0.000001}_{10^{-6}} = \underbrace{100}_{=10^{-4-(-6)}=10^2}$$

Powers of Powers of 10

We can use the multiplication and division rules to raise powers of 10 to other powers. For example:

$$(10^4)^3 = 10^4 \times 10^4 \times 10^4 = 10^{4+4+4} = 10^{12}$$

Note that we get the same result by simply multiplying the two powers:

$$(10^4)^3 = 10^{4 \times 3} = 10^{12}$$

Adding and Subtracting Powers of 10

There is no shortcut for adding or subtracting powers of 10, as there is for multiplication or division. The values must be written in longhand notation. For example:

$$10^6 + 10^2 = 1{,}000{,}000 + 100 = 1{,}000{,}100$$

$$10^8 + 10^{-3} = 100{,}000{,}000 + 0.001 = 100{,}000{,}000.001$$

$$10^7 - 10^3 = 10{,}000{,}000 - 1000 = 9{,}999{,}000$$

Summary

We summarize our findings as follows, using n and m to represent any numbers:

- To *multiply* powers of 10, *add* exponents:

$$10^n \times 10^m = 10^{n+m}$$

- To *divide* powers of 10, *subtract* exponents:

$$\frac{10^n}{10^m} = 10^{n-m}$$

- To *raise* powers of 10 to other powers, multiply exponents:

$$(10^n)^m = 10^{n \times m}$$

historical note

People around the world use the same base 10 *(decimal)* counting system. This may seem natural because we have 10 fingers, but various ancient cultures used base 2, base 3, base 5, base 20, and other bases. Vestiges of other bases remain in our language. For example, we have 60 seconds in a minute and 60 minutes in an hour because the ancient Babylonians used base 60. Similarly, a dozen is 12 and a gross is $12 \times 12 = 144$, presumably because base 12 was once common in northern Europe.

by the way...

Technically, pounds are a unit of *weight* and kilograms are a unit of *mass*, so the given conversions between pounds and kilograms are valid only on Earth. That is, on Earth, a 50-kilogram astronaut weighs about 110 pounds. In Earth orbit, the astronaut still has a mass of 50 kilograms, but has a weight of zero (weightless).

EXAMPLE 3 *Using Metric Prefixes*

a. Convert 2759 centimeters to meters.
b. How many nanoseconds are in a microsecond?

Solution

a. Table 2.5 shows that *centi* means 10^{-2}, so 1 cm = 10^{-2} m or, equivalently, 1 m = 100 cm. Thus, 2759 centimeters is the same as

$$2759 \text{ cm} \times \frac{1 \text{ m}}{100 \text{ cm}} = 27.59 \text{ m}$$

b. We compare the quantities by dividing the longer time (microsecond) by the shorter time (nanosecond):

$$\frac{1 \text{ μs}}{1 \text{ ns}} = \frac{10^{-6} \text{ s}}{10^{-9} \text{ s}} = 10^{-6-(-9)} = 10^{-6+9} = 10^3$$

A microsecond is 10^3, or 1000, times longer than a nanosecond, so there are 1000 nanoseconds in a microsecond. ■

Metric–USCS Conversions

We carry out conversions between metric and USCS units like any other unit conversions. Table 2.6 lists a few handy conversion factors. It's useful to memorize approximate conversions, particularly if you plan to travel internationally or if you work with metric units in sports or business. For example, if you remember that a kilometer is about 0.6 mile, you will know that a 10-kilometer road race is about 6 miles. Similarly, if you remember that a meter is about 10% longer than a yard, you'll know that a 100-meter race is about the same as a 110-yard race.

TABLE 2.6 USCS–Metric Conversions

USCS TO METRIC	METRIC TO USCS
1 in. = 2.540 cm	1 cm = 0.3937 in.
1 ft = 0.3048 m	1 m = 3.28 ft
1 yd = 0.9144 m	1 m = 1.094 yd
1 mi = 1.6093 km	1 km = 0.6214 mi
1 lb = 0.4536 kg	1 kg = 2.205 lb
1 fl oz = 29.574 mL	1 mL = 0.03381 fl oz
1 qt = 0.9464 L	1 L = 1.057 qt
1 gal = 3.785 L	1 L = 0.2642 gal

EXAMPLE 4 *Gas Price per Liter*

At a gas station in Mexico, the price of gasoline is 8 pesos per liter. What is the price in dollars per gallon? Use the currency exchange rate in Table 2.1.

Solution We use a chain of conversions to convert from pesos to dollars and then from liters to gallons. From Table 2.1, the currency conversion is $0.1073 per peso; from Table 2.6, there are 3.785 liters per gallon.

$$\frac{8 \text{ peso}}{1 \text{ L}} \times \frac{\$0.1073}{1 \text{ peso}} \times \frac{3.785 \text{ L}}{1 \text{ gal}} \approx \frac{\$3.25}{1 \text{ gal}}$$

The price of the gasoline is about $3.25 per gallon.

EXAMPLE 5 Square Kilometers to Square Miles

How many square kilometers are in one square mile?

Solution We square both sides of the conversion factor, 1 mi = 1.6093 km.

$$(1 \text{ mi})^2 = (1.6093 \text{ km})^2 \quad \Rightarrow \quad 1 \text{ mi}^2 = 2.5898 \text{ km}^2$$

One square mile is 2.5898 square kilometers.

Temperature Units: Fahrenheit, Celsius, and Kelvin

Three temperature scales are commonly used today (Figure 2.6). The **Fahrenheit** scale is used in the United States; it is defined so that water freezes at 32°F and boils at 212°F. Internationally, temperature is usually measured on the **Celsius** scale, which places the freezing point of water at 0°C and the boiling point at 100°C. In science, temperature is usually measured on the **Kelvin** scale, which is the same as the Celsius scale except for its zero point. A temperature of 0 K is the coldest possible temperature, known as **absolute zero**; it is approximately −273.15°C or −459.67°F.

We can find formulas for converting between the three scales by studying Figure 2.6. Because the Kelvin and Celsius scales are the same except for the 273.15° difference in their zero points, we simply add 273.15 to a Celsius temperature to get a Kelvin temperature. Next, note that the Fahrenheit scale has 180° (212°F − 32°F = 180°F) between the freezing and boiling points of water, whereas the Celsius scale has only 100° between these points. Each Celsius degree therefore represents a temperature change equivalent to 1.8 Fahrenheit degrees. Thus, we convert from Celsius to Fahrenheit by multiplying by 1.8, or $\frac{9}{5}$, and then adding 32 to account for the difference in the zero points. We do the reverse—subtract 32 and then divide by 1.8 (which is the same as multiplying by $\frac{5}{9}$)—to convert from Fahrenheit to Celsius.

FIGURE 2.6

Temperature Scale: 212 °F / 100 °C / 373.15 K — Water boils; 32 °F / 0 °C / 273.15 K — Water freezes; −459.67 °F / −273.15 °C / 0 K — Absolute zero. Fahrenheit, Celsius, Kelvin.

historical note

The Fahrenheit scale was invented by German-born scientist Gabriel Fahrenheit in 1714. He originally defined the scale with 0°F representing the coldest temperature he could create with a mixture of ice and salt, and 100°F representing a rough estimate of human body temperature. The Celsius scale was invented by Swedish astronomer Anders Celsius in 1742. The Kelvin scale was invented in 1848 by British physicist William Thomson, later known as Lord Kelvin.

TECHNICAL NOTE

Algebraically, we write these formulas as

$$C = \frac{F - 32}{1.8}$$

$$= \frac{5}{9}(F - 32)$$

$$F = (C \times 1.8) + 32$$

$$= \frac{9}{5}C + 32$$

$$K = C + 273.15$$

where C, F, and K are the temperatures in Celsius, Fahrenheit, and Kelvin, respectively.

Temperature Conversions

From Celsius to Fahrenheit: Multiply by 1.8 (or $\frac{9}{5}$); then add 32.

From Fahrenheit to Celsius: Subtract 32; then divide by 1.8 (or multiply by $\frac{5}{9}$).

From Celsius to Kelvin: Add 273.15.

From Kelvin to Celsius: Subtract 273.15.

Time out to think

The local weather report says that tomorrow's temperature will be 59°, but does not specify whether it is in Celsius or Fahrenheit. Can you tell which it is? How?

EXAMPLE 6 *Paris Weather Report*

You'll be in Paris tomorrow, and the forecast calls for a temperature of 30°C. Convert to Fahrenheit. Will you need a winter jacket?

Solution We convert from Celsius to Fahrenheit by first multiplying by 1.8 and then adding 32:

$$(30 \times 1.8) + 32 = 54 + 32 = 86$$

The forecast calls for 86°F, so you won't need a winter jacket. ∎

EXAMPLE 7 *Human Body Temperature*

Average human body temperature is 98.6°F. What is it in Celsius and Kelvin?

Solution We convert from Fahrenheit to Celsius by subtracting 32 and then dividing by 1.8:

$$\frac{98.6 - 32}{1.8} = \frac{66.6}{1.8} = 37.0$$

Human body temperature is 37°C. We find the Kelvin equivalent by adding 273.15, which makes it 37 + 273.15 = 310.15 K. ∎

UNITS OF ENERGY AND POWER

We pay energy bills to power companies, we use energy from gasoline to run our cars, and we argue about whether nuclear energy is a sensible alternative to fossil fuels. But what *is* energy?

Broadly speaking, energy is what makes matter move or heat up. We need energy from food to keep our hearts beating, to maintain our body temperatures, and to walk or run. A car needs energy to move the pistons in its engine, which turn the wheels. A light bulb needs energy to generate the heat that makes its filament glow.

For Americans, the most familiar energy unit is the *Calorie*, used to measure the energy our bodies can draw from food. A typical adult uses about 2500 Calories of energy each day. The international metric unit of energy is the **joule**; outside the United States, most foods and beverages are labeled with their energy content in joules (or kilojoules) rather than Calories.

The words *energy* and *power* are often used together, but they are not quite the same: Power is the *rate* at which energy is used, which means it has units of energy divided by time. The most common unit of power is the **watt**, defined as 1 joule per second.

TECHNICAL NOTE
One calorie (lowercase c) is the energy needed to raise the temperature of 1 gram of water by 1° Celsius. A food Calorie (uppercase C) is 1000 calories, which is equivalent to 4184 joules.

> **Energy and Power**
>
> **Energy** is what makes matter move or heat up. The international metric unit of energy is the **joule.**
>
> **Power** is the *rate* at which energy is used. The international metric unit of energy is the **watt,** defined as
>
> $$1 \text{ watt} = 1 \frac{\text{joule}}{\text{s}}$$

If you check an electric utility bill, you'll probably find that it refers to electric usage in units of **kilowatt-hours.** The hyphen in this unit tells us that it means 1 kilowatt × 1 hour, which is equivalent to 1000 watts × 1 hour. Replacing 1000 watts with 1000 joule/s, we find

$$1 \text{ kilowatt-hour} = \frac{1000 \text{ joule}}{1 \text{ s}} \times 1 \text{ hr} \times \frac{60 \text{ min}}{1 \text{ hr}} \times \frac{60 \text{ s}}{1 \text{ min}} = 3{,}600{,}000 \text{ joule}$$

by the way...
If you purchase a gas appliance, such as a gas stove or a kerosene heater, its energy requirements may be labeled in *British thermal units,* or BTUs. One BTU is equivalent to 1055 joules.

> **DEFINITION**
>
> A **kilowatt-hour** is a unit of energy:
>
> $$1 \text{ kilowatt-hour} = 3.6 \text{ million joules}$$

EXAMPLE 8 *Operating Cost of a Light Bulb*

Suppose your utility company charges 8¢ per kilowatt-hour of electricity. How much does it cost to keep a 100-watt light bulb on for a week?

Solution One watt is one joule per second, so a 100-watt light bulb consumes energy at a rate of 100 joules per second. The energy used by the bulb in a week is 100 joule/s × 1 week. Using a chain of conversions to convert a week to seconds, we have

$$\frac{100 \text{ joule}}{1 \text{ s}} \times 1 \text{ week} \times \frac{7 \text{ day}}{1 \text{ week}} \times \frac{24 \text{ hr}}{1 \text{ day}} \times \frac{60 \text{ min}}{1 \text{ hr}} \times \frac{60 \text{ s}}{1 \text{ min}} = 60{,}480{,}000 \text{ joule}$$

Next, we convert this result to kilowatt-hours, remembering that 1 kilowatt-hour is 3,600,000 joules:

$$60{,}480{,}000 \text{ joule} \times \frac{1 \text{ kilowatt-hour}}{3{,}600{,}000 \text{ joule}} = 16.8 \text{ kilowatt-hour}$$

The light bulb uses 16.8 kilowatt-hours of energy in a week. At a price of 8¢ per kilowatt-hour, the total cost is

$$16.8 \text{ kilowatt-hr} \times 8 \frac{¢}{\text{kilowatt-hr}} = 134.4¢ \approx \$1.34$$

The electricity to keep the 100-watt light bulb on all week costs about $1.34 (assuming a price of 8¢ per kilowatt-hour).

Time out to think

Check a utility bill (yours or a friend's). Is the electricity usage metered in units of kilowatt-hours (often abbreviated kWh)? If not, what units are used? If so, what is the price per kilowatt-hour?

practical matters

SAVE MONEY AND SAVE THE EARTH

You can save both money and energy by replacing standard (incandescent) light bulbs with more efficient bulbs, such as compact fluorescent bulbs. The compact fluorescent bulbs save energy because virtually all the energy they use goes into making light, whereas most of the energy used by standard bulbs goes into heat (that's why standard bulbs get so hot when they are turned on). As a result, a compact fluorescent bulb needs much less energy to generate the same light as a standard bulb. For example, a typical 25-watt compact fluorescent produces the same light as a 100-watt standard bulb.

To see how much you can save, consider a 100-watt light bulb that is left on all the time. At an energy cost of 8¢ per kilowatt-hour, this light bulb costs $1.34 to operate for a week (see Example 8), which translates into about $70 per year. Replacing the 100-watt bulb with a 25-watt compact fluorescent means your annual energy cost will be only one-fourth (25%) as much, or $17.50. Thus, the light bulb replacement saves some $52.50 per year in energy costs while providing just as much light. This savings more than makes up for the fact that compact fluorescent bulbs are more expensive than standard bulbs. (Compact fluorescent bulbs also last typically ten times longer than standard bulbs.)

Of course, you probably don't (or shouldn't) leave 100-watt light bulbs on all the time, so your savings for a single light bulb will be less than calculated above. Nevertheless, your savings may still be quite substantial. And, while you save money by switching to more efficient bulbs, you also reduce total energy usage, which helps protect the environment.

EXAMPLE 9 Pedal Power

You are riding an exercise bicycle at a fitness center. The readout states that you are using 500 Calories per hour. Are you generating enough power to light a 100-watt bulb (1 Calorie = 4184 joules)?

Solution We use a chain of conversions to go from Calories per hour to joules per second:

$$\frac{500 \text{ Cal}}{1 \text{ hr}} \times \frac{4184 \text{ joule}}{1 \text{ Cal}} \times \frac{1 \text{ hr}}{60 \text{ min}} \times \frac{1 \text{ min}}{60 \text{ s}} \approx 581 \frac{\text{joule}}{\text{s}}$$

Your pedaling generates energy at a rate of 581 joules per second, which is a power of 581 watts—enough to light five (almost six) 100-watt bulbs.

Review Questions

1. Briefly describe the origin and use of common units in the U.S. customary system.
2. Briefly describe the origin and use of metric units. What are the basic metric units of length, mass, time, and volume? How are the metric prefixes used?
3. Using examples, show how to convert between the Fahrenheit, Celsius, and Kelvin temperature scales.
4. What is *energy*? List at least three common units of energy. Under what circumstances do the different units tend to be used?
5. What is the difference between *energy* and *power*? What are the standard units for power?

In the News

1. **Everyday Metric.** Describe three ways that you use metric units in your everyday life.
2. **Should the United States Go Metric?** Discuss the pros and cons of having the United States switch fully to the metric system. Do you think it will ever happen?
3. **Energy.** Look for a news article concerning energy or power. What units are used to describe energy or power? Summarize the article and explain the meaning of the units.
4. **Utility Bill.** Analyze a utility bill. Explain all the units shown, and determine the relative costs of different energy uses. What changes would you recommend if the recipient of the bill wanted to lower energy costs?

Exercises

Powers of 10. Exercises 1–2 use skills covered in the Brief Review on p. 91.

1. Evaluate each of the following.
 a. $10^6 \times 10^5$ b. $10^4 \times 10^{-3}$
 c. $\dfrac{10^6}{10^5}$ d. $\dfrac{10^4}{10^{-3}}$
 e. $10^{-3} \times 10^{-4}$ f. $\dfrac{10^{-7}}{10^{-9}}$
 g. $10^6 + 10^5$ h. $10^7 - 10^2$

2. Evaluate each of the following.
 a. $10^{12} \times 10^9$ b. $10^6 \times 10^{-8}$
 c. $\dfrac{10^{15}}{10^{12}}$ d. $\dfrac{10^{-7}}{10^4}$
 e. $10^{-4} \times 10^{-6}$ f. $\dfrac{10^{-10}}{10^{-14}}$
 g. $10^8 + 10^5$ h. $10^4 - 10^{-1}$

USCS Units. Exercises 3–4 involve conversions within the U.S. customary system.

3. a. What is your height in inches?
 b. A gallon of water weighs about 128 ounces. How many pounds is that?
 c. You have a $\frac{1}{2}$-gallon milk jug. How many liquid pints can it hold? How many dry pints can it hold?
 d. If 150 million bushels of wheat are traded on one day, how many cubic inches are traded? how many cubic feet?

4. a. What is your weight in ounces (avoirdupois)? in tons?
 b. A boat is moving at 30 knots (nautical miles per hour). What is its speed in miles per hour?
 c. Most soda cans contain 12 fluid ounces. How many cubic inches do they contain?
 d. A small city produces 500,000 cubic feet of garbage per week. If all of this garbage were stacked neatly (in a nice vertical pile) on a 100-yard by 60-yard football field, how high would the pile be (in feet)?

Metric Prefixes. For each pair of units in Exercises 5–6, state how much larger or smaller the first unit is than the second.

5. a. millimeter, meter
 b. gram, kilogram
 c. square millimeter, square kilometer

6. a. milliliter, deciliter
 b. kilometer, micrometer
 c. cubic meter, cubic centimeter

USCS–Metric Conversions. Convert each measurement in Exercises 7–8 to the units specified.

7. a. 10 meters to feet
 b. 880 yards to kilometers
 c. 20 gallons to liters
 d. 5 milliliters to cubic inches
 e. 1200 square feet to square meters

8. a. 105 centimeters to yards
 b. 150 pounds to kilograms
 c. 100 kilometers per hour to miles per hour
 d. 5.5 grams per cubic centimeter to pounds per cubic foot
 e. 25 miles per hour to kilometers per hour

Celsius-Fahrenheit Conversions. In Exercises 9–10, convert Celsius temperatures into Fahrenheit or Fahrenheit temperatures into Celsius.

9. a. 45°F b. 20°C
 c. −15°C d. −30°C
 e. 70°F

10. a. −8°C b. 15°F
 c. 15°C d. 75°F
 e. 20°F

Celsius-Kelvin Conversions. In Exercises 11–12, convert Celsius temperatures into Kelvin or Kelvin temperatures into Celsius.

11. a. 50 K b. 240 K
 c. 10°C

12. a. −40°C b. 400 K
 c. 125°C

Sensible or Ridiculous? In Exercises 13–20, determine whether each statement is sensible or ridiculous, and explain why.

Example 1: I ate 2 meters of apples at lunch. *Solution:* The statement is ridiculous. A meter is a unit of length, so talking about "meters of apples" makes no sense.

Example 2: My brother is 4 meters tall. *Solution:* The statement is ridiculous. A meter is slightly longer than a yard, so 4 meters is slightly more than 12 feet; no one is 12 feet tall.

13. I drank 2 liters of water today.
14. I know a professional bicyclist who weighs 300 kilograms.
15. Today I drove along the interstate at 100 kilometers per hour.
16. I know someone who can run 35 liters per second.
17. A guy on our track team can high jump 7 meters.
18. My friend ran 10,000 meters in less than an hour.
19. The book I sent you weighs 3 milligrams.
20. My car's gas tank holds 12 meters of gasoline.
21. **The Metric Mile.** In track and field, the 1500-meter race is sometimes called the *metric mile*.
 a. Compare the metric mile to a USCS mile. How much longer or shorter is it, by percentage?

b. Look up the current men's and women's world records for the mile. If you assume that the runners maintain the same pace for the metric mile, what should their times be for the metric mile? Compare the predicted records at this pace to the actual current world records for the metric mile.

22. **Metric Tools.** Many tools come in both USCS and metric standards. In a standard socket set, the smallest USCS subdivision is $\frac{1}{16}$ inch; the smallest metric subdivision is 0.5 millimeter. Are the tools interchangeable? Explain.

23. **Tallest Mountain?** Mauna Kea, the highest mountain on the island of Hawaii, rises 13,796 feet above sea level. It extends an additional 18,200 feet from sea level to its base on the ocean floor. How tall is Mauna Kea from its base to its peak, in feet, miles, and kilometers? Compare its total extent to the height of Mt. Everest above sea level (29,023 ft). Would it be fair to call Mauna Kea the highest mountain in the world? Why or why not?

24. **The Cullinan Diamond and the Star of Africa.** The largest single rough diamond ever found, the Cullinan diamond, weighed 3106 carats; it was used to cut the world's largest diamond gem, the Star of Africa (530.2 carats), which is part of the British crown jewels collection. How much did the Cullinan diamond weigh in milligrams? in pounds? How much does the Star of Africa weigh in milligrams? in pounds?

Gems and Gold. Exercises 25–28 use karats and carats (see the Practical Matters box on p. 89).

25. You find a nugget that is 25% gold. What is its purity in karats?

26. You purchase a 14-karat gold chain that weighs 15 grams. How much gold have you purchased (in grams)? Bonus: At the current price of gold, how much is the gold in the chain worth?

27. How much does a 2.5-carat diamond weigh in grams?

28. Is it possible to have jewelry made of 30-karat gold? Why or why not?

Electric Bills. For the electric bills described in Exercises 29–30, do the following:

 a. Determine your total electrical energy use, in joules.

 b. Determine your average power use, in watts.

 c. Assuming the power company generated the energy by burning oil, calculate the amount of oil needed to provide the energy shown on your bill. Give your answer in both liters and gallons. (Hint: Burning 1 liter of oil releases 12 million joules of energy.)

29. Your electric bill states that you used 1250 kilowatt-hours of energy in June.

30. Your electric bill states that you used 970 kilowatt-hours of energy in September.

31. **Basketball Power.** You burn 800 Calories while playing in a basketball game for an hour. What is your average power during the game, in watts? Is it enough to keep a 100-watt light bulb shining? Explain.

32. **Aerobics Power.** You burn 500 Calories while doing an aerobics class for 45 minutes. What is your average power during the class, in watts? Is it enough to keep a 100-watt light bulb shining? Explain.

33. **Refrigerator Cost.** Your refrigerator uses an average power of 350 watts, and your utility company charges 8¢ per kilowatt-hour of energy. How much does it cost to run your refrigerator for a year? Explain.

34. **Hair Dryer Cost.** You have an 1800-watt hair dryer, which you use for an average of 10 minutes per day. Your utility company charges 9¢ per kilowatt-hour of energy. How much does it cost to run the hair dryer each day? each year?

35. **Compact Fluorescent Light Bulbs.** You replace a 100-watt standard light bulb with a 25-watt compact fluorescent bulb that supplies the same light. Over a 10,000-hour life (typical for a compact fluorescent bulb), how much energy, in kilowatt-hours, do you save? If electricity costs 8¢ per kilowatt-hour, how much money do you save?

36. **Human Wattage.** Suppose you require 2500 food Calories per day (which is about average).
 a. What is your average power, in watts? Compare your answer to the wattage of some familiar appliance.
 b. How much energy, in joules, do you require from food in a year? Counting all forms of energy (such as gasoline, electricity, and energy for heating), the average U.S. citizen consumes about 400 billion joules of energy each year. Compare this value to the energy needed from food alone.

37. **Coal Power Plant.** A new coal-burning power plant can generate 1 gigawatt (billion watts) of power. Burning 1 kilogram of coal yields about 450 kilowatt-hours of energy. How much energy, in kilowatt-hours, can the plant generate each month? How much coal, in kilograms, is needed by this power plant each month? If a typical home uses 1000 kilowatt-hours per month, how many homes can this power plant supply with energy?

38. **Nuclear Power Plant.** Operating at full capacity, the Fort St. Vrain Nuclear Power Station in Colorado can generate 330 megawatts of power. Nuclear fission of 1 kilogram of uranium (in the form of uranium-235) releases 16 million kilowatt-hours of energy. How much energy, in kilowatt-hours, can the plant generate each month? How much uranium, in kilograms, is needed by this power plant each month? If a typical home uses 1000 kilowatt-hours per month, how many homes can this power plant supply with energy?

Solar Energy. Use the following facts in Exercise 39–40: Solar (photovoltaic) cells convert sunlight directly into electricity. If solar cells were 100% efficient, they would generate about 1000 watts of power per square meter of surface area when exposed to direct sunlight. With lower efficiency, they generate proportionally less power; for example, 10% efficient cells generate 100 watts of power in direct sunlight.

39. Suppose a 1-square-meter panel of solar cells has an efficiency of 20% and receives the equivalent of 6 hours of direct sunlight per day. How much energy, in joules, can it produce each day? What average power, in watts, does the panel produce?

40. Suppose you want to supply 1 kilowatt of power to a house (the average household power requirement) by putting solar panels on its roof. For the solar cells described in Exercise 39, how many square meters of solar panels would you need? Assume that you can make use of the average power from the solar cells (by, for example, storing the energy in batteries until it is needed).

41. **Project: Personal Energy Audit.** Do a thorough electrical energy audit of your home, apartment, or dormitory. That is, determine the energy used in a typical month by each of your electrical appliances, and your total energy usage per month for electricity. Based on the cost of electricity in your area, calculate the average monthly cost for your electricity. After studying your findings, propose at least three energy conservation strategies that you would consider implementing to save money and energy. Discuss the pros and cons of each strategy.

Web Projects

Find useful links for Web projects on the text Web site: www.aw.com/bennett-briggs

1. **Metric History.** Research some aspect of the history of the metric system, such as how it came to be adopted around the world or how various units have been scientifically defined. Write a short report on your findings.

2. **Energy Issues.** The question of how we will continue to meet our energy needs is one of the most important issues of our time. Research one aspect of this issue using information available at the Web site for the U.S. Energy Information Administration. Write a short report on your findings.

UNIT 2C

PROBLEM-SOLVING GUIDELINES AND HINTS

So far in this chapter, we've focused on how units help us solve problems. But if you look back through the examples we've studied, you'll see that we've actually done much more. In a few cases, we've used pictures or diagrams to help us understand a problem. We've studied each problem in order to find a good strategy for solving it, and we've carefully interpreted our results to make sure they had the correct units and made sense. You may have noticed that while every problem is different, the *process* of problem solving always has a few common features.

The four-step process on the next page provides a useful guide to problem solving. Study it carefully and look back to see how these steps were applied in earlier examples. Keep in mind that the four steps offer *general* advice; following them will not automatically lead to a solution. In this sense, problem solving is more of an art than a science, and the four steps are more like instructions for painting than a recipe for the Mona Lisa. Note also that while these steps apply to virtually all problem solving, it is not necessary (or particularly useful) to write them out step by step in solutions.

> *historical note*
>
> These four steps are modified from a process in *How to Solve It,* by George Polya (Princeton University Press, Princeton, N.J., 1957). First published in 1945, this book has sold more than 1 million copies and is available in at least 17 different languages.

Time out to think

Choose two examples from Units 2A and 2B, and identify how each of the four steps was carried out in those examples. More generally, explain how these four steps are implicit in the solutions to the examples we've studied, even though we have not written them out in a step-by-step list.

Using the four-step process will help you become better organized. But the only sure way to become more creative and improve your problem solving is through practice. We now offer several hints about problem solving, each illustrated with an example. The hints themselves relate more to *attitudes* about problem solving than to actual techniques. Indeed, their primary purpose is to help you develop a "mind set" that is conducive to enjoyable and successful problem solving.

HINT 1: THERE MAY BE MORE THAN ONE ANSWER

How can society best reduce the total amount of greenhouse gases emitted into the atmosphere? We won't even attempt to answer this question, but it should make the point that no *single* best answer may be available. Indeed, many different political and economic strategies could yield similar reductions in greenhouse gas emissions.

Most people recognize that policy questions do not have unique answers, but the same is true of many mathematical problems. For example, both $x = 4$ and $x = -4$ are solutions to the equation $x^2 = 16$. Without further information and

A Four-Step Problem-Solving Process

Step 1: Understand the problem. Be sure that you understand the nature of the problem. For example:

- Think about the context of the problem (that is, how it relates to other problems in the real world) to gain insight into its purpose.
- Make a list or table of the specific information given in the problem, including units for numerical data.
- Draw a picture or diagram to help you make sense of the problem.
- Restate the problem in different ways to clarify its question.
- Make a mental or written model of the solution, into which you will fill in details as you work through the problem.

Step 2: Devise a strategy for solving the problem. Finding an appropriate strategy requires creativity, organization, and experience. In seeking a strategy, try any or all of the following:

- Obtain needed information that is not provided in the problem statement, using recall, estimation, or research.
- Make a list of possible strategies and hints that will help you select your overall strategy.
- Map out your strategy with a flow chart or diagram.

Step 3: Carry out your strategy, and revise it if necessary. In this step you are likely to use analytical and computational tools. As you work through the mathematical details of the problem, be sure to do the following:

- Keep an organized, neat, and written record of your work, which will be helpful if you later need to review your solution.
- Double-check each step so that you do not risk carrying errors through to the end of your solution.
- Constantly reevaluate your strategy as you work; if you find a flaw in your strategy, return to step 2 and create a revised strategy.

Step 4: Look back to check, interpret, and explain your result. Although you may be tempted to think you have finished after you find a result in step 3, this final step is the most important. After all, a result is useless if it is wrong or misinterpreted or cannot be explained to others. Always do all of the following:

- Be sure that your result makes sense; for example, be sure that it has the expected units, that its numerical value is sensible, and that it is a reasonable answer to the original problem.
- Once you are sure that your result is reasonable, recheck your calculations or find an independent way of checking the result.
- Identify and understand potential sources of uncertainty in your result.
- Write your solution clearly and concisely, including discussion of any relevant uncertainties or assumptions.
- Consider and discuss any pertinent implications of your result.

context, we have no way to determine whether both solutions are valid for a particular problem. Nonunique solutions often occur because not enough information is available to distinguish among a variety of possibilities.

EXAMPLE 1 *Traffic Counters*

A traffic counter is a device designed to count the number of vehicles passing along a street. It usually is a thin black tube stretched across a street or highway, and connected to a "brain box" at the side of the road. The device registers one "count" each time a set of wheels (that is, wheels on a single axle) rolls over the tube. A normal automobile registers two counts: one for the front wheels and one for the rear wheels. A light truck with three axles registers three counts. A large semitrailer truck might register four or five counts.

Suppose that, during a one-hour period, a particular traffic counter registers 35 counts on a residential street on which only two-axle vehicles (cars) and three-axle vehicles (light trucks) are allowed. How many cars and light trucks passed over the traffic counter?

Solution Let's proceed by trial and error. For example, we can rule out the possibility of 12 light trucks, because at 3 counts per truck, 12 light trucks would yield $12 \times 3 = 36$ counts—more than the 35 registered. However, 11 light trucks would yield only $11 \times 3 = 33$ counts. Because a car registers 2 counts, 11 light trucks and 1 car would produce the 35 counts. We've found a solution, but it is not the only one. Note that 9 light trucks and 4 cars also is a solution: The 9 trucks yield $9 \times 3 = 27$ counts, and the four cars yield $4 \times 2 = 8$ counts, for a total of $27 + 8 = 35$ counts.

In fact, six different combinations of cars and light trucks can produce a total count of 35. One of these combinations represents the actual number of cars and trucks that passed over the traffic counter during the hour. However, without further information, we cannot determine which of the six solutions represents the actual traffic flow.

Copyright ©1998 by Sidney Harris.

HINT 2: THERE MAY BE MORE THAN ONE STRATEGY

Just as we must often admit more than one right answer, we should also expect more than one strategy for finding an answer. Given how rich and diverse mathematics and the human mind are, it would be unreasonable to expect everyone to follow the same path to a solution. As illustrated in the following example, an efficient strategy can save a lot of time and work.

EXAMPLE 2 *Jill and Jack's Race*

Jill and Jack ran a 100-meter race. Jill won by 5 meters; that is, Jack had run only 95 meters when Jill crossed the finish line. They decide to race again, but this time Jill starts 5 meters behind the starting line. Assuming that both runners run at the same pace as before, who will win?

Solution Strategy 1 One approach to this problem is analytical—we analyze each race quantitatively. We were not told how fast either Jill or Jack ran, so we can choose some reasonable numbers. For example, we might assume that Jill ran the 100 meters in the first race in 20 seconds. In that case, her pace was 100 m ÷ 20 s = 5 m/s, or 5 meters per second. Because Jack ran only 95 meters in the same 20 seconds, his pace was 95 m ÷ 20 s = 4.75 m/s, or 4.75 meters per second.

For the second race, Jill must run 105 meters (because she starts 5 meters behind the starting line) to Jack's 100 meters. We predict their times by dividing their respective race distances by their speeds from the first race:

$$\text{Jill:} \quad 105 \text{ m} \div 5 \frac{\text{m}}{\text{s}} = 105 \text{ m} \times \frac{1 \text{ s}}{5 \text{ m}} = 21 \text{ s}$$

$$\text{Jack:} \quad 100 \text{ m} \div 4.75 \frac{\text{m}}{\text{s}} = 100 \text{ m} \times \frac{1 \text{ s}}{4.75 \text{ m}} \approx 21.05 \text{ s}$$

Thus, Jill will win the second race by a slim margin.

Strategy 2 Although the analytical method works, we can use a much more intuitive and direct solution. We simply note that Jill runs 100 meters in the same time that Jack runs 95 meters. Therefore, in the second race, Jill will pull even with Jack 95 meters from the starting line. In the remaining 5 meters, Jill's faster speed will allow her to pull away and win. Note how this insight avoids the calculations needed in Strategy 1. ■

HINT 3: USE APPROPRIATE TOOLS

You don't need a computer to check your tab in a restaurant, and you don't need calculus to find the area of a rectangular room. For any given task there is an appropriate level of power that is needed, and it is a matter of style and efficiency neither to underestimate nor to overestimate that level. You usually will have a choice of tools to use in any problem. Choosing the ones most suited to the job will make your task much easier.

EXAMPLE 3 *The Cars and the Canary*

Two cars, 120 miles apart, begin driving toward each other on a long straight highway. One car travels 20 miles per hour and the other 40 miles per hour (Figure 2.7 on the following page). At the same time a canary, starting on one car, flies back and forth between the two cars as they approach each other. If the canary flies always 150 miles per hour and turns around instantly at each car, how far has it flown when the cars collide?

FIGURE 2.7 Set-up for the cars and canary problem.

Solution Because the problem asks "how far," we might be tempted to calculate the *distance* traveled by the canary on each back and forth trip between the cars. However, these trips get shorter as the cars approach each other and we would have to add up all the individual distances. In principle, we would need to add up an *infinite* number of ever-smaller distances—a task that involves the mathematics of calculus.

But note what happens if we focus on *time* rather than distance. The cars will collide with each other at a speed of 60 mi/hr (because one car is traveling at 20 mi/hr and the other is traveling in the opposite direction at 40 mi/hr). Because they initially are 120 miles apart, they will collide in precisely 2 hours:

$$120 \text{ mi} \div 60 \frac{\text{mi}}{\text{hr}} = 120 \text{ mi} \times \frac{1 \text{ hr}}{60 \text{ mi}} = 2 \text{ hr}$$

The canary is flying at a speed of 150 miles per hour, so in 2 hours it will fly

$$2 \text{ hr} \times 150 \frac{\text{mi}}{\text{hr}} = 300 \text{ mi}$$

The canary will fly 300 miles before the cars collide. We could have found the answer with calculus, but why bother when we were able to do it with just multiplication and division?

HINT 4: CONSIDER SIMPLER, SIMILAR PROBLEMS

Sometimes you are confronted with a problem that at first seems daunting. Our fourth hint is to consider a simpler, but similar, problem. The insight gained from solving the easier problem may help you understand the original problem.

EXAMPLE 4 *Coffee and Milk*

Suppose you have two cups in front of you: One holds coffee and one holds milk (Figure 2.8 on page 107). You take a teaspoon of milk from the milk cup and stir it into the coffee cup. Next, you take a teaspoon of the mixture in the coffee cup and put it back into the milk cup. After the two transfers, there will be either (1) more coffee in the milk cup than milk in the coffee cup, (2) less coffee in the milk cup than milk in the coffee cup, or (3) equal amounts of coffee in the milk cup and milk in the coffee cup. Which of these three possibilities is correct?

Solution A cup of either milk or coffee contains something like a *trillion trillion* molecules of liquid. Clearly, it would be difficult to visualize how such enormous numbers of molecules mix together, let alone to calculate the result. However, the essence

thinking about...
ZENO'S PARADOX

The Greek philosopher Zeno of Elea (c. 460 B.C.) posed several paradoxes that defied solution for thousands of years. (A *paradox* is a situation or statement that seems to violate common sense or to contradict itself.) One paradox begins with an imaginary race between the warrior Achilles and a slow-moving tortoise. The tortoise is given a head start, but our common sense says that the swift Achilles will soon overtake the tortoise and win.

Zeno suggested a different way to look at the race. Suppose that, as shown in the figure, Achilles starts from point P0 and the tortoise starts from P1. During the time it takes Achilles to reach P1, the slow-moving tortoise will move ahead a little bit to P2. While Achilles continues on to P2, the tortoise will move ahead to P3. And so on. Thus, Achilles must cover an infinite set of ever-smaller distances to catch the tortoise (that is, from P0 to P1, from P1 to P2, etc.). From this point of view, it seems that Achilles will never catch the tortoise.

This paradox puzzled philosophers and mathematicians for more than 2000 years. Its resolution depends on a key insight that became clear only with the invention of calculus in the 17th century: It does not necessarily require an infinite amount of time to cover an infinite set of distances. For example, imagine that the infinite set of distances covered by Achilles begins with 1 mile, then $\frac{1}{2}$ mile, then $\frac{1}{4}$ mile, and so on. Then the total distance he covers, in miles, is

$$1 + \frac{1}{2} + \frac{1}{4} + \frac{1}{8} + \frac{1}{16} + \frac{1}{32} + \frac{1}{64} + \frac{1}{128} + \frac{1}{256}$$
$$+ \frac{1}{512} + \frac{1}{1024} + \frac{1}{2048} + \cdots$$

This sum is called an **infinite series** because it is the sum of an infinite number of terms. Note what happens when we add just the first four terms, then the first eight terms, and then the first twelve terms.

$$1 + \frac{1}{2} + \frac{1}{4} + \frac{1}{8} = 1.875$$

$$1 + \frac{1}{2} + \frac{1}{4} + \frac{1}{8} + \frac{1}{16} + \frac{1}{32} + \frac{1}{64} + \frac{1}{128} = 1.9921875$$

$$1 + \frac{1}{2} + \frac{1}{4} + \frac{1}{8} + \frac{1}{16} + \frac{1}{32} + \frac{1}{64} + \frac{1}{128} + \frac{1}{256}$$
$$+ \frac{1}{512} + \frac{1}{1024} + \frac{1}{2048} = 1.99951171875$$

(You can confirm these results with a calculator.) If you continue to add more terms in this infinite series, you will find that the sum gets closer and closer to 2, but never exceeds it. It is possible to prove deductively that the sum of this infinite series is 2. So if the fractions represent distances in miles run by Achilles, then he runs a total distance of only 2 miles (even though this sum involves an infinite set of distances). Clearly, it won't take him long to run a finite distance of 2 miles, so he will pass the slower tortoise and win the race.

(a)

While Achilles runs from P0 to P1... ...tortoise runs to P2

(b)

Achilles runs from P1 to P2... ...but tortoise moves on to P3

FIGURE 2.8 Set-up for the coffee and milk problem.

of this problem is the *mixing* of two things. So one approach is to try a similar mixing problem that is much easier: mixing two piles of marbles.

Suppose that the black pile in Figure 2.9 has ten black marbles and represents the coffee. The white pile has ten white marbles and represents the milk. In this simpler problem, we can represent the first transfer—the teaspoon of milk into the coffee cup—by moving two white marbles to the black pile. This leaves the white pile with just eight white marbles, while the black pile now has ten black and two white marbles.

FIGURE 2.9 The white and black piles of marbles represent the milk and coffee, respectively. Moving two white marbles to the black pile represents putting a teaspoon of milk in the coffee.

Representing the second transfer—of the teaspoon from the coffee cup into the milk cup—involves taking two marbles from the black pile and putting them in the white pile. We can then ask a question analogous to the original question: Are there more black marbles in the white pile or white marbles in the black pile?

Because the marbles represent molecules that mix thoroughly, the two marbles for the second transfer must be drawn at random. This opens three possible cases: The two marbles in the second transfer can be either both black, both white, or one of each. But, as shown in Figure 2.10, in all three cases we end up with the same number of white marbles in the black pile as black marbles in the white pile. By analogy,

FIGURE 2.10 When we transfer two marbles back to the white pile, the two marbles may be both black (left), both white (center), or one black and one white (right). In all three cases, we end up with the same number of white marbles in the black pile as black marbles in the white pile.

we have the answer to our original question: After the two transfers, the amounts of coffee in the milk cup and milk in the coffee cup are equal.

The only remaining step is to confirm that the simpler problem is a reasonable representation of the real problem. Our choice of using two marbles to represent a teaspoon was arbitrary. If we redo this example with transfers of one, three, or any other number of marbles, we will find the same result: We always end up with the same number of black marbles in the white pile as white marbles in the black pile. Starting with ten marbles in each pile also was arbitrary; the conclusion remains the same if we start with twenty, fifty, or a trillion trillion marbles. Because molecules can be thought of as tiny marbles for the purpose of this problem, the real problem has no essential differences from the marble problem. ■

Time out to think

Most people are surprised by the result of the coffee and milk problem. Are you? Now that you know the solution, can you give a simple explanation of the real problem that would satisfy surprised friends?

HINT 5: CONSIDER EQUIVALENT PROBLEMS WITH SIMPLER SOLUTIONS

Replacing a problem with a similar, simpler problem can reveal essential insights about a problem, as we've just seen. However, "similar" is not good enough when we need a numerical solution. In that case a useful approach to a difficult problem is to look for an *equivalent* problem. An equivalent problem will have the same numerical solution but may be easier to solve.

EXAMPLE 5 *A Coiled Wire*

Let's consider a problem of a type often encountered by plumbers, electricians, and engineers: measuring or wrapping a wire around a cylindrical pipe. Suppose that eight turns of a wire are wrapped around a pipe with a length of 20 centimeters and a circumference of 6 centimeters. What is the length of the wire?

Solution This problem is difficult because it involves three-dimensional geometry. However, we can convert it into an equivalent problem by imagining that we cut the pipe along its length and unfold the pipe and wire into a flat rectangle (Figure 2.11 on the following page). The width of the rectangle is the 6-centimeter circumference of the pipe, and its length is the 20-centimeter length of the pipe.

Now, instead of dealing with wire wrapped around a three-dimensional pipe, we have a simple rectangle with eight diagonal segments of wire. The total length of the eight segments must be the length of the wire in the original problem. The height of each triangle is $\frac{1}{8}$ of the length of the rectangle, or 20 cm ÷ 8 = 2.5 cm. The base of each triangle is the 6-centimeter width of the rectangle. The Pythagorean theorem tells us that

$$\text{base}^2 + \text{height}^2 = \text{hypotenuse}^2 \quad \text{or} \quad \text{hypotenuse} = \sqrt{\text{base}^2 + \text{height}^2}$$

Substituting the 6-centimeter base and 2.5-centimeter height yields

$$\text{hypotenuse} = \sqrt{(6 \text{ cm})^2 + (2.5 \text{ cm})^2} = 6.5 \text{ cm}$$

FIGURE 2.11 We want to know the length of wire coiled around a pipe (upper left). We can simplify the problem by imagining that we cut the pipe lengthwise (lower left). When the pipe is pressed flat, the wire now lays in straight segments, each of which is the hypotenuse of a right triangle.

Thus, the length of each wire segment is 6.5 centimeters, and the total length of the wire is 8 × 6.5 cm = 52 cm. Note how much easier it was to solve this equivalent problem than the original one.

HINT 6: APPROXIMATIONS CAN BE USEFUL

Another useful strategy is to make problems easier by using approximations. Most real problems involve approximate numbers to begin with, so an approximation often is good enough for a final answer. In other cases, an approximation will reveal the essential character of a problem, making it easier to reach an exact solution. Approximations also provide a useful check: If you come up with an "exact solution" that isn't close to the approximate one, something may have gone wrong.

EXAMPLE 6 A Bowed Rail

Imagine a mile-long bar of metal such as a railroad rail. Suppose that the rail was anchored on both ends (a mile apart) and that, on a hot day, its length expanded by 1 foot. If the added length caused the rail to bow upward in a circular arc as shown in Figure 2.12a, about how high would the center of the rail rise above the ground?

Solution Because the added length is short compared to the original length, we can approximate the curved rail with two straight lines (Figure 2.12b). We now have two right triangles and the Pythagorean theorem applies. The bases of the two right triangles together give the original rail length of 1 mile, so each base is $\frac{1}{2}$ mile long. The two hypotenuses together represent the expanded length of 1 mile + 1 foot, so each hypotenuse is $\frac{1}{2}$ mile + $\frac{1}{2}$ foot long. Because there are 5280 feet in a mile, $\frac{1}{2}$ mile

FIGURE 2.12 (a) If a rail lengthens while remaining anchored at its ends, it will bow upward as shown. (b) We can find the approximate height of the bow by ignoring its curvature and assuming that each side makes a triangle.

equals 2640 feet. The height of the rail off the ground is approximately the height of the triangles:

$$\text{height of triangle} = \sqrt{(2640.5 \text{ ft})^2 - (2640 \text{ ft})^2} \approx 51.4 \text{ ft}$$

According to our approximation, the center of the rail would rise more than 50 feet off the ground! Because a triangle will stick up higher than a curve with the same base and length, the actual height would be less than what we found with the approximation; an exact solution shows that the top of the curved rail would be about 48 feet off the ground. Because real rails don't bow 48 feet upward on hot days, the assumptions in the original question must not have been very realistic.

Time out to think

Are you surprised by the answer to Example 6? How does the use of the approximation tell you that the original question contained at least some unrealistic assumptions? Which assumptions do you think were unrealistic?

HINT 7: TRY ALTERNATIVE PATTERNS OF THOUGHT

Try to avoid rigid patterns of thought that tend to suggest the same ideas and methods over and over again. Instead, approach every problem with an openness that allows innovative ideas to percolate. In its most wondrous form, this approach is typified by what Martin Gardner, a well-known popularizer of mathematics, calls "aha!" problems. These are problems whose best solution involves a penetrating insight that reduces the problem to its essential parts.

EXAMPLE 7 *China's Population Policy*

In an effort to reduce population growth, in 1978 China instituted a policy that allows only one child per family. One unintended consequence has been that, because of a cultural bias toward sons, China now has many more young boys than girls (presumably because of selective abortions or, in some cases, infanticide). To solve this problem, some people have suggested replacing the one-child policy with a *one-son* policy. That is, if a family's first child is a boy, the family has met its limit of children; but if the first child is a girl, the family can have additional children until one is a boy.

Suppose that the one-son policy were implemented and birth rates returned to their natural levels (half boys and half girls). Compared to the one-child policy, how would this affect the overall birth rate and the numbers of boys and girls?

Solution Most people initially approach this problem by considering the families. Half the families would have a boy as their first and only child. Of the remaining families, half would have a boy as their second child (reaching their one-son limit), while the other half would have a girl and go on to a third child. Half of these families would have a boy on this third try, while the other half would continue on after a third girl. And so on.

However, a moment of insight makes the problem much easier. Neither policy, by itself, affects the natural probability that any child will be a boy or a girl. If the one-son policy restores natural birth rates, then equal numbers of boys and girls must be born. Moreover, nearly all families would have exactly one boy under this policy. Because there must be equal numbers of boys and girls overall, there must also be one girl per family, on average. Thus, the average number of children per family must be two (one boy and one girl) under a one-son policy. In summary, a one-son policy would lead to equal numbers of boys and girls but, compared to a one-child policy, would double the average number of children per family from one to two.

by the way...
Hospital birth records show that boys and girls are born in nearly, but not precisely, equal numbers: Roughly 106 boys are born for every 100 girls. However, males have higher mortality rates than females at every age, so the numbers even out in adulthood and women outnumber men in old age.

Time out to think

China occupies roughly the same amount of land as the United States, but has more than four times as many people. Given these circumstances, do you think that the one-child policy is a good idea? Would a one-son policy be better? Defend your opinions.

HINT 8: DO NOT SPIN YOUR WHEELS
Finally, everyone has had the experience of getting "bogged down" with a problem. When your wheels are spinning, let up on the gas! Often the best strategy in problem solving is to put a problem aside for a few hours or days. You may be amazed at what you see (and what you overlooked) when you return to it.

Review Questions

1. Describe the four basic steps of problem solving.
2. Summarize the strategic hints for problem solving given in this unit, with an example of the meaning of each one.

In the News

1. **Multiple Solutions.** Find an example of a real problem for which, because of insufficient data, we cannot distinguish between two or more possible solutions. The problem might come from a news report or from your own experiences. What additional data would be useful?
2. **Multiple Strategies.** Find an example of a real problem that could potentially be solved by two or more competing strategies. The problem might come from a news report or from your own experiences. Describe each strategy. Which one do you think is better? Why?
3. **Novel Solution.** Find a news report concerning a problem in business or science that was solved by a surprising method. Describe the method and why it was useful.

4. **China's Population Policy.** Look for a recent news report concerning consequences of China's population policies. Does the news report affect your overall opinion of China's policy? Explain.

Exercises

Traffic Counters. Refer to Example 1 for Exercises 1–2. Be sure to look for patterns to simplify your work.

1. **a.** Find, by trial and error, all six solutions to the problem.

 b. Suppose that the same traffic counter registers 41 counts during another period of time. Find all possible combinations of two- and three-axle vehicles that might have passed.

2. **a.** A similar traffic counter is placed on a major highway, where vehicles with two, three, and four axles can travel (that is, vehicles that register 2, 3, or 4 counts). Suppose that, during a 5-minute period late at night, the counter registers 10 counts. Find all possible combinations of two-, three-, and four-axle vehicles that might have passed that point.

 b. Suppose that the traffic counter (from part a) registers 20 counts during a 1-minute period of heavy traffic. Find all possible combinations of two-, three-, and four-axle vehicles that might have passed.

More on Jack and Jill. Refer to Example 2 for Exercises 3–4.

3. Suppose that Jill's time in the first race was 10 seconds. Using strategy 1, determine by how much she would win the second race.

4. Hack and Quill race 200 meters and Hack wins by 10 meters. They race a second time, with Hack starting 10 meters behind the starting line. Who wins the second race? Explain.

5. **Cars and Canary.** Refer to Example 3. Two cars, 150 kilometers apart, begin driving toward each other on a long, straight highway. One car travels 80 kilometers per hour and the other 100 kilometers per hour. At the same time a canary, starting on one car, flies back and forth between the two cars as they approach each other. If the canary flies 120 kilometers per hour and spends no time to turn around at each car, how far has it flown when the cars collide?

6. **Cars and Canary.** Repeat Exercise 5, but assume the canary flies at only 60 kilometers per hour.

7. **Mixing Marbles.** Refer to Example 4. Consider the case in which each pile initially has fifteen marbles. Suppose that on the first transfer three black marbles are moved to the white pile. On the second transfer, any three marbles are taken from the white pile and put into the black pile. Demonstrate, with diagrams and words, that you will always end up with as many white marbles in the black pile as black marbles in the white pile.

8. **Mixing Marbles.** Repeat Exercise 7, but for the case of piles of twenty marbles and transfers of four marbles at a time.

9. **A One-Son Policy.** To convince yourself that a one-son policy would lead to an average of two children per family, with equal numbers of boys and girls, do the following. Suppose that 10,000 families are having children according to the one-son policy. Describe the general makeup of all of the families (that is, start with the fact that 5000 families have a boy as their first and therefore only child, and continue on). Use this process to show that the average number of children is two and that boys and girls are equal in number.

10. **Demographic Imbalance.** Suppose that, as the current generation matures, China's population of young adults has more men than women by a ratio of 118 to 100. With 400 million young adults in China, how many men will be unable to find a spouse?

Puzzle Problems. Exercises 11–25 are puzzle problems that require careful thinking and can help your problem-solving skills. Hint: Look for "aha!" solutions.

11. It takes you 30 seconds to walk from the first (ground) floor of a building to the third floor. How long will it take to walk from the first floor to the sixth floor (at the same pace, assuming that all floors have the same height)?

12. Reuben says, "Two days ago I was 20 years old. Later next year I will be 23 years old." How is this possible?

13. There are three kinds of apples all mixed up in a basket. How many apples must you draw (without looking) from the basket to be sure of getting at least two of one kind?

14. "Brothers and sisters I have none, but that man's father is my father's son." Who is "that man"?

15. "I am the brother of the blind fiddler, but brothers I have none." How can this be?

16. A woman bought a horse for $500 and then sold it for $600. She bought it back for $700 and then sold it again for $800. How much did she gain or lose on these transactions?

17. Three boxes of fruit are labeled *Apples,* *Oranges,* and *Apples and Oranges.* Each label is wrong. By selecting just

one fruit from just one box, how can you determine the correct labeling of the boxes?

18. Each of ten large barrels is filled with golf balls that all look alike. The balls in nine of the barrels weigh 1 ounce and the balls in one of the barrels weigh 2 ounces. With only *one* weighing on a scale, how can you determine which barrel contains the heavy golf balls?

19. A woman is traveling with a wolf, a goose, and a mouse. She must cross a river in a boat that will hold only herself and one other animal. If left to themselves, the wolf will eat the goose and the goose will eat the mouse. How many crossings are required to get all four creatures across the river alive?

20. How do you measure 9 minutes with a 7-minute and a 4-minute hourglass? Assume the hourglasses can only measure 7-minute and 4-minute intervals and cannot be used to measure other time intervals (for example, you can't tell when 2 minutes have gone by).

21. A 150-foot rope is suspended at its two ends from the tops of two 100-foot flagpoles. The lowest point of the rope is 25 feet from the ground. What is the distance between the flagpoles?

22. You are considering buying 12 gold coins that look alike, but you have been told that one of them is a heavy counterfeit. How can you find the heavy coin in three weighings on a balance scale?

23. Suppose you have 40 blue socks and 40 brown socks in a drawer. How many socks must you take from the drawer (without looking) to be sure of getting a pair of the same color?

24. Alma, Bess, Cleo, and Dina visited Elf on Saturday. Alma visited at 8:00, Bess visited at 9:00, Cleo visited at 10:00, and Dina visited at 11:00. Some clues: (1) The times may be either A.M. or P.M. (2) At least one woman visited Elf between Alma and Bess. (3) Alma did not visit Elf before both Cleo and Dina. (4) Cleo did not visit Elf between Bess and Dina. Who visited Elf last?

25. Three prisoners know that the jailer has three white hats and two red hats. The jailer gives one hat to each prisoner and says, "If you can deduce the color of your own hat, you will be freed." Each prisoner can see the hats of the other two prisoners, but not his own. The first prisoner says, "I cannot tell the color of my hat." The second prisoner says, "I cannot tell the color of my hat." The third prisoner is blind, but he is freed. What color hat does he have and how did he know?

Projects: Real Problems in the Real World. Exercises 26–34 describe complex problems that do not have single, straightforward solutions. For each, describe how you would apply the four-step problem-solving process described in the text (without actually carrying out the process to obtain a solution). You may choose either to enumerate the four steps or to describe your process in essay form. Either way, be sure that you list as many relevant factors as possible and discuss sources of uncertainty associated with each. Also, describe how you would work with these factors to find a solution. Conclude by describing your overall impression of whether the problem can be solved and whether any solution would likely generate controversy.

26. You are asked to calculate the cost of installing enough bike racks on campus to solve a bicycle parking problem.

27. You want to know how many new faculty members would be needed, and the total cost to the university, to make sure that all classes have twenty or fewer students (to replace large lecture classes with smaller classes).

28. You want to know whether having a top-quality football program means more money for academics at a university.

29. You want to figure out how much taxes would have to be increased to provide public school teachers with twice their present salaries.

30. You decide that, in the interest of the environment, you will convert your home heating and hot water system to solar power. How much will this conversion cost or save over the next ten years?

31. Suppose that China and India decide to use their extensive coal reserves to supply energy to their populations at the same per capita level as in the United States. How much carbon dioxide would be added to the atmosphere?

32. Are automobile insurance companies gouging drivers? Suppose that you want to figure out whether they are justified in raising insurance rates as rapidly as they have during the past few years.

33. A large city is facing a severe shortage of water. How should water be rationed?

34. Suppose that a city added new bus routes and handed out free bus passes. How many people would give up driving in favor of the bus? How much money, overall, would this cost or save the city?

35. **Project: Textbook Analysis.** Although research shows that most adults today have difficulty with "story problems," we might hope that the next generation will have less difficulty. Find a current textbook in mathematics

THE FAR SIDE BY GARY LARSON

Hell's library

The Far Side ® by Gary Larson © 1987 FarWorks, Inc. All Rights Reserved. Used with permission.

that is used at the upper elementary school level (grades 4–6). Read through the "story problems" in the textbook. Write a critical analysis of the problems and conclude with an opinion as to whether the problems make mathematics meaningful.

Web Projects

Find useful links for Web projects on the text Web site: www.aw.com/bennett-briggs

1. **China's Population.** Find current statistics regarding China's population, such as its total population and average number of children per family, and projections of its future population. Based on what you find, does it appear that China is slowing its rate of population growth? Explain.

2. **Polya and Problem Solving.** Many Web sites discuss George Polya's work and other problem-solving strategies. Explore a few such sites. Write a one- or two-paragraph summary of something you learn about problem solving that you think would be useful to many people.

CHAPTER 2 SUMMARY

UNIT	KEY TERMS	KEY IDEAS AND SKILLS
2A	units conversion factor currency conversions	**Identify** units with key words *per*, *square*, or *cube* or with hyphens. **Write** conversion factors in three equivalent forms. **Work** with units to check answers and help solve problems.
2B	U.S. customary system (USCS) metric system (SI) Celsius, Fahrenheit, Kelvin absolute zero energy units: Calorie, joule, kilowatt-hour power units: 1 watt = 1 joule/s	**Understand and convert** basic USCS and metric units. **Know** metric prefixes. **Know** temperature units and convert between °F, °C, and Kelvin. **Understand** energy units and power units. Energy is what makes matter move or heat up. Power is the *rate* at which energy is used.
2C	four-step problem-solving process	**Apply** eight strategic hints for problem solving.

3

NUMBERS IN THE REAL WORLD

The concept of number is the obvious distinction between the beast and man. Thanks to number, the cry becomes song, noise acquires rhythm, the spring is transformed into a dance, force becomes dynamic, and outlines figures.

—Joseph de Maistre,
19th-century French philosopher

Life is filled with numbers that may at first seem incomprehensible: a population measured in billions of people, national budgets measured in trillions of dollars, and distances ranging from nanometers to light-years. Moreover, the numbers that we encounter nearly always involve uncertainties. In this chapter, we will discuss the use and interpretation of numbers we encounter in our daily lives.

UNIT 3A Uses and Abuses of Percentages: Percentages can be surprisingly subtle and difficult to interpret. We will study both uses and abuses of percentages.

UNIT 3B Putting Numbers in Perspective: Many of the numbers we hear daily are extremely large or small. We will look at several techniques for giving meaning to such numbers.

UNIT 3C Dealing with Uncertainty: We will discuss ways of dealing with the inevitable uncertainty in numbers heard daily in the news. We will also consider the types of errors that affect measured numbers.

UNIT 3D How Numbers Deceive: Polygraphs, Mammograms, and More: Numbers may seem straightforward, but they can be deceiving unless we interpret them carefully. We will study several interesting cases that relate to everyday issues.

UNIT 3A

USES AND ABUSES OF PERCENTAGES

One of the most common ways of communicating quantitative information is with percentages. Percentages may seem pretty basic; after all, a percentage is simply a fraction. But percentages can be surprisingly subtle, as illustrated by the quotation in Figure 3.1. For example, what does it mean to say "up 44%, to 10.4%"? In this unit, we will investigate some of the uses and abuses of percentages.

Smoking Rates

Teenage smoking rates are still lower than in the 1970s. But the percentage of twelfth graders who smoked daily last year jumped 20 percent since 1991, to 22 percent, according to the most recent edition of the University of Michigan's Monitoring the Future Survey, an annual study widely followed by tobacco researchers. The rate among tenth graders jumped 45 percent, to 18.3 percent, and the rate for eighth graders is up 44 percent, to 10.4 percent.

FIGURE 3.1 SOURCE: *New York Times*, April 20, 1997.

THREE WAYS OF USING PERCENTAGES

Consider the following statements from news reports:

- A total of 13,000 newspaper employees, 2.6% of the newspaper work force, lost their jobs.
- Cisco stock rose 5.7% last week, to $81.75.
- High definition television sets (HDTVs) have 125% more resolution than conventional TV, but cost 400% more.

On close examination, each statement uses a percentage in a different way. The first uses a percentage to express a *fraction* of the total work force. The second uses a percentage to describe a *change* in stock price. The third uses percentages to *compare* the resolutions and costs of televisions.

Using Percentages as Fractions

Percent is just a fancy way of saying "divided by 100," so *P*% simply means *P*/100. For example, 10.4% means 10.4/100, or 0.104. Therefore, if 10.4% of eighth-graders smoke and there are 100,000 eighth-graders, then the number who smoke is 10.4% *of* 100,000, or

$$10.4\% \times 100{,}000 = 0.104 \times 100{,}000 = 10{,}400$$

Note that the word *of* implies multiplication. We've found that if 10.4% of 100,000 eighth-graders smoke, there are 10,400 smokers.

a brief review . . .

PERCENTAGES

The words *per cent* simply mean "per 100" or "divided by 100." We use the symbol % as a shorthand for the words *per cent*. For example, we read 50% as "50 percent" and its meaning is

$$50\% = \frac{50}{100} = 0.5$$

More generally, for a number *P*,

$$P\% = \frac{P}{100}$$

For example:

$$100\% = \frac{100}{100} = 1 \qquad 200\% = \frac{200}{100} = 2$$

$$350\% = \frac{350}{100} = 3.5$$

Note that multiplying a number by 100% does not change its meaning, because 100% is just another way of writing 1. For example, multiplying 1.25 by 100% gives

$$1.25 \times 100\% = 125\%$$

This shows that 125% is just another way of writing 1.25. These basic ideas lead to the following rules for converting between percentages and decimals or common fractions.

- **To convert a percentage to a common fraction:** Replace the % symbol with division by 100; simplify the fraction if necessary.

 Example: $25\% = \dfrac{25}{100} = \dfrac{1}{4}$

- **To convert a percentage to a decimal:** Drop the % symbol and move the decimal point two places to the left (that is, divide by 100).

 Example: $25\% = 0.25$

- **To convert a decimal to a percentage:** Move the decimal point two places to the right (that is, multiply by 100) and add the % symbol.

 Example: $0.43 = 43\%$

- **To convert a common fraction to a percentage:** First convert the common fraction to decimal form; then convert the decimal to a percentage.

 Example: $\dfrac{1}{5} = 0.2 = 20\%$

EXAMPLE 1 *Newspaper Survey*

A newspaper reports that 44% of 1069 people surveyed said that the president is doing a good job. How many said the president is doing a good job?

Solution The 44% represents the fraction of respondents who said the president is doing a good job. Because *of* indicates multiplication, we find

$$44\% \times 1069 = 0.44 \times 1069 = 470.36 \approx 470$$

About 470 people said the president is doing a good job. We rounded the answer to 470 because 470.36 people would not make sense in this context. ■

Using Percentages to Describe Change

Percentages are often used to describe how a quantity changes with time. As an example, suppose the population of a town was 10,000 in 1970 and 15,000 in 2000. We can express the change in population in two basic ways:

- Because the population rose by 5000 people (from 10,000 to 15,000), we say that the **absolute change** in the population was 5000 people.
- Because the increase of 5000 people was 50% of the starting population of 10,000, we say that the **relative change** in the population was 50%, or 0.5.

In general, calculating an absolute or relative change always involves two numbers: a starting number, or **reference value,** and a **new value.** Once we identify these two values, we calculate the absolute and relative change with the formulas in the following box.

Absolute and Relative Change

The **absolute change** describes the actual increase or decrease from a reference value to a new value:

$$\text{absolute change} = \text{new value} - \text{reference value}$$

The **relative change** describes the size of the absolute change in comparison to the reference value:

$$\text{relative change} = \frac{\text{absolute change}}{\text{reference value}} = \frac{\text{new value} - \text{reference value}}{\text{reference value}}$$

The absolute and relative changes are *positive* if the new value is greater than the reference value and *negative* if the new value is less than the reference value.

The above formula gives the relative change as a fraction. Following the usual rules for converting fractions to percentages, we can write the relative change as a **percent change** by multiplying it by 100%. For example, a relative change of 0.5 means a percent change of $0.5 \times 100\% = 50\%$. As Example 2 (below) shows, this idea leads to the following important rules:

- When a quantity *doubles* in value, its relative change is 1, or 100%.
- When a quantity *triples* in value, its relative change is 2, or 200%.
- When a quantity *quadruples* in value, its relative change is 3, or 300%. And so on.

EXAMPLE 2 Stock Price Rise

During a 6-month period, Nokia's stock doubled in price from $25 to $50. What were the absolute and relative changes in the stock price?

Solution The reference value is the starting stock price of $25 and the new value is the later stock price of $50. The absolute change is the difference between the new and starting stock prices:

$$\text{absolute change} = \text{new value} - \text{reference value}$$
$$= \$50 - \$25 = \$25$$

The absolute change divided by the reference value gives the relative change:

$$\text{relative change} = \frac{\text{absolute change}}{\text{reference value}} = \frac{\$25}{\$25} = 1 = 100\%$$

That is, the doubling of the stock price means a relative increase in value of 100%.

Time out to think

In your own words, explain why doubling in value means a relative increase of 1, or 100% (as opposed to, say, 200%). Extend this thinking to explain why tripling means a 200% increase, and so on.

EXAMPLE 3 World Population Growth

World population in 1950 was 2.6 billion. By the beginning of 2000, it had reached 6.0 billion. Describe the absolute and relative change in world population from 1950 to 2000.

Solution The reference value is the 1950 population of 2.6 billion and the new value is the 2000 population of 6.0 billion.

$$\text{absolute change} = \text{new value} - \text{reference value}$$
$$= 6.0 \text{ billion} - 2.6 \text{ billion} = 3.4 \text{ billion}$$

$$\text{relative change} = \frac{\text{new value} - \text{reference value}}{\text{reference value}}$$
$$= \frac{6.0 \text{ billion} - 2.6 \text{ billion}}{2.6 \text{ billion}} \approx 1.3 = 130\%$$

World population increased by 3.4 billion people, or by 130%, from 1950 to 2000.

by the way...
According to estimates made by the U.S. Census Bureau, world population hit 6 billion on July 19, 1999. According to United Nations estimates, the 6 billion mark was passed on October 12, 1999.

EXAMPLE 4 Depreciating a Computer

You bought a computer three years ago for $2000. Today, it is worth only $600. Describe the absolute and relative change in the computer's value.

Solution The reference value is the original price of $2000 and the new value is its current worth of $600. The absolute change in the computer's value is

$$\text{absolute change} = \text{new value} - \text{reference value}$$
$$= \$600 - \$2000 = -\$1400$$

The negative sign tells us that the computer's current worth is $1400 *less than* the price you paid three years ago. The relative change is

$$\text{relative change} = \frac{\text{new value} - \text{reference value}}{\text{reference value}}$$

$$= \frac{\$600 - \$2000}{\$2000}$$

$$= -0.7 = -70\%$$

Again, the negative sign tells us that the computer is now worth 70% *less than* it was three years ago. ∎

Using Percentages for Comparisons

Percentages are also commonly used to compare pairs of numbers. Suppose we want to compare the price of a $50,000 Mercedes to the price of a $40,000 Cadillac. The difference between the Mercedes and Cadillac prices is

$$\$50,000 - \$40,000 = \$10,000$$

That is, the Mercedes costs $10,000 *more than* the Cadillac. We can also express this difference as a percentage of the Cadillac's price:

$$\frac{\$10,000}{\$40,000} = 0.25 = 25\%$$

Thus, in relative terms, the Mercedes costs 25% *more than* the Cadillac.

Because we are *comparing to* the Cadillac price, we say that the Cadillac price is the **reference value;** notice that the reference value follows the word *than*. The Mercedes price is the **compared value.** We can now define the absolute and relative *difference* between the quantities, much as we defined absolute and relative change.

Absolute and Relative Difference

The **absolute difference** describes the actual amount by which the compared value is larger or smaller than the reference value; it is the difference between these two values:

$$\text{absolute difference} = \text{compared value} - \text{reference value}$$

The **relative difference** describes the size of the absolute difference in comparison to the reference value:

$$\text{relative difference} = \frac{\text{absolute difference}}{\text{reference value}}$$

$$= \frac{\text{compared value} - \text{reference value}}{\text{reference value}}$$

The absolute and relative differences are *positive* if the compared value is greater than the reference value and *negative* if the compared value is less than the reference value.

Like the relative change formula, the relative difference formula above gives a fraction. We can convert the answer to a **percent difference** by multiplying it by 100%.

That's essentially all there is to comparisons, except for one subtlety: There's no particular reason why we chose to compare to the Cadillac price. We can just as easily compare to the Mercedes price. In that case, the Mercedes price is the reference value and the Cadillac price is the compared value. The absolute difference is

$$\text{Cadillac price} - \text{Mercedes price} = \$40{,}000 - \$50{,}000 = -\$10{,}000$$

The negative sign tells us that the Cadillac costs $10,000 *less than* the Mercedes. The relative difference is

$$\frac{\text{absolute difference}}{\text{Mercedes price}} = \frac{-\$10{,}000}{\$50{,}000} = -0.20 = -20\%$$

The negative sign indicates that the Cadillac costs 20% *less than* the Mercedes.

We now have two ways to express the relative difference:

- The Mercedes costs 25% more than the Cadillac; or
- the Cadillac costs 20% less than the Mercedes.

$50,000 Mercedes +25% / −20% $40,000 Cadillac

TECHNICAL NOTE
With a little algebra, it is possible to prove the following rule: If quantity A is $P\%$ more than quantity B, then quantity B is $\dfrac{100}{100 + P}$ percent less than quantity A.

Both statements are correct, yet they contain *different* percentages. Thus, we see the importance of keeping careful track of the reference and compared values.

EXAMPLE 5 *Russian and American Life Expectancy*

Life expectancy for American men is about 73 years, while life expectancy for Russian men is about 59 years. Compare the life expectancy of American men to that of Russian men in absolute and relative terms.

Solution We are told to compare the American life expectancy to the Russian life expectancy. Therefore, we use American life expectancy as the compared value and Russian life expectancy as the reference value.

$$\begin{aligned}
\text{absolute difference} &= \text{compared value} - \text{reference value} \\
&= 73 \text{ years} - 59 \text{ years} \\
&= 14 \text{ years} \\
\text{relative difference} &= \frac{\text{compared value} - \text{reference value}}{\text{reference value}} \\
&= \frac{73 \text{ years} - 59 \text{ years}}{59 \text{ years}} \\
&\approx 0.24 = 24\%
\end{aligned}$$

by the way...
Alcoholism is one reason for the low Russian life expectancy. On average, Russians drink nearly 10 gallons of vodka per person per year. An estimated 2.3% of all Russians are alcoholics, and some 1 million Russians die each year from alcohol-related illnesses.

The life expectancy of American men is 14 years greater in absolute terms and 24% greater in relative terms than the life expectancy of Russian men.

EXAMPLE 6 *Pay Comparison*

Average pay for full-time wage earners varies from state to state. As of 1998, Connecticut ranked first in average pay, at $38,895 per person. South Dakota had the lowest average pay, at $21,645 per person. Compare average pay in South Dakota to that in Connecticut in both absolute and relative terms.

Solution Based on the wording of the statement, we use South Dakota pay as the compared value and Connecticut pay as the reference value.

$$\text{absolute difference} = \text{compared value} - \text{reference value}$$
$$= \$21{,}645 - \$38{,}895$$
$$= -\$17{,}250$$

$$\text{relative difference} = \frac{\text{compared value} - \text{reference value}}{\text{reference value}}$$
$$= \frac{\$21{,}645 - \$38{,}895}{\$38{,}895}$$
$$\approx -0.444 = -44.4\%$$

In absolute terms, average pay in South Dakota is $17,250 *less than* average pay in Connecticut. In relative terms, average pay in South Dakota is about 44% less than average pay in Connecticut. (You might want to verify that this also means that average pay in Connecticut is 79.7% more than average pay in South Dakota.)

OF VERSUS MORE THAN

Percentage statements have subtly different meanings depending on whether they contain the words *more than* (or *less than*) or *of*. Consider a population that triples in size from 200 to 600. There are two equivalent ways in which we can state this change with percentages:

- Using *more than:* The new population is 200% *more than* the original population. Here we are stating the *relative change* in the population; we find the increase of 200% by using the relative change formula.
- Using *of:* The new population is 300% *of* the original population. In this case, we are stating a *ratio* of the new population to the original population. The ratio is 3 to 1, or 300%, because the new population is three times the size of the original population.

Time out to think

Use the relative change formula to confirm that a population which grows from 200 to 600 has increased in size by 200%.

Notice that the percentages in the *of* and *more than* statements are related by 300% = 100% + 200%. Generalizing, we find the relationships in the following box.

by the way . . .

Although South Dakota has the lowest average pay, Mississippi ranks lowest in per capita (per person) income. Average pay is the total amount of wages earned in the state divided by the total number of wage earners. Per capita income is all income earned, divided by the total number of people, employed or not. Per capita income is a better measure of poverty or wealth, which is why Mississippi is generally considered the poorest state. But jobs there pay better, on average, than those in South Dakota.

> **Of versus More Than (or Less Than)**
> - If the compared value is *P%* *more than* the reference value, it is $(100 + P)\%$ *of* the reference value.
> - If the compared value is *P%* *less than* the reference value, it is $(100 - P)\%$ *of* the reference value.

For example, 40% *more than* the reference value is $(100 + 40)\% = 140\%$ *of* the reference value, and 40% *less than* the reference value is $(100 - 40)\% = 60\%$ *of* the reference value.

EXAMPLE 7 Salary Difference

Carol earns 50% more than William. How many times larger is her income than his?

Solution We use the rule that *P%* *more than* means $(100 + P)\%$ *of*; since Carol's income is 50% more than William's, we set $P = 50$. Thus, Carol's income is 150% of William's income. Remembering that 150% = 1.5, we know that Carol earns 1.5 times as much as William. ■

a brief review...

WHAT IS A RATIO?

Suppose we want to compare two quantities, such as the prices of a BMW that costs $80,000 and a Honda that costs $20,000. We can, of course, find the absolute or relative difference in the quantities. But another way to make the comparison is by dividing, which tells us the *ratio* of the two quantities. In this case, the two quantities are $80,000 and $20,000, so their ratio is

$$\frac{\$80,000}{\$20,000} = \frac{4}{1} = 4$$

We can state this result in three equivalent ways:

- The ratio of the price of the BMW to that of the Honda is 4 to 1. (The ratio can also be stated as 4, but "4 to 1" is more common.)
- The price of the BMW is 4 times the price of the Honda.
- The price of the Honda is $\frac{1}{4}$ (or 25%) of the price of the Mercedes.

Notice how the units of *dollars* canceled on the top and bottom of the fraction. This is a general characteristic of ratios: Because comparisons make sense only when we compare quantities with the same units, ratios always end up without units.

Example: The average density of the Earth is about 5.5 grams per cubic centimeter and the average density of Saturn is about 0.7 gram per cubic centimeter. What is the ratio of their densities?

Solution: We divide Earth's density by Saturn's density:

$$\frac{\text{average density of Earth}}{\text{average density of Jupiter}} = \frac{5.5 \text{ g/cm}^3}{0.7 \text{ g/cm}^3} \approx 8$$

The ratio of the Earth's average density to Saturn's average density is about 8 to 1. Alternatively, we can say that Earth's density is about 8 times that of Saturn, or that Saturn's density is about $\frac{1}{8}$ that of Earth. Again, notice how the units canceled, leaving the ratio without units.

TECHNICAL NOTE

The *of* versus *more than* rule works well for statements with percentages, but may fail if we convert the percentages to ordinary fractions or numbers. For example, the statement "50 is 100% more than 25" makes sense, but the statement "50 is 1 more than 25" does not make sense—even though 100% = 1. Similarly, most people would say that "50 is two times larger than 25," but it is not correct to say that "50 is 200% larger than 25" (it is 200% *of* 25, or 100% more than 25). Because ordinary language can be ambiguous, it's important to examine context.

EXAMPLE 8 *Sale!*

A store is having a "25% off" sale. How does an item's sale price compare to its original price?

Solution The "25% off" means that an item's sale price is 25% *less than* its original price. Thus, the sale price is (100 − 25)% = 75% *of* the original price. For example, if the original price is $100, the sale price is $75.

Time out to think

One store advertises "30% off everything!" Another store advertises "Sale prices are 30% of original prices!" Which store is having the bigger sale? Explain.

PERCENTAGES OF PERCENTAGES

Percentage comparisons can be particularly confusing when the compared and reference values *themselves* are percentages. Suppose your bank increases the interest rate on your savings account from 3% to 4%. It's tempting to say that the interest rate increased by 1%, but this statement is ambiguous at best. The interest rate increased by 1 *percentage point,* but the relative change in the interest rate was

$$\frac{\text{new value} - \text{reference value}}{\text{reference value}} = \frac{4\% - 3\%}{3\%} = 0.33 = 33\%$$

Note that we identify the new rate as the new value and the old rate as the reference value. Thus, you can say that the bank raised your interest rate by 33%, even though the actual rate increased by only 1 percentage point. More specifically, the *absolute change* in the interest rate was 1 percentage point, while the *relative change* in the interest rate was 33%.

> **Percentage Points versus %**
>
> When you see a change or difference expressed in *percentage points,* you can assume it is an *absolute* change or difference. If it is expressed with the % sign or the word *percent,* it should be a *relative* change or difference.

EXAMPLE 9 *Newspaper Readership Declines*

According to a *Time Magazine* story,

> The percentage of adults reading daily newspapers fell from 78% in 1970 to 64% in 1995.

Describe the change in newspaper readership.

Solution The drop in readership from 78% to 64% represents a decline of 14 percentage points. This is the *absolute change* in the percentage of adults who read a daily newspaper. The *relative change* in readership is

$$\frac{\text{new value} - \text{reference value}}{\text{reference value}} = \frac{64\% - 78\%}{78\%} = -0.18 = -18\%$$

The negative sign indicates a *decrease* in readership. Thus, we can say that readership dropped by 18% in relative terms or by 14 percentage points in absolute terms. ∎

EXAMPLE 10 *Care in Wording*

Assume that 40% of the registered voters in Carson City are Republicans. Read the following questions carefully, and give the most appropriate answers.

a. The percentage of voters registered as Republicans is 25% higher in Freetown than in Carson City. What percentage of the registered voters in Freetown are Republicans?
b. The percentage of voters registered as Republicans is 25 percentage points higher in Freetown than in Carson City. What percentage of the registered voters in Freetown are Republicans?

If you can't convince them, confuse them.
—HARRY S TRUMAN

Solution

a. We interpret the 25% as a relative difference, and 25% of 40% is 10% (because 0.25 × 0.40 = 0.10). We add this value to the Carson City percentage to find that the percentage of registered Republicans in Freetown is 40% + 10% = 50%.
b. We interpret 25 *percentage points* as an absolute difference, so we add this value to the percentage of Republicans in Carson City. The percentage of registered Republicans in Freetown is 40% + 25% = 65%. ∎

SOLVING PERCENTAGE PROBLEMS

Consider this statement:

 Retail prices are 25% more than wholesale prices.

If you knew a wholesale price, how would you calculate the retail price? One way to begin is by translating the *more than* statement into an *of* statement. The statement becomes:

 Retail prices are (100 + 25)% = 125% of wholesale prices.

Replacing *of* with multiplication, we write this statement mathematically as

$$\text{retail price} = 125\% \times \text{wholesale price}$$

With this equation, we can find a retail price from a wholesale price. For example, if the wholesale price is $10, the retail price is 125% × $10 = 1.25 × $10 = $12.50.

Now suppose we want to find a wholesale price from a retail price. In this case, we must divide both sides of the equation by 125%:

$$\frac{\text{retail price}}{125\%} = \frac{\cancel{125\%} \times \text{wholesale price}}{\cancel{125\%}}$$

We then interchange the left and right sides to write the result:

$$\text{wholesale price} = \frac{\text{retail price}}{125\%}$$

For example, if the retail price is $15, the wholesale price is $15/1.25 = $12. We can now generalize our results.

> **Solving Percentage Problems**
>
> If the compared value is $P\%$ *more than* the reference value, then
>
> $$\text{compared value} = (100 + P)\% \times \text{reference value}$$
>
> and
>
> $$\text{reference value} = \frac{\text{compared value}}{(100 + P)\%}$$
>
> If the compared value is *less than* the reference value, then use $(100 - P)$ instead of $(100 + P)$ in the above equations.

EXAMPLE 11 Tax Calculations

a. You purchase a shirt with a labeled (pre-tax) price of $17. The local sales tax rate is 5%. What is your final cost (including tax)?
b. Your receipt shows that you paid $19.26 for a DVD movie, tax included. The local sales tax rate is 7%. What was the labeled (pre-tax) price of the DVD?

Solution

a. The final cost of an item is its labeled price plus the sales tax. For a tax rate of $P\% = 5\%$, the final cost is

$$\text{final cost} = \text{labeled price} + (5\% \times \text{labeled price})$$

In other words, the final cost is 5% more than the labeled price. Changing from a *more than* statement to an *of* statement, we have

$$\text{final cost} = (100 + 5)\% \times \text{labeled price} = 105\% \times \text{labeled price}$$

When we substitute the given labeled price of $17, the final cost is

$$\text{final cost} = 105\% \times \$17 = 1.05 \times \$17 = \$17.85$$

The final cost for the shirt is $17.85.

b. This time the tax rate is $P\% = 7\%$, so the relation between labeled price and final cost is

$$\text{final cost} = (100 + 7)\% \times \text{labeled price} = 107\% \times \text{labeled price}$$

We are given the final cost of $19.26. We find the labeled price by dividing both sides by 107%:

$$\text{labeled price} = \frac{\text{final cost}}{107\%} = \frac{\$19.26}{107\%} = \frac{\$19.26}{1.07} = \$18.00$$

The labeled price of the DVD was $18.

EXAMPLE 12 *Up 44%, to 10.4%*

Consider the following statement from the quote in Figure 3.1:

The [smoking] rate for eighth graders is up 44 percent, to 10.4 percent.

What was the previous smoking rate for eighth-graders?

by the way...

Percentages are often called *rates*. For example, a sales tax of 6% is called a *tax rate* of 6%; if 15% of high school students smoke, we say that the *smoking rate* among high school students is 15%.

Solution The "10.4 percent" means that 10.4% of all eighth-graders now smoke. The "44 percent" expresses the relative change from the previous smoking rate to the new rate. It tells us that the new rate is 44% *more than* the previous rate, which means it is (100 + 44)% = 144% *of* the previous rate:

$$\text{new rate} = 144\% \times \text{previous rate}$$

We solve for the previous rate by dividing both sides by 144% = 1.44:

$$\text{previous rate} = \frac{\text{new rate}}{144\%} = \frac{\text{new rate}}{1.44}$$

Now we find the previous rate by substituting 10.4% for the new rate:

$$\text{previous rate} = \frac{10.4\%}{1.44} \approx 7.2\%$$

The previous smoking rate for eighth-graders was 7.2%. Note that we left the numerator as a percentage in the above equation (rather than converting it to decimal form) so that our answer would come out as a percentage.

ABUSES OF PERCENTAGES

Because percentages can be so subtle, many people misuse percentages—sometimes inadvertently and sometimes deliberately. In the rest of this unit, we'll investigate a few common abuses of percentages.

Beware of Shifting Reference Values

Consider the following situation: Because of losses by your employer, you agree to accept a temporary 10% pay cut. Your employer promises to give you a 10% pay raise after six months. Will the pay raise restore your original salary?

We can answer this question by assuming some arbitrary number for your original weekly pay, such as $200. A 10% pay cut means that your pay will decrease by 10% of $200, or $20, so your weekly pay after the cut is

$$\$200 - \$20 = \$180$$

The subsequent raise increases your pay by 10% of $180, or $18, making your weekly pay

$$\$180 + \$18 = \$198$$

DILBERT reprinted by permission of United Feature Syndicate, Inc.

Thus, the 10% pay cut followed by the 10% pay raise leaves you with *less* money than you started with. This result arises because the reference value for the calculations shifted during the problem: It was $200 in the first calculation and $180 in the second.

EXAMPLE 13 *Shifting Investment Value*

A stockbroker offers the following defense to angry investors: "I admit that the value of your investments fell 60% during my first year on the job. This year, however, their value has increased by 75%, so you are now 15% ahead!" Evaluate the stockbroker's defense.

Solution Imagine that you began with an investment of $1000. During the first year, your investment lost 60% of its value, or $600, leaving you with $400. During the second year, your investment gained 75% of $400, or 0.75 × $400 = $300. Thus, at the end of the second year, your investment was worth $400 + $300 = $700. This is *less* than your original investment of $1000 and certainly not a 15% gain overall. We can trace the problem with the stockbroker's defense to a shifting reference value: It was $1000 for the first calculation and $400 for the second. ■

EXAMPLE 14 *Tax Cuts*

A politician promises, "If elected, I will cut your taxes by 20% for each of the first three years of my term, for a total cut of 60%." Evaluate the promise.

Solution The politician neglected the effects of the shifting reference value. A cut of 20% in each of three years will *not* make an overall cut of 60%. To see what really happens, suppose that you currently pay $1000 in taxes. The following table shows how your taxes change over the three years.

Year	Tax paid in previous year	20% of previous year tax	New tax this year
1	$1000	$200	$800
2	$800	$160	$640
3	$640	$128	$512

Over three years, your taxes decline by $1000 − $512 = $488. But this is only 48.8% of $1000, so over the three years your tax bill declines by 48.8%, not by 60%. ■

Less than Nothing

We often see numbers that represent large "more than" percentages. For example, a price of $40 is 300% more than a price of $10. However, in most cases it is not possible to have a "less than" percentage that is greater than 100%. To see why, consider an advertisement claiming that replacing standard light bulbs with energy-efficient ones would use "200 percent less energy." If you think about it, you'll realize that such a savings is impossible. If the new light bulbs used 100% less energy, they'd be using no energy at all. Thus, the only way they could use 200% less would be if they actually *produced* energy. Clearly, whoever wrote the advertisement made a mistake.

by the way...
The light bulb example related here was used as the title story in the book *200% of Nothing* by A. K. Dewdney (Wiley, 1993). The book contains many other interesting stories of misuses of numbers.

EXAMPLE 15 Impossible Sale

A store advertises that it will take "150% off" the price of all merchandise. What should happen when you go to the counter to buy a $500 item?

Solution If the price were 100% off, the item would be free. Thus, if the price is 150% off, the store should *pay you* half the item's cost, or $250. More likely, the store manager did not understand percentages.

Time out to think

Can an athlete give a 110% effort? Can a glass of juice have 110% of the minimum daily requirement for vitamin C? Explain.

Don't Average Percentages

Suppose you got 70% of the questions correct on a midterm exam and 90% correct on the final exam. Can we conclude that you answered 80% of all the questions correctly? It might be tempting to say yes—after all, 80% is the average (mean) of 70% and 90%. But it would be wrong, unless both tests happened to have the same number of questions.

To see why, let's say that the midterm had 10 questions and the final had 100 questions, for a total of 110 questions. Your 70% score on the midterm means you answered 7 questions correctly, while your 90% score on the final means you answered 90 questions correctly. Thus, on the two exams combined, you answered 97 out of 110 questions correctly, which is 88.2% (because 97/110 = 0.882). This is much higher than the 80% "average" of the two individual exam percentages.

This example carries a very important lesson: As a general rule, *you should never average percentages.*

> **by the way ...**
> We usually think of the average of two numbers as their sum divided by 2. Technically, this type of average is called a *mean*.

EXAMPLE 16 Batting Average

In baseball, a player's batting average represents the percentage of at-bats in which he got a hit. For example, a batting average of .350 means the player got a hit 35% of the times he batted. Suppose a player had a batting average of .200 during the first half of the season and .400 during the second half of the season. Can we conclude that his batting average for the entire season was .300 (the average of .200 and .400)? Why or why not? Give an example that illustrates your reasoning.

Solution No. For example, suppose he had 300 at-bats during the first half of the season and 200 at-bats during the second half, for a total of 500 at-bats. His first-half batting average of .200 means he got hits on 20% of his 300 at-bats, or $0.2 \times 300 = 60$ hits. His second-half batting average of .400 means he got hits on 40% of his 200 at-bats, or $0.4 \times 200 = 80$ hits. For the season, he got a total of $60 + 80 = 140$ hits in his 500 at-bats. Thus, his season batting average was $140/500 = 28\%$, or .280—not the .300 found by averaging his first-half and second-half batting percentages. (In fact, the only case in which his season average would be .300 is if he had precisely the same number of at-bats in each half of the season.)

Review Questions

1. Describe the three basic uses of percentages. Give a sample statement that uses percentages in each of the three ways.
2. Distinguish between absolute and relative change. Give an example that illustrates how we calculate a relative change.
3. Distinguish between an absolute and relative difference. Give an example that illustrates how we calculate a relative difference.
4. Explain the difference between the key words *of* and *more than* when dealing with percentages. How are their meanings related?
5. Explain the difference between the terms *percent* (%) and *percentage points*. Give an example of how they can differ for the same situation.
6. Suppose a compared value is $P\%$ more than a reference value. How can you find the compared value if you are given the reference value? How can you find the reference value if you are given the compared value? Give examples.
7. Suppose the stock market declines by 20% one week, then gains 20% the next week. Has it returned to its original value? Explain.
8. Is it possible for a person to be 200% shorter than another person? Explain.
9. Give an example to explain why, in general, it is not legitimate to average percentages.

In the News

1. **Percentages.** Find three recent news reports that quote percentages. In each case, describe the use of the percentage (as a fraction, to describe change, or for comparison) and explain its context.
2. **Percentage Change.** Find a recent news report that quotes a percentage change. Describe the meaning of the change.
3. **Abuse of Percentages.** Find a news article or report in which the use of a percentage is either suspicious or wrong. If possible, clarify or correct the statement.

Exercises

Review of Fractions. Exercises 1–2 use skills covered in the Brief Review on p. 119.

1. Express the following numbers in three forms: fraction, decimal, and percentage.

 a. $\frac{1}{4}$ b. 0.45 c. 0.33333… d. 23%
 e. $\frac{1}{8}$ f. 1.34 g. 0.65 h. 99%

2. Express the following numbers in three forms: fraction, decimal, and percentage.

 a. 120% b. 0.652 c. 2.34 d. 0.98
 e. 45% f. 1.01 g. $\frac{3}{8}$ h. 125%

Review of Ratios. Exercises 3–4 use skills covered in the Brief Review on p. 125. Each exercise gives a pair of quantities, A and B. Compare them in three ways. First, find the ratio of A to B. Then, find the ratio of B to A. Finally, complete this statement: A is _____ percent of B.

3. a. $A = 10$ and $B = 2$ b. $A = 150$ and $B = 25$
 c. $A = 550$ and $B = 50$ d. $A = 45$ and $B = 550$
 e. $A = 75$ and $B = 480$ f. $A = 33$ and $B = 123$

4. a. A is the 7.5 million population of North Carolina and B is the 16.5 million population of Texas.
 b. A is the 2.8 million population of Chicago and B is the 0.75 million population of Baltimore.
 c. A is the 359,000 births recorded in the United States in a recent July and B is the 290,000 births recorded in the United States in a recent January.
 d. A is the average U.S. family size of 3.58 in 1970 and B is the average U.S. family size of 3.18 in 1998.
 e. A is the median price of $23,000 for a single-family house in the United States in 1970 and B is the median price of $131,000 for a single-family house in the United States in 1998.

Percentages as Fractions. Each statement in Exercises 5–6 gives two numbers. Express the first number as a percentage of the second.

5. a. 25 women in a room of 113 people
 b. 345 blooming tulips in a field of 398 tulips
 c. 1234 people who voted for the losing candidate out of 3009 voters

6. a. 38 men at a convention of 236 people

 b. 23 purple jelly beans in a jar of 123 jelly beans

 c. 2345 people who voted for the winning candidate out of 4023 voters

Percentage Change. Exercises 7–13 each describe measurements made at two different times. Find the absolute and relative change (as a percentage) in each case.

7. The number of daily newspapers in the United States was 2226 in 1900 and 1509 in 1997.

8. The annual number of deaths from cardiovascular disease in the United States decreased from 1,008,000 in 1970 to 945,000 in 1998.

9. Consumer debt in the United States increased from $160 billion in 1974 to $850 billion in 1995.

10. The percentage of all bachelor's degrees awarded to women increased from 44% in 1972 to 54% in 1992.

11. The percentage of single-parent households in the United States increased from 12% (of all households) in 1970 to 30% (of all households) in 1995.

12. The annual per capita consumption of alcohol in the form of beer in the United States increased from 0.73 gallon in 1940 to 1.26 gallons in 1994.

13. The total number of libraries in the United States was 28,638 in 1980 and 33,004 in 1997.

14. **Salary Comparison.** John's salary grew from $20,000 in 1998 to $28,000 in 2001. Mary's salary grew from $25,000 to $35,000 in the same period. Whose salary grew more in absolute terms? in relative terms? Explain.

Percentage Difference. Exercises 15–20 each state two measurements. Find the absolute and relative difference (as a percentage) in each case. Assume that the first quantity is the compared value and the second quantity is the reference value.

15. The daily circulation of the *Wall Street Journal* is about 1.77 million (the largest in the country). The daily circulation of the *New York Times* is about 1.07 million (the third largest in the country).

16. In 1996, Ford sold 401,049 Taurus cars and Honda sold 382,298 Accord cars.

17. In 1996, France ranked as the number one tourist destination with about 66 million international arrivals. The United States ranked second with about 48 million international arrivals.

18. In 1996, Saudi Arabia produced 8.2 million barrels of crude oil per day, while the United States produced 6.4 million barrels per day.

19. In 1995, 504,000 males had open-heart surgery in the United States, while 209,000 females had the operation.

20. The life expectancy in Canada is 79.1 years (highest in the world in 1997), while the life expectancy in the United States is 76.0 years.

Of **vs.** ***More Than.*** Fill in the blanks in Exercises 21–24. Briefly explain your reasoning in each case.

21. Jack weighs 40% more than Jill, so his weight is ____% of Jill's weight.

22. The area of Norway is 24% more than the area of Colorado, so Norway's area is ____% of Colorado's area.

23. The population of Montana is 20% less than the population of New Hampshire, so Montana's population is ____% of New Hampshire's population.

24. Henry earns 45% less than Ingrid, so his salary is ____% of Ingrid's salary.

Prices and Sales. Fill in the blanks or answer the questions in Exercises 25–28.

25. The wholesale cost of a car is 50% less than its retail cost. Therefore, the retail cost is ____ times the wholesale cost.

26. A store is having a 15% off sale. Therefore, a sale price is ____ times an item's original price.

27. The retail cost of a TV is 40% more than its wholesale cost. Therefore, the retail cost is ____ times the wholesale cost.

28. A store is having a 40% off sale. What is the sale price for an item with a regular cost of $80?

Percentages of Percentages. Exercises 29–32 describe changes in which the measurements themselves are percentages. Express each change in two ways: (1) as an absolute difference in terms of *percentage points* and (2) as a relative difference in terms of %.

29. The percentage of high school seniors using alcohol decreased from 68.2% in 1975 to 52.7% in 1997.

30. The percentage of the world's population living in developed countries decreased from 27% in 1970 to 20% in 1996.

31. The five-year survival rate for Caucasians for all forms of cancer increased from 39% in the 1960s to 60% in the 1990s.

32. The five-year survival rate for Blacks for all forms of cancer increased from 27% in the 1960s to 44% in the 1990s.

33. **Care in Wording.** Assume that 30% of city employees in Carson City ride the bus to work. Consider the following two statements:

 - The percentage of city employees who ride the bus to work is 10% higher in Freetown than in Carson City.
 - The percentage of city employees who ride the bus to work is 10 percentage points higher in Freetown than in Carson City.

 For each case, state the percentage of city employees in Freetown who ride the bus to work. Briefly explain why the two statements have different meanings.

34. **Ambiguous News.** The average annual precipitation on Mt. Washington, New Hampshire, is 90 inches. During one particularly wet year, different news reports carried the following statements.

 - The precipitation this year is 200% of normal.
 - The precipitation this year is 200% above normal.

 What does each of these statements imply about the precipitation during this year? Do the two statements have the same meaning? Explain.

35. **Sales Tax.** You purchase a bicycle with a labeled (pre-tax) price of $699. The local sales tax rate is 7.6%. What is the final cost?

36. **Sales Tax.** The final cost of your new shoes is $107.69. The local sales tax rate is 6.2%. What was the labeled (pre-tax) price?

37. **Sales Tax.** You purchase a car with a labeled (pre-tax) price of $17,600. The local sales tax rate is 6.5%. What is the final cost?

38. **Sales Tax.** The final cost of your new HDTV is $3,706.30. The local sales tax rate is 8.1%. What was the labeled (pre-tax) price?

39. **Tip.** Your dinner bill is $18.75. You leave $22. What percent tip did you leave? (Ignore taxes in this question.)

40. **Tip.** Your dinner bill is $42.50. You leave $50. What percent tip did you leave? (Ignore taxes in this question.)

41. **Smoking Rates.** Consider the statement "The smoking rate among tenth-graders jumped 45 percent, to 18.3 percent." What percentage of tenth-graders smoked before the increase?

42. **Smoking Rates.** Consider the statement: "The smoking rate among twelfth-graders jumped 20 percent, to 22 percent." What percentage of twelfth-graders smoked before the increase?

Solving News Percentages. Each of Exercises 43–48 has a quote from a *Time Magazine* article. Answer the question that follows the quote.

43. "Almost 75% of the $100 billion in U.S. trade with Mexico is delivered by truck and most of that cargo travels through Texas."

 Question: What is the value of U.S. trade with Mexico (in dollars) that is delivered by truck?

44. "In 1991 just 12% of the U.S. population was considered obese; by 1998 it was 20%."

 Question: What is the percentage increase in the proportion of the U.S. population considered obese?

45. "Bagel sales have reached $3 billion and are growing 20% annually according to Lehman Brothers."

 Question: At this growth rate, how much will be spent on bagels next year?

46. "With containerization there has been a 98.5% reduction in the work needed to unload a timber ship since 1970, down to just 8 man-days."

 Question: How many man-days did it take to unload a timber ship in 1970?

47. "Consumer debt has soared 39% in the last five years and now exceeds $1 trillion."

 Question: What (approximately) was consumer debt five years ago?

48. "In a new TIME/CNN poll, 80% of working people say they're not worried about losing their jobs, up 16% since last winter."

Question: What percentage of people said they were not worried last winter?

Shifting Reference Value. In each of Exercises 49–52, determine whether the concluding statement is true or false. If it is false, state the true change. Explain clearly.

49. Sue went on a weight loss program and lost 10% of her weight. However, she then gained 10% (from her lowest weight). In the end, her weight was unchanged.

50. You receive a pay raise of 5%, then receive a pay cut of 5%. After the two changes in pay, your salary is unchanged.

51. Your property taxes increase 5% in one year and increase 6% the next year. The total increase in taxes over the two years is 11%.

52. A high school reports that its students' SAT scores were down by 20% for one year. The next year, however, they rose by 30%. The high school principal announces, "Overall, test scores have improved by 10% over the past two years."

Sensible? Decide whether each claim in Exercises 53–58 makes any sense at all. If it does, explain how. If not, explain why not.

53. The price of cellular service has decreased 150% in the last ten years.

54. Mary is 100% shorter than Vivian.

55. The price of real estate in this city has increased 450% in the last ten years.

56. Super Oats have 100% more calcium than Regular Oats.

57. Your computer is 250% faster than mine.

58. Your computer is 250% slower than mine.

59. **Average Percentages.** You are a teacher. Your first-period class, with 25 students, had a mean score of 86% on the midterm exam. Your second-period class, with 30 students, had a mean score of 84% on the same exam. Does it follow that the mean score for both classes is 85%? Explain.

60. **Average Percentages.** A player has a batting average over many games of 0.400. In his next game, he goes 2 for 4, which is a batting average of 0.500 for the game. Does it follow that his new batting average is $(0.400 + 0.500)/2 = 0.450$? Explain.

Analyzing Percentage Statements. For Exercises 61–64, decide if the statement is true or false. Explain.

61. The class is 60% women and 10% of the women have blond hair, so blond-haired women comprise $60\% \times 10\% = 6\%$ of the class.

62. The class is 40% men and 20% of the men are bald, so bald men comprise $40\% \times 20\% = 80\%$ of the class.

63. 70% of the hotels have a restaurant and 20% have a swimming pool, so 90% of the hotels in town have a restaurant or a pool.

64. 40% of the class drives to school and 15% takes a bus there, so 45% of the class neither drives nor takes a bus.

Web Project

Find useful links for Web Projects on the text Web site: www.aw.com/bennett-briggs

1. **Web News Percentages.** Go to the Web site for a newspaper and look for at least three stories with percentages. Explain the percentages in each case.

UNIT 3B

PUTTING NUMBERS IN PERSPECTIVE

What is a billion dollars or how much is a trillion dollars? For many people, these numbers are just words—so large that they hardly seem to mean anything. But we hear numbers like these every day, and we cannot truly understand the major issues of our time unless we have some understanding of the numbers that are involved. In this unit, we will study several techniques for putting large or small numbers into a perspective that gives them real meaning.

WRITING LARGE AND SMALL NUMBERS

Before we can work with large and small numbers, we need to have a way of writing them that makes calculations easy. We do this by using powers of 10 (see the Brief Review in Unit 2B). For example, we know that a billion is ten to the ninth power, or 10^9. We can therefore write the number 6 billion as 6×10^9. More generally, we write each number in the form of a number *between* 1 and 10 multiplied by a power of 10. For example, we write 420 as 4.2×10^2 (because $10^2 = 100$ and $4.2 \times 100 = 420$) and 0.00067 as 6.7×10^{-4} (because $10^{-4} = 0.0001$ and $6.7 \times 0.0001 = 0.00067$). This format for writing numbers is called **scientific notation.**

> **DEFINITION**
>
> **Scientific notation** is a format in which a number is expressed as a number *between* 1 and 10 multiplied by a power of 10.

by the way...
The only drawback to scientific notation is that it makes extremely large or small numbers deceptively easy to write. For example, it's easy to write 10^{80}, and it doesn't look so incredibly large on paper—but it is a number larger than the total number of atoms in the known universe.

Another advantage of scientific notation is that it makes it easy to *approximate* answers without a calculator. For example, we can quickly approximate the answer to 5795×326 by rounding 5795 to 6000 and 326 to 300. Writing the rounded numbers in scientific notation, we then see that

$$5795 \times 326 \approx (6 \times 10^3) \times (3 \times 10^2) = 18 \times 10^5 = 1,800,000$$

Because the exact answer is 1,889,170, this approximation provides a good estimate. This technique can be especially useful for checking answers (see Example 2 below).

EXAMPLE 1 Numbers in Scientific Notation

Rewrite each of the following statements using scientific notation.

 a. The U.S. federal debt is about $5,800,000,000,000.
 b. The diameter of a hydrogen nucleus is about 0.000 000 000 000 001 meter.

Solution Notice how much easier it is to read the numbers with scientific notation.

 a. The U.S. federal debt is about 5.8×10^{12}, or $5.8 trillion.
 b. The diameter of a hydrogen nucleus is about 1×10^{-15} meter.

by the way...
A million is a thousand thousand, or 10^6. But above the millions, the same words have different meanings in different countries. In the United States, a *billion* is a thousand million, or 10^9, and a trillion is a thousand billion, or 10^{12}. But in Great Britain and Germany, a billion is a million million, or 10^{12}, and a trillion is a million billion, or 10^{18}. We will use only the U.S. meanings in this book.

a brief review...

WORKING WITH SCIENTIFIC NOTATION

Converting to Scientific Notation

To convert a number from ordinary notation to scientific notation:

Step 1. Move the decimal point to come after the *first* nonzero digit.

Step 2. For the power of 10, use the number of places the decimal point moves; the power is *positive* if the decimal point moves to the left and *negative* if it moves to the right.

Examples:

$$3042 \xrightarrow{\text{decimal moves 3 places to left}} 3.042 \times 10^3$$

$$0.00012 \xrightarrow{\text{decimal moves 4 places to right}} 1.2 \times 10^{-4}$$

$$226 \times 10^2 \xrightarrow{\text{decimal moves 2 places to left}} (2.26 \times 10^2) \times 10^2 = 2.26 \times 10^4$$

Converting from Scientific Notation

To convert a number from scientific notation to ordinary notation:

Step 1. The power of 10 indicates how many places to move the decimal point; move it to the *right* if the power of 10 is positive and to the *left* if it is negative.

Step 2. If moving the decimal point creates any open places, fill them with zeros.

Examples:

$$4.01 \times 10^2 \xrightarrow{\text{move decimal 2 places to right}} 401$$

$$3.6 \times 10^6 \xrightarrow{\text{move decimal 6 places to right}} 3{,}600{,}000$$

$$5.7 \times 10^{-3} \xrightarrow{\text{move decimal 3 places to left}} 0.0057$$

Multiplying or Dividing with Scientific Notation

Multiplying or dividing numbers in scientific notation simply requires operating on the powers of 10 and the other parts of the number separately. *Examples:*

$$(6 \times 10^2) \times (4 \times 10^5) = (6 \times 4) \times (10^2 \times 10^5)$$
$$= 24 \times 10^7$$
$$= 2.4 \times 10^8$$

$$\frac{4.2 \times 10^{-2}}{8.4 \times 10^{-5}} = \frac{4.2}{8.4} \times \frac{10^{-2}}{10^{-5}}$$
$$= 0.5 \times 10^{-2-(-5)}$$
$$= 0.5 \times 10^3$$
$$= 5 \times 10^2$$

Note that, in both examples, we first found an answer in which the number multiplied by a power of 10 was *not* between 1 and 10. We then followed the process for converting the final answer into scientific notation.

Addition and Subtraction with Scientific Notation

In general, we must write numbers in ordinary notation before adding or subtracting. *Examples:*

$$(3 \times 10^6) + (5 \times 10^2) = 3{,}000{,}000 + 500$$
$$= 3{,}000{,}500$$
$$= 3.0005 \times 10^6$$

$$(4.6 \times 10^9) - (5 \times 10^8) = 4{,}600{,}000{,}000 - 500{,}000{,}000$$
$$= 4{,}100{,}000{,}000$$
$$= 4.1 \times 10^9$$

When both numbers have the *same* power of 10, we can factor out the power of 10 first. *Examples:*

$$(7 \times 10^{10}) + (4 \times 10^{10}) = (7 + 4) \times 10^{10}$$
$$= 11 \times 10^{10}$$
$$= 1.1 \times 10^{11}$$

$$(2.3 \times 10^{-22}) - (1.6 \times 10^{-22}) = (2.3 - 1.6) \times 10^{-22}$$
$$= 0.7 \times 10^{-22}$$
$$= 7.0 \times 10^{-23}$$

EXAMPLE 2 *Checking Answers with Approximations*

You and a friend are doing a rough calculation of how much garbage New City residents produce every day. You estimate that, on average, each of the 8 million residents produces 1.8 pounds, or 0.0009 ton, of garbage each day. Thus, the total amount of garbage is

$$8{,}000{,}000 \text{ persons} \times 0.0009 \frac{\text{ton}}{\text{person}}$$

Your friend quickly presses calculator buttons and tells you that the answer is 225 tons. Without using your calculator, determine whether this answer is reasonable.

Solution You can write 8 million as 8×10^6, which is nearly 10^7. You can write 0.0009 as 9×10^{-4}, which is nearly 10^{-3}. Thus, the product should be approximately

$$10^7 \times 10^{-3} = 10^{7-3}$$
$$= 10^4$$
$$= 10{,}000$$

Clearly, your friend's answer of 225 tons is too small. This simple approximation technique provided a useful check, even though it did not tell us the exact answer. (You should confirm that the exact answer is 7200 tons.)

GIVING MEANING TO NUMBERS

Now that we have a method for writing large and small numbers, we can move toward our goal of putting numbers in perspective. As for problem solving generally, there's no single recipe for giving meaning to numbers. Nevertheless, a few techniques are particularly useful. Here, we will study three of them: putting numbers in perspective through estimation, through comparisons, and through scaling.

Perspective Through Estimation

Making estimates is one of the best techniques for putting numbers into perspective. For example, how high is 1000 feet? For most people, the quantity 1000 feet has little meaning by itself. However, we know that each story in a building is typically about 10 feet from floor to ceiling. Therefore, we can put 1000 feet in perspective by thinking of it as the approximate height of a 100-story building.

Keep in mind that estimates are not meant to be exact. For example, the 102-story Empire State Building stands 1250 feet tall (not counting its needle), which means our estimate of 1000 feet for a 100-story building was off by about 250 feet, or 25%, in this case. But if we started with no idea about what 1000 feet looks like, this 25% difference is not very important—picturing a 100-story building still gives us a good sense of the quantity 1000 feet.

Often, estimates are useful even if they only get us within a very broad range of an exact value. For example, you can infer a lot about the character of a town or city simply by knowing whether its population is in the ten thousands or the millions—the former is the size of a small college town and the latter is a huge city. Estimates that give only a broad range of values, such as "in the millions," are called **order of magnitude** estimates.

by the way...

To make it *convincingly* taller than all other buildings in the world at the time, the architects of the Empire State Building designed a 200-foot cap and needle for the top of the building. The original purpose of the needle was to serve as a docking station for air ships. While the needle was never used for this purpose, it has served as a good radio and TV antenna.

DEFINITION

An **order of magnitude** estimate specifies only a broad range of values, such as "in the ten thousands" or "in the millions."

In ordinary language, we must interpret the term *order* in the way that seems appropriate to its context. For example, in describing the United States population, we might say that it is *of order 300 million,* because we know that it is closer to 300 million than to 200 million or 400 million. In contrast, when astronomers say that the number of stars in a galaxy is *of order 100 billion,* they mean that it could be anywhere within about a factor of 10 of this number—that is, anywhere between about 10 billion and 1 trillion.

TECHNICAL NOTE
In science, the term *order of magnitude* refers specifically to powers of 10. For example, 10^{23} is said to be five orders of magnitude larger than 10^{18}, because it is 10^5 times larger ($10^{23} = 10^{18} \times 10^5$).

EXAMPLE 3 *Order of Magnitude of Ice Cream Spending*

Make an order of magnitude estimate of total annual spending on ice cream in the United States.

Solution We begin by devising a strategy for the problem. Clearly, we can calculate total annual ice cream spending if we know how much the average person spends each year; we simply multiply the per person spending by the population to find the total spending. We can find the per person annual spending by multiplying the average cost of a serving of ice cream by the average number of servings a person eats in a year. Putting these ideas together, we can find the total annual ice cream spending with the following equation:

$$\underbrace{\text{total annual spending}}_{\text{units of } \frac{\$}{\text{yr}}} = \underbrace{\text{servings per person per year}}_{\text{units of } \frac{\text{servings}}{\text{person} \times \text{yr}}} \times \underbrace{\text{price per serving}}_{\text{units of } \frac{\$}{\text{serving}}} \times \underbrace{\text{population}}_{\text{units of persons}}$$

Notice how the units work out, telling us that our equation will indeed give us the units of dollars per year that we expect for the answer.

Now we need a value for each term in the equation. We can reasonably guess that an average person has something like 50 servings of ice cream per year (which means about one a week). The price of ice cream is certainly of order $1 per serving (as opposed to, say, 10¢ or $10). And the U.S. population is of order 300 million (3×10^8) people.

$$\text{total annual spending} = 50 \frac{\text{servings}}{\text{person} \times \text{yr}} \times \frac{\$1}{\text{serving}} \times (3 \times 10^8 \text{ persons})$$

$$= \frac{\$1.5 \times 10^{10}}{1 \text{ yr}}$$

Total annual spending on ice cream in the United States is of order 1.5×10^{10}, or $15 billion, per year. The actual number could be up to a few times larger or smaller; for example, actual spending might be anywhere between about $5 billion and $50 billion. Nevertheless, we've learned a lot—before we started, we had no idea how

by the way...

Calculations such as the one here for ice cream, in which we find a basic equation by looking at the units and then use the equation for an order of magnitude estimate, are often called *back-of-the-envelope calculations*. The term reflects the idea that the calculation could fit on the back of an envelope and is informal.

much Americans spend on ice cream, and now we know that they spend billions of dollars on ice cream each year.

Time out to think

Suppose you work for a company that distributes ice cream to stores throughout the United States. Market research tells you that a $5 million advertising campaign could give you an additional 5% of the total U.S. market for ice cream. Based on the estimate in Example 3, would you expect the advertising campaign to be worth its cost? Explain.

Perspective Through Comparisons

A second general way to put numbers in perspective is by making comparisons. For example, consider $100 billion, which is roughly the wealth of the world's richest individuals. It's easy to say a number like 100 billion, but how big is it? Let's think of it in terms of counting: Suppose you were asked to count $100 billion in $1 bills; how long would it take? Clearly, if we assume you can count 1 bill each second, it would take 100 billion seconds. We can put 100 billion (10^{11}) seconds in perspective by converting to years, using a chain of conversions:

$$10^{11} \text{ s} \times \left(\frac{1 \text{ min}}{60 \text{ s}}\right) \times \left(\frac{1 \text{ hr}}{60 \text{ min}}\right) \times \left(\frac{1 \text{ day}}{24 \text{ hr}}\right) \times \left(\frac{1 \text{ yr}}{365 \text{ days}}\right) \approx 3171 \text{ yr}$$

We see that 100 billion seconds is equivalent to 3171 years. In other words, you would need *more than three thousand years* just to count $100 billion in $1 bills. And that assumes that you never take a break: no sleeping, no eating, and absolutely no dying!

Comparisons are particularly useful in dealing with units that are relatively unfamiliar, such as energy units. Table 3.1 lists various energies that you can use in comparisons. For example, we immediately see that U.S. annual energy consumption is about $\frac{1}{5}$ of world annual energy consumption.

TABLE 3.1 Selected Energy Comparisons

ITEM	ENERGY (JOULES)
Energy released by metabolism of 1 average candy bar	1×10^6
Energy needed for 1 hour of running (adult)	4×10^6
Energy released by burning 1 liter of oil	1.2×10^7
Electrical energy used in an average home daily	5×10^7
Energy released by burning 1 kilogram of coal	1.6×10^9
Energy released by fission of 1 kilogram of uranium-235	5.6×10^{13}
Energy released by fusion of hydrogen in 1 liter of water	7×10^{13}
U.S. annual energy consumption	1×10^{20}
World annual energy consumption	5×10^{20}
Annual energy generation of Sun	1×10^{34}

NOTE: Consumption data for 2000, from the U.S. Energy Information Administration.

Using your CALCULATOR

Scientific Notation

Most calculators have a special key for entering scientific notation. The key, usually labeled exp (short for "exponent") or EE (short for "enter exponent"), is used to enter a power of 10. For example, enter the number 3.5×10^6 with the key sequence

3.5 [EE] 6

You should think of the EE (or exp) as "times 10 to the power entered next." Thus, to enter a "pure" power of 10, such as 10^5, you must use the key sequence for 1×10^5:

1 [EE] 5

Beware of two common mistakes: First, some people try to enter 10^5 by pressing 10 [EE] 5, but this key sequence actually means 10×10^5, or 10^6. Second, some people try to enter a number like 7×10^6 by pressing 7 [EE] 10 [EE] 6, which actually means $7 \times 10 \times 10^6$, or 7×10^7. The correct way to enter 7×10^6 is 7 [EE] 6.

EXAMPLE 4 U.S. versus World Energy Consumption

Compare the U.S. population to the world population and U.S. energy consumption to world energy consumption. What does this tell you about energy usage by Americans?

Solution The world population is of order 6 billion (6×10^9) and the U.S. population is of order 300 million (3×10^8). Comparing the two quantities, we find

$$\frac{\text{U.S. population}}{\text{world population}} \approx \frac{3 \times 10^8}{6 \times 10^9} = \frac{1}{2} \times 10^{-1} = \frac{1}{20}$$

The U.S. population is roughly $\frac{1}{20}$ of the world's population but, as shown in Table 3.1, the United States uses roughly $\frac{1}{5}$ of the world's energy. This tells us that Americans use much more energy per person (about four times as much) than the world average.

EXAMPLE 5 Fusion Power

No one has yet succeeded in creating a commercially viable way to produce energy through nuclear fusion. However, suppose that at some time in the future we are able to build fusion power plants that are safe and cost-efficient. If we could extract all the hydrogen from water and use it for fusion, how much water would we need each minute to meet U.S. energy needs?

Solution First, we find the amount of water needed each year by dividing the U.S. annual energy consumption by the amount of energy available through fusion of 1 liter of water (from Table 3.1):

$$\frac{10^{20} \; \cancel{\text{joule}}}{7 \times 10^{13} \frac{\cancel{\text{joule}}}{\text{L}}} = 1.4 \times 10^6 \; \text{L}$$

That is, we would need about 1.4 million liters of water *per year*. Now we convert 1.4 million liters per year to units of liters per minute; note how the units work out.

$$\frac{1.4 \times 10^6 \; \text{L}}{\cancel{\text{yr}}} \times \frac{1 \; \cancel{\text{yr}}}{365 \; \cancel{\text{days}}} \times \frac{1 \; \cancel{\text{day}}}{24 \; \cancel{\text{hr}}} \times \frac{1 \; \cancel{\text{hr}}}{60 \; \text{min}} \approx \frac{2.7 \; \text{L}}{\text{min}}$$

In other words, it would take less than 3 liters of water per minute—which is less than 1 gallon per minute—to meet all U.S. energy needs through nuclear fusion. This is somewhat less than the rate at which water flows from a typical kitchen faucet. So if we could simply attach a nuclear fusion reactor to your kitchen faucet, we could stop producing and importing oil, remove all the hydroelectric dams, shut down all the coal-burning power plants, and still have energy to spare.

Time out to think

Scientists have been working to achieve viable nuclear fusion power plants for decades, but so far without success. Nevertheless, it's conceivable that nuclear fusion will eventually become a realistic source of energy. Based on what you've learned in Example 5, how do you think this would change our world?

by the way . . .
Nuclear fission means splitting large atomic nuclei into smaller ones, while nuclear fusion means combining small nuclei into larger ones. All current nuclear reactors involve fission of uranium or plutonium. In contrast, nuclear bombs generate most of their destructive force through fusion of hydrogen into helium. The sun also generates energy by fusion.

by the way . . .
Scientists working on fusion power actually use a rare form of hydrogen called *deuterium*. (Ordinary hydrogen nuclei contain just a single proton; deuterium contains a proton and a neutron.) About 1 in 50,000 hydrogen atoms is deuterium, so the flow rate needed to power the U.S. would be 50,000 times higher than that calculated in this example—but still only about the flow rate of a small creek.

Perspective Through Scaling

A third general technique for giving meaning to numbers uses *scaling* or scale models. You are probably familiar with three common ways of expressing scales:

- *Verbally:* A scale can be described in words such as "One centimeter represents one kilometer" or, more simply, as "1 cm = 1 km."
- *Graphically:* A marked miniruler can show the scale visually (Figure 3.2).
- *As a ratio:* We can state the ratio of a scaled size to an actual size. For example, there are 100,000 centimeters in a kilometer (because there are 100 centimeters in 1 meter and 1000 meters in 1 kilometer). Thus, a scale where 1 centimeter represents 1 kilometer can be described as a scale ratio of 1 to 100,000 (or 1/100,000).

Scales have many uses other than maps. Architects and engineers build scale models to visualize their plans. *Timelines* represent the scale of time; a given distance along a timeline represents a certain number of years. *Time-lapse photography* and computer simulations allow us to represent large blocks of time in short periods. For example, in a 4-second time-lapse video clip that shows 24 hours of weather, each second of the video represents 6 hours of real time. Similarly, a computer simulation of continental drift might represent 1 billion years of geographical change in just 1 minute.

FIGURE 3.2 The miniruler acts as the map scale.

EXAMPLE 6 Scale Ratio

A city map states, "One inch represents one mile." What is the scale ratio for this map?

Solution We find the ratio by converting 1 mile into inches:

$$1 \text{ mi} \times 5280 \frac{\text{ft}}{\text{mi}} \times 12 \frac{\text{in.}}{\text{ft}} = 63{,}360 \text{ in.}$$

One inch on the map represents 1 mile, or 63,360 inches. Therefore, the scale ratio for this map is 1 to 63,360, meaning that real distances are 63,360 times larger than distances on the map. Note that, as should always be the case, the scale ratio has no units.

EXAMPLE 7 Earth and Sun

The distance from the Earth to the Sun is about 150 million kilometers. The diameter of the Sun is about 1.4 million kilometers and the diameter of the Earth is about 12,760 kilometers. Put these numbers in perspective by using a scale model of the solar system with a 1 to 10 billion scale.

Solution The scale ratio tells us that real sizes and distances are 10 billion (10^{10}) times larger than the sizes and distances in the model. Therefore, we find scaled sizes and distance by dividing the actual values by 10^{10}. For the Earth–Sun distance, we find

$$\text{scaled Earth–Sun distance} = \frac{\text{actual distance}}{10^{10}} = \frac{1.5 \times 10^8 \text{ km}}{10^{10}}$$

$$= 1.5 \times 10^{-2} \text{ km} \times 10^3 \frac{\text{m}}{\text{km}} = 15 \text{ m}$$

Note that, in the last step, we converted the distance from kilometers to meters because it is easier to understand "15 meters" than "0.015 kilometer." We find the scaled Sun and Earth diameters similarly, this time converting the units to centimeters and millimeters, respectively:

$$\text{scaled Sun diameter} = \frac{\text{actual Sun diameter}}{10^{10}} = \frac{1.4 \times 10^6 \text{ km}}{10^{10}}$$

$$= 1.4 \times 10^{-4} \text{ km} \times 10^5 \frac{\text{cm}}{\text{km}} = 14 \text{ cm}$$

$$\text{scaled Earth diameter} = \frac{\text{actual Earth diameter}}{10^{10}} = \frac{1.276 \times 10^4 \text{ km}}{10^{10}}$$

$$= 1.276 \times 10^{-6} \text{ km} \times 10^6 \frac{\text{mm}}{\text{km}} = 1.276 \text{ mm}$$

The model Sun, at 14 centimeters in diameter, is roughly the size of a grapefruit. The model Earth, at about 1.3 millimeters in diameter, is about the size of the ball point in a pen, and the distance between them is 15 meters. At this scale, it is easy to visualize relative sizes and distances.

Time out to think

Find a grapefruit or similar-sized ball and the ball point from a pen. Set them 15 meters apart to represent the Sun and Earth a 1 to 10 billion scale. How does a scale model like this one make it easier to understand our solar system? Discuss.

by the way . . .

The Earth's distance from the Sun varies between a minimum of 147.1 million kilometers each January and 152.1 million kilometers each July. The Earth's *average* distance from the Sun, about 149.6 million kilometers, is called an *astronomical unit,* or AU.

EXAMPLE 8 — Distances to the Stars

The distance from the Earth to the nearest stars besides the Sun (the three stars of the *Alpha Centauri* system) is about 4.3 light-years. On the 1 to 10 billion scale of Example 7, how far are these stars from the Earth? Note: A light-year is the distance that light can travel in one year; 1 light-year = 9.5×10^{12} km.

Solution Because a light-year is about 9.5×10^{12} kilometers, 4.3 light-years is equivalent to

$$4.3 \text{ light-years} \times \frac{9.5 \times 10^{12} \text{ km}}{1 \text{ light-year}} = 4.1 \times 10^{13} \text{ km}$$

We divide this actual distance by 10 billion to find the scaled distance to the nearest stars:

$$\text{scaled distance} = \frac{\text{actual distance}}{10^{10}} = \frac{4.1 \times 10^{13} \text{ km}}{10^{10}}$$
$$= 4.1 \times 10^3 \text{ km} = 4100 \text{ km}$$

The distance to even the nearest stars on this scale is more than 4000 kilometers, or approximately the distance across the United States.

Time out to think

Suppose that an Earth-like planet is orbiting a nearby star. Based on the results of Examples 7 and 8, discuss the challenge of trying to detect such a planet.

EXAMPLE 9 — Timeline

Human civilization, at least since the time of ancient Egypt, is on the order of 5000 years old. The age of the Earth is on the order of 5 billion years. Suppose we use the length of a football field, or about 100 meters, as a timeline to represent the age of the Earth. If we put the birth of the Earth at the start of the timeline, how far from the line's end would human civilization begin?

Solution First, we compare the 5000-year history of human civilization to the 5 billion-year age of the Earth:

$$\frac{5000 \text{ yr}}{5 \text{ billion yr}} = \frac{5 \times 10^3 \text{ yr}}{5 \times 10^9 \text{ yr}} = 10^{-6}$$

That is, 5000 years is about 10^{-6}, or one *millionth*, of the age of the Earth. One millionth of a 100-meter (10^2 m) timeline is

$$10^{-6} \times 10^2 \text{ m} = 10^{-6+2} \text{ m} = 10^{-4} \text{ m} \times \frac{10^3 \text{ mm}}{1 \text{ m}} = 0.1 \text{ mm}$$

In the last step, we converted from meters to millimeters. On a timeline where the Earth's history stretches the length of a football field, human civilization shows up only in the final tenth of a millimeter.

by the way...

Light travels so fast that it could circle the Earth nearly 8 times in just one second, so you can imagine that light can go a long way in a year. In fact, light can go a light-year in a year, which is where this unit gets its name. Be sure to note that a light-year is a unit of *distance*, not of time. The distance is easy to remember as being about 10 trillion kilometers (6 trillion mi).

by the way...

According to modern science, the Earth and solar system formed from the collapse of a large cloud of interstellar gas. Meteorites represent some of the first rocks that formed as the cloud collapsed. Radioactive dating shows the oldest meteorites to be 4.6 billion years old, so scientists conclude that the Earth formed 4.6 billion years ago.

PUTTING IT ALL TOGETHER: CASE STUDIES

We've studied several techniques for putting numbers in perspective, but in many cases we gain more perspective by using two or more techniques together. Sometimes we need additional creativity to think of ways to make sense of a particular number. The following case studies illustrate a few more of the many ways of putting numbers in perspective. As you study them, ask whether the numbers now have more meaning and how else you might give meaning to the numbers.

CASE STUDY *How Big Is a University?*

Consider a large university with 25,000 students. The number 25,000 is small compared to many of the numbers we've dealt with in this chapter, but it still takes thought to put it in perspective. Here's one way: Imagine that the new university president wants to get to know the students. She proposes to meet for lunch with groups of 5 students at a time. Is it possible for her to have lunch with *all* the students?

To answer, let's assume she holds the lunch meetings five days per week. With five students at each lunch, she will meet $5 \times 5 = 25$ students per week. Since there are 25,000 students at the school, it would take her 1000 weeks to have lunch with everyone. If we now assume that she has the lunches for 50 weeks each year (out of the total of 52 weeks in a year), it would take her 1000 weeks ÷ 50 weeks/yr = 20 years to meet all the students. But after only 4 years, most of the 25,000 students would have graduated and been replaced by a new group of 25,000 students, so she could never keep up. Thus, it would *not* be possible to meet all the students at these small group lunches. The lesson from this example is that while 25,000 people might not sound like much, it is far more than you can reasonably get to know in 4 years—or even a lifetime.

Time out to think

Today, a typical Congressional Representative represents about 650,000 constituents. Is it possible for a Representative to campaign by going door to door to everyone in his or her district? Explain.

CASE STUDY *What Is a Billion Dollars?*

One way to put $1 billion in perspective is to ask a question like "How many people can you employ with $1 billion per year?" Let's take $25,000 as a typical starting salary for a college graduate. Further, let's assume that it costs a business an additional $25,000 per year to have a $25,000 employee on the payroll (costs for office space, computer services, health insurance, and other benefits). At $50,000 per employee, $1 billion would allow a business to hire

$$\frac{\$1 \text{ billion}}{\$50{,}000 \text{ per employee}} = \frac{\$10^9}{\$5 \times 10^4/\text{employee}} = 2 \times 10^4 \text{ employees}$$

One billion dollars per year could support a work force of some 20,000 employees.

Another way to put $1 billion in perspective also points out how different numbers can be, even when they sound similar (like million, billion, and trillion). Suppose you become a sports star and earn a salary of $1 million per year. How long

A billion here, a billion there; soon you're talking real money.
—SENATOR EVERETT DIRKSEN

would it take you to earn a billion dollars? We simply divide $1 billion by your salary of $1 million/year:

$$\frac{\$1 \text{ billion}}{\$1 \text{ million/yr}} = \frac{\$10^9}{\$10^6/\text{yr}} = 10^3 \text{ yr} = 1000 \text{ yr}$$

Even at a salary of $1 million per year, earning a billion dollars would take a thousand years.

CASE STUDY *The Scale of the Atom*

Everything is made from *atoms*, which consist of a nucleus (made from *protons* and *neutrons*) surrounded by a "cloud" of *electrons*. A typical atom has a diameter of about 10^{-10} meter (as defined by its electron cloud), while its nucleus is about 10^{-15} meter in diameter. Can we put these numbers in perspective?

Let's begin with the atom itself. Because its diameter is 10^{-10} meter, or one *ten-billionth* of a meter, we could fit 10 billion (10^{10}) atoms in a line along a meter stick. A centimeter is 1/100 (or $1/10^2$) of a meter, so we could fit $10^{10}/10^2 = 10^8$, or 100 million, atoms in a 1-centimeter line. Thus, if we could shrink people down to the size of atoms, the roughly 300 million people in the United States could fit in a line only about 3 centimeters—just over an inch—long.

Next, let's compare the diameter of the atom to the diameter of its nucleus:

$$\frac{\text{diameter of atom}}{\text{diameter of nucleus}} = \frac{10^{-10} \text{ m}}{10^{-15} \text{ m}} = 10^{-10-(-15)} = 10^5$$

> **by the way...**
> The size of an atom is determined by the size of its electron cloud, but most of the mass of an atom is contained in its nucleus. Because the nucleus is so tiny compared to the atom itself, we have the surprising fact that atoms consist mostly of empty space!

The atom itself is about 10^5, or 100,000, times as large as its nucleus. Thus, if we made a scale model of an atom in which the nucleus were the size of a marble (1 cm), the atom would have a diameter of about 100,000 centimeters, which is 1 kilometer, or more than half a mile.

CASE STUDY *Until the Sun Dies*

We can hope that living things will flourish on Earth until the Sun dies, which astronomers estimate will be in about 5 billion (5×10^9) years. How long is 5 billion years? First, let's compare the Sun's remaining lifetime to a long human lifetime of 100 years:

$$\frac{5 \times 10^9 \text{ yr}}{100 \text{ yr}} = 5 \times 10^7$$

The Sun's remaining lifetime is equivalent to about 5×10^7, or 50 million, human lifetimes. We can add perspective by dividing a human lifetime of 100 years by 50 million; we convert the result from years to minutes:

$$\frac{100 \text{ yr}}{5 \times 10^7} = 2 \times 10^{-6} \text{ yr} \times \frac{365 \text{ days}}{1 \text{ yr}} \times \frac{24 \text{ hr}}{1 \text{ day}} \times \frac{60 \text{ min}}{1 \text{ hr}} \approx 1 \text{ min}$$

Thus, a human lifetime in comparison to the Sun's remaining life is roughly the same as *one minute* in comparison to a human lifetime.

How about human creations? The Egyptian pyramids are often described as "eternal." But they are slowly eroding because of wind, rain, air pollution, and the impact of tourists; all traces of them will have vanished within a few million years. If we estimate the lifetime of the pyramids as 5 million (5×10^6) years, then

$$\frac{\text{Sun's remaining lifetime}}{\text{lifetime of pyramids}} \approx \frac{5 \times 10^9 \text{ yr}}{5 \times 10^6 \text{ yr}} = 10^3 = 1000$$

A few million years may seem like a long time, but the Sun's remaining lifetime is a thousand times longer.

On a more somber note, we can gain perspective on the Sun's remaining 5 billion years by considering evolutionary time scales. During the past century, we have acquired sufficient technology and power to destroy human life totally, if we so choose. However, even if we make that unfortunate choice, some species (including many insects) are likely to survive. Would another intelligent species ever emerge on the Earth? There is no way to know, but we can look to the past for guidance. It's been 65 million years since the dinosaurs were wiped out in what many scientists believe was a fairly sudden cataclysm caused by the impact of an asteroid. To keep the numbers simple, imagine that another intelligent species could emerge some 50 million years after a human extinction. Then, if they also destroyed themselves, another species might emerge 50 million years after that, and so on. If this pattern continued, how many intelligent species would eventually emerge on Earth? We simply divide the Sun's remaining 5 billion years by the 50 million years between successive intelligent species:

$$\frac{5 \text{ billion years}}{50 \text{ million years}} = \frac{5 \times 10^9 \text{ years}}{5 \times 10^7 \text{ years}} = 100$$

Even at 50 million years per shot, the Earth would have time for some 100 more intelligent species to emerge. Perhaps one of those species will not destroy itself, and descendants of the Earth might move on to other star systems by the time the Sun finally dies. Perhaps it will be our species.

I met a traveller from an antique land
Who said: Two vast and trunkless legs of stone
Stand in the desert . . . Near them, on the sand,
Half sunk, a shattered visage lies, whose frown,
And wrinkled lip, and sneer of cold command,
Tell that its sculptor well those passions read
Which yet survive, stamped on these lifeless things,
The hand that mocked them, and the heart that fed:
And on the pedestal these words appear:
'My name is Ozymandias, king of kings:
Look on my works, ye Mighty, and despair!'
Nothing beside remains. Round the decay
Of that colossal wreck, boundless and bare,
The lone and level sands stretch far away.

—PERCY BYSSHE SHELLEY, *Ozymandias*

Review Questions

1. Briefly describe scientific notation. How is it useful for writing large and small numbers? How is it useful for making approximations?

2. Explain how we can use estimation to put numbers in perspective. Give an example.

3. What is an *order of magnitude* estimate? Explain why such an estimate can be useful even though it may be as much as 10 times too large or too small.

4. Explain how we can use comparisons to put numbers in perspective. Give an example.

5. Describe three common ways of expressing the scale of a map or model. How would you show a scale of 1 cm = 100 km graphically? How would you describe it as a ratio?

6. Explain how we can use scaling to put numbers in perspective. Give an example.

7. Suppose that the Sun were the size of a grapefruit. How big and how far away would the Earth be on this scale? How far would the nearest stars (besides the Sun) be?

8. Describe several ways of putting each of the following in perspective: the size of a large university; $1 billion; the size of an atom; the Sun's remaining lifetime.

In the News

1. **Large Numbers.** Search today's newspaper for as many instances of numbers larger than 100,000 as you can find. Briefly explain the context within which each large number is used.

2. **Perspective in the News.** Find an example in the recent news in which a reporter uses a technique to put a number in perspective. Describe the example. Do you think the technique is effective? Can you think of a better way to put the number in perspective? Explain.

3. **Putting Numbers in Perspective.** Find at least two examples of very large or small numbers in recent news reports. Use a technique of your choosing to put each number into perspective in a way that you believe most people would find meaningful.

Exercises

Review of Scientific Notation. Exercises 1–6 use skills covered in the Brief Review on p. 137.

1. Convert each of the following numbers from scientific to ordinary notation and write its name.

 Example: $2 \times 10^3 = 2000 =$ two thousand
 a. 5×10^6
 b. 7×10^9
 c. 2×10^{-2}
 d. 8×10^{11}
 e. 1×10^{-7}
 f. 9×10^{-4}

2. Convert each of the following numbers from scientific to ordinary notation and write its name.
 a. 7.0×10^3
 b. 1.5×10^9
 c. 3.0906×10^3
 d. 2.3×10^6
 e. 2.2×10^{-4}
 f. 7.06×10^{-1}

3. Write each of the following numbers in scientific notation.
 a. 600
 b. 0.9
 c. 50,000
 d. 0.003
 e. 0.0005
 f. 70,000,000,000

4. Write each of the following numbers in scientific notation.
 a. 64,598
 b. 0.92
 c. 500.098
 d. 0.002×10^6
 e. 250 million
 f. $-23,800$

5. Do the following operations without a calculator and show your work clearly. Be sure to express answers in scientific notation. You may round your answers to two decimal places (as in 3.2×10^5).
 a. $(3 \times 10^4) \times (8 \times 10^5)$
 b. $(6.3 \times 10^2) + (1.5 \times 10^1)$
 c. $(9 \times 10^3) \times (5 \times 10^{-7})$
 d. $(4.4 \times 10^{99}) \div (2.0 \times 10^{11})$

6. Do the following operations without a calculator and show your work clearly. Be sure to express answers in scientific notation. You may round your answers to two decimal places (as in 3.2×10^5).
 a. $(8 \times 10^{12}) \div (4 \times 10^9)$
 b. $(7.5 \times 10^{21}) \div (1.5 \times 10^{13})$
 c. $(3.2 \times 10^{22}) \div (1.6 \times 10^{-14})$
 d. $(6 \times 10^{10}) - (5 \times 10^9)$

They Don't Look That Different! In Exercises 7–8, compare each pair of numbers. By what factor is one number larger than the other?

Example: 10^6 is 10^2, or 100, times larger than 10^4.

7. a. $10^{35}, 10^{26}$
 b. $10^{17}, 10^{27}$
 c. 1 billion, 1 million
 d. 7 trillion, 7 thousand
 e. $2 \times 10^{-6}, 2 \times 10^{-9}$
 f. $6.1 \times 10^{27}, 6.1 \times 10^{29}$

8. a. 250 million, 5 billion
 b. $9.3 \times 10^2, 3.1 \times 10^{-2}$
 c. $10^{-8}, 2 \times 10^{-13}$
 d. $3.5 \times 10^{-2}, 7 \times 10^{-8}$
 e. 1 thousand, 1 thousandth
 f. $10^{12}, 10^{-9}$

Using Scientific Notation. Rewrite each statement in Exercises 9–14 with a number in scientific notation.

9. Corporate profits in the United States in 1996 were 632 billion dollars.

10. Consumer debt (loans and credit cards) in the United States in 1994 was 800 billion dollars.

11. The hard drive on my computer has a capacity of 5.4 gigabytes. (Hint: Recall that the prefix *giga* means 1 billion.)

12. In 1993, cardiovascular diseases (heart diseases and strokes) caused 739,860 deaths in the United States.

13. The area of the Earth's surface is 509,600,000 square kilometers.

14. In 1990, Americans used 2,626,165 gigawatt-hours of electricity; express this quantity in watt-hours. (Hint: Recall that the prefix *giga* means 1 billion.)

Approximation with Scientific Notation. In Exercises 15–16, make an estimate of the answer without a calculator, showing your method of estimation. Then do the exact calculation (with a calculator if necessary) and describe how well your approximation technique worked.

15. a. $20,000 \times 100$
 b. $9642 \div 31$
 c. $-12.5 \times 11,890$
 d. 250 million $\times$ 40
 e. 7.453×291
 f. $6,570,999 \div 32.7$

16. a. 5.6 billion $\div$ 200
 b. 4 trillion $\div$ 260 million
 c. $9000 \times 54,986$
 d. 3 billion $\div$ 25,000
 e. 5987×341
 f. $43 \div 765$

Perspective Through Estimation. Make estimates as needed to answer Exercises 17–20.

17. Which is bigger, the height of a 10-story apartment building or the length of a football field? By how much?
18. Could a person walk across the United States (New York to California) in a year? If not, about how long would it take?
19. Which is more, the number of miles Americans fly each year or the number of miles Americans drive each year?
20. Which holds more people, a football stadium or 10 movie theaters?

Orders of Magnitude Estimates. For each quantity described in Exercises 21–27, make an order of magnitude estimate. Be sure to explain your estimates clearly.

21. The total amount of money spent each year by Americans in going to the movies
22. The total annual spending on magazines by students at your school
23. The number of breaths you take in a week
24. The amount of water you drink in a year
25. The total number of words in this textbook
26. The amount of gasoline an average adult uses per year
27. The total amount you spend on food in a year
28. **CO_2 Emissions.** For every gallon of gasoline burned by an automobile, approximately 10.2 kilograms of carbon dioxide are emitted into the atmosphere. Estimate the total amount of carbon dioxide added to the atmosphere by all automobile travel in the United States over the past year.

Making Numbers Understandable. Restate the facts in Exercises 29–36 as indicated.

29. There are approximately 4 million births in the United States in a year. Express this quantity in births per minute.
30. There are approximately 2.2 million marriages in the United States each year. Express this quantity in marriages per day.
31. There are approximately 132 million births worldwide in a year. Express this quantity in births per minute.
32. There are roughly 30,000 suicides in the United States each year. Express this quantity in suicides per day.
33. American businesses spent about $220 billion on all forms of advertising in 2000. Express this quantity as the height of a stack of $1 bills. (Hint: Measure the thickness of a stack of ten $1 bills.)
34. The national debt is about $5.8 trillion. Express this quantity as the height of a stack of $1 bills.
35. Approximately 720,000 Americans die of a heart attack each year. Express this quantity in heart attacks per day.
36. In 2000, approximately 1.7 million college students received bachelor's degrees. Express this number of students in terms of the number of football stadiums (capacity 70,000) that could be filled by these graduates.

Energy Comparisons. Use Table 3.1 to answer Exercises 37–44.

37. How many average candy bars would you have to eat to supply the energy needed for four hours of running?
38. How many liters of oil are required to supply the electrical energy needs of an average home for a month?
39. Compare the energy released by burning 1 kilogram of coal to that released by fission of 1 kilogram of uranium-235.
40. Compare the energy released by burning 1 liter of oil to that released by fusion of the hydrogen in 1 liter of water.
41. If you could generate energy by fusing the hydrogen in ordinary tap water, how much water would you need to generate the electrical energy used daily by a typical home?
42. If you could generate energy by fusing the hydrogen in ordinary tap water, how much water would you need to meet all the world's current energy consumption?
43. How many kilograms of uranium would be required to supply the energy needs of the United States for one year using fission?
44. Suppose that we could somehow capture all the energy released by the Sun for just one second. Would this energy be enough to supply U.S. energy needs for a year? Explain.
45. **Scale Ratio.** What is the scale ratio for each of the following?
 a. 1 centimeter on the map represents 1 kilometer on the ground.
 b. 2 inches on the map represents 0.5 mile on the ground.
 c. 5 cm (map) = 100 km (actual)
 d. 1 ft (map) = 100 m (actual)

46. **Scale Ratio.** What is the scale ratio for each of the following?
 a. 1 inch on the map represents 1 mile on the ground.
 b. 3 inches on the map represents 2 miles on the ground.
 c. 10 cm (map) = 10 km (actual)
 d. 1 ft (map) = 1 km (actual)

47. **Scale Model Solar System.** The following table gives size and distance data for all the planets. Calculate the scaled size and distance for each planet using a 1 to 10 billion scale model solar system. Give your results in table form. Then write one or two paragraphs that describe your findings in words and give perspective to the size of our solar system.

Planet	Diameter	Average distance from Sun
Mercury	4880 km	57.9 million km
Venus	12,100 km	108.2 million km
Earth	12,760 km	149.6 million km
Mars	6790 km	227.9 million km
Jupiter	143,000 km	778.3 million km
Saturn	120,000 km	1427 million km
Uranus	52,000 km	2870 million km
Neptune	48,400 km	4497 million km
Pluto	2260 km	5900 million km

48. **Universal Timeline.** Suppose the universe is 15 billion years old, and you represent its age with a timeline 100 meters long. What distance along the timeline represents 1 billion years? Written human history extends back about 10,000 years. How far would that be on the timeline?

49. **The Amazing Amazon.** The February 1995 issue of *National Geographic* contains the following statement:

 Dropping less than two inches per mile after emerging from the Andes, the Amazon drains a sixth of the world's runoff into the ocean. One day's discharge at its mouth—4.5 trillion gallons—could supply all U.S. households for five months.

 Based on this statement, determine how much water an average U.S. household uses each month. Does this answer seem reasonable? Explain any estimates you make.

50. **Wood for Energy?** A total of about 180,000 terawatts of solar power reaches the Earth's surface, of which about 0.06% is used by plants for photosynthesis. Of the energy that goes to photosynthesis, about 1% ends up stored in plant matter (including wood). (Recall that 1 watt = 1 joule/s; 1 terawatt = 10^{12} watts.)

 a. Calculate the total amount of energy that becomes stored in plant matter each second.
 b. Suppose that power stations generated electricity by burning plant matter. If all the energy stored in plants could be converted to electricity, what average power, in terawatts, would be possible? Would it be enough to meet world electricity demand, which is of order 10 terawatts?
 c. Based on your answer to b, can you draw any conclusions about why humans depend on fossil fuels, such as oil and coal, which are the remains of plants that died long ago? Explain.

51. **Stellar Corpses: White Dwarfs and Neutron Stars.** A few billion years from now, after exhausting its nuclear engines, the Sun will become a type of remnant star called a *white dwarf*. It will still have nearly the same mass (about 2×10^{30} kg) as the Sun today, but its radius will be only about that of the Earth (about 6400 km).

 a. Calculate the average density of the white dwarf in units of kilograms per cubic centimeter.
 b. What is the mass of a teaspoon of material from the white dwarf? (Hint: A teaspoon is about 4 cubic centimeters.) Compare this mass to the mass of something familiar (for example, a person, a car, a tank).
 c. A neutron star is a type of stellar remnant compressed to even greater densities than a white dwarf. Suppose that a neutron star has a mass 1.4 times the mass of the Sun but a radius of only 10 kilometers. What is its density? Compare the mass of 1 cubic centimeter of neutron star material to the total mass of Mt. Everest (about 5×10^{10} kg).

Sampling Problems. Sampling techniques can be used to estimate physical quantities. To estimate a large quantity, you might measure a representative small sample and find the total quantity by "scaling up." To estimate a small quantity, you might measure several of the small quantities together and "scale down." Carry out the necessary measurements and use sampling techniques to answer Exercises 52–55.

Example: How thick is a sheet of a paper?

Solution: One way to estimate the thickness of a sheet of paper is to measure the thickness of a ream (500 sheets) of paper. A particular ream was 7.5 centimeters thick. Thus, a sheet of paper from this ream was 7.5 cm ÷ 500 = 0.015 cm thick, or 0.15 millimeter.

52. How much does a sheet of paper weigh?

53. How thick is a penny? a nickel? a dime? a quarter? Would you rather have your height stacked in pennies, nickels, dimes, or quarters? Explain.

54. How much does a grain of sand weigh? How many grains of sand are in a typical playground sand box?

55. How many stars are visible in the sky on the clearest, darkest nights? How could astronomers estimate the total number of stars in the universe?

Web Projects

Find useful links for Web Projects on the text Web site: www.aw.com/bennett-briggs

1. **Energy Comparisons.** Using data available from the Energy Information Administration Web site, choose a few measures of U.S. or world energy consumption or production. Make comparisons that put these numbers in perspective.

2. **Nuclear Fusion.** Learn about the current state of research into building commercially viable fusion power plants. What obstacles must still be overcome? Do you think fusion power will be a reality in your lifetime? Explain.

3. **Scale Model Solar System.** Visit the Web site for the Voyage scale model solar system on the National Mall in Washington, DC. Take the virtual tour, and write a brief report on what you learn.

4. **Richest People.** Find the net worth of the world's three richest people. Put these monetary values in perspective through any techniques you wish.

UNIT 3C

DEALING WITH UNCERTAINTY

After decades of running huge deficits, the United States federal government finally recorded a budget surplus in 1998. Politicians and economists cheered, as the government collected $69 billion more than it spent. Yet news reports just a year earlier had predicted a deficit, not a surplus, of $121 billion. In other words, the original projection for the 1998 budget turned out to be off by $190 billion ($121 billion + $69 billion = $190 billion).

We can draw two important lessons from this budget example. First, government projections of future budgets are notoriously inaccurate, even when made just a year ahead of time. Second, although economists were aware of the huge uncertainties in the earlier budget projection, most news reports didn't mention the uncertainties at all—they simply said the government would run a $121 billion deficit. Thus, we see that numbers in the news are often stated as though they hold truth when, in fact, they are quite uncertain.

In this unit, we will discuss ways of dealing with uncertainty. We'll discuss the types of errors that can occur in measurements, estimates, and projections. We'll also discuss how to account for possible errors when we state results. Once you understand errors, you'll be able to assess the reliability of numbers that you hear in the news.

None of us really understands what's going on with all these numbers.

—DAVID STOCKMAN, budget director for President Reagan, 1981

SIGNIFICANT DIGITS

The first consideration in dealing with uncertainty involves how we write measured numbers. For example, suppose you measure your weight to be 132 pounds on a scale that can be read only to the nearest pound. Saying that you weigh 132.00 pounds would be misleading, because it would incorrectly imply that you know your weight to the nearest *hundredth* of a pound, rather than to the nearest pound. In other words, 132 pounds and 132.00 pounds do *not* have the same meaning.

The digits in a number that represent actual measurements and therefore have meaning are called **significant digits.** For example, 132 pounds has 3 significant digits and implies a measurement to the nearest pound, while 132.00 pounds has 5 significant digits and implies a measurement to the nearest hundredth of a pound.

Note that zeros are significant when they represent actual measurements, but not when they serve only to locate the decimal point. We assume that the zeros in 132.00 pounds are significant because there was no reason to include them unless they represented an actual measurement. In contrast, we assume that the zeros in 600 centimeters are *not* significant, because they serve only to tell us that the decimal point comes to their right; rewriting 600 centimeters as 6 meters makes it easier to see that only the 6 is a significant digit.

The only subtlety in counting significant digits arises when we cannot be sure whether zeros are truly significant. For example, suppose your professor states that there are 200 students in your class. Without further information, you have no way to know whether she means exactly 200 students or roughly 200. We can avoid this kind of ambiguity by writing numbers in scientific notation. In that case, zeros appear only when they are significant. For example, an enrollment of 2×10^2 implies a measurement to the nearest hundred students, while 2.00×10^2 implies exactly 200 students.

SUMMARY When Are Digits Significant?

Nonzero digits	always significant
Zeros that follow a nonzero digit *and* lie to the right of the decimal point (as in 4.20 or 3.00)	always significant
Zeros *between* nonzero digits (as in 4002 or 3.06) or other significant zeros (such as the first zero in 30.0)	always significant
Zeros to the *left* of the first nonzero digit (as in 0.006 or 0.00052)	never significant
Zeros to the *right* of the last nonzero digit but before the decimal point (as in 40,000 or 210)	not significant unless stated otherwise

EXAMPLE 1 *Counting Significant Digits*

State the number of significant digits and the implied meaning of the following numbers.

 a. a time of 11.90 seconds
 b. a length of 0.000067 meter
 c. a population reported as 240,000
 d. a population reported as 2.40×10^5

Solution

 a. The number 11.90 seconds has 4 significant digits and implies a measurement to the nearest 0.01 second.

a brief review...

ROUNDING

The basic process of rounding numbers takes just two steps.

Step 1. Decide which decimal place (e.g., tens, ones, tenths, or hundredths) is the smallest that should be kept.

Step 2. Look at the number in the next place to the *right* (for example, if rounding to tenths, look at hundredths). If the value in the next place is *less than 5*, round *down*; if it is *5 or greater*, round *up*.

For example, the number 382.2593 is given to the nearest ten thousandth. It can be rounded in the following ways:

382.2593 rounded to the nearest thousandth is 382.259.
382.2593 rounded to the nearest hundredth is 382.26.
382.2593 rounded to the nearest tenth is 382.3.
382.2593 rounded to the nearest one is 382.
382.2593 rounded to the nearest ten is 380.
382.2593 rounded to the nearest hundred is 400.

(Some statisticians use a more complex rounding rule if the value in the next column is exactly 5: They round up if the last digit being kept is odd and down if it is even. We won't use this rule in this book.)

b. The number 0.000067 meter has 2 significant digits and implies a measurement to the nearest 0.000001 meter; note that we can rewrite this number as 67 micrometers, showing clearly that it has only 2 significant digits.

c. We assume that the zeros are not significant in 240,000 people. Thus, the number has 2 significant digits and implies a measurement to the nearest 10,000 people.

d. The number 2.40×10^5 has 3 significant digits. Although this number means 240,000, we know that in this case the leftmost zero is significant, so it implies a measurement to the nearest 1000 people. ∎

EXAMPLE 2 *Rounding with Significant Digits*

For each of the following operations, give your answer with the specified number of significant digits.

 a. 7.7 mm × 9.92 mm; give your answer with 2 significant digits
 b. 240,000 × 72,106; give your answer with 4 significant digits

Solution

 a. 7.7 mm × 9.92 mm = 76.384 mm². Because we are asked to give the answer with 2 significant digits, we round to 76 mm².
 b. 240,000 × 72,106 = 1.730544×10^{10}. Because we are asked to give the answer with 4 significant digits, we round to 1.731×10^{10}. ∎

UNDERSTANDING ERRORS

We are now ready to deal with errors themselves. We'll begin by classifying errors into two types. Then we'll discuss how to quantify the sizes of errors and how to report final results to account for errors.

Types of Error: Random and Systematic

Broadly speaking, measurement errors fall into two categories: random errors and systematic errors. The difference between them can best be illustrated with an example.

Suppose you work in a pediatric office and use a digital scale to weigh babies. If you've ever worked with babies, you know that they usually aren't very happy about being put on a scale. Their thrashing and crying tends to shake the scale, making the readout jump around. For the case shown in Figure 3.3a, you could equally well record the baby's weight as anything between about 14.5 and 15.0 pounds. We say that the shaking of the scale introduces a **random error** because any particular measurement may be either too high or too low.

Now suppose you have weighed babies all day with the scale in Figure 3.3b. At the end of the day, you notice that the scale reads 1.2 pounds when there is nothing on it. In that case, every measurement you made was too high by 1.2 pounds. This type of error is called a **systematic error** because it is caused by an error in the measurement *system*—that is, an error that consistently (systematically) affects all measurements.

A systematic error affects all measurements in the same way, such as making them all too high or all too low. Thus, if you discover a systematic error, you can go back and adjust the affected measurements. In contrast, the unpredictable nature of random errors makes it impossible to correct for them. However, you can minimize the effects of random errors by making many measurements and averaging them. For example, if you measure the baby's weight ten times, your measurements will probably be too high in some cases and too low in others. Thus, when you average the ten measurements, you are likely to get a value that better represents the true weight.

FIGURE 3.3 (a) The baby's motion introduces random errors. (b) The scale reads 1.2 pounds when empty, introducing a systematic error that makes all measurements 1.2 pounds too high.

Two Types of Measurement Error

Random errors occur because of random and inherently unpredictable events in the measurement process.

Systematic errors occur when there is a problem in the measurement system that affects all measurements in the same way, such as making them all too low or too high by a particular amount.

Time out to think

Call the local phone number that gives the current time. How far off is your clock or watch? Describe possible sources of random and systematic errors in your time-keeping.

EXAMPLE 3 *Errors in Global Warming Data*

Scientists studying global warming need to know how the average temperature of the entire Earth, or the *global average temperature,* has changed with time. Consider two difficulties (among many others) in trying to interpret historical temperature data

from the early 20th century: (1) temperatures were measured with simple thermometers and the data were recorded by hand; and (2) most temperature measurements were recorded in or near urban areas, which tend to be warmer than surrounding rural areas because of heat released by human activity. Discuss whether each of these two difficulties produces random or systematic errors, and consider the implications of these errors.

Solution The first difficulty involves *random errors* because people undoubtedly made occasional errors in reading the thermometers, in calibrating the thermometers, and in recording temperature readings. There is no way to predict whether any individual reading is correct, too high, or too low. However, if there are several readings for the same region on the same day, averaging these readings can minimize the effects of the random errors.

The second difficulty involves a *systematic error* because the excess heat in urban areas always causes the temperature reading to be higher than it would be otherwise. If the researchers could determine how much this systematic error affected the temperature readings, they could correct the data for this problem.

> *by the way ...*
> The fact that urban areas tend to be warmer than they would be in the absence of human activity is often called the *urban heat island effect*. Major causes of this effect include heat released by burning fuel in automobiles, homes, and industry and the fact that pavement and large masonry buildings tend to retain heat from sunlight.

EXAMPLE 4 *The Census*

The Constitution of the United States mandates a census of the population every 10 years. The United States Census Bureau conducts the census by distributing household surveys through the mail or through personal visits. Suggest several sources of both random and systematic error in the census.

Solution The surveys are useful if they are properly filled out and recorded. However, some people make errors when they complete the surveys, and Census Bureau employees may make errors when they enter the data into their computers. These kinds of errors are random errors because too few or too many people may end up being counted for any individual household.

Systematic errors arise from problems in the surveying process. For example, it is very difficult to deliver surveys to the homeless, and it is difficult to count undocumented aliens, who may be reluctant to reveal their presence in the United States. Thus, the homeless and undocumented aliens tend to be systematically undercounted. Other systematic errors lead to overcounts. For example, college students often get counted both by their parents and in the households where they reside at school, and children of divorced parents are sometimes counted in both parents' households.

Although there is no way to determine the overall effects of random errors, statistical studies can estimate the effects of systematic errors. For example, the 2000 census originally counted about 281.4 million people. However, follow-up statistical studies suggested that the surveys had missed about 7.6 million people and counted 4.3 million people twice. Overall, then, the census undercounted the population by over 3 million people.

Time out to think

The question of whether the Census Bureau should be allowed to adjust its "official" count on the basis of statistical surveys is very controversial. The Constitution calls for an "actual enumeration" of the population (Article 1, Section 2, Subsection 2). Do you believe that this

wording precludes or allows the use of statistical surveys in the official count? Defend your opinion. Also discuss reasons why Democrats tend to favor the use of sampling methods while Republicans tend to oppose it.

Size of Errors: Absolute versus Relative

Besides wanting to know whether an error is random or systematic, we often want to know something about the overall size of the error. For example, is the error big enough to be of concern or small enough to be unimportant? Let's explore the concept of error size.

Suppose you go to the store and buy what you think is 6 pounds of hamburger, but because the store's scale is poorly calibrated, you actually get only 4 pounds. You'd probably be upset by this 2-pound error. Now suppose you buy a car which the owner's manual says weighs 3132 pounds, but you find that it really weighs 3130 pounds. The error is the same 2 pounds as before, but this error probably doesn't seem very important for the car. This simple example shows that the importance of an error can depend on its context.

Speaking more technically, we say that the 2-pound error in the measurement of the hamburger and the car is the absolute error. The **absolute error** describes how far the measured value lies from the true value. The **relative error** compares the size of the absolute error to the true value. The relative error for the hamburger is fairly large because the absolute error of 2 pounds is half the true weight of 4 pounds. In this case, the relative error is 2/4, or 50%. In contrast, the relative error for the car is quite small, because 2 pounds is very small in comparison to the car's true weight of 3130 pounds. Here, the relative error is 2/3130, which is about 0.0006, or 0.06%. Note the similarity between the ideas of absolute and relative error and those of absolute and relative change or difference (see Unit 3A).

Absolute and Relative Error

The **absolute error** describes how far a measured value lies from the true value; it is the difference between the measured value and the true value:

$$\text{absolute error} = \text{measured value} - \text{true value}$$

The **relative error** compares the size of the error to the true value:

$$\text{relative error} = \frac{\text{absolute error}}{\text{true value}} = \frac{\text{measured value} - \text{true value}}{\text{true value}}$$

The absolute and relative errors are *positive* when the measured value is greater than the true value and *negative* when the measured value is less than the true value. Note that the above formula gives the relative error as a fraction, which can be converted to a percentage.

EXAMPLE 5 *Absolute and Relative Error*

Find the absolute and relative error in each case.

a. Your true weight is 125 pounds, but a scale says you weigh 130 pounds.

b. The government claims that a program costs $49.0 billion, but an audit shows that the true cost is $50.0 billion.

Solution

a. The measured value is the scale reading of 130 pounds, and the true value is your true weight of 125 pounds. The absolute and relative errors are

$$\text{absolute error} = \text{measured value} - \text{true value}$$
$$= 130 \text{ lb} - 125 \text{ lb}$$
$$= 5 \text{ lb}$$
$$\text{relative error} = \frac{\text{absolute error}}{\text{true value}} = \frac{5 \text{ lb}}{125 \text{ lb}} = 0.04 = 4\%$$

The measured weight is too high by 5 pounds, or 4%.

b. The measured value is the claimed cost of $49 billion, and the true value is the true cost of $50 billion. The absolute and relative errors are

$$\text{absolute error} = \$49.0 \text{ billion} - \$50.0 \text{ billion}$$
$$= -\$1.0 \text{ billion}$$
$$\text{relative error} = \frac{\$49 \text{ billion} - \$50 \text{ billion}}{\$50 \text{ billion}}$$
$$= -0.02 = -2\%$$

The claimed cost is too low by $1 billion, or 2%.

And now for some temperatures around the nation: 58, 72, 85, 49, 77.

—GEORGE CARLIN, *comedian*

Describing Results: Accuracy and Precision

Once a measurement is reported, we should evaluate it to see whether it is believable in light of any potential errors. In particular, we should consider two key ideas about any reported value: its *accuracy* and its *precision*. The terms are often used interchangeably in English, but mathematically they are different.

The goal of any measurement is to obtain a value that is as close as possible to the *true value*. The closer the measured value lies to the true value, the greater is the **accuracy** of the measurement. If a census says that the population of your home town is 72,453 but the true population is 96,000, then the census report is not very accurate. In contrast, if a company projects sales of $7.30 billion and true sales turn out to be $7.32 billion, we would say that projection is quite accurate.

The **precision** of a number describes the amount of detail in a measurement; that is, it describes how *precisely* a number is measured. For example, a distance given as 2.345 kilometer is more precise than a distance given as 2.3 kilometer, because the first number gives detail to the nearest 0.001 kilometer, whereas the second number gives detail only to the nearest 0.1 kilometer. Similarly, an income of $45,678.90 has greater precision than one of $46,000 because the first income is precise to the nearest penny whereas the second income is precise only to the nearest thousand dollars.

by the way . . .

The Mars Climate Orbiter was supposed to go into orbit of Mars in late 1999. NASA engineers controlled the spacecraft from Earth by sending precise computer instructions telling it when and for how long to fire its rockets. However, the instructions were sent in English units (pounds) and the spacecraft software interpreted them in metric units (kilograms). Result: The very precise instructions were actually quite inaccurate, sending the $160 million spacecraft so close to Mars that it burned up in the Martian atmosphere.

> **DEFINITIONS**
>
> **Accuracy** describes how closely a measurement approximates a true value. An accurate measurement is very close to the true value.
>
> **Precision** describes the amount of detail in a measurement.

DILBERT reprinted by permission of United Feature Syndicate, Inc.

EXAMPLE 6 Accuracy and Precision in Your Weight

Suppose you actually weigh 102.4 pounds. The scale at the doctor's office, which can be read only to the nearest quarter pound, says that you weigh $102\frac{1}{4}$ pounds. The scale at the gym, which gives a digital readout to the nearest 0.1 pound, says that you weigh 100.7 pounds. Which scale is more *precise*? Which is more *accurate*?

Solution The scale at the gym is more *precise*, because it gives your weight to the nearest tenth of a pound whereas the doctor's scale gives your weight only to the nearest quarter pound. However, the scale at the doctor's office is more *accurate* because its value is closer to your true weight.

Time out to think

In Example 6, we needed to know your *true* weight in order to determine which scale was more accurate. But how would we know your true weight? Can we ever be *sure* that we know your true weight? Explain.

CASE STUDY Does the Census Measure the True Population?

Upon completing the 2000 census, the U.S. Census Bureau reported a population of 281,421,906 (as of April 1, 2000), thereby implying an exact count of everyone living in the United States. Unfortunately, such a precise count could not possibly be accurate.

Even in principle, the only way to get an exact count of the number of people living in the United States would be to count everyone *instantaneously*. Otherwise, the count would be off because the number of people changes very rapidly. An average of about eight births and four deaths occur every minute in the United States, and a new immigrant enters the United States about every three minutes on average. Moreover, random and systematic errors can easily make the census inaccurate by a few million people (see Example 4).

Thus, in reality, the Census Bureau did *not* know the exact population and only found an estimate of the population to within a few million people. Given these facts, it was fundamentally dishonest to report the population as an exact count. In this

case, a more honest report would have used much less precision—for example, stating the population as "about 280 million." (In fairness to the Census Bureau, their detailed reports do explain the uncertainties in the population count, but these uncertainties are rarely mentioned by the press.)

Summary: Dealing with Errors

The ideas we've covered in this section are a bit technical, but very important to understanding measurements and errors. Let's briefly summarize how the ideas relate to one another.

- Errors can occur in many ways, but generally can be classified as one of two basic types: random errors or systematic errors.
- Whatever the source of an error, its size can be described in two different ways: as the absolute size of the error or as the relative size of the error.
- Once a measurement is reported, we can evaluate it in terms of its accuracy and its precision.

Mistakes are the portals of discovery.
—JAMES JOYCE

COMBINING MEASURED NUMBERS

Suppose you live in a city with a population of 300,000. One day, your best friend moves to your city to share an apartment with you. What is the population of your city now?

You might be tempted to add your friend to the city's population, making the new population 300,001. However, the number 300,000 has only 1 significant digit, implying that the population is known only to the nearest 100,000 people. The number 300,001 has 6 significant digits, which implies that you know the population *exactly*. Clearly, your friend's move cannot change the fact that the population is known only to the nearest 100,000. Thus, the population is still 300,000, despite the addition of your friend.

As this example illustrates, we must be very careful when we combine measured numbers, to avoid giving answers with more certainty than they deserve. In scientific or statistical work, researchers conduct careful analyses to determine how to combine numbers properly. For our purposes, however, we can use two simple rounding rules.

Combining Measured Numbers

Rounding rule for addition or subtraction: An answer obtained by adding or subtracting measured numbers should be rounded to the same precision as the *least precise* number in the problem.

Rounding rule for multiplication or division: An answer obtained by multiplying or dividing measured numbers should be rounded to the same number of *significant digits* as the number in the problem with the *fewest significant digits*.

Note: You should do the rounding only after completing all the operations, not during intermediate steps.

EXAMPLE 7 Combining Measured Numbers

a. A book written in 1962 states that the oldest Mayan ruins are 2000 years old. How old are they now?
b. The government in a town of 82,000 people plans to spend $41.5 million this year. Assuming all this money must come from taxes, what average amount must the city collect from each resident?

Solution

a. A book written in 1962 is about 40 years old, so we might be tempted to add 40 years to 2000 years to get 2040 years for the age of the ruins. However, 2000 years is the least precise of the two numbers: It is precise only to the nearest 1000 years, while 40 years is precise to the nearest 10 years. Thus, the answer also should be precise only to the nearest 1000 years:

$$\underbrace{2000 \text{ yr}}_{\text{precise to nearest 1000}} + \underbrace{40 \text{ yr}}_{\text{precise to nearest 10}} = \underbrace{2040 \text{ yr}}_{\text{must round to nearest 1000}} \approx \underbrace{2000 \text{ yr}}_{\text{correct final answer}}$$

To the precision with which we know the age of the ruins, they are still 2000 years old despite the 40-year age of the book.

b. We find the average tax by dividing the $41.5 million, which has 3 significant digits, by the population of 82,000, which has 2 significant digits. Thus, the answer should have 2 significant digits:

$$\underbrace{\$41,500,000}_{\text{3 significant digits}} \div \underbrace{82,000 \text{ persons}}_{\text{2 significant digits}} = \underbrace{\$506.10 \text{ per person}}_{\substack{\text{must round to} \\ \text{2 significant digits}}} \approx \underbrace{\$510 \text{ per person}}_{\text{correct final answer}}$$

The average resident must pay about $510 in taxes.

Review Questions

1. What are significant digits? How can you tell whether zeros are significant?
2. Distinguish between *random errors* and *systematic errors*. Give an example of how each type might affect measurements of weight. How can we minimize the effects of random errors? How can we account for the effects of a systematic error?
3. Distinguish between the *absolute error* and the *relative error* in a measurement. Give an example in which the absolute error is large but the relative error is small and another example in which the absolute error is small but the relative error is large.
4. Distinguish between *accuracy* and *precision*. Give an example of a measurement that is precise but inaccurate and another example that is accurate but imprecise.
5. Why can it be misleading to give measurements with more precision than is justified by the measurement process?
6. State the rounding rule for adding or subtracting measured numbers. Give an example of its use.
7. State the rounding rule for multiplying or dividing measured numbers. Give an example of its use.

In the News

1. **Random and Systematic Errors.** Find a recent news report that gives a measured quantity (for example, a report of population, average income, or the number of homeless people). Write a short description of how the quantity was measured, and briefly describe any likely sources of either random or systematic errors. Overall, do you think that the reported measurement was accurate? Why or why not?

2. **Absolute and Relative Errors.** Find a recent news report that describes some mistake in a measured, estimated, or projected number (for example, a budget projection that turned out to be incorrect). In words, describe the size of the error in terms of both absolute error and relative error.

3. **Accuracy and Precision.** Find a recent news article that causes you to question accuracy or precision. For example, the article might report a figure with more precision than you think is justified, or it may cite a figure that you know is inaccurate. Write a one-page summary of the report and explain why you question its accuracy or precision (or both).

4. **Uncertainty in the News.** Look through the newspapers from the past week for articles that mention numbers in any context. Find at least two numbers from each of the following sections: national/international news, local news, sports, and business. In each case, describe the number and its context, and discuss any uncertainty that you think is associated with the number.

5. **Real-World Rounding.** Find at least three different examples of numbers that were rounded in news stories. Explain how and why each number was rounded. (Hint: You might consider baseball batting averages, times in athletic events, or the stock market.)

Exercises

Review of Rounding. Exercises 1–2 use skills covered in the Brief Review on p. 153.

1. Round the following numbers to the nearest whole number.

 a. 2.4 b. 14.500 c. 779.49
 d. 4.1999 e. 13.5001 f. 234.5
 g. 1999.5 h. 88.71 i. −13.998

2. Round the following numbers to the nearest thousandth, tenth, ten, and hundred.

 a. 2365.98521 b. 322354.09005 c. 6000
 d. $\frac{34}{3}$ e. 78.555 f. 0.45232768
 g. −12.1 h. −850.7654 i. −12.001

3. **Counting Significant Digits.** State the number of significant digits and the implied meaning in the given numbers.

 a. 96.2 km/hr b. 100.020 seconds
 c. 0.00098 mm d. 0.0002020 meter
 e. 300,000 f. 3.0000×10^5

4. **Counting Significant Digits.** State the number of significant digits and the implied meaning in the given numbers.

 a. 401 people b. 200.0 liters c. 1.00098 mm
 d. 0.000202 meter e. 550,000,000 f. 5.500×10^8

Sources of Errors. For each measurement described in Exercises 5–12, briefly describe any likely sources of random errors and any likely sources of systematic errors.

5. A count of sport utility vehicles passing through a busy intersection during a 20-minute period

6. The population of a county in Mississippi, according to a census count

7. The average income of 25 people, found by checking their tax returns

8. The average income of 25 people, found by asking them their incomes during interviews conducted at a supermarket

9. Weights of sandwich meats, measured in a delicatessen using a well-calibrated scale

10. Weights of adults, measured by weighing them with their clothes on

11. The numbers of popped kernels in "large" boxes of popcorn at a movie theater

12. Times in a swimming meet

13. **Tax Audit.** A tax auditor reviewing a tax return looks for several kinds of problems, including these two: (1) mistakes made in entering or calculating numbers on the tax return and (2) places where the taxpayer reported income dishonestly. Discuss whether each problem involves random or systematic errors.

14. **AIDS Epidemic.** Researchers trying to study the progression of the AIDS epidemic need to know how many people are suffering from AIDS, which they can try to determine by studying medical records. Two of the many problems they face in this research are that (1) some people who are suffering from AIDS are misdiagnosed as having other diseases, and vice versa, and (2) some people

with AIDS never seek medical help and therefore do not have medical records. Discuss whether each problem involves random or systematic errors.

15. **Safe Air Travel.** Before taking off, a pilot is supposed to set the aircraft altimeter to the elevation of the airport. A pilot leaves from Denver (altitude 5280 ft) with her altimeter set to 2500 feet. Explain how this affects the altimeter readings throughout the flight. What kind of error is this?

16. **Lumber Yard.** A lumber yard is cutting boards that are supposed to be 4 feet long. Each board is cut by an employee who first measures and marks the 4-foot length. Later, careful measurements show that the average length of the boards was indeed 4 feet, but that most of the boards were actually slightly longer or shorter than 4 feet. What type of measurement error is involved in this situation? Explain.

Absolute and Relative Errors. In Exercises 17–24, find the absolute and relative errors.

17. The price of a gift you purchase by mail-order is supposed to be $18.50, but instead you are billed for $19.00.

18. The label on a bag of dog food says "20 pounds," but the true weight is only 18 pounds.

19. The bank is supposed to charge you interest of $65 on your loan, but instead charges you only $48.

20. You cut a piece of linen to a length of 48.5 inches, but the instructions call for a piece that is 48 inches long.

21. The recipe calls for $5\frac{1}{2}$ cups of flour, but you measure $5\frac{1}{4}$ cups of flour.

22. The meeting is supposed to last one hour, but it lasts one hour and five minutes.

23. The official distance to Longmont is 12 miles, but your odometer reads 11.7 miles.

24. The diameter of a ball bearing is supposed to be 0.1 centimeter, but instead is 0.2 centimeter.

Accuracy and Precision. Exercises 25–28 each describe a pair of measurements. In each case, state which measurement is more precise and which measurement is more accurate. Explain.

25. Your true height is 62.50 inches. A tape measure that can be read to the nearest $\frac{1}{8}$ inch gives your height as $62\frac{3}{8}$ inches. A new laser device at the doctor's office that gives readings to the nearest 0.05 inch gives your height as 62.90 inches.

26. Your true height is 62.50 inches. A tape measure that can be read to the nearest $\frac{1}{8}$ inch gives your height as $62\frac{5}{8}$ inches. A new laser device at the doctor's office that gives readings to the nearest 0.05 inch gives your height as 62.50 inches.

27. Your weight is 52.55 kilograms. A scale at a health clinic that gives weight measurements to the nearest half kilogram gives your weight as 53 kilograms. A digital scale at the gym that gives readings to the nearest 0.01 kilogram gives your weight as 52.88 kilograms.

28. Your weight is 52.55 kilograms. A scale at a health clinic that gives weight measurements to the nearest half kilogram gives your weight as $52\frac{1}{2}$ kilograms. A digital scale at the gym that gives readings to the nearest 0.01 kilogram gives your weight as 51.48 kilograms.

Believable Facts? Exercises 29–36 give statements of "fact" taken from official sources. For each statement, briefly discuss possible sources of error in the measurement. Then in light of the precision with which the measurement is given, discuss whether you think the fact is believable.

29. The population of the United States in 1860 was 31,443,321.

30. The number of deaths worldwide due to coronary heart disease in 1997 was 7.2 million.

31. The population of Tokyo in 2015 will be 28.7 million.

32. The Petrona Towers in Kuala Lumpur are 1483 feet tall, making them the world's tallest buildings as of 2000.

33. The average maximum temperature in January in Lagos, Nigeria is 88 degrees.

34. The number of deaths due to AIDS in the United States as of 1997 was 390,242.

35. The number of cell phone users in the United States in 1997 was 55,312,293.

36. Worldwide in 1998, there were 1694 officially declared endangered or threatened species.

Calvin and Hobbes by Bill Watterson

CALVIN AND HOBBES © 1991 Watterson. Reprinted with permission of UNIVERSAL PRESS SYNDICATE. All rights reserved.

Combining Measured Numbers. Use the appropriate rounding rules to answer Exercises 37–40 with the correct precision or correct number of significant digits.

37. Subtract the volumes 140 liters and 1.09 liters.
38. Add the times 2 hours, 37 minutes and 1 hour, 22 minutes, 15 seconds.
39. Multiply the weights 9.7 kilograms and 165 kilograms.
40. Divide the distance 110 kilometers by the time 55 minutes.
41. As you drive down the freeway, a sign tells you that it is 36 miles to the downtown. Your destination lies 2.2 miles beyond city hall (which is downtown). How much farther do you have to drive?
42. At the hardware store, you buy a 50-kilogram bag of sand. You also buy 1.25 kilograms of nails. What is the total weight of your purchases?
43. The government in a city of 480,000 people plans to spend $112.4 million on a transportation project. Assuming all this money must come from taxes, what average amount must the city collect from each resident?
44. Representatives in Congress serve an average of 630,000 constituents. If a political consultant recommends campaign spending of $4.25 per constituent for an upcoming election, how much should a representative spend in total?

Web Projects

Find useful links for Web Projects on the text Web site: www.aw.com/bennett-briggs

1. **The 2000 Census.** Go to the Web site for the Census Bureau and find the results from the 2000 census. According to the census, what was the population of the United States in 2000? How does the Census Bureau estimate the population in non-census years?

2. **Census Controversies.** Use the Library of Congress's "Thomas" Web site to find out about any pending legislation concerning the collection or use of census data. If you find more than one legislative bill pending, choose one to study in depth. Summarize the proposed legislation, and briefly discuss arguments both for and against it.

3. **Global Warming.** Find current estimates regarding how much the Earth has warmed over the past century and how much human activity contributes to that warming. Based on what you learn, write a one- to two-page essay about the uncertainty involved in understanding how human activity is contributing to global warming.

UNIT 3D

HOW NUMBERS DECEIVE: POLYGRAPHS, MAMMOGRAMS, AND MORE

The government administers polygraph tests ("lie detectors") to new applicants for sensitive security jobs. The polygraph tests are reputed to be 90% accurate; that is, they catch 90% of the people who are lying and validate 90% of the people who are truthful. Most people therefore guess that only 10% of the people who fail the polygraph tests have been falsely identified as lying. In fact, the actual percentage of false accusations can be *much* higher—in some cases, more than 90% of those who fail the test have been falsely accused! How can this be?

We'll discuss the answer soon, but first let's draw the important moral of this story: Numbers may not lie, but they can be deceiving if we are not very careful when we interpret them. In this unit, we'll discuss several common ways in which numbers can deceive.

BETTER IN EACH CASE, BUT WORSE OVERALL

Suppose a pharmaceutical company creates a new treatment for acne. To decide whether the new treatment is better than an older treatment, the company conducts a test. The old treatment is given to 90 patients, while the new treatment is given to 110 patients. Here are the results after four weeks of treatment:

- Among patients with mild acne, 10 received the old treatment and 90 received the new treatment. Among the 10 receiving the old treatment, 2 were cured, for a 20% cure rate. Among the 90 receiving the new treatment, 30 were cured, for a 33% cure rate. Thus, the new treatment had a higher cure rate among patients with mild acne.
- Among patients with severe acne, 80 received the old treatment and 20 received the new treatment. Among the 80 receiving the old treatment, 40 were cured, for a 50% cure rate. Among the 20 receiving the new treatment, 12 were cured, for a 60% cure rate. Thus, the new treatment had a higher cure rate among patients with severe acne.

Notice that the new treatment had a higher cure rate both for patients with mild acne and for patients with severe acne. Is it fair for the company to claim that their new treatment is better than the old treatment?

At first, this might seem to make sense. But let's look at the overall results:

- Among the total of 90 patients receiving the old treatment, 42 were cured— 2 out of 10 with mild acne and 40 out of 80 with severe acne. Thus, the overall cure rate for the old treatment was $42/90 = 46.7\%$.
- Among the total of 110 patients receiving the new treatment, 42 were cured—30 out of 90 with mild acne and 12 out of 20 with severe acne. Thus, the overall cure rate for the new treatment was $42/110 = 38.2\%$.

Overall, the old treatment had the higher cure rate, despite the fact that the new treatment had a higher rate for both mild and severe acne cases.

This example illustrates that it is possible for something to appear better in each of two or more group comparisons, but actually be worse overall. This situation is an example of **Simpson's paradox.** If you look carefully, you'll see that it occurs because of the way in which the overall results were divided into unequally sized groups (in this case, the groups were mild acne patients and severe acne patients). Simpson's paradox arises surprisingly often and is one way in which numbers can deceive unless they are examined with great care.

> *historical note*
> Simpson's paradox gets its name from Edward Simpson, who described it in 1951. However, the same idea was actually described much earlier, by Scottish statistician George Yule around 1900.

EXAMPLE 1 Who Played Better?

Table 3.2 gives the shooting performance of two players in a basketball game during the two halves.

TABLE 3.2 Basketball Shots

PLAYER	First Half BASKETS	First Half ATTEMPTS	First Half PERCENT	Second Half BASKETS	Second Half ATTEMPTS	Second Half PERCENT
Shaq	4	10	40%	3	4	75%
Kobe	1	4	25%	7	10	70%

Shaq had a higher shooting percentage both in the first half (40% to 25%) and in the second half (75% to 70%). Can Shaq claim that he had the better game?

Solution No, and we can see why by looking at the overall game statistics. Shaq made a total of 7 baskets (4 in the first half and 3 in the second half) on 14 shots (10 in the first half and 4 in the second half), for an overall shooting percentage of $7/14 = 50\%$. Kobe made a total of 8 baskets on 14 shots, for an overall shooting percentage of $8/14 = 57.1\%$. Thus, even though Shaq had a higher shooting percentage in both halves, Kobe had a better overall shooting percentage for the game.

IS IT DISCRIMINATION?—AND OTHER CASES OF SIMPSON'S PARADOX

Variations on Simpson's paradox arise in many situations. One famous case involves an investigation of possible gender discrimination in graduate programs at the University of California, Berkeley. Examining graduate admissions for 1973, university officials found that male applicants were being admitted at a significantly higher rate than female applicants. Assuming that male and female applicants were equally qualified, the university appeared to be engaging in discrimination against women.

The university began investigating individual departments to see which ones were responsible for the discrimination. Table 3.3 shows the results for the six departments with the largest numbers of graduate applicants. (University policy requires that departments not be named, so we refer to them as Departments A through F.)

| TABLE 3.3 | Applicants and Admissions for Six Graduate Programs at UC Berkeley, 1973 |||||||
|---|---|---|---|---|---|---|
| | **Men** ||| **Women** |||
| DEPARTMENT | APPLIED | ADMITTED | PERCENT | APPLIED | ADMITTED | PERCENT |
| A | 825 | 512 | 62% | 108 | 89 | 82% |
| B | 560 | 353 | 63% | 25 | 17 | 68% |
| C | 325 | 120 | 37% | 593 | 202 | 34% |
| D | 417 | 138 | 33% | 375 | 131 | 35% |
| E | 191 | 53 | 28% | 393 | 94 | 24% |
| F | 373 | 22 | 6% | 341 | 24 | 7% |
| Total | 2691 | 1198 | 44.5% | 1835 | 557 | 30.4% |

SOURCE: P. Bickel and J. W. O'Connell, *Science*, Vol. 187, 1975, pp. 398–404.

by the way...

A National Science Foundation study once found that, on average, women in science and engineering earned only 77% as much as men. But when the data were examined within each field, in every case the average salary for women was at least 92% of the average salary for men. The explanation for this case of Simpson's paradox: Women were more highly represented in life sciences and social sciences, which tend to be lower paying than physical sciences and engineering.

Note that the admissions rates (percentages) for women were actually higher than those for men in all but Departments C and E, and the rates were quite close in those cases. Nevertheless, the totals confirm a much higher overall admission rate for men than for women. Thus, we see Simpson's paradox in action again: While women were admitted at a significantly lower rate overall, no individual department was guilty of this practice.

Time out to think

Look closely at the data in Table 3.3. Would it be reasonable to conclude that Department A discriminates against men? Why or why not?

EXAMPLE 2 *Does Smoking Make You Live Longer?*

In the early 1970s, a medical study in England involved many adult residents from a district called Wickham. Twenty years later, a follow-up study looked at the survival rates of the people from the original study. The follow-up study found the following surprising results (D. R. Appleton, J. M. French, and M. P. Vanderpump, *American Statistician*, Vol. 50, 1996, pp. 340–341):

- Among the adults who smoked, 24% died during the 20 years since the original study.
- Among the adults who did not smoke, 31% died during the 20 years since the original study.

Do these results suggest that smoking can make you live longer?

Solution Not necessarily, because the given results don't tell us the ages of the smokers and nonsmokers. It turned out that, in the original study, the nonsmokers were older on average than the smokers. Thus, the higher death rate among nonsmokers simply reflected the fact that death rates tend to increase with age. When the results were broken into age groups, they clearly showed that for any given age group, nonsmokers had a higher 20-year survival rate than smokers. That is, 55-year-old smokers were less likely to reach age 75 than 55-year-old nonsmokers, and so on. Rather than suggesting that smoking prolongs life, a careful study of the data showed just the opposite.

DOES A POSITIVE MAMMOGRAM MEAN CANCER?

Although we often associate tumors with cancers, many tumors are not cancers. Medically, any kind of abnormal swelling or tissue growth is considered a tumor. A tumor caused by cancer is said to be *malignant* (or *cancerous*); all others are said to be *benign*. When a medical practitioner spots a tumor, one of the first steps in treatment is determining whether the tumor is benign or malignant.

With that background, imagine you are a doctor or nurse treating a patient who has been referred to you because she has a breast tumor. The patient will be understandably nervous, but you can give her some comfort by telling her that only about 1 in 100 breast tumors turn out to be malignant. But, just to be safe, you order a mammogram to determine whether her tumor is one of the 1% that are malignant.

Now, suppose the mammogram comes back positive; that is, it indicates the tumor is likely to be malignant. Mammograms are not perfect, so even the positive result does not necessarily mean that your patient has breast cancer. More specifically, let's assume that the mammogram screening is 85% accurate: It will correctly identify 85% of malignant tumors as malignant and 85% of benign tumors as benign. When you tell your patient that her mammogram was positive, what should you tell her about the chance that she actually has cancer?

Because the mammogram screening is 85% accurate, most people guess that the positive result means that the patient probably has cancer. Studies have shown that in this situation most doctors also believe this to be the case, and would tell the patient to be prepared for cancer treatment. But a more careful analysis shows otherwise. In fact, the chance that the patient has cancer is still quite small—about 5%. We can see why by analyzing some numbers.

Consider a study in which mammograms are given to 10,000 women with breast tumors. Assuming that 1% of tumors are malignant, 1% × 10,000 = 100 of the women actually have cancer; the remaining 9900 women have benign tumors. Table 3.4 summarizes the mammogram results. Note the following:

- The mammogram screening correctly identifies 85% of the 100 malignant tumors as malignant. Thus, it gives positive (malignant) results for 85 of the malignant tumors; these cases are called **true positives.** In the other 15 malignant cases, the result is negative, even though the women actually have cancer; these cases are **false negatives.**

TECHNICAL NOTE
The mammogram example and the polygraph example that follow illustrate cases in which conditional probabilities (discussed in Unit 6B) lead to confusion. The proper way of handling conditional probabilities was discovered by the Reverend Thomas Bayes (1702–1761) and is often called *Bayes rule.*

by the way...
Most cancer tests have different accuracy rates for malignant and benign tumors. For example, the accuracy rate for malignant tumors (called the *sensitivity*) might be 80% and the accuracy rate for benign tumors (called the *specificity*) might be 90%.

TABLE 3.4 Summary of Results for 10,000 Mammograms (in which 100 of the women have malignant tumors and 9900 have benign tumors)

	TUMOR IS MALIGNANT	TUMOR IS BENIGN	TOTAL
POSITIVE MAMMOGRAM	85 *true positives*	1485 *false positives*	1570
NEGATIVE MAMMOGRAM	15 *false negatives*	8415 *true negatives*	8430
TOTAL	100	9900	10,000

- The mammogram screening correctly identifies 85% of the 9900 benign tumors as benign. Thus, it gives negative (benign) results for 85% × 9900 = 8415 of the benign tumors; these cases are **true negatives.** The rest of the 9900 women with benign tumors (9900 − 8415 = 1485 women) get positive results in which the mammogram incorrectly suggests their tumors are malignant; these cases are **false positives.**

Overall, the mammogram screening gives positive results to 85 women who actually have cancer and to 1485 women who do *not* have cancer. The total number of positive results is 85 + 1485 = 1570. Because only 85 of these are true positives (the rest are false positives), the chance that a positive result really means cancer is only 85/1570 = 0.054, or 5.4%. Therefore, when your patient's mammogram comes back positive, you should reassure her that there's still only a small chance that she has cancer.

EXAMPLE 3 False Negatives

Suppose you are a doctor seeing a patient with a breast tumor. Her mammogram comes back negative. Based on the numbers in Table 3.4, what is the chance that she has cancer?

Solution For the 10,000 cases summarized in Table 3.4, the mammograms are negative for 15 women with cancer and for 8415 women with benign tumors. The total number of negative results is 15 + 8415 = 8430. Thus, the fraction of women with false negatives is 15/8430 = 0.0018, or slightly less than 2 in 1000. In other words, the chance that a woman with a negative mammogram has cancer is only about 2 in 1000.

Time out to think

While the chance of cancer with a negative mammogram is small, it is not zero. Therefore, it might seem like a good idea to biopsy all tumors, just to be sure. However, biopsies involve surgery, which means they can be painful and expensive, among other things. Given these facts, do you think that biopsies should be routine for all tumors? Should they be routine for cases of positive mammograms? Defend your opinion.

POLYGRAPHS AND DRUG TESTS

We're now ready to return to the question asked at the beginning of this unit, about how a 90% accurate polygraph test can lead to a surprising number of false accusations. You may already have guessed that the explanation is very similar to that found in the case of the mammograms.

Suppose the government gives the polygraph test to 1000 applicants for sensitive security jobs. Further suppose that 990 of these 1000 people tell the truth on their polygraph test, while only 10 people lie. For a test that is 90% accurate, we should find the following results:

- Of the 10 people who lie, the polygraph should correctly identify 90%, meaning that 9 fail the test (they are identified as liars) and 1 passes.
- Of the 990 people who tell the truth, the polygraph should correctly identify 90%, meaning that 90% × 990 = 891 truthful people pass the test and the other 10% × 990 = 99 people fail the test.

by the way...

The accuracy of mammograms is improving with technology. In 2000, studies showed the overall accuracy rate to be over 91%, and some experimental technologies appeared to achieve accuracy rates of over 98%. The accuracy is lower, however, for women receiving hormone replacement therapy, which is quite common among women over 40.

by the way...

A polygraph, often called a "lie detector," measures a variety of bodily functions including heart rate, skin temperature, and blood pressure. Polygraph operators look for subtle changes in these functions that typically occur when people lie. However, polygraph results have never been allowed as evidence in criminal proceedings. First of all, 90% accuracy is far too low for justice. In addition, studies show that polygraphs are easily fooled by people who train to beat them.

Figure 3.4 summarizes these results. The total number of people who fail the test is 9 + 99 = 108. Of these, only 9 were actually liars; the other 99 were falsely accused of lying. That is, 99 out of 108, or 99/108 = 91.7%, of the people who fail the test were actually telling the truth.

The percentage of people who are falsely accused in any real situation depends on both the accuracy of the test and the proportion of people who are lying. Nevertheless, for the numbers given here, we have an astounding result: Assuming the government rejects applicants who fail the polygraph test, then almost 92% of the rejected applicants were actually being truthful and may have been highly qualified for the jobs.

```
                        1000 applicants
                       /              \
                  10 lie              990 tell truth
                 /     \              /           \
   Polygraph does    Polygraph catches   Polygraph         Polygraph
   not catch         90% × 10 = 9 liars  falsely accuses   finds 90% × 990
   10% × 10 = 1 liar                     10% × 990 = 99    = 891 truthful

                        9 + 99 = 108 accused of lying,
                        but 99 of these are falsely accused
```

FIGURE 3.4 A tree diagram summarizes results of a 90% accurate polygraph test for 1000 people, of whom only 10 are lying.

Time out to think

Imagine that you are falsely accused of a crime. The police suggest that, if you are truly innocent, you should agree to take a polygraph test. Would you do it? Why or why not?

EXAMPLE 4 High School Drug Testing

All athletes participating in a regional high school track and field championship must provide a urine sample for a drug test. Those who fail are eliminated from the meet and suspended from competition for the following year. Studies show that, at the laboratory selected, the drug tests are 95% accurate. Assume that 4% of the athletes actually use drugs. What fraction of the athletes who fail the test are falsely accused and therefore suspended without cause?

Solution The easiest way to answer this question is by using some sample numbers. Suppose there are 1000 athletes in the meet. We are told to assume that 4%, or 40 athletes, actually use drugs; the remaining 960 athletes do not use drugs. In that case, the 95% accurate drug test should return the following results:

- 95% of the 40 athletes who use drugs, or 0.95 × 40 = 38 athletes, fail the test. The other 2 athletes who use drugs pass the test.
- 95% of the 960 athletes who do not use drugs pass the test, but 5% of these 960 athletes fail. Thus, the number of athletes who fail despite not using drugs is 5% × 960 = 48.

The total number of athletes who fail the test is 38 + 48 = 86. But 48 of these athletes who fail the test, or 48/86 = 56%, are actually nonusers. Despite the 95% accuracy of the drug test, more than half of the suspended students are innocent of drug use.

POLITICAL MATH

Another type of deception occurs when two sides in a debate argue with different sets of numbers. Some of the classic cases arise in the politics of taxes and tax cuts.

The two charts in Figure 3.5 both purport to show the effects of the *same* proposed tax cut. Figure 3.5a reflects numbers supplied by Congressional Republicans, who supported the tax cut. It suggests that the tax cut would benefit families of all incomes similarly in percentage terms, with slightly greater benefits going to middle-income families than to the poor or the rich. Figure 3.5b reflects numbers supplied by Democrats, who opposed the cut. This chart suggests that the benefits would go disproportionately to the rich. If we assume that neither side is lying outright, how can they make such remarkably different claims? The answer lies in how each side chose to calculate "benefits" of the tax cut.

The Republicans calculated the *average* tax cut that would be received by families in each group. For example, the last bar in Figure 3.5a shows that families with income over $200,000 would get an average tax cut of 2.9%. Of course, a 2.9% tax cut means many more actual dollars to someone paying a lot in taxes than to someone paying a little. For example, someone who pays $100,000 in taxes would save $2900 with a 2.9% tax cut, while someone paying $1000 in taxes would save only $29.

The Democrats calculated the percentage of *total* benefits that would go to each group. For example, the last bar in Figure 3.5b shows that families with an income over $200,000 would receive 28.1% of the total benefits from the tax cut. But this does *not* mean that these families get a 28.1% tax cut. To see why, consider the effects of an across-the-board tax cut of the same percentage for everyone. Because families with incomes above $200,000 pay more than one-fourth of the total income taxes collected by the U.S. government, these families would receive about one-fourth of the total benefit of *any* across-the-board cut. For example, if the government collected $1 trillion in taxes, a 10% across-the-board cut would mean savings of $100 billion for taxpayers as a whole. In that case, because one-fourth of the $1 trillion was paid by families with income over $200,000, one-fourth of the $100 billion in savings would go to these same

FIGURE 3.5 SOURCE: *The New York Times*, April 7, 1995.

families. Thus, these families would get 25% of the *benefits* of the tax cut, even though their actual tax cut would be only 10%.

Which side was being more fair? Neither, really. The Republicans chose to neglect the fact that most of the total tax savings would go to the wealthy, while the Democrats chose to neglect the fact that the wealthy already pay most of the taxes. Unfortunately, this type of "selective truth" is very common when it comes to numbers, bringing home the necessity of listening carefully to quantitative arguments.

EXAMPLE 5 *A Cut or an Increase?*

Government spending for a popular education program was $100 million last year. When Congress prepares its budget for next year, spending for the program is slated to rise to $102 million. Lobbyists immediately go into action. Those who support the program complain about the planned cut in the program. Those who oppose the program complain that it is being increased when it should be cut. Is one side lying? Explain. Assume that the Consumer Price Index is expected to rise by 3% over the next year.

Solution Neither side is lying. It's easiest to see for those who oppose the program: Clearly, spending *is* rising, since it is going from $100 million to $102 million. However, the opponents are neglecting the effects of inflation. If the Consumer Price Index rises by 3%, then this year's $100 million is equivalent to $103 million next year. Those who support the program say it is being cut because the increase to $102 million is not enough to keep up with inflation. In other words, $102 million next year will not buy as much as $100 million this year, so the program is being cut in the sense that it will not be able to do as much next year.

Review Questions

1. What is Simpson's paradox? Give an example to show how it can make something look better in individual cases yet be worse overall.

2. Describe an example that illustrates how Simpson's paradox can suggest discrimination in overall numbers even when no discrimination occurs.

3. Briefly explain why a positive result on a cancer test such as a mammogram does not necessarily mean that a patient has cancer.

4. Explain how it is possible for a very accurate polygraph or drug test to result in a large proportion of false accusations.

5. Give an example explaining how politicians on both sides of an issue can use numbers to support their case without actually lying.

In the News

1. **Simpson's Paradox.** Find a recent news article that gives data whose interpretation may be complicated by Simpson's paradox (such as an article about possible discrimination). Explain what additional information you would need to reach a definitive conclusion.

2. **Polygraphs.** Find a recent article in which someone or some group proposes a polygraph test to determine whether a person is being truthful. In light of what you know about polygraph tests, do you think the results will be meaningful? Why or why not?

3. **Drug Testing and Athletes.** Find a new report concerning drug testing of athletes. Summarize how the testing is being used, and discuss whether the testing is reliable.

4. **Tax Cut.** Find information about a recently proposed tax cut. According to its proponents, who will benefit most from the cut? Who will benefit most according to its opponents? Discuss the numerical data cited by each side. Which side do you think has the stronger argument? Why?

Exercises

1. **Batting Percentages.** The table on the next page shows the batting records of two baseball players in the first half (first 81 games) and last half of a season.

First Half

Player	Hits	At-bats	Batting average
Josh	50	150	.333
Jude	10	50	.200

Second Half

Player	Hits	At-bats	Batting average
Josh	35	70	.500
Jude	70	150	.467

Who had the higher batting average in the first half of the season? Who had the higher batting average in the second half? Who had the higher overall batting average? Explain how these results illustrate Simpson's paradox.

2. **Passing Percentages.** The table below shows the passing records of two rival quarterbacks in the first half and second half of a football game.

First Half

Player	Completions	Attempts	Percent
Allan	8	20	40%
Abner	2	6	33%

Second Half

Player	Completions	Attempts	Percent
Allan	3	6	50%
Abner	12	25	48%

Who had the higher completion percentage in the first half? Who had the higher completion percentage in the second half? Who had the higher overall completion percentage? Explain how these results illustrate Simpson's paradox.

3. **Test Scores.** The table below shows eighth-grade mathematics test scores in Nebraska and New Jersey. The scores are separated according to the race of the students. Also shown are the state averages for all races.

	White	Nonwhite	Average for all races
Nebraska	281	250	277
New Jersey	283	252	272

National Assessment of Educational Progress scores for 1992, from *Chance*, Spring 1999.

a. Which state had the higher scores in both racial categories? Which state had the higher overall average across both racial categories?

b. Explain how a state could score lower in both categories and still have a higher overall average.

c. Now consider the table below that gives the percentages of whites and nonwhites in each state. Use these percentages to verify that the overall average test score in Nebraska is 277, as claimed in the first table.

	White	Nonwhite
Nebraska	87%	13%
New Jersey	66%	34%

d. Use the racial percentages to verify that the overall average test score in New Jersey is 272, as claimed in the first table.

e. Explain briefly, in your own words, how Simpson's paradox appeared in this case.

4. **Test Scores.** Consider the following table comparing the grade point average (GPA) and mathematics SAT scores of high school students in 1988 and 1998.

	% students		SAT score		
GPA	1988	1998	1988	1998	Change
A+	4	7	632	629	−3
A	11	15	586	582	−4
A−	13	16	556	554	−2
B	53	48	490	487	−3
C	19	14	431	428	−3
Overall average			504	514	+10

Cited in *Chance*, Vol. 12, No. 2, 1999, from data in *New York Times*, September 2, 1999.

a. In general terms, how did the SAT scores of the students in the five grade categories change between 1988 and 1998?

b. How did the overall average SAT score change between 1988 and 1998?

c. How is this an example of Simpson's paradox?

5. **Tuberculosis Deaths.** The following table shows deaths due to tuberculosis (TB) in New York City and Richmond, Virginia, in 1910.

	New York	
Race	Population	TB deaths
White	4,675,000	8400
Nonwhite	92,000	500
Total	74,767,000	8900

	Richmond	
Race	Population	TB deaths
White	81,000	130
Nonwhite	47,000	160
Total	128,000	290

SOURCE: Cohen and Nagel, *An Introduction to Logic an Scientific Method*, Harcourt, Brace and World, 1934.

a. Compute the death rates for whites, nonwhites, and all residents in New York City.

b. Compute the death rates for whites, nonwhites, and all residents in Richmond.

c. Explain why this is an example of Simpson's paradox and explain how the paradox arises.

6. **Weight Training.** Two cross-country running teams participated in a (hypothetical) study in which a fraction of each team used weight training to supplement a running workout. The remaining runners did not use weight training. At the end of the season, the mean improvement in race times (in seconds) was recorded in the table below.

	Weight training	No weight training	Team average
Gazelles	10	2	6.0
Cheetahs	9	1	6.2

Describe how Simpson's paradox arises in this table. Resolve the paradox by finding the percentage of each team that used weight training.

7. **Hiring Statistics.** (This problem is based on an example in "Ask Marilyn," *Parade Magazine,* April 28, 1996.) A company decided to expand, so it opened a factory, generating 455 jobs. For the 70 white-collar positions, 200 males and 200 females applied. Of the females who applied, 20% were hired, while only 15% of the males were hired. Of the 400 males applying for the blue-collar positions, 75% were hired, while 85% of the 100 females who applied were hired. How does looking at the white-collar and blue-collar positions separately suggest a hiring preference for women? Do the overall data support the idea that the company hires women preferentially? Explain why this is an example of Simpson's paradox and how the paradox can be resolved.

8. **Drug Trials.** (This problem is based on an example in "Ask Marilyn," *Parade Magazine,* April 28, 1996.) A company runs two trials of two treatments for an illness. In the first trial, Treatment A cures 20% of the cases (40 out of 200) and Treatment B cures 15% of the cases (30 out of 200). In the second trial, Treatment A cures 85% of the cases (85 out of 100) and Treatment B cures 75% of the cases (300 out of 400). Which treatment had the better cure rate in the two trials individually? Which treatment had the better overall cure rate? Explain why this is an example of Simpson's paradox and how the paradox can be resolved.

9. **Basketball Records.** Consider the following hypothetical basketball records for Spelman and Morehouse Colleges.

	Spelman College	Morehouse College
Home games	10 wins, 19 losses	9 wins, 19 losses
Away games	12 wins, 4 losses	56 wins, 20 losses

a. Give numerical evidence to support the claim that Spelman College has a better team than Morehouse College.

b. Give numerical evidence to support the claim that Morehouse College has a better team than Spelman College.

c. Which claim do you think makes more sense? Why?

10. **Better Drug.** Two drugs, A and B, were tested on a total of 2000 patients, half of whom were women and half of

whom were men. Drug A was given to 900 patients, and Drug B to 1100 patients. The results appear in the table below.

	Women	Men
Drug A	5 of 100 cured	400 of 800 cured
Drug B	101 of 900 cured	196 of 200 cured

a. Give numerical evidence to support the claim that Drug B is more effective than Drug A.

b. Give numerical evidence to support the claim that Drug A is more effective than Drug B.

c. Which claim do you think makes more sense? Why?

11. **Polygraph Test.** Suppose that a polygraph is 90% accurate (it will correctly detect 90% of people who are lying and it will correctly detect 90% of people who are telling the truth). The 2000 employees of a company are given a polygraph test during which they are asked whether they use drugs. All of them deny drug use, when, in fact, 1% of the employees actually use drugs. Assume that anyone whom the polygraph operator finds untruthful is accused of lying.

a. Verify that the entries in the table below follow from the given information. Explain each entry.

	Users	Non-users	Total
Test finds employee lying	18	198	216
Test finds employee truthful	2	1782	1784
Total	20	1980	2000

b. How many employees are accused of lying? Of these, how many were actually lying and how many were telling the truth? What percentage of those accused of lying were falsely accused?

c. How many employees are found truthful? Of these, how many were actually truthful? What percentage of those found truthful really were truthful?

12. **Disease Test.** Suppose a test for a disease is 80% accurate for those who have the disease (true positives) and 80% accurate for those who do not have the disease (true negatives). Within a sample of 4000 patients, the incidence rate of the disease is the national average, which is 1.5%.

a. Verify that the entries in the table below follow from the information given and that the overall incidence rate is 1.5%. Explain.

	Disease	No disease	Total
Test positive	48	788	836
Test negative	12	3152	3164
Total	60	3940	4000

b. Of those with the disease, what percentage test positive?

c. Of those who test positive, what percentage have the disease? Compare this result to the one in part b and explain why they are different.

d. Suppose a patient tests positive for the disease. As a doctor using this table, how would you describe the patient's chance of actually having the disease? Compare this figure to the overall incidence rate of the disease.

13. **HIV Risks.** The New York State Department of Health estimates a 10% rate of HIV for the "at risk" population and a 0.3% rate for the general population. Tests for HIV are 95% accurate in detecting both true negatives and true positives. Random selection of 5000 "at risk" people and 20,000 people from the general population results in the following table.

	"At risk" population	
	Test positive	Test negative
Infected	475	25
Not infected	225	4275

	General population	
	Test positive	Test negative
Infected	57	3
Not infected	997	18,943

a. Verify that incidence rates for the general and "at risk" populations are 0.3% and 10%, respectively. Also verify that detection rates for the general and "at risk" populations are 95%.

b. Consider a patient in the "at risk" category. Of those with HIV, what percentage test positive? Of those who test positive, what percentage have HIV? Explain why these two percentages are different.

c. Suppose a patient in the "at risk" category tests positive for the disease. As a doctor using this table, how would you describe the patient's chance of actually having the disease? Compare this figure to the overall incidence rate of the disease.

d. Consider a patient in the general population. Of those with HIV, what percentage test positive? Of those who test positive, what percentage have HIV? Explain why these two percentages are different.

 e. Suppose a patient in the general population tests positive for the disease. As a doctor using this table, how would you describe the patient's chance of actually having the disease? Compare this figure to the overall incidence rate of the disease.

14. **Political Math.** Government spending for a popular housing program was $1 billion last year. When Congress prepares its budget for next year, spending for the program is slated to rise to $1.01 billion. Assume that the Consumer Price Index is expected to rise by 3% over the next year. Those who support the program complain that the program is being cut. Those who oppose the program complain that the program is being increased. Explain each position.

Web Projects

Find useful links for Web Projects on the text Web site: www.aw.com/bennett-briggs

1. **Polygraph Arguments.** Visit Web sites devoted to either opposing or supporting the use of polygraph tests. Summarize the arguments on both sides, specifically noting the role that false negative rates play in the discussion.

2. **Drug Testing.** Explore the issue of drug testing either in the workplace or for athletic competitions. Discuss the legality of drug testing in these settings and the accuracy of the tests that are commonly conducted.

3. **Cancer Screening.** Investigate recommendations concerning routine screening for some type of cancer (for example, breast cancer, prostate cancer, colon cancer). Explain how the accuracy of the screening test is measured. How is the test useful? How can its results be misleading?

CHAPTER 3 SUMMARY

UNIT	KEY TERMS	KEY IDEAS AND SKILLS
3A	absolute change relative change absolute difference relative difference	**Absolute change** = new value − reference value **Relative change** = $\dfrac{\text{absolute change}}{\text{reference value}}$ = $\dfrac{\text{new value} - \text{reference value}}{\text{reference value}}$ **Absolute difference** = compared value − reference value **Relative difference** = $\dfrac{\text{absolute difference}}{\text{reference value}}$ = $\dfrac{\text{compared value} - \text{reference value}}{\text{reference value}}$ **Understand** the *of* versus *more than* rule. **Understand** percentage points versus %. **Solve** percentage problems. **Identify** common abuses of percentages.
3B	scientific notation order of magnitude	**Write** and interpret numbers in scientific notation. **Put numbers in perspective** through estimation, including order of magnitude estimates comparisons scaling
3C	significant digits systematic error relative error accuracy precision	**Distinguish significant digits** from nonsignificant zeros. **Identify and distinguish** between random and systematic errors. **Absolute error** = measured value − true value **Relative error** = $\dfrac{\text{absolute error}}{\text{true value}}$ = $\dfrac{\text{measured value} - \text{true value}}{\text{true value}}$ **Distinguish** between accuracy and precision. **Apply rounding rules** for combining approximate numbers. Addition/subtraction: Round to the precision of the *least precise* number in the problem. Multiplication/division: Round to the same number of significant digits as in the number in the problem with the fewest significant digits.
3D	Simpson's paradox test results: false positive true positive false negative true negative	**Understand and give examples** of Simpson's paradox. **Understand and give examples** of how mammograms and polygraphs can lead to surprising results.

4

FINANCIAL MANAGEMENT

A fool and his money are soon parted.
—ENGLISH PROVERB

Managing your personal finances is a complex task in the modern world. If you are like most Americans, you already have a bank account and at least one credit card. You may also have student loans, a home mortgage, and various investment plans. In this chapter, we discuss key issues in personal financial management, including compound interest, savings, and loans.

UNIT 4A The Power of Compounding: If you put $100 in a piggy bank, many years from now it will still be $100. But if you invest $100 well, it may grow to much more. We'll study how money grows with compound interest.

UNIT 4B Savings Plans: One way to save is to make regular deposits of, say, $100 per month. We'll calculate the future value of such savings plans and study their applications.

UNIT 4C Loan Payments, Credit Cards, and Mortgages: Nearly everyone has some type of loan. We'll calculate monthly payments and explore loan issues.

UNIT 4A

THE POWER OF COMPOUNDING

On July 18, 1461, King Edward IV of England borrowed the modern equivalent of $384 from New College of Oxford. The King soon paid back $160, but never repaid the remaining $224. The debt was forgotten for 535 years. Upon its rediscovery in 1996, a New College administrator wrote to the Queen of England asking for repayment, with interest. Assuming an interest rate of 4% per year, he calculated that the college was owed $290 billion.

This example illustrates what is sometimes called the "power of compounding": the remarkable way that money grows when interest continues to accumulate year after year. In the New College case, there is no clear record of a promise to repay the debt with interest, and even if there were, the Queen might not feel obliged to pay a debt that had been forgotten for more than 500 years. But anyone can take advantage of compound interest simply by opening a savings account. With patience, the results may be truly astonishing.

> *by the way...*
> Of course, the New College administrator did not seriously believe that the Queen would pay $290 billion. However, he suggested a compromise of assuming a 2% per year interest rate, in which case the college was owed only $8.9 million. This, he said, would be enough to pay for a modernization project at the College. The Queen has not yet paid.

SIMPLE VERSUS COMPOUND INTEREST

Imagine that you deposit $1000 in Honest John's Money Holding Service, which promises to pay 5% interest each year. At the end of the first year, Honest John's sends you a check for

$$5\% \times \$1000 = 0.05 \times \$1000 = \$50$$

You also get $50 at the end of the second and third years. Over the 3 years, you receive total interest of

$$3 \times \$50 = \$150$$

Your original $1000 has grown in value to $1150. Honest John's method of payment represents **simple interest,** in which interest is paid only on your actual investment, or **principal.**

Now, suppose that you place the $1000 in a bank account that pays the same 5% interest once a year. But instead of paying you the interest directly, the bank adds the interest to your account. At the end of the first year, the bank deposits $50 interest into your account, raising your balance to $1050. At the end of the second year, the bank again pays you 5% interest. This time, however, the 5% interest is paid on the balance of $1050, so it amounts to

$$5\% \times \$1050 = 0.05 \times \$1050 = \$52.50$$

Adding this $52.50 raises your balance to

$$\$1050 + \$52.50 = \$1102.50$$

This is the new balance on which your 5% interest is computed at the end of the third year. So your third interest payment is

$$5\% \times \$1102.50 = 0.05 \times \$1102.50 = \$55.13$$

Therefore, your balance at the end of the third year is

$$\$1102.50 + \$55.13 = \$1157.63$$

Despite identical interest rates, you end up with $7.63 more if you use the bank instead of Honest John's. The difference comes about because the bank pays you *interest on the interest* as well as on the original principal. This type of interest payment is called **compound interest.**

> **DEFINITIONS**
>
> The **principal** in financial formulas is the balance upon which interest is paid.
>
> **Simple interest** is interest paid only on the original principal, and not on any interest added at later dates.
>
> **Compound interest** is interest paid both on the original principal and on all interest that has been added to the original principal.

EXAMPLE 1 *Savings Bond*

While banks almost always pay compound interest, bonds usually pay simple interest. Suppose you invest $1000 in a savings bond that pays simple interest of 10% per year. How much total interest will you receive in 5 years? If the bond paid compound interest, would you receive more or less total interest? Explain.

Solution With simple interest, every year you receive the same interest payment of 10% × $1000 = $100. Thus, you receive a total of $500 in interest over 5 years. With compound interest, you receive *more* than $500 in interest because the interest each year is calculated on your growing balance rather than on your original principal. For example, because your first interest payment of $100 raises your balance to $1100, your next compound interest payment is 10% × $1100 = $110, which is more than the simple interest payment of $100. For the same interest rate, compound interest always raises your balance faster than simple interest.

THE COMPOUND INTEREST FORMULA

Let's return to King Edward's debt to the New College. We can calculate the amount owed to the College by pretending that the $224 he borrowed was deposited into an interest-bearing account for 535 years. Let's assume, as did the New College administrator, that the interest rate was 4% per year. For each year, we can calculate the interest and the new balance with interest; the first three columns of Table 4.1 show these calculations for 4 years.

To find the total balance, we could continue the calculations to 535 years. Fortunately, there's a much easier way. The 4% annual interest rate means that each end-of-year balance is 104% of, or 1.04 times, the previous year's balance. Thus, as shown in the last column of Table 4.1, we can get each balance as follows:

- The balance at the end of 1 year is the original principal times 1.04:

$$\$224 \times 1.04 = \$232.96$$

TABLE 4.1 Calculating Compound Interest

AFTER N YEARS	INTEREST	BALANCE	OR EQUIVALENTLY
1 year	4% × $224 = $8.96	$224 + $8.96 = $232.96	$224 × 1.04 = $232.96
2 years	4% × $232.96 = $9.32	$232.96 + $9.32 = $242.28	$224 × $(1.04)^2$ = $242.28
3 years	4% × $242.28 = $9.69	$242.28 + $9.69 = $251.97	$224 × $(1.04)^3$ = $251.97
4 years	4% × $251.97 = $10.08	$251.97 + $10.08 = $262.05	$224 × $(1.04)^4$ = $262.05

- The balance at the end of 2 years is the 1-year balance times 1.04:

$$\$224 \times 1.04 \times 1.04 = \$224 \times (1.04)^2 = \$242.28$$

- The balance at the end of 3 years is the 2-year balance times 1.04:

$$\$224 \times 1.04 \times 1.04 \times 1.04 = \$224 \times (1.04)^3 = \$251.97$$

Continuing the pattern, we find that the balance after N years is the original principal times 1.04 raised to the Nth power. For example, the balance after $N = 10$ years is

$$\$224 \times (1.04)^{10} = \$331.57$$

We can generalize this result as the **compound interest formula.**

The Compound Interest Formula (General Form)

Suppose you invest a principal P in an account that earns compound interest. Each time interest is paid, your balance increases by an interest rate i (note that i should be expressed as a decimal rather than a percentage). After interest has been paid N times, your accumulated balance A is

$$A = P \times (1 + i)^N$$

by the way...
Financial planners often call the principal, P, the *present value* (PV) of the money and the accumulated amount, A, the *future value* (FV). This terminology is also used on many financial calculators and software packages.

In the New College case, the interest is paid at a rate of $i = 4\% = 0.04$, and it is paid a total of $N = 535$ times (once a year). Thus, the accumulated balance over the 535 years is

$$\begin{aligned} A = P \times (1 + i)^N &= \$224 \times (1 + 0.04)^{535} \\ &= \$224 \times (1.04)^{535} \\ &= \$224 \times 1{,}296{,}691{,}085 \\ &\approx \$2.9 \times 10^{11} = \$290 \text{ billion} \end{aligned}$$

Using Your **CALCULATOR**

Most calculators have a key for raising to powers, labeled y^x or $\wedge$ or something similar. For example, calculate 1.04^{535} by pressing 1.04 y^x 535 $=$ or 1.04 $\wedge$ 535.

As the administrator claimed, a 4% interest rate means the Queen owes about $290 billion.

COMPOUND INTEREST AS EXPONENTIAL GROWTH

The New College case demonstrates the remarkable way in which money can grow with compound interest. Figure 4.1 shows how the value of the New College debt rises during the first 100 years, assuming a starting value of $224 and an interest rate

of 4% per year. Note that while the value rises slowly at first, it rapidly accelerates, so in later years the value grows by much more each year than it did during earlier years.

This rapid growth is a hallmark of what we generally call *exponential growth*. You can see how exponential growth gets its name by looking again at the general compound interest formula:

$$A = P \times (1 + i)^N$$

Because the principal P and the interest rate i have fixed values for any particular compound interest calculation, the growth of the accumulated value A depends only on N (the number of times interest has been paid) and N appears in the *exponent* of the calculation.

Exponential growth is one of the most important topics in mathematics, with applications that include population growth, resource depletion, and radioactivity. We will study exponential growth in much more detail in Chapter 7. In this chapter, we focus only on its applications in finance.

FIGURE 4.1 The value of the debt in the New College case during the first 100 years. The original value in year 0 is $224, and the interest rate is 4% per year. Note that the value rises much more rapidly in later years than in earlier years—a hallmark of exponential growth.

CALCULATIONS FOR DIFFERENT COMPOUNDING PERIODS

We can use the general form of the compound interest formula for any compound interest problem. However, working with the formula is often easier if we rewrite it to fit the problem at hand. In this section, we'll restate and use the compound interest formula for three common cases: interest paid once a year (annually), interest paid two or more times per year, and interest paid continuously at all times.

Compound Interest Paid Once a Year

We can make the compound interest formula specific to any particular case by rewriting it with appropriate replacements for the interest rate i and the number of payments N. For interest paid once a year, the interest rate is the **annual percentage rate,** or **APR.** The total number of interest payments is simply the number of years, which we call Y. We can therefore rewrite the compound interest formula in the following form.

Compound Interest Formula for Interest Paid Once a Year

$$A = P \times (1 + \text{APR})^Y$$

where

P = starting principal
APR = annual percentage rate (as a decimal)
Y = number of years
A = acumulated balance after Y years

As in the general compound interest formula, we must write the interest rate APR as a decimal; for example, an annual interest rate of 4% means APR = 0.04.

EXAMPLE 2 New College Debt at 2%

If the interest rate is 2%, calculate the amount due to New College using

 a. simple interest **b.** compound interest

Solution

a. At a rate of 2%, the simple interest due each year is

$$2\% \times \$224 = 0.02 \times \$224 = \$4.48$$

Over 535 years, the total interest due would be

$$535 \times \$4.48 = \$2396.80$$

Adding this to the original loan principal of $224 gives the payoff amount of

$$\$224 + \$2396.80 = \$2620.80$$

b. To find the amount due with compound interest, we set the annual interest rate to APR = 2% = 0.02 and the number of years to $Y = 535$. Then we use the formula for compound interest paid once a year:

$$\begin{aligned} A = P \times (1 + \text{APR})^Y &= \$224 \times (1 + 0.02)^{535} \\ &= \$224 \times (1.02)^{535} \\ &= \$224 \times 39{,}911 \\ &\approx \$8.94 \times 10^6 \end{aligned}$$

The amount due with compound interest is about $8.94 million—far higher than the amount due with simple interest. You should note the remarkable effects of small changes in the compound interest rate. Here, we found that a 2% compound interest rate leads to $8.94 million after 535 years. Earlier, we found that a 4% interest rate for the same 535 years leads to $290 billion—which is more than 30,000 times larger than $8.94 million. Figure 4.2 contrasts the values of the New College debt during the first 100 years at interest rates of 2% and 4%. Note that the rates don't make much difference for the first few years, but over time the higher rate becomes far more valuable. ■

FIGURE 4.2 This figure contrasts the debt in the New College case during the first 100 years at interest rates of 2% and 4%. The original value in year 0 is $224. Note that the higher interest rate has an ever-greater impact at later times.

Time out to think

Suppose the interest rate for the New College debt were 3%. Without calculating, do you think the value after 535 years would be halfway between the values at 2% and 4% or closer to one or the other of these values? Now, check your guess by calculating the value at 3%. What happens at an interest rate of 6%? Briefly discuss why small changes in the interest rate can lead to large changes in the accumulated value.

EXAMPLE 3 *Mattress Investments*

Your grandfather put $100 under his mattress 50 years ago. If he had instead invested it in a bank account paying 3.5% interest compounded yearly (roughly the average U.S. rate of inflation during that period), how much would it be worth now?

Solution The starting principal is $P = \$100$. The annual percentage rate is APR = 3.5% = 0.035. The number of years is $Y = 50$. So the accumulated balance is

$$A = P \times (1 + \text{APR})^Y = \$100 \times (1 + 0.035)^{50}$$
$$= \$100 \times (1.035)^{50}$$
$$= \$558.49$$

Invested at a rate of 3.5%, the $100 would be worth over $550 today. Unfortunately, the $100 was put under a mattress, so it still has a face value of only $100.

Compound Interest Paid More Than Once a Year

Suppose you deposit $1000 into a bank that pays interest at an annual percentage rate of APR = 8%. If the interest is paid all at once at the end of a year, you'll receive interest of

$$8\% \times \$1000 = 0.08 \times \$1000 = \$80$$

Thus, your year-end balance will be $1000 + $80 = $1080.

Now, assume instead that the bank pays the interest *quarterly*, or four times a year (that is, once every 3 months). The quarterly interest rate is one-fourth of the annual interest rate, or

$$\text{quarterly interest rate} = \frac{\text{APR}}{4} = \frac{8\%}{4} = 2\% = 0.02$$

Table 4.2 shows how quarterly compounding affects the $1000 starting principal during the first year.

TABLE 4.2 Quarterly Interest Payments

AFTER N QUARTERS	INTEREST PAID	NEW BALANCE
1st quarter (3 months)	2% × $1000 = $20	$1000 + $20 = $1020
2nd quarter (6 months)	2% × $1020 = $20.40	$1020 + $20.40 = $1040.40
3rd quarter (9 months)	2% × $1040.40 = $20.81	$1040.40 + $20.81 = $1061.21
4th quarter (1 full year)	2% × $1061.21 = $21.22	$1061.21 + $21.22 = $1082.43

Note that the year-end balance with quarterly compounding ($1082.43) is *greater* than the year-end balance with interest paid all at once ($1080). That is, when interest is compounded more than once a year, the balance increases by *more* than the APR in 1 year.

We can find the same results with the general compound interest formula. In this case, the starting principal is $P = \$1000$. The quarterly payments have an interest rate of $i = \text{APR}/4 = 0.02$. In 1 year, interest is paid $N = 4$ times. Thus, the accumulated balance at the end of 1 year is

$$A = P \times (1 + i)^N = \$1000 \times (1 + 0.02)^4 = \$1000 \times (1.02)^4 = \$1082.43$$

We see that if interest is paid quarterly, the interest rate at each payment is $i = \text{APR}/4$. Generalizing, if interest is paid n times per year, the interest rate at each payment is $i = \text{APR}/n$. The total number of times that interest is paid after Y years is $N = nY$. (Note that n and N have different meanings: Lower-case n is the number of times interest is paid *each year*, while upper-case N is the total number of interest payments.) Substituting these expressions for i and N into the general compound interest formula, we get a formula for interest paid more than once each year.

> **Compound Interest Formula for Interest Paid n Times per Year**
>
> $$A = P\left(1 + \frac{\text{APR}}{n}\right)^{(nY)}$$
>
> where
>
> P = starting principal
> APR = annual percentage rate (as a decimal)
> n = number of compounding periods per year
> Y = number of years (may be a fraction)
> A = accumulated balance after Y years

Time out to think

Confirm that substituting $n = 1$ into the formula for interest paid n times per year gives you the formula for interest paid once a year. Explain why this should be true.

EXAMPLE 4 *Monthly Compounding at 6%*

You deposit $5000 in a bank account that pays an APR of 6% and compounds interest monthly. How much money will you have after 5 years? Compare this amount to the amount you'd have if interest were paid only once each year.

Solution The starting principal is $P = \$5000$ and the interest rate is $\text{APR} = 0.06$. Monthly compounding means that interest is paid $n = 12$ times a year, and we are considering a period of $Y = 5$ years. We put these values into the compound interest formula to find the accumulated balance, A.

$$A = P \times \left(1 + \frac{\text{APR}}{n}\right)^{(nY)} = \$5000 \times \left(1 + \frac{0.06}{12}\right)^{(12 \times 5)}$$
$$= \$5000 \times (1.005)^{60}$$
$$= \$6744.25$$

For interest paid only once each year, we find the balance after 5 years by using the formula for compound interest paid once a year:

$$A = P \times (1 + \text{APR})^Y = \$5000 \times (1 + 0.06)^5$$
$$= \$5000 \times (1.06)^5$$
$$= \$6691.13$$

Using Your CALCULATOR The Compound Interest Formula (for Interest Paid More Than Once per Year)

You can do compound interest calculations on any calculator that has a y^x or similar key for raising to powers. Here's a five-step procedure that will work on most scientific calculators, along with an example in which $P = \$1000$, APR $= 8\% = 0.08$, $Y = 10$ years, and $n = 12$ (monthly compounding). With a graphing calculator, you may be able to do the calculation more directly by using the parentheses keys; some business calculators have built-in functions for calculating compound interest in a single step.
Note: It is very important that you not round any answers until you have completed all the calculations.

		In general	Example	Display
Starting Formula:		$A = P \times \left(1 + \dfrac{APR}{n}\right)^{(nY)}$	$\$1000 \times \left(1 + \dfrac{0.08}{12}\right)^{(12 \times 10)}$	———
Step 1.	Multiply factors in exponent.	$n \; \boxed{\times} \; Y \; \boxed{=}$	$12 \; \boxed{\times} \; 10 \; \boxed{=}$	120.
Step 2.	Store product in memory (or write down).	$\boxed{\text{Store}}$	$\boxed{\text{Store}}$	120.
Step 3.	Add terms 1 and APR/n.	$1 \; \boxed{+} \; APR \; \boxed{\div} \; n \; \boxed{=}$	$1 \; \boxed{+} \; 0.08 \; \boxed{\div} \; 12 \; \boxed{=}$	1.0066666667
Step 4.	Raise result to power in memory.	$\boxed{y^x} \; \boxed{\text{Recall}} \; \boxed{=}$	$\boxed{y^x} \; \boxed{\text{Recall}} \; \boxed{=}$	2.219640235
Step 5.	Multiply result by P.	$\boxed{\times} \; P \; \boxed{=}$	$\boxed{\times} \; \$1000 \; \boxed{=}$	2219.640235

With the calculation complete, you can round to the nearest cent, writing the answer as $2219.64. Finally, because it's easy to push the wrong buttons by accident, you should always check the calculation (preferably twice) and make sure your answer makes sense.

After 5 years, monthly compounding gives you a balance of $6744.25 while annual compounding gives you a balance of $6691.13. That is, monthly compounding earns $6744.25 − $6691.13 = $53.12 more, even though the APR is the same in both cases. ∎

Annual Percentage Yield (APY)
We've seen that in 1 year, money grows by *more* than the APR when interest is compounded more than once a year. For example, we found that with quarterly compounding and an 8% APR, a $1000 principal increases to $1082.43 in 1 year. This represents a relative increase of 8.243%:

$$\text{relative increase} = \frac{\text{absolute increase}}{\text{starting principal}} = \frac{\$82.43}{\$1000} = 0.08243 = 8.243\%$$

This relative increase over 1 year is called the **annual percentage yield (APY)**. Note that the APY of 8.243% is greater than the APR of 8%.

> **DEFINITION**
>
> The **annual percentage yield (APY)** is the actual percentage by which a balance increases in 1 year. It is equal to the APR if interest is compounded annually. It is greater than the APR if interest is compounded more than once a year.

by the way...
Banks usually list both the annual percentage rate (APR) and the annual percentage yield (APY). The APY is what your money really earns and is the more important number when you are comparing interest rates. Banks are required by law to state the APY on interest-bearing accounts. The APY is sometimes called the *effective yield,* or simply the *yield.*

EXAMPLE 5 *More Compounding Means a Higher Yield*

You deposit $1000 into an account with APR = 8%. Find the annual percentage yield with *monthly* compounding and with *daily* compounding.

Solution The easiest way to find the annual percentage yield is by finding the balance at the end of 1 year. We have $P = \$1000$, APR = 8% = 0.08, $Y = 1$ year. For monthly compounding, we set $n = 12$. Thus, at the end of 1 year, the accumulated balance with monthly compounding is

$$A = P \times \left(1 + \frac{\text{APR}}{n}\right)^{(nY)} = \$1000 \times \left(1 + \frac{0.08}{12}\right)^{(12 \times 1)}$$
$$= \$1000 \times (1.006666667)^{12} = \$1083.00$$

Note that your balance increases by $83.00, so the annual percentage yield is

$$\text{APY} = \text{relative increase in 1 year} = \frac{\$83.00}{\$1000} = 0.083 = 8.3\%$$

With monthly compounding, the annual percentage yield is 8.3%.

Daily compounding means that interest is paid $n = 365$ times per year. At the end of 1 year, your accumulated balance with daily compounding is

$$A = P \times \left(1 + \frac{\text{APR}}{n}\right)^{(nY)} = \$1000 \times \left(1 + \frac{0.08}{365}\right)^{(365 \times 1)}$$
$$= \$1000 \times (1.000219178)^{365} = \$1083.28$$

Your balance increases by $83.28, so the annual percentage yield is

$$\text{APY} = \text{relative increase in 1 year} = \frac{\$83.28}{\$1000} = 0.08328 = 8.328\%$$

With daily compounding, the annual percentage yield is 8.328%, which is slightly higher than the APY for monthly compounding.

TECHNICAL NOTE

Most banks divide the APR by 360, rather than 365, when calculating the interest rate and APY for daily compounding. Thus, the results found here may not agree exactly with actual bank results.

Continuous Compounding

Suppose that interest were compounded more often than daily—say, every second or every trillionth of a second. How would this affect the annual yield?

Let's examine what we've found so far for APR = 8%. If interest is compounded annually (once a year), the annual yield is simply APY = APR = 8%. With quarterly compounding, we found APY = 8.243%. With monthly compounding, we found APY = 8.300%. With daily compounding, we found APY = 8.328%. Clearly, more frequent compounding means a higher APY (for a given APR).

However, notice that the change gets smaller as the frequency of compounding increases. For example, changing from annual compounding ($n = 1$) to quarterly compounding ($n = 4$) increases the APY quite a bit, from 8% to 8.243%. In contrast, going from monthly ($n = 12$) to daily ($n = 365$) compounding increases the APY only slightly, from 8.300% to 8.328%. You probably won't be surprised to learn that the APY can't get much bigger than it already is for daily compounding.

Table 4.3 shows the APY for various compounding periods and Figure 4.3 is a graph of the results. As expected, the annual yield does *not* grow indefinitely. Instead, it approaches a *limit* that must be very close to the APY of 8.3287068% found for

TABLE 4.3 Annual Yield (APY) for APR = 8% with Various Numbers of Compounding Periods (n)

n	APY	n	APY
1	8.0000000%	1000	8.3283601%
4	8.2432160%	10,000	8.3286721%
12	8.2999507%	1,000,000	8.3287064%
365	8.3277572%	10,000,000	8.3287067%
500	8.3280135%	1,000,000,000	8.3287068%

FIGURE 4.3 The annual yield (APY) for APR = 8% depends on the number of times interest is compounded per year.

n = 1 billion. In other words, even if we could compound *infinitely many times* per year, the annual yield would not go much above 8.3287068%. Compounding infinitely many times per year is called **continuous compounding,** and it represents the best possible compounding for a particular APR. With continuous compounding, the compound interest formula takes the following special form.

Compound Interest Formula for Continuous Compounding

$$A = P \times e^{(APR)(Y)}$$

where

P = starting principal
APR = annual percentage rate (as a decimal)
Y = number of years
A = accumulated balance after Y years

The number e is a special irrational number with a value of $e \approx 2.71828$. You can compute e to a power with the e^x key on your calculator.

> **historical note**
>
> Like the number π that arises so frequently in mathematics, the number e is one of the universal constants of mathematics. It appears in countless applications, most importantly to describe exponential growth and decay processes. The notation e was proposed by the Swiss mathematician Leonhard Euler in 1727. Like π, the number e is not only an irrational number, but also a transcendental number.

Time out to think

Look for the $\boxed{e^x}$ key on your calculator. Use it to enter e^1 and thereby verify that $e \approx 2.71828$.

EXAMPLE 6 *Continuous Compounding*

You deposit $100 in an account with an APR of 8% and continuous compounding. How much will you have after 10 years?

Solution We have $P = \$100$, APR = 8% = 0.08, and $Y = 10$ years of continuous compounding.

$$A = P \times e^{(APR \times Y)} = \$100 \times e^{(0.08 \times 10)}$$
$$= \$100 \times e^{0.8}$$
$$= \$222.55$$

Your balance will be $222.55 after 10 years. Note: Be sure you can get the above answer by using the $\boxed{e^x}$ on your calculator. ■

PLANNING AHEAD WITH COMPOUND INTEREST

Suppose you have a new baby and want to make sure that you'll have $100,000 for his or her college education. Assuming your baby will start college in 18 years, how much money should you deposit now?

If we know the interest rate, this problem is simply a "backwards" compound interest problem. We start with the amount A needed after 18 years and then calculate the necessary starting principal, P. The following two examples illustrate the calculations.

EXAMPLE 7 *College Fund at 7%*

Suppose you could make a single deposit in an investment with an interest rate of APR = 7%, compounded annually, and leave it there for the next 18 years. How much would you have to deposit now to realize $100,000 after 18 years?

Solution We know the interest rate (APR = 0.07), the number of years of compounding ($Y = 18$), and the amount desired after 18 years ($A = \$100,000$). We want to find the starting principal, P, that must be deposited now. We therefore solve the compound interest formula (for interest paid once a year) for P, by dividing both sides by $(1 + APR)^Y$.

$$\underbrace{A = P \times (1 + APR)^Y}_{\text{compound interest formula (interest paid once a year)}} \xrightarrow[\text{then interchange left and right sides}]{\text{divide both sides by } (1 + APR)^Y} P = \frac{A}{(1 + APR)^Y}$$

> **by the way . . .**
>
> The process of finding the principal (present value) that must be deposited today to yield some particular future amount is called *discounting* by financial planners.

Now we substitute the given values for A, APR, and Y.

$$P = \frac{A}{(1 + APR)^Y} = \frac{\$100,000}{(1 + 0.07)^{18}} = \frac{\$100,000}{(1.07)^{18}} = \$29,586.39$$

Depositing $29,586 now will yield the desired $100,000 in 18 years—assuming that the 7% APR doesn't change and that you make no withdrawals or additional deposits. ■

a brief review...

THREE BASIC RULES OF ALGEBRA

In prior units, we've already encountered several instances in which we needed to solve an equation by adding, subtracting, multiplying, or dividing both sides by the same quantity. These operations will be useful in this unit and the rest of the book, so it is important to review the basic rules.

Three Basic Rules
The following three rules can always be used:

1. We can add or subtract the same quantity on both sides of an equation.
2. We can multiply or divide both sides of an equation by the same quantity, as long as we do not multiply or divide by zero.
3. We can interchange the left and right sides of an equation; that is, if $x = y$, it is also true that $y = x$.

Adding and Subtracting
The following examples show how adding to or subtracting from both sides can help solve equations that involve unknowns.

Example: Solve the equation $x - 9 = 3$ for x.
Solution: We isolate x by adding 9 to both sides:

$$x - 9 + 9 = 3 + 9 \quad \rightarrow \quad x = 12$$

Example: Solve the equation $y + 6 = 2y$ for y.
Solution: Because we have y on the left side and $2y$ on the right, we can isolate y on the right by subtracting y from both sides:

$$y + 6 - y = 2y - y \quad \rightarrow \quad 6 = y$$

We interchange the left and right sides, writing the answer as $y = 6$.

Example: Solve the equation $8q - 17 = p + 4q - 2$ for p.
Solution: We isolate p by subtracting $4q$ from both sides while also adding 2 to both sides:

$$8q - 17 - 4q + 2 = p + 4q - 2 - 4q + 2$$
$$\downarrow$$
$$4q - 15 = p$$

We interchange the left and right sides, writing the answer more simply as $p = 4q - 15$. Note that the equation is solved for p, but we cannot state a numerical value for p until we know the value of q.

Multiplying and Dividing
When we cannot isolate a variable by addition or subtraction alone, we may need to multiply or divide both sides of an equation by the same thing.

Example: Solve the equation $4x = 24$ for x.
Solution: We isolate x by dividing both sides by 4:

$$\frac{4x}{4} = \frac{24}{4} \quad \rightarrow \quad x = 6$$

Example: Solve the equation $\frac{3z}{4} - 2 = 10$ for z.
Solution: First, we isolate the term containing z by adding 2 to both sides:

$$\frac{3z}{4} - 2 + 2 = 10 + 2 \quad \rightarrow \quad \frac{3z}{4} = 12$$

Now we multiply both sides by $\frac{4}{3}$:

$$\frac{3z}{4} \times \frac{4}{3} = 12 \times \frac{4}{3} \quad \rightarrow \quad z = 16$$

Example: Solve the equation $7w = 3s + 5$ for s.
Solution: We isolate the term containing s by subtracting 5 from both sides:

$$7w - 5 = 3s + 5 - 5 \quad \rightarrow \quad 7w - 5 = 3s$$

Next we divide both sides by 3 to isolate s; to write the final answer more simply, we also switch the left and right sides after dividing.

$$\frac{7w - 5}{3} = \frac{3s}{3} \quad \rightarrow \quad s = \frac{7w - 5}{3}$$

EXAMPLE 8 College Fund at 10%, Compounded Monthly

Repeat Example 7, but with an interest rate of APR = 10% and monthly compounding. Compare the results.

Solution This time we must solve for *P* in the compound interest formula for interest paid more than once a year.

$$\underbrace{A = P \times \left(1 + \frac{APR}{n}\right)^{(nY)}}_{\text{compound interest formula (interest paid n times per year)}} \xrightarrow[\text{then interchange left and right sides}]{\text{divide both sides by }\left(1 + \frac{APR}{n}\right)^{(nY)}} P = \frac{A}{\left(1 + \frac{APR}{n}\right)^{nY}}$$

We substitute the given interest rate (APR = 0.10), the number of years (*Y* = 18), and the balance after 18 years (*A* = $100,000). With monthly compounding, we have *n* = 12.

$$P = \frac{A}{\left(1 + \frac{APR}{n}\right)^{(nY)}} = \frac{\$100{,}000}{\left(1 + \frac{0.10}{12}\right)^{(12 \times 18)}} = \frac{\$100{,}000}{(1.008333333)^{216}} = \$16{,}653.64$$

With a 10% APR and monthly compounding, you can reach $100,000 in 18 years by depositing about $16,650 today. This is just over half the $29,586 that you must deposit to reach the same goal with an interest rate of 7% (compounded annually). ∎

Time out to think

Aside from long-term government bonds, it is extremely difficult to find investments with a constant interest rate for 18 years. Nevertheless, financial planners often make such assumptions when exploring investment options. Explain why such calculations can be useful, despite the fact that you can't be sure of a steady interest rate.

Review Questions

1. What is the difference between simple interest and compound interest? Why do you end up with more money with compound interest?

2. Explain how New College could claim that a debt of $224 from 535 years ago is worth $290 billion today. How does this show the "power of compounding"?

3. State the general form of the compound interest formula, and define *A*, *P*, *i*, and *N*. Give an example that uses this formula.

4. State the compound interest formula for interest paid once a year. Define APR and *Y*. Briefly explain how this formula is related to the general compound interest formula.

5. State the compound interest formula for interest paid more than once a year. Define *n*, and explain how this formula is related to the general compound interest formula.

6. What is an annual percentage yield (APY)? Explain why, for a given APR, the APY is higher if the interest is compounded more frequently.

7. What is continuous compounding? How does the APY for continuous compounding compare to, say, the APY for daily compounding? Explain the use of the formula for continuous compounding.

8. Give an example of a situation in which you might want to solve the compound interest formula to find the principal *P* that must be invested now to yield a particular amount *A* in the future.

In the News

1. **Bank Advertisement.** Find two bank advertisements that refer to compound interest rates. Explain the terms in each advertisement. Which bank offers the better deal? Explain.

2. **Power of Compounding.** In an advertisement or article about an investment, find a description of how money has grown (or will grow) over a period of many years. What is the annual yield listed? How does the value of the account change?

Exercises

Algebra Review. Exercises 1–2 use skills covered in the Brief Review on p. 189.

1. Solve the following equations for the unknown quantity.
 a. $x - 4 = 6$
 b. $y + 5 = 10$
 c. $2z = 12$
 d. $3x = 15$
 e. $2x - 5 = 13$
 f. $3n + 4 = 13$
 g. $4x + 8 = 24$
 h. $5 - 2w = 9$

2. Solve the following equations for the unknown quantity.
 a. $x - 8 = 2$
 b. $x + 7 = 4$
 c. $4y = 28$
 d. $5n = 100$
 e. $5z - 5 = 35$
 f. $3x + 4 = 13 - 6x$
 g. $s + 4 = 24 - 4s$
 h. $6 - 2y = 9 - y$

Simple vs. Compound Interest. Exercises 3–4 describe two similar, but not identical, investment accounts. Make a table that shows the performance of both accounts for 5 years. The table should list the amount of interest earned each year and the balance in each account. Compare the balances after 5 years in absolute and relative (percentage) terms.

3. Yancy invests $500 in an account that earns simple interest at an annual rate of 5% per year. Samantha invests $500 in a savings account with annual compounding at a rate of 5% per year.

4. Trevor invests $1000 in an account that earns simple interest at an annual rate of 6% per year. Kendra invests $1000 in a savings account with annual compounding at a rate of 6% per year.

Compound Interest. In Exercises 5–10, use the compound interest formula to determine the accumulated balance after the stated period. Assume that interest is compounded annually.

5. $2000 is invested at an APR of 3% for 10 years.
6. $10,000 is invested at an APR of 5% for 20 years.
7. $30,000 is invested at an APR of 7% for 25 years.
8. $3000 is invested at an APR of 4% for 12 years.
9. $10,000 is invested at an APR of 6% for 25 years.
10. $40,000 is invested at an APR of 8.5% for 30 years.

Small Rate Differences. Exercises 11–12 describe two similar, but not identical, investment accounts. In each case, compare the balances after 10 years and after 30 years in absolute and relative (percentage) terms. Briefly discuss the effects of the small difference in interest rates.

11. Chang invests $500 in a savings account that earns 3.5% compounded annually. Kio invests $500 in a different savings account that earns 3.75% compounded annually.

12. José invests $1500 in a savings account that earns 5.6% compounded annually. Marta invests $1500 in a different savings account that earns 5.7% compounded annually.

Compounding More Than Once a Year. In Exercises 13–20, use the compound interest formula for compounding more than once a year to determine the accumulated balance after the stated period.

13. A $1000 deposit at an APR of 5.5% with monthly compounding for 10 years
14. A $2000 deposit at an APR of 3% with daily compounding for 5 years
15. A $5000 deposit at an APR of 7.3% with quarterly compounding for 20 years
16. A $10,000 deposit at an APR of 6.2% with monthly compounding for 5 years
17. A $1000 deposit at an APR of 7% with monthly compounding for 15 years
18. A $3000 deposit at an APR of 5% with daily compounding for 10 years
19. A $5000 deposit at an APR of 6.2% with quarterly compounding for 30 years
20. A $15,000 deposit at an APR of 7.8% with monthly compounding for 15 years

Annual Percentage Yield (APY). Find the annual percentage yield (APY) for the banks described in Exercises 21–24.

21. A bank offers an APR of 6.5% compounded daily.
22. A bank offers an APR of 4.5% compounded monthly.
23. A bank offers an APR of 3.25% compounded monthly.
24. A bank offers an APR of 7.25% compounded quarterly.

25. **Comparing Annual Yields.** Consider an account with an APR of 6.6%. Find the APY with quarterly compounding, monthly compounding, and daily compounding. Comment on how changing the compounding period affects the annual yield.

26. **Comparing Annual Yields.** Consider an account with an APR of 8%. Find the APY with quarterly compounding, monthly compounding, and daily compounding. Comment on how changing the compounding period affects the annual yield.

27. **Rates of Compounding.** Compare the accumulated balance in two accounts that both start with an initial deposit of $1000. Both accounts have an APR of 5.5%, but one account compounds interest annually while the other account compounds interest daily. Make a table that shows the interest earned each year and the accumulated balance in both accounts for the first 10 years. Compare the balance in the accounts, in percentage terms, after 10 years. Round all figures to the nearest dollar.

28. **Understanding Annual Percentage Yield (APY).**

 a. Explain why APR and APY are the same with annual compounding.

 b. Explain why APR and APY are different with daily compounding.

 c. Does APY depend on the starting principal, P? Why or why not?

 d. How does APY depend on the number of compoundings during a year, n? Explain.

Continuous Compounding. In Exercises 29–34, use the compound interest formula for continuous compounding to determine the accumulated balance after 1 year, 5 years, and 20 years. Also find the APY for each account.

29. A $1000 deposit in an account with an APR of 4%

30. A $2000 deposit in an account with an APR of 5%

31. A $10,000 deposit in an account with an APR of 6%

32. A $3000 deposit in an account with an APR of 4%

33. A $5000 deposit in an account with an APR of 6.5%

34. A $500 deposit in an account with an APR of 7%

35. **Comparing Investment Plans.** Bernard deposits $1600 in a savings account that compounds interest annually at an APR of 4%. Carla deposits $1400 in a savings account that compounds interest daily at an APR of 5%. Who will have the higher accumulated balance after 5 years and after 20 years? Explain.

36. **Comparing Investment Plans.** Brian invests $1600 in an account with annual compounding and an APR of 5.5%. Celeste invests $1400 in an account with continuous compounding and an APR of 5.2%. Determine who has the higher accumulated balance after 5 years and after 20 years. Discuss the effect of the APR and the compounding period.

Planning Ahead with Compounding. For Exercises 37–40, suppose you start saving today for a $10,000 down payment that you plan to make on a house in 10 years. Assume that you make no deposits into the account after your initial deposit. For each account described, how much would you have to deposit now to reach your $10,000 goal in 10 years?

37. An account with annual compounding and an APR of 9%

38. An account with quarterly compounding and an APR of 9%

39. An account with monthly compounding and an APR of 9%

40. An account with continuous compounding and an APR of 9%

41. **College Fund.** You want to have a $100,000 college fund in 18 years. How much will you have to deposit now under each of the following scenarios?

 a. An APR of 6%, compounded daily

 b. An APR of 7.5%, compounded continuously

 c. An APR of 11%, compounded monthly

42. **Retirement Fund.** You want to accumulate $75,000 for your retirement in 35 years. You have two choices: Plan A is an account with annual compounding and an APR of 5%. Plan B is an account with continuous compounding and an APR of 4.5%. How much of an investment does each plan require to reach your goal?

43. **Your Bank Account.** Find the current APR, the compounding period, and the claimed APY for your personal savings account. (Choose just one account if you have more than one, or pick a rate from a nearby bank if you don't have an account.)

 a. Calculate the APY on your account. Does your calculation agree with the APY claimed by the bank? Explain.

 b. Suppose you receive a gift of $10,000 and place it in your account. If the interest rate never changes, how much will you have in 10 years?

 c. Suppose you find another account that offers interest at an APR that is 2 percentage points higher than yours, with the same compounding period. For the $10,000 deposit, how much will you have after 10 years? Briefly discuss how this result compares to the result from part b.

44. **Continuous Compounding.** Explore continuous compounding by answering the following questions.

 a. For an APR of 12%, make a table similar to Table 4.3 in which you display the APY for $n = 1, 4, 12, 365, 500, 1000$.

 b. Find the APY for continuous compounding at an APR of 12%.

 c. Show the results of parts a and b on a graph similar to Figure 4.3.

 d. In words, compare the APY with continuous compounding to the APY with other types of compounding.

 e. You deposit $500 in an account with an APR of 12%. With continuous compounding, how much money will you have at the end of 1 year? at the end of 5 years?

45. **Finding Time Periods.** Use a calculator and possibly some trial and error to answer the following questions.

 a. How long will it take your money to triple at an APR of 8% compounded annually?

 b. How long will it take your money to grow by 50% at an APR of 7% compounded annually?

 c. You deposit $1000 in an account that pays an APR of 7% compounded annually. How long will it take for your balance to reach $100,000?

Web Projects

Find useful links for Web Projects on the text Web site: www.aw.com/bennett-briggs

1. **Compound Interest Calculators.** Although you know how to calculate balances with the compound interest formula, the Web has many compound interest calculators. Find such a calculator on the Web. Experiment with various APRs, initial deposits, and compounding periods to determine if the Web calculator is accurate. Note and discuss any terms that are new or different from those you encountered in this unit.

2. **Money Stretcher.** The Money Stretcher is a Web-based tutorial on compound interest. Read this short article and comment on its accuracy, given what you have learned in this unit.

3. **Rate Comparisons.** Find a Web site that compares interest rates available for ordinary savings accounts at different banks. What is the range of rates currently being offered? What is the best deal? How does your own bank account compare?

UNIT 4B

SAVINGS PLANS

Suppose you want to save money for retirement, for your child's college expenses, or for some other reason. You could deposit a lump sum of money today and let it grow through the power of compound interest. But what if you don't have a large lump sum to start such an account?

For most people, a more realistic way to save is by depositing smaller amounts on a regular basis. For example, you might put $50 a month into savings. Such long-term **savings plans** are so popular that many have special names—and some even get special tax treatment. Popular types of savings plans include Individual Retirement Accounts (IRAs), 401(k) plans, Keogh plans, and employee pension plans.

> **by the way...**
> Financial planners call any series of equal, regular payments an *annuity*. Thus, savings plans are a type of annuity, as are loans that you pay with equal monthly payments.

THE SAVINGS PLAN FORMULA

We can study savings plans with a simple example. Suppose you deposit $100 into your savings plan at the end of each month. Further suppose that your plan pays monthly interest at an annual rate of APR = 12%, or 1% per month.

- You begin with $0 in the account. At the end of month 1, you make the first deposit of $100.
- At the end of month 2, you receive the monthly interest on the $100 already in the account, which is 1% × $100 = $1. In addition, you make your monthly deposit of $100. Thus, your balance at the end of month 2 is

$$\underbrace{\$100}_{\text{prior balance}} + \underbrace{\$1.00}_{\text{interest}} + \underbrace{\$100}_{\text{new deposit}} = \$201.00$$

- At the end of month 3, you receive 1% interest on the $201 already in the account, or 1% × $201 = $2.01. Adding your monthly deposit of $100, you have a balance at the end of month 3 of

$$\underbrace{\$201.00}_{\text{prior balance}} + \underbrace{\$2.01}_{\text{interest}} + \underbrace{\$100}_{\text{new deposit}} = \$303.01$$

Table 4.4 continues these calculations through 6 months.

TABLE 4.4 Savings Plan Calculations

END OF...	PRIOR BALANCE	INTEREST ON PRIOR BALANCE	END-OF-MONTH DEPOSIT	NEW BALANCE
Month 1	$0	$0	$100	$100
Month 2	$100	1% × $100 = $1	$100	$201
Month 3	$201	1% × $201 = $2.01	$100	$303.01
Month 4	$303.01	1% × $303.01 = $3.03	$100	$406.04
Month 5	$406.04	1% × $406.04 = $4.06	$100	$510.10
Month 6	$510.10	1% × $510.10 = $5.10	$100	$615.20

NOTE: The last column shows the new balance at the end of each month, which is the sum of the prior balance, the interest, and the end-of-month deposit.

In principle, we could extend this table indefinitely—but it would take a lot of work! Fortunately, there's a much easier way: the **savings plan formula.**

Savings Plan Formula (Regular Payments)

$$A = \text{PMT} \times \frac{\left[\left(1 + \frac{\text{APR}}{n}\right)^{(nY)} - 1\right]}{\left(\frac{\text{APR}}{n}\right)}$$

where

A = accumulated savings plan balance
PMT = regular payment (deposit) amount
APR = annual percentage rate (as a decimal)
n = number of payment periods per year
Y = number of years

TECHNICAL NOTE
This formula assumes the same payment and compounding periods. For example, if payments are made monthly, interest also is calculated and paid monthly. If the compounding period is different from the payment period, replace the term APR/n by the effective yield for each payment period.

EXAMPLE 1 Using the Savings Plan Formula

Use the savings plan formula to calculate the balance after 6 months for an APR of 12% and monthly payments of $100.

Solution We have monthly payments of PMT = $100, annual interest rate of APR = 0.12, $n = 12$ because the payments are made monthly, and $Y = \frac{1}{2}$ because 6 months is a half year. Using the savings plan formula, we can find the balance after 6 months:

$$A = \text{PMT} \times \frac{\left[\left(1 + \frac{\text{APR}}{n}\right)^{(nY)} - 1\right]}{\left(\frac{\text{APR}}{n}\right)}$$

$$= \$100 \times \frac{\left[\left(1 + \frac{0.12}{12}\right)^{(12 \times 1/2)} - 1\right]}{\left(\frac{0.12}{12}\right)}$$

$$= \$100 \times \frac{[(1.01)^6 - 1]}{0.01} = \$615.20$$

Note that this answer agrees with Table 4.4.

by the way...
A savings plan in which payments are made at the end of each month is called an *ordinary annuity*. A plan in which payments are made at the beginning of each period is called an *annuity due*. In both cases, the accumulated amount, A, at some future date is called the *future value* of the annuity. The formulas in this unit apply only to ordinary annuities.

Using Your CALCULATOR The Savings Plan Formula

There are many ways to do the savings plan formula on your calculator. On a graphing calculator, you may be able to do the calculations directly if you use the parentheses keys; some business calculators have built-in functions that allow you to make savings plan calculations in a single step. However, the following procedure will work on most scientific calculators. The example uses numbers from Example 2 (below): $n = 12$ payments (monthly payments), PMT $= \$100$, APR $= 8\% = 0.08$, and $Y = 35$ years. *It is very important that you not round any number until the end of the calculation.*

	In general	Example	Display
Starting Formula:	$A = \text{PMT} \times \left[\dfrac{\left(1 + \dfrac{\text{APR}}{n}\right)^{(nY)} - 1}{\dfrac{\text{APR}}{n}}\right]$	$\$100 \times \left[\dfrac{\left(1 + \dfrac{0.08}{12}\right)^{(12 \times 35)} - 1}{\dfrac{0.08}{12}}\right]$	—
Step 1. Multiply factors in exponent.	$n \; \boxed{\times} \; Y \; \boxed{=}$	$12 \; \boxed{\times} \; 35 \; \boxed{=}$	420.
Step 2. Store product in memory (or write down).	$\boxed{\text{Store}}$	$\boxed{\text{Store}}$	420.
Step 3. Add terms 1 and APR/n.	$1 \; \boxed{+} \; \text{APR} \; \boxed{\div} \; n \; \boxed{=}$	$1 \; \boxed{+} \; 0.08 \; \boxed{\div} \; 12 \; \boxed{=}$	1.0066666667
Step 4. Raise result to power in memory.	$\boxed{y^x} \; \boxed{\text{Recall}} \; \boxed{=}$	$\boxed{y^x} \; \boxed{\text{Recall}} \; \boxed{=}$	16.2925499
Step 5. Subtract 1 from result.	$\boxed{-} \; 1 \; \boxed{=}$	$\boxed{-} \; 1 \; \boxed{=}$	15.2925499
Step 6. Store result in memory (or write down).	$\boxed{\text{Store}}$	$\boxed{\text{Store}}$	15.2925499
Step 7. Compute denominator, then take its reciprocal.	$\text{APR} \; \boxed{\div} \; n \; \boxed{=} \; \boxed{1/x}$	$0.08 \; \boxed{\div} \; 12 \; \boxed{=} \; \boxed{1/x}$	150.
Step 8. Multiply by result in memory and payment.	$\boxed{\times} \; \boxed{\text{Recall}} \; \boxed{\times} \; \text{PMT} \; \boxed{=}$	$\boxed{\times} \; \boxed{\text{Recall}} \; \boxed{\times} \; 100 \; \boxed{=}$	229,388.249

With the calculation complete, you can round to the nearest cent, writing the answer as $229,388.25. Be sure to check the calculation.

EXAMPLE 2 *Retirement Plan*

At age 30, Michelle starts an IRA to save for retirement. She deposits $100 at the end of each month. If she can count on an APR of 8%, how much will she have when she retires 35 years later at age 65? Compare the IRA's value to the total amount of her deposits over this time period.

Solution We use the savings plan formula for payments of PMT $= \$100$, an interest rate of APR $= 0.08$, and $n = 12$ for monthly deposits. The balance after $Y = 35$ years is

$$A = \text{PMT} \times \dfrac{\left[\left(1 + \dfrac{\text{APR}}{n}\right)^{(nY)} - 1\right]}{\left(\dfrac{\text{APR}}{n}\right)} = \$100 \times \dfrac{\left[\left(1 + \dfrac{0.08}{12}\right)^{(12 \times 35)} - 1\right]}{\left(\dfrac{0.08}{12}\right)}$$

$$= \$100 \times \dfrac{[(1.006666667)^{420} - 1]}{0.006666667}$$

$$= \$229,388.25$$

Because 35 years is 420 months (35 × 12 = 420), the total amount of her deposits over the 35 years is

$$420 \text{ months} \times \$100/\text{month} = \$42{,}000$$

She will deposit a total of $42,000 over 35 years. However, thanks to compounding, her IRA will have a balance of almost $230,000—*more than five times* the amount of her contributions. ∎

PLANNING AHEAD WITH SAVINGS PLANS

Most people start savings plans with a particular goal in mind, such as saving enough for retirement or enough to buy a new car in a couple of years. For planning ahead, the important question is this: Given a financial goal (the total amount, A, desired after a certain number of years), what regular payments are needed to reach the goal? The following two examples show how the calculations work.

> *by the way ...*
> The lump sum deposit that would give you the same end result as regular payments into a savings plan is called the *present value* of the savings plan.

EXAMPLE 3 *College Savings Plan at 7%*

You want to build a $100,000 college fund in 18 years by making regular, end-of-month deposits. Assuming an APR of 7%, calculate how much you should deposit monthly. How much of the final value comes from actual deposits and how much from interest?

Solution The goal is to accumulate $A = \$100{,}000$ over $Y = 18$ years. The interest rate is APR = 0.07 and monthly payments mean $n = 12$. We want to calculate the required monthly payments, PMT. We therefore need to solve the savings plan formula for PMT. We begin with the savings plan formula:

$$A = \text{PMT} \times \frac{\left[\left(1 + \frac{\text{APR}}{n}\right)^{(nY)} - 1\right]}{\left(\frac{\text{APR}}{n}\right)}$$

To isolate PMT, we multiply both sides by $\left(\frac{\text{APR}}{n}\right)$ and divide both sides by $\left[\left(1 + \frac{\text{APR}}{n}\right)^{(nY)} - 1\right]$. You should confirm the following result:

$$\text{PMT} = \frac{A \times \frac{\text{APR}}{n}}{\left[\left(1 + \frac{\text{APR}}{n}\right)^{(nY)} - 1\right]}$$

Using Your CALCULATOR

On most calculators, it is easiest to calculate the denominator first, then take its reciprocal and multiply by the other terms.

Now we substitute the given values for A, APR, n, and Y.

$$\text{PMT} = \frac{A \times \frac{\text{APR}}{n}}{\left[\left(1 + \frac{\text{APR}}{n}\right)^{(nY)} - 1\right]} = \frac{\$100{,}000 \times \frac{0.07}{12}}{\left[\left(1 + \frac{0.07}{12}\right)^{(12 \times 18)} - 1\right]}$$

$$= \frac{\$100{,}000 \times 0.005833333}{[(1.005833333)^{216} - 1]}$$

$$= \$232.17$$

Assuming the APR remains 7%, monthly payments of $232.17 will give you $100,000 after 18 years. During that time, you deposit a total of

$$18 \text{ yr} \times \frac{12 \text{ mo}}{\text{yr}} \times \frac{\$232.17}{\text{mo}} = \$50{,}148.72$$

Just over half of the $100,000 comes from your actual deposits; the rest is the result of compound interest. ■

Time out to think

Suppose you want a $100,000 college fund in 18 years and you are counting on an APR of 7%. In Example 3 above, we found that you could reach your goal with monthly deposits of about $232. In Unit 4A (Example 7), we found that you could reach the same goal with a lump sum deposit of $29,586 today. Discuss the circumstances under which you might choose either the lump sum or the savings plan.

EXAMPLE 4 A Comfortable Retirement

You would like to retire 25 years from now, and you would like to have a retirement fund from which you can draw an income of $50,000 per year—forever! How can you do it? Assume a constant APR of 9%.

Solution You can achieve your goal by building a retirement fund that is large enough to earn $50,000 per year *from interest alone*. In that case, you can withdraw the interest for your living expenses while leaving the principal untouched (for your heirs!). The principal will then continue to earn the same $50,000 interest year after year (assuming there is no change in interest rates).

What balance do you need to earn $50,000 annually from interest? Since we are assuming an APR of 9%, the $50,000 must be 9% = 0.09 of the total balance. That is,

$$\$50{,}000 = 0.09 \times (\text{total balance})$$

Dividing both sides by 0.09, we find

$$\text{total balance} = \frac{\$50{,}000}{0.09} = \$555{,}556$$

In other words, with a 9% APR, a balance of about $556,000 allows you to withdraw $50,000 per year without ever reducing the principal.

Let's assume you will try to accumulate this balance of $A = \$556{,}000$ by making regular, monthly deposits into a savings plan. We have APR = 0.09, $n = 12$ (for monthly deposits), and $Y = 25$ years. As in Example 3, we calculate the required monthly deposits by using the savings plan formula solved for PMT.

$$\text{PMT} = \frac{A \times \frac{\text{APR}}{n}}{\left[\left(1 + \frac{\text{APR}}{n}\right)^{(nY)} - 1\right]} = \frac{\$556{,}000 \times \frac{0.09}{12}}{\left[\left(1 + \frac{0.09}{12}\right)^{(12 \times 25)} - 1\right]}$$

$$= \frac{\$556{,}000 \times 0.0075}{[(1.0075)^{300} - 1]}$$

$$= \$495.93$$

If you deposit about $500 per month over the next 25 years, you will achieve your retirement goal—assuming you can count on a 9% APR. Although saving $500 per month may seem like a lot, it can be easier than it sounds thanks to special tax treatment for retirement plans. ∎

> **by the way ...**
> An account that provides a permanent source of income without reducing its principal is called an *endowment*. Many charitable foundations are endowments: They spend each year's interest (or a portion of the interest) on charitable activities, while leaving the principal untouched to earn interest again in future years. Of course, the *value* of a particular dollar amount tends to decline with time, because inflation reduces the value of a dollar.

TOTAL AND ANNUAL RETURN

In the examples so far, we've assumed that you get a constant interest rate for a long period of time. In reality, interest rates usually vary over time. Consider a case in which you invest a starting principal of $1000 and it grows to $1500 in 5 years. Although the interest rate may have varied during the 5 years, we can still describe the change in both total and annual terms.

Your **total return** is the relative change in the investment value over the 5-year period (see Unit 3A for a discussion of relative change):

$$\text{total return} = \frac{\text{new value} - \text{starting principal}}{\text{starting principal}} = \frac{\$1500 - \$1000}{\$1000} = 0.5$$

The total return on this investment is 50% over 5 years.

Your **annual return** is the average annual rate at which your money grew over the 5 years. That is, it is the constant annual percentage yield (APY) that would give the same result in 5 years. In this case, the annual return is about 8.5%. We can see why by using the compound interest formula for interest paid once a year. We set the interest rate (APR) to the annual yield of APY = 8.5% = 0.085 and the number of years to $Y = 5$. The compound interest formula confirms that a starting principal $P = \$1000$ grows to about $A = \$1500$ in 5 years:

$$A = P \times (1 + \text{APY})^Y = \$1000 \times (1 + 0.085)^5 = \$1503.66$$

In this case, we "guessed" the APY, then confirmed our guess with the compound interest formula. We can calculate the annual return more directly by solving the compound interest formula above for APY. The algebra is shown in the Brief Review on pp. 200–201. The result is summarized in the box on p. 202.

a brief review...

ALGEBRA WITH POWERS AND ROOTS

As we have seen, powers and roots are often used in financial calculations. A review of these operations may be helpful.

Basics of Powers
A number raised to the nth power is that number multiplied by itself n times (n is called an *exponent*). For example:

$$2^1 = 2 \quad\quad 2^2 = 2 \times 2 = 4 \quad\quad 2^3 = 2 \times 2 \times 2 = 8$$

A number to the *zero* power is defined to be 1. For example,

$$2^0 = 1$$

Negative powers are the reciprocals of the corresponding positive powers. For example:

$$5^{-2} = \frac{1}{5^2} = \frac{1}{5 \times 5} = \frac{1}{25} \quad\quad 2^{-3} = \frac{1}{2^3} = \frac{1}{2 \times 2 \times 2} = \frac{1}{8}$$

Power Rules
In the following rules, x represents a number being raised to a power, and n and m are exponents; note that these rules work only when all powers involve the same number x. (See also A Brief Review on p. 91 [Unit 2B].)

- To multiply powers of the same number, *add* the exponents:

$$x^n \times x^m = x^{m+n}$$

Example: $2^3 \times 2^2 = 2^{3+2} = 2^5 = 32$

- To divide powers of the same number, *subtract* the exponents:

$$\frac{x^n}{x^m} = x^{n-m}$$

Example: $\frac{5^3}{5^2} = 5^{3-2} = 5^1 = 5$

- When a power is raised to another power, *multiply* the exponents:

$$(x^n)^m = x^{n \times m}$$

Example: $(2^2)^3 = 2^{2 \times 3} = 2^6 = 64$

Basics of Roots
Finding a root is the reverse of raising a number to a power. Second roots, or square roots, are written with a number under the root symbol $\sqrt{}$. More generally, we indicate an nth root by writing a number under the symbol $\sqrt[n]{}$. For example:

$$\sqrt{4} = 2 \text{ because } 2^2 = 2 \times 2 = 4$$
$$\sqrt[3]{27} = 3 \text{ because } 3^3 = 3 \times 3 \times 3 = 27$$
$$\sqrt[4]{16} = 2 \text{ because } 2^4 = 2 \times 2 \times 2 \times 2 = 16$$
$$\sqrt[6]{1{,}000{,}000} = 10 \text{ because } 10^6 = 1{,}000{,}000$$

EXAMPLE 5 *Mutual Fund Gain*

You invest $3000 in the Clearwater mutual fund. Over 4 years, your investment grows in value to $8400. What are your total and annual returns for the 4-year period?

Solution You have a starting principal $P = \$3000$ and a later accumulated value $A = \$8400$. Thus, your total and annual returns for the 4 years are

$$\text{total return} = \frac{(A - P)}{P} = \frac{(\$8400 - \$3000)}{\$3000}$$
$$= 1.8 = 180\%$$

and

Roots as Fractional Powers

The nth root of a number is the same as the number raised to the $1/n$ power. That is,

$$\sqrt[n]{x} = x^{1/n}$$

For example:

$$64^{1/3} = \sqrt[3]{64} = 4^{1/2} = \sqrt{4} = 2$$
$$1{,}000{,}000^{1/6} = \sqrt[6]{1{,}000{,}000} = 10$$

Power and Root Algebra

The following two rules hold true for working with equations:

1. We can raise both sides of an equation to the same power.
2. We can take the same root of both sides of an equation, which is equivalent to raising both sides to the same fractional power. (Note: This process may produce both positive and negative roots; we consider only positive roots here.)

Example: Find the positive solution of the equation $x^4 = 16$.
Solution: We isolate x by raising both sides to the $1/4$ power:

$$(x^4)^{1/4} = 16^{1/4} \;\rightarrow\; x^{4 \times 1/4} = 16^{1/4} \;\rightarrow\; x = 16^{1/4} = 2$$

Therefore, one solution of the equation is $x = 2$. Note that, in the last step, we recognized that $x^{4 \times 1/4} = x^1 = x$.

Example: Solve the equation $A = P \times (1 + \text{APY})^Y$ for APY.
Solution: First, we isolate the term containing APY by dividing both sides of the equation by P:

$$\frac{A}{P} = (1 + \text{APY})^Y$$

Next, we raise both sides of the equation to the $1/Y$ power, then simplify the right side:

$$\left(\frac{A}{P}\right)^{1/Y} = [(1 + \text{APY})^Y]^{1/Y}$$
$$= \underbrace{(1 + \text{APY})^{Y \times 1/Y}}_{\text{This step from the rule } (x^n)^m = x^{n \times m}} = \underbrace{1 + \text{APY}}_{\text{This step from the rule } x^{n/n} = x^1 = x}$$

Finally, we isolate APY by subtracting 1 from both sides:

$$\left(\frac{A}{P}\right)^{1/Y} - 1 = 1 + \text{APY} - 1 \;\rightarrow\; \text{APY} = \left(\frac{A}{P}\right)^{1/Y} - 1$$

(In the last step, we interchanged the left and right sides.) To get the formula in the box on p. 202, we replace APY by *annual return*; this replacement is valid because the annual return is the same as the constant APY that would lead to an accumulated balance A.

$$\text{annual return} = \left(\frac{A}{P}\right)^{(1/Y)} - 1 = \left(\frac{\$8400}{\$3000}\right)^{(1/4)} - 1 = \sqrt[4]{2.8} - 1 = 0.294 = 29.4\%$$

Your total return is 1.8, or 180%, meaning that the value of your investment after 4 years is 1.8 times its original value. Your annual return is 0.294, or 29.4%, meaning that your investment has grown by an average of 29.4% each year.

Note that you should check your answer for the annual return by using the compound interest formula. If you use the annual return as the APY, the compound interest formula should give you the correct accumulated value. In this case,

$$A = P \times (1 + \text{APY})^Y = \$3000 \times (1 + 0.294)^4 = \$8411.21$$

This is very close to the correct value of $8400; the slight difference is due to rounding when we calculated the annual return. ■

Using Your CALCULATOR

Raising a number to the $1/Y$ power is the same as taking the Yth root. The key for taking a root will be labeled something like $\sqrt[y]{x}$ or $x^{1/y}$. For example, on many calculators you would calculate $\sqrt[4]{2.8}$ by pressing 2.8 $\sqrt[y]{x}$ 4 $=$.

> **Total and Annual Return**
>
> Consider an investment that grows from an original principal P to a later accumulated balance A.
>
> The **total return** is the relative change in the investment value:
>
> $$\text{total return} = \frac{(A - P)}{P}$$
>
> The **annual return** is the average annual percentage yield (APY) that would give the same overall growth. The formula is
>
> $$\text{annual return} = \left(\frac{A}{P}\right)^{(1/Y)} - 1$$
>
> where Y is the investment period in years.

EXAMPLE 6 *Investment Loss*

You purchased shares in NewWeb.com for $2000. Three years later, you sold them for $1100. What were your total return and annual return on this investment?

Solution You had a starting principal $P = \$2000$ and a later accumulated value $A = \$1100$. Thus, your total and annual returns for the 3 years were

$$\text{total return} = \frac{(A - P)}{P} = \frac{(\$1100 - \$2000)}{\$2000} = -0.45$$

and

$$\text{annual return} = \left(\frac{A}{P}\right)^{(1/Y)} - 1 = \left(\frac{\$1100}{\$2000}\right)^{(1/3)} - 1 = \sqrt[3]{0.55} - 1 = -0.18$$

The returns are negative because you *lost* money on this investment. Your total return was -0.45, or -45%, meaning that your investment lost 45% of its original value. Your annual return was -0.18, or -18%, meaning that your investment lost an average of 18% of its value each year.

Review Questions

1. What is a savings plan? Explain the savings plan formula.
2. Give an example of a situation in which you might want to solve the savings plan formula to find the payments, PMT, required to achieve some goal.
3. Distinguish between the total return and the annual return on an investment. How do we calculate the annual return? Give an example.

In the News

1. **Advertised Investment.** Find an advertisement for an investment plan. Describe some of the cited benefits of the plan. Using what you learned in this unit, identify at least one possible drawback of the plan.

Exercises

Review of Powers and Roots. Exercises 1–4 use skills covered in the Brief Review on pp. 200–201.

1. Evaluate the following expressions; express the answer in simplest terms.
 a. 4^3 b. $2^3 \times 2^5$ c. $3^6 \div 3^2$
 d. 4^{-2} e. $4^5 \times 4^{-2}$ f. $5^3 \div 5^{-4}$

2. Evaluate the following expressions; express the answer in simplest terms.
 a. $25^{1/2}$ b. $27^{1/3}$ c. $81^{1/2} \times 3^2$
 d. $16^{-1/2}$ e. $64^{-1/3}$ f. $36^{1/2} \div 16^{1/4}$

3. Solve the following equations for the unknown, x.
 a. $x^2 = 100$ b. $x^3 = 27$ c. $x^{1/3} = 2$ d. $x^{1/5} = 2$

4. Solve the following equations for the unknown, x.
 a. $(x-2)^2 = 36$ b. $x^3 + 3 = 30$
 c. $(x/2)^2 = 25$ d. $x^2 - 4 = 32$

Investment Plans. Use the savings plan formula in Exercises 5–8.

5. You set up an IRA (individual retirement account) with an APR of 8% at age 25. At the end of each month, you deposit $50 in the account. How much will the IRA contain when you retire at age 65? Compare that amount to the total amount of deposits made over the time period.

6. A friend creates an IRA with an APR of 8.25%. She starts the IRA at age 25 and deposits $50 per month. How much will her IRA contain when she retires at age 65? Compare that amount to the total amount of deposits made over the time period.

7. You put $200 per month in an investment plan that pays an APR of 7%. How much money will you have after 18 years? Compare this amount to the total amount of deposits made over the time period.

8. You put $200 per month in an investment plan that pays an APR of 7.5%. How much money will you have after 18 years? Compare this amount to the total amount of deposits made over the time period.

Who Comes Out Ahead? Exercises 9–12 each describe two savings plans. Compare the balances in the two plans after 10 years. Who deposits more money in each case? Who comes out ahead in each case? Comment on any lessons about savings plans that you find in the results. (Assume that, for each plan, the payment and compounding periods are the same, so the savings plan formula is valid.)

9. Yolanda deposits $100 per month in an account with an APR of 5%, while Zach deposits $1200 at the end of each year in an account with an APR of 5%.

10. Polly deposits $50 per month in an account with an APR of 6%, while Quint deposits $40 per month in an account with an APR of 6.5%.

11. Juan deposits $200 per month in an account with an APR of 6%, while Maria deposits $2500 at the end of each year in an account with an APR of 6.5%.

12. George deposits $40 per month in an account with an APR of 7%, while Harvey deposits $150 per quarter in an account with an APR of 7.5%.

Investment Planning. Use the savings plan formula in Exercises 13–16.

13. You intend to create a college fund for your baby. If you can get an APR of 7.5% and want the fund to have a value of $150,000 after 18 years, how much should you deposit monthly?

14. At age 35 you start saving for retirement. If your investment plan pays an APR of 9% and you want to have $2 million when you retire in 30 years, how much should you deposit monthly?

15. You want to purchase a new car in 3 years and expect the car to cost $10,000. Your bank offers a plan with a

guaranteed interest rate of APR = 5.5% if you make regular monthly deposits. How much should you deposit each month to end up with $10,000 in 3 years?

16. At age 20 you start saving for retirement. If your investment plan pays an APR of 8% and you want to have $5 million when you retire in 45 years, how much should you deposit monthly?

Comparing Investment Plans. Suppose you want to accumulate $50,000 for your child's college fund within the next 15 years. Explain fully whether the investment plans in Exercises 17–20 will allow you to reach your goal.

17. You deposit $50 per month into an account with an APR of 7%.
18. You deposit $75 per month into an account with an APR of 7%.
19. You deposit $100 per month into an account with an APR of 6%.
20. You deposit $75 per month into an account with an APR of 10%.
21. **Comfortable Retirement.** Suppose you are 30 years old and would like to retire at age 60. Furthermore, you would like to have a retirement fund from which you can draw an income of $50,000 per year—forever! How can you do it? Assume a constant APR of 8%.
22. **Very Comfortable Retirement.** Suppose you are 25 years old and would like to retire at age 65. Furthermore, you would like to have a retirement fund from which you can draw an income of $200,000 per year—forever! How can you do it? Assume a constant APR of 8%.

Total and Annual Returns. In Exercises 23–29, compute the total and annual returns on the described investment.

23. Five years after buying 100 shares of XYZ stock for $55 per share, you sell the stock for $10,300.
24. You pay $8000 for a municipal bond. When it matures after 20 years, you receive $12,500.
25. Ten years after purchasing shares in a mutual fund for $5500, you sell them for $11,300.
26. Three years after paying $4500 for shares in a startup company, you sell the shares for $2500 (at a loss).
27. Five years after paying $5000 for shares in a new company, you sell the shares for $3000 (at a loss).
28. Three years after buying 200 shares of XYZ stock for $25 per share, you sell the stock for $8500.
29. Eight years after purchasing shares in a mutual fund for $7500, you sell them for $12,600.

30. **Death and the Maven (A True Story).** In December 1995, 101-year-old Anne Scheiber died and left $22 million to Yeshiva University. This fortune was accumulated through shrewd and patient investment of a $5000 nest egg over the course of 50 years. In turning $5000 into $22 million, what were her total and annual returns?

31. **Personal Savings Plan.** Describe something for which you would like to save money right now. How much do you need to save? How long do you have to save it? Based on these needs, calculate how much you should deposit each month in a savings plan to meet your goal. For the interest rate, use the highest rate currently available at local banks.

Web Projects

Find useful links for Web Projects on the text Web site: www.aw.com/bennet-briggs

1. **Investment Tracking.** Choose three stocks that you think would make good investments. Imagine that you invest $100 in each of these nine investments. Use the Web to track the value of your investment portfolio over the next 5 weeks. Based on the portfolio value at the end, find your return for the 5-week period. Which investments fared the best, and which did most poorly?

2. **Financial Web Sites.** Visit one of the many financial news and advising Web sites. Describe the services offered by the Web site. Explain whether, as an active or prospective investor, you find the Web site useful.

UNIT 4C

LOAN PAYMENTS, CREDIT CARDS, AND MORTGAGES

Do you have a credit card? Do you have a loan for your car? Do you have student loans? Do you own a house? Chances are that you owe money for at least one of these purposes. If so, you not only have to pay back the money you borrowed, but also have to pay interest on the money that you owe. In this unit, we will begin by studying the basic ideas of loans, then apply these ideas to common loans including credit cards and mortgages.

LOAN BASICS

Suppose you borrow $1200 at an annual interest rate of APR = 12%, or 1% per month. At the end of the first month, you owe interest in the amount of

$$1\% \times \$1200 = \$12$$

If you paid *only* this $12 in interest, you'd still owe $1200. That is, the total amount of the loan, called the loan **principal**, would still be $1200. In that case, you'd owe the same $12 in interest the next month. In fact, if you paid only the interest from one month to the next, the loan would never be paid off and you'd have to pay $12 per month forever.

If you hope to make progress in paying off the loan, you need to pay part of the principal as well as interest. For example, suppose that you pay $200 toward your loan principal each month, plus the current interest. At the end of the first month, you'd pay $200 toward principal *plus* $12 for the 1% interest you owe, making a total payment of $212. Because you've paid $200 toward principal, your new loan principal would be $1200 − $200 = $1000.

At the end of the second month, you'd again pay $200 toward principal and 1% interest. But this time the interest is on the $1000 that you still owe. Thus, your interest payment would be 1% × $1000 = $10, making your total payment $210. Table 4.5 shows how the calculations continue until the loan is paid off after 6 months.

> **Loan Basics**
>
> For any loan, the **principal** is the amount of money owed at any particular time. Interest is charged on the loan principal. To pay off a loan, you must gradually pay down the principal. Thus, in general, every payment should include all the interest you owe plus some amount that goes toward paying off the principal.

TABLE 4.5 Payments and Loan Principal for a $1200 Loan with Principal Paid Off at a Constant $200 per Month

END OF...	PRIOR PRINCIPAL	INTEREST ON PRIOR PRINCIPAL	PAYMENT TOWARD PRINCIPAL	TOTAL PAYMENT	NEW PRINCIPAL
Month 1	$1200	1% × $1200 = $12	$200	$212	$1000
Month 2	$1000	1% × $1000 = $10	$200	$210	$800
Month 3	$800	1% × $8000 = $8	$200	$208	$600
Month 4	$600	1% × $6000 = $6	$200	$206	$400
Month 5	$400	1% × $4000 = $4	$200	$204	$200
Month 6	$200	1% × $2000 = $2	$200	$202	$0

Installment Loans

For the case illustrated in Table 4.5, your total payment decreases from month to month because of the declining amount of interest that you owe. There's nothing inherently wrong with this method of paying off a loan, but most people prefer to pay the *same* total amount each month because it makes planning a budget easier. A loan that you pay off with equal regular payments is called an **installment loan** (or *amortized loan*).

Suppose you wanted to pay off your $1200 loan with 6 equal monthly payments. How much should you pay each month? Because the payments in Table 4.5 vary between $202 and $212, it's clear that the equal monthly payments must lie somewhere in this range. The exact amount is not obvious, but we can calculate it with the **loan payment formula.**

Loan Payment Formula (Installment Loans)

$$\text{PMT} = \frac{P \times \left(\dfrac{\text{APR}}{n}\right)}{\left[1 - \left(1 + \dfrac{\text{APR}}{n}\right)^{(-nY)}\right]}$$

where

P = starting loan principal (amount borrowed)
PMT = regular payment amount
APR = annual percentage rate
n = number of payment periods per year
Y = loan term in years

In our current example, the starting loan principal is P = $1200, the interest rate is APR = 12%, the loan term is $Y = \frac{1}{2}$ year (6 months), and monthly payments mean n = 12. The loan payment formula gives

$$\text{PMT} = \frac{P \times \left(\dfrac{\text{APR}}{n}\right)}{\left[1 - \left(1 + \dfrac{\text{APR}}{n}\right)^{(-nY)}\right]} = \frac{\$1200 \times \left(\dfrac{0.12}{12}\right)}{\left[1 - \left(1 + \dfrac{0.12}{12}\right)^{(-12 \times 1/2)}\right]}$$

$$= \frac{\$1200 \times (0.01)}{\left[1 - (1 + 0.01)^{-6}\right]}$$

$$= \frac{\$12}{1 - 0.942045235}$$

$$= \$207.06$$

The monthly payments would be $207.06, which, as we expected, is between $200 and $212.

Using Your CALCULATOR The Loan Payment Formula

As with other formulas in this chapter, there are many ways to do loan calculations on your calculator. Graphing or business calculators may make the calculations easier. Here is a procedure that will work on most scientific calculators. The example uses $P = \$1200$, APR = 12%, $n = 12$ (monthly payments), and $Y = \frac{1}{2}$ year (6 months). *It is important that you not round any numbers until the last step.*

		In general	Example	Display
Starting Formula:		$\text{PMT} = \dfrac{P \times \left(\dfrac{\text{APR}}{n}\right)}{\left[1 - \left(1 + \dfrac{\text{APR}}{n}\right)^{(-nY)}\right]}$	$\dfrac{\$1200 \times \left(\dfrac{0.12}{12}\right)}{\left[1 - \left(1 + \dfrac{0.12}{12}\right)^{-(12 \times 1/2)}\right]}$	___
Step 1.	Multiply factors in exponent.	n $\boxed{+/-}$ $\boxed{\times}$ Y $\boxed{=}$	12 $\boxed{+/-}$ $\boxed{\times}$ 1 $\boxed{\div}$ 2 $\boxed{=}$	$-6.$
Step 2.	Store product in memory (or write down).	$\boxed{\text{Store}}$	$\boxed{\text{Store}}$	$-6.$
Step 3.	Add denominator terms 1 and APR/n.	1 $\boxed{+}$ APR $\boxed{\div}$ n $\boxed{=}$	1 $\boxed{+}$ 0.12 $\boxed{\div}$ 12 $\boxed{=}$	1.01
Step 4.	Raise result to power in memory.	$\boxed{y^x}$ $\boxed{\text{Recall}}$ $\boxed{=}$	$\boxed{y^x}$ $\boxed{\text{Recall}}$ $\boxed{=}$	0.942045235
Step 5.	Subtract result from 1 by making result negative and adding 1.	$\boxed{+/-}$ $\boxed{+}$ 1 $\boxed{=}$	$\boxed{+/-}$ $\boxed{+}$ 1 $\boxed{=}$	0.057954765
Step 6.	Denominator is now complete; take its reciprocal.	$\boxed{1/x}$	$\boxed{1/x}$	17.25483667
Step 7.	Multiply result by factors P and APR/n.	$\boxed{\times}$ P $\boxed{\times}$ APR $\boxed{\div}$ n $\boxed{=}$	$\boxed{\times}$ 1200 $\boxed{\times}$ 0.12 $\boxed{\div}$ 12 $\boxed{=}$	207.0580401

With the calculation complete, you can round to the nearest cent, writing the answer as $207.06. Be sure to check the calculation.
*The $\boxed{+/-}$ key is used on scientific calculators to change the sign of a number.

Because the loan principal is gradually paid down with the installment payments, the interest due each month must also decline gradually. Thus, because the payments remain the same, the amount paid toward principal each month gradually rises. We therefore have the general relationship between principal and interest summarized in the following box.

> **Principal and Interest for Installment Loans**
>
> The portions of installment loan payments going toward principal and toward interest vary as the loan is paid down. Early in the loan term, the portion going toward interest is relatively high and the portion going toward principal is relatively low. As the term proceeds, the portion going toward interest gradually decreases and the portion going toward principal gradually increases.

EXAMPLE 1 *Student Loan*

Suppose you have student loans totaling $7500 when you graduate from college. The interest rate is APR = 9% and the loan term is 10 years. What are your monthly payments? How much will you pay over the lifetime of the loan? What is the total interest you will pay on the loan?

Solution The starting loan principal is $P = \$7500$, the interest rate is APR = 0.09, the loan term is $Y = 10$ years, and $n = 12$ for monthly payments. We use the loan payment formula to find the monthly payments:

$$\text{PMT} = \frac{P \times \left(\frac{\text{APR}}{n}\right)}{\left[1 - \left(1 + \frac{\text{APR}}{n}\right)^{(-nY)}\right]} = \frac{\$7500 \times \left(\frac{0.09}{12}\right)}{\left[1 - \left(1 + \frac{0.09}{12}\right)^{(-12 \times 10)}\right]}$$

$$= \frac{\$7500 \times (0.0075)}{[1 - (1.0075)^{-120}]}$$

$$= \frac{\$56.25}{[1 - 0.407937305]}$$

$$= \$95.01$$

Your monthly payments are $95.01. Over the 10-year term, your total payments will be

$$10 \text{ yr} \times 12 \frac{\text{mo}}{\text{yr}} \times \frac{\$95.01}{\text{mo}} = \$11{,}401.20$$

Of this amount, $7500 pays off the principal. The rest, or $11,401 − $7500 = $3901, represents interest payments.

TECHNICAL NOTE
Because we assume the compounding period is the same as the payment period and because we round payments to the nearest cent, the calculated payments may differ slightly from actual payments.

EXAMPLE 2 *Principal and Interest Payments*

For the loan in Example 1, calculate the portions of your payments that go to principal and to interest during the first 3 months.

Solution The monthly interest rate is APR/12 = 0.09/12 = 0.0075. For a $7500 starting loan principal, the interest due at the end of the first month is

$$0.0075 \times \$7500 = \$56.25$$

Your monthly payment (calculated in Example 1) is $95.01. We've found that the interest due is $56.25, so the rest, or $95.01 − $56.25 = $38.76, goes to principal. Thus, after your first payment, your new loan principal is

$$\$7500 - \$38.76 = \$7461.24$$

Table 4.6 continues the same calculations for months 2 and 3. Note that, as expected, the interest payment gradually decreases and the payment toward principal gradually increases. But also note that, for these first 3 months of a 10-year loan, more than half of each payment goes toward interest. We could continue this table through the life of the loan, but it's generally easier to use software that finds principal and interest payments with built-in functions.

> **by the way...**
> A table of principal and interest payments over the life of a loan is called an *amortization schedule*. Most banks will provide an amortization schedule for any loan you are considering.

TABLE 4.6 Interest and Principal Portions of Payments on a $7500 Loan (10-year term, APR = 9%)

END OF...	INTEREST = 0.0075 × BALANCE	PAYMENT TOWARD PRINCIPAL	NEW PRINCIPAL
Month 1	0.0075 × $7500 = $56.25	$95.01 − $56.25 = $38.76	$7500 − $38.76 = $7461.24
Month 2	0.0075 × $7461.24 = $55.96	$95.01 − $55.96 = $39.05	$7461.24 − $39.05 = $7422.19
Month 3	0.0075 × $7422.19 = $55.67	$95.01 − $55.67 = $39.34	$7422.19 − $39.34 = $7382.85

Time out to think

In a case such as the student loan in Examples 1 and 2, many people are surprised to find that more than half of their early loan payments goes to interest when the annual interest rate is only 9%. By referring to Table 4.6, explain why this is the case. How will the payments toward principal and interest compare toward the end of the loan?

Choices of Rate and Term

You'll usually have several choices of interest rate and loan term when seeking a loan. For example, a bank might offer a 3-year car loan at 8%, a 4-year loan at 9%, and a 5-year loan at 10%. You'll pay less total interest with the shorter-term, lower-rate loan, but this loan will have the highest monthly payments. Thus, you'll have to evaluate your choices and make the decision that is best for your personal situation.

EXAMPLE 3 *Choice of Auto Loans*

You need a $6000 loan to buy a used car. Your bank offers a 3-year loan at 8%, a 4-year loan at 9%, and a 5-year loan at 10%. Calculate your monthly payments and total interest over the loan term with each option.

Solution Let's begin with the 3-year loan at 8%. The starting loan principal is $P = \$6000$, the interest rate is APR $= 0.08$, the loan term is $Y = 3$ years, and $n = 12$ monthly payments. Your monthly payments would be

$$\text{PMT} = \frac{P \times \left(\dfrac{\text{APR}}{n}\right)}{\left[1 - \left(1 + \dfrac{\text{APR}}{n}\right)^{(-nY)}\right]} = \frac{\$6000 \times \left(\dfrac{0.08}{12}\right)}{\left[1 - \left(1 + \dfrac{0.08}{12}\right)^{(-12 \times 3)}\right]}$$

$$= \frac{\$40}{[1 - (1.006666667)^{-36}]}$$

$$= \$188.02$$

Three years is 36 months, so your payments would total $36 \times \$188.02 = \6768.72. Of this total, $6000 pays off your principal, so the total interest is the remaining $768.72.

For the 4-year loan at 9%, we repeat the calculations with APR $= 0.09$ and $Y = 4$ years:

$$\text{PMT} = \frac{P \times \left(\dfrac{\text{APR}}{n}\right)}{\left[1 - \left(1 + \dfrac{\text{APR}}{n}\right)^{(-nY)}\right]} = \frac{\$6000 \times \left(\dfrac{0.09}{12}\right)}{\left[1 - \left(1 + \dfrac{0.09}{12}\right)^{(-12 \times 4)}\right]} = \$149.31$$

Your total payments over 4 years, or 48 months, would be $48 \times \$149.31 = \7166.88. After we subtract the $6000 that goes to principal, the total interest is $1166.88.

For the 5-year loan, we set APR $= 0.1$ and $Y = 5$ years:

$$\text{PMT} = \frac{P \times \left(\dfrac{\text{APR}}{n}\right)}{\left[1 - \left(1 + \dfrac{\text{APR}}{n}\right)^{(-nY)}\right]} = \frac{\$6000 \times \left(\dfrac{0.1}{12}\right)}{\left[1 - \left(1 + \dfrac{0.1}{12}\right)^{(-12 \times 5)}\right]} = \$127.48$$

Your total payments over 5 years, or 60 months, would be $60 \times \$127.48 = \7648.80. After we subtract the $6000 that goes to principal, the total interest is $1648.80. As we expected, the monthly payments are lower with the longer-term loans, but the total interest is higher.

> *by the way...*
> You should always watch out for financial scams, especially when borrowing money. Keep in mind what is sometimes called the first rule of finance: *If it sounds too good to be true, it probably is!*

Time out to think

Consider your own current financial situation. If you needed a $6000 car loan, which option from Example 3 would you choose? Why?

CREDIT CARDS

Credit card loans differ from installment loans in that you are not required to pay off your balance in any set period of time. Instead, you are required to make only a minimum monthly payment that generally covers all the interest but very little principal. As a result, it takes a very long time to pay off your credit card loan if you make only the minimum payments. Therefore, if you wish to pay off your loan in a particular amount of time, you should use the loan payment formula to calculate the necessary payments.

A word of caution: Most credit cards have very high interest rates compared to other types of loans. As a result, it is easy to get into financial trouble if you get overextended with credit cards. The trouble is particularly bad if you miss your payments. In that case, you will probably be charged a late fee that is added to your principal, thereby increasing the amount of interest due the next month. With the interest charges operating like compound interest in reverse, failure to pay on time can put a person into an ever-deepening financial hole.

EXAMPLE 4 *Credit Card Debt*

You have a credit card balance of $2300 with an annual interest rate of 21%. You decide to pay off your balance over 1 year. How much will you need to pay each month? Assume you make no further credit card purchases.

Solution Your starting loan principal is $P = \$2300$, the interest rate is APR = 0.21, and monthly payments mean $n = 12$. Because you want to pay off the loan in 1 year, we set $Y = 1$. The required payments are

$$\text{PMT} = \frac{P \times \left(\frac{\text{APR}}{n}\right)}{\left[1 - \left(1 + \frac{\text{APR}}{n}\right)^{(-nY)}\right]} = \frac{\$2300 \times \left(\frac{0.21}{12}\right)}{\left[1 - \left(1 + \frac{0.21}{12}\right)^{(-12 \times 1)}\right]} = \$214.16$$

You must pay $214.16 per month to pay off the balance in 1 year. ■

Time out to think

Continuing Example 4, suppose you can get a personal loan at a bank at an annual interest rate of 10%. Should you take this loan and use it to pay off your $2300 credit card debt? Why or why not?

EXAMPLE 5 *A Deepening Hole*

Paul has gotten into credit card trouble. He has a balance of $9500 and just lost his job. His credit card company charges interest of APR = 21%, compounded daily. Suppose the credit card company allows him to suspend his payments until he finds a new job—but continues to charge interest. If it takes him a year to find a new job, how much will he owe when he starts his new job?

Solution Because Paul is not making payments during the year, this is not a loan payment problem. Instead, it is a compound interest problem, in which Paul's balance of $9500 grows at an annual rate of 21%, compounded daily. We use the compound interest formula with a starting balance of $P = \$9500$, APR = 0.21, $Y = 1$ year, and $n = 365$ (for daily compounding). At the end of the year, his loan balance will be

$$A = P \times \left(1 + \frac{\text{APR}}{n}\right)^{(nY)} = \$9500 \times \left(1 + \frac{0.21}{365}\right)^{(365 \times 1)} = \$11{,}719.23$$

During his year of unemployment, interest alone will make Paul's credit card balance grow from $9500 to over $11,700, an increase of more than $2200. Clearly, this increase will only make it more difficult for Paul to get back on his financial feet. ■

by the way . . .

About three-fourths of American households have at least one credit card, and their average credit card balance is about $8000 (as of 2000). The average credit card interest rate is about 17%—far higher than the interest rate on most other consumer loans.

by the way . . .

The U.S. population of 280 million holds a total of over 500 million VISA, MasterCard, and American Express cards, plus another 800 million store credit cards and debit cards. Americans charge more than $1 trillion each year to their major credit cards and have a total revolving credit card debt of about $600 billion (2001 statistics).

practical matters

AVOIDING CREDIT CARD TROUBLE

Most adults have credit cards for good reason. Used properly, credit cards offer many conveniences: They are safer and easier to carry than cash, they offer monthly statements that list everything charged to the card, and they can be used as ID to rent a car. But credit card trouble can compound quickly, and many people get into financial trouble as a result. A few simple guidelines can help you avoid credit card trouble.

- Use only one credit card. People who accumulate balances on several cards often lose track of their overall debt. A lost wallet or purse means more credit cards that must be canceled.
- If possible, pay off your balance in full each month. Then there's no chance of getting into a financial hole.
- If you plan to pay off your balance in full each month, be sure that your credit card offers an interest-free "grace period" on purchases (usually of about 1 month) so that you will not have to pay any interest.
- Compare the interest rate and annual fee (if any) of your credit card and others. Fees and rates differ greatly among credit cards, so be sure you are getting a good deal. In particular, if you carry a balance, look for a card with a relatively low interest rate.
- When choosing a credit card, watch out for teaser rates. These are low interest rates that are offered for a short period, such as 6 months, after which the card reverts to very high rates.
- Never use your credit card for a cash advance except in an emergency, because nearly all credit cards charge both fees and high interest rates for cash advances; in addition, most credit cards charge interest immediately on cash advances, even if there is a grace period on purchases. When you need cash, get it directly from your own bank account by cashing a check or using an ATM card.
- If you own a home, consider replacing a common credit card with a home equity credit line. You'll generally get a lower interest rate, and the interest may be tax deductible.
- If you find yourself in a deepening financial hole, consult a financial advisor right away. A good place to start is with the National Foundation for Credit Counseling (www.nfcc.org). The longer you wait, the worse off you'll be in the long run.

MORTGAGES

One of the most popular types of installment loans is designed specifically to help you buy a home—a home **mortgage.** Mortgage interest rates generally are lower than interest rates on other types of loans because your home itself serves as a payment guarantee. If you fail to make your payments, the lender (usually a bank or mortgage company) can take possession of your home and sell it to recover the amount loaned to you.

There are several considerations in getting a home mortgage. First, the lender will probably require a **down payment,** typically 10% to 20% of the purchase price. Then the lender will loan you the rest of the money needed to purchase the home. Most lenders also charge fees, or **closing costs,** at the time you take out a loan. Closing costs can be substantial and may vary significantly between lenders, so you should be sure that you understand them. In general, there are two types of closing costs:

- Direct fees, such as fees for getting the home appraised and checking your credit history, for which the lender charges a fixed dollar amount. These fees typically range from a few hundred dollars to well over one thousand dollars.
- Fees charged as **points,** where each point is 1% of the loan amount. Many lenders divide points into two categories: an "origination fee" that is charged

by the way...

The idea of a mortgage contract originated in early British real estate law. The curious word *mortgage* comes from Latin and old French. It literally means "dead pledge."

on all loans and "discount points" that vary for loans with different rates. For example, a lender might charge an origination fee of 1 point (1%) on all loans, then offer you a choice of adding 1 discount point for a loan at 8% or 2 discount points for a loan at 7.75%. Despite their different names, there is no essential difference between an origination fee and discount points.

As always, you should watch out for any fine print that may affect the cost of your loan. For example, you should check to make sure that there are no *prepayment penalties* if you decide to pay off your loan early. Most people pay off mortgages early, either because they sell the home or because they decide to refinance the loan to get a better interest rate or to change their monthly payments.

> **Mortgage Basics**
>
> If you are seeking a home mortgage, be sure to keep the following considerations in mind as you compare lenders:
>
> - What interest rate and down payment are required for the loan?
> - What closing costs will be charged? Be sure you identify all closing costs, including origination fees and discount points, since different lenders may quote their fees differently.
> - Watch out for fine print, such as prepayment penalties, that may make the loan more expensive than it seems on the surface.

Fixed Rate Mortgages

The simplest type of home loan is a **fixed rate mortgage**, in which you are guaranteed that the interest rate will not change over the life of the loan. Most fixed rate loans have a term of either 15 or 30 years, with lower interest rates on the shorter-term loans. We can calculate payments on fixed rate loans with the loan payment formula.

EXAMPLE 6 *Fixed Rate Payment Options*

You need a loan of $100,000 to buy your new home. The bank offers a choice of a 30-year loan at an APR of 8% or a 15-year loan at 7.5%. Compare your monthly payments and total loan cost under the two options. Assume that the closing costs are the same in both cases and therefore do not affect the choice.

Solution The starting loan principal is $P = \$100{,}000$ and we set $n = 12$ for monthly payments. For the 30-year loan, we have APR = 0.08 and $Y = 30$. The monthly payments are

$$\text{PMT} = \frac{P \times \left(\frac{\text{APR}}{n}\right)}{\left[1 - \left(1 + \frac{\text{APR}}{n}\right)^{(-nY)}\right]} = \frac{\$100{,}000 \times \left(\frac{0.08}{12}\right)}{\left[1 - \left(1 + \frac{0.08}{12}\right)^{(-12 \times 30)}\right]} = \$733.76$$

by the way...

The average mortgage in the United States is paid off after 7 years, usually because the home is sold.

Over the 30-year life of the loan, your total payments are

$$30 \text{ yr} \times \frac{12 \text{ mo}}{\text{yr}} \times \frac{\$733.76}{\text{mo}} \approx \$264{,}150$$

For the 15-year loan, we have APR = 0.075 and $Y = 15$. The monthly payments are

$$\text{PMT} = \frac{P \times \left(\dfrac{\text{APR}}{n}\right)}{\left[1 - \left(1 + \dfrac{\text{APR}}{n}\right)^{(-nY)}\right]} = \frac{\$100{,}000 \times \left(\dfrac{0.075}{12}\right)}{\left[1 - \left(1 + \dfrac{0.075}{12}\right)^{(-12 \times 15)}\right]} = \$927.01$$

Over the 15-year life of the loan, your total payments are

$$15 \text{ yr} \times \frac{12 \text{ mo}}{\text{yr}} \times \frac{\$927.01}{\text{mo}} \approx \$166{,}860$$

Note that the payments of $927.01 on the 15-year loan are almost $200 higher than the payments of $733.76 on the 30-year loan. However, the 15-year loan saves you almost $100,000 in total payments. Thus, the 15-year loan saves you a lot in the long run, but it's a good plan only if you are confident that you can afford the additional $200 per month that it will cost you for the next 15 years. (See Example 9 for an alternative payment strategy.)

EXAMPLE 7 *Closing Costs*

Great Bank offers a $100,000, 30-year, 8% fixed rate loan with closing costs of $500 plus 2 points. Big Bank offers a lower rate of 7.9%, but with closing costs of $1000 plus 2 points. Evaluate the two options.

Solution In Example 6, we calculated the payments on the 8% loan to be $733.76. At the lower 7.9% rate, the payments are

$$\text{PMT} = \frac{P \times \left(\dfrac{\text{APR}}{n}\right)}{\left[1 - \left(1 + \dfrac{\text{APR}}{n}\right)^{(-nY)}\right]} = \frac{\$100{,}000 \times \left(\dfrac{0.079}{12}\right)}{\left[1 - \left(1 + \dfrac{0.079}{12}\right)^{(-12 \times 30)}\right]} = \$726.81$$

Thus, you'll save about $7 per month with Big Bank's lower interest rate. Now we must consider the difference in closing costs. Both banks charge the same 2 points, so this portion of the closing costs won't affect your decision. (Note, however, that 2 points means 2% of the $100,000 loan, which is a $2000 fee!) But you must consider Big Bank's extra $500 in direct fees.

The choice comes down to this: Big Bank costs you an extra $500 now, but saves you $7 per month in payments. Dividing $500 by $7 per month, we find the time it will take to recoup the extra $500:

$$\frac{\$500}{\$7/\text{mo}} = 71.4 \text{ mo} \approx 6 \text{ yr}$$

Thus, it will take you about 6 years to save the extra $500 that Big Bank charges up front. Unless you are sure that you will be staying in your house (and keeping the same loan) for much more than 6 years, you probably should go with the lower closing costs at Great Bank, even though your monthly payments will be slightly higher.

by the way . . .

When you pay points, most lenders give you a choice between paying them up front or folding them into the loan. For example, if you pay 2 points on a $100,000 loan, your choice is to pay $2000 up front or to make the loan amount $102,000 rather than $100,000. For the examples in this book, we assume that you pay the points up front.

EXAMPLE 8 Points Decision

Continuing Example 7, suppose you've decided to go with Great Bank's lower closing costs. You learn that Great Bank actually offers two options for 30-year loans: an 8% interest rate with 2 points or a 7.5% rate with 4 points. Evaluate your options.

Solution We already know that the monthly payments for the 8% loan are $733.76. For the 7.5% loan, we have $P = \$100{,}000$, $n = 12$, and $Y = 30$. Setting APR = 7.5% = 0.075, we find that the monthly payments are

$$\text{PMT} = \frac{P \times \left(\frac{\text{APR}}{n}\right)}{\left[1 - \left(1 + \frac{\text{APR}}{n}\right)^{(-nY)}\right]} = \frac{\$100{,}000 \times \left(\frac{0.075}{12}\right)}{\left[1 - \left(1 + \frac{0.075}{12}\right)^{(-12 \times 30)}\right]} = \$699.21$$

The 7.5% loan lowers the monthly payments by $733.76 − $699.21 = $34.55. However, this loan has 2 additional points in closing costs, which means 2% of your $100,000 loan, or $2000. Thus, you must decide whether it is worth an extra $2000 up front for a monthly savings of just under $35. Let's calculate how long it will take to make up the added up-front costs:

$$\frac{\$2000}{\$34.55/\text{mo}} = 57.9 \text{ mo}$$

This is not quite 5 years. If you think it's likely that you will sell or refinance within 5 years, you should not pay the extra points. However, if you expect to keep the loan for a long time, the added points might be worth it. For example, if you keep the loan for the full 30 years (360 months), you'll save 360 × $34.55 = $12,438 in monthly payments over the life of the loan—far more than the extra $2000 you pay for the lower rate today.

> *by the way...*
> Most financial analysts say that you should pay additional up-front costs only if they will be made up in savings on your monthly payments within 2 to 3 years.

Prepayment Strategies

Because of the long term, the early payments on a mortgage tend to be almost entirely interest. For example, Figure 4.4 shows the portion of each payment going to principal

FIGURE 4.4 Portions of monthly payments going to principal and interest over the life of a 30-year, $100,000 loan at 8%.

and interest for a 30-year, $100,000 loan at 8%. In addition, the total interest paid on mortgages is often much more than the principal. In Example 6, we found that the total payments for this $100,000 loan would be about $264,000—more than $2\frac{1}{2}$ times the starting principal!

Clearly, you can save a lot if you can reduce your interest payments. One way to do this is to pay extra toward the principal, particularly early in the term. For example, suppose you pay an extra $100 toward principal in the first monthly payment of your $100,000 loan; that is, instead of paying the required $734 (see Example 6), you pay $834. Because you've reduced your loan balance by $100, you will save the compounded value of this $100 over the rest of the 30 year loan term—which is nearly $1100. In other words, paying an extra $100 in the first month saves you about $1100 in interest over the 30 years.

EXAMPLE 9 *An Alternative Strategy*

An alternative strategy to the mortgage options in Example 6 is to take the 30-year loan at 8%, but to try to pay it off in 15 years by making larger payments than are required. How much would you have to pay each month? Discuss the pros and cons of this strategy.

Solution To reflect paying off an 8% loan in 15 years, we set APR = 0.08 and $Y = 15$; we still have $P = \$100{,}000$ and $n = 12$. The monthly payments are

$$\text{PMT} = \frac{P \times \left(\frac{\text{APR}}{n}\right)}{\left[1 - \left(1 + \frac{\text{APR}}{n}\right)^{(-nY)}\right]} = \frac{\$100{,}000 \times \left(\frac{0.08}{12}\right)}{\left[1 - \left(1 + \frac{0.08}{12}\right)^{(-12 \times 15)}\right]} = \$955.65$$

In Example 6, we found that the 30-year loan requires payments of $733.76. Thus, to pay off the loan in 15 years, you must make payments that are more than the minimum required by $955.65 − $733.76 = $221.89 per month.

Note that this payment is about $30 per month *more* than the payment of $927.01 required with the 15-year loan (see Example 6), because the 15-year loan had a lower interest rate. Clearly, if you *know* you're going to pay off the loan in 15 years, you should take the lower-interest 15-year loan. However, taking the 30-year loan has one advantage: Because your required payments are only $733.76, you can always drop back to this level if you find it difficult to afford the extra needed to pay off the loan in 15 years.

Time out to think

Consider two options for paying off a loan in 15 years: taking out a 15-year loan or taking out a 30-year loan and making an extra principal payment each month. Assuming that you would like to pay off the loan in 15 years, how would you decide which strategy is better for you?

Adjustable Rate Mortgages

A fixed rate mortgage is advantageous for you because your monthly payments never change. However, it poses a risk to the lender. Imagine that you take out a fixed, 30-year loan of $100,000 from Great Bank at an 8% interest rate. Initially, the loan may seem like a good deal for Great Bank. But suppose that, 2 years later, prevailing interest rates have risen to 10%. If Great Bank still had the $100,000 that it lent to you, it could lend it out to someone else at this higher 10% rate. Instead, it's stuck with the

by the way . . .
Although it may seem strange at first, making prepayments on a home loan is not always a good idea even though it reduces the total payments. For example, if you have $200 per month to spare, you might choose to invest it rather than pay down the loan. If the investment return is greater than the effective loan interest rate, you will come out ahead. For home mortgages, where tax benefits can make your effective interest rate much lower than the actual rate (because of the mortgage interest deduction), it may be relatively easy to come out ahead by investing.

8% rate that you are paying. In effect, Great Bank loses potential future income if prevailing rates rise substantially and you have a fixed rate loan.

Lenders can lessen the risk of rising interest rates by charging higher rates for longer-term loans. That is why rates generally are higher for 30-year loans than for 15-year loans. But an even lower-risk strategy for the lender is an **adjustable rate mortgage (ARM),** in which the interest rate you pay changes whenever prevailing rates change. Because of the reduced long-term risk to lenders, ARMs generally have much lower initial interest rates than fixed rate loans. For example, a bank offering an 8% rate on a fixed 30-year loan might offer an ARM that begins at 6%. Most ARMs guarantee their starting interest rate for the first 6 months or 1 year, but interest rates in subsequent years move up or down to reflect prevailing rates. Most ARMs also include a *rate cap* that cannot be exceeded. For example, if your ARM begins at an interest rate of 6%, you may be promised that your interest rate can never go higher than a rate cap of 12%. Making a decision between a fixed rate loan and an ARM can be one of the most important financial decisions of your life.

> **by the way ...**
> Watch out for "teaser" rates on adjustable rate mortgages. Under normal circumstances, your rate on an ARM rises only if prevailing interest rates rise. However, some lenders offer low teaser rates—rates below the prevailing rates—for the first few months of an ARM. Teaser rates are certain to rise as soon as the teaser period is over. Thus, while teaser rates may be attractive, the longer-term policies of the ARM are far more important.

EXAMPLE 10 *Rate Approximations for ARMs*

You have a choice between a 30-year fixed rate loan at 8% and an ARM with a first-year rate of 5%. Neglecting compounding and changes in principal, estimate your monthly savings with the ARM during the first year on a $100,000 loan. Suppose that the ARM rate rises to 11% by the fourth year. How will your payments be affected?

Solution Because mortgage payments are mostly interest in the early years of a loan, we can make approximations by pretending that the principal remains unchanged. For the 8% fixed rate loan, the interest on the $100,000 loan for the first year will be approximately 8% × $100,000 = $8000. With the 5% loan, your first-year interest will be approximately 5% × $100,000 = $5000. Thus, the ARM will save you about $3000 in interest during the first year, which means a monthly savings of about $3000 ÷ 12 = $250.

By the fourth year, when rates reach 11%, the situation is reversed. The rate on the ARM is now 3 percentage points above the rate on the fixed rate loan. Instead of saving $250 per month, you'd be paying $250 per month *more* on the ARM than on the 8% fixed rate loan. Moreover, if interest rates remain high on the ARM, you will continue to make these high payments for many years to follow. Thus, while ARMs reduce risk for the lender, they add risk for the borrower. ■

The Mortgage Interest Tax Deduction

One final but very important consideration concerning the affordability of a mortgage is its potential tax benefits. Under current tax law, most people are allowed to deduct from their income any interest (but not principal) paid on a home mortgage. If you are able to get the full benefit of this deduction, it can be quite significant. For example, if your tax rate is 33%, the deduction effectively returns to you one-third of the amount of your mortgage interest payments. However, not everyone can get the full benefit of the mortgage interest tax deduction. (It depends, for example, on whether you are better off itemizing deductions or taking the standard deduction.) Because the tax consequences of a mortgage can be so significant, be sure you understand these consequences before making any final decisions about a home mortgage.

Review Questions

1. Suppose you pay only the interest on a loan. Will the loan ever be paid off? Why not?
2. What is an *installment loan?* Explain the meaning and use of the loan payment formula.
3. Explain, in general terms, how the portions of installment loan payments going to principal and interest change over the life of the loan.
4. Suppose that you need a loan of $10,000 and are offered a choice of a 3-year loan at 7% interest or a 5-year loan at 8% interest. Discuss the pros and cons of each choice.
5. How do credit card loans differ from ordinary installment loans? Why are credit card loans particularly dangerous?
6. What is a *mortgage?* What is a *down payment* on a mortgage? Explain how closing costs, including points, can affect loan decisions.

In the News

1. **Mortgage Rates.** Find advertisements in the newspaper for two different home mortgages companies. Using the ideas of this unit, evaluate the terms of loans from each company and decide which company you would use for a home mortgage.
2. **Credit Card Statement.** Look carefully at the terms of financing explained on your most recent credit card statement. Explain all the important terms, including the interest rates that apply, annual fees, and grace periods.
3. **Bank Rates.** Find the interest rates that your bank (or another local bank) charges for different types of loans, such as auto loans, personal loans, and home mortgages. Why do you think the rates are different in the different cases?

Exercises

1. **Loan Terminology.** You borrowed $40,000 at an APR of 7%, which you are paying off with monthly payments of $310 for 20 years.
 a. Clearly identify each of the following: the starting loan principal, the annual interest rate, the number of payments per year, the loan term, and the payment amount.
 b. How many payments will you make in total? What total amount will you pay over the full term of the loan?
 c. Of the total amount you pay, how much will go toward principal and how much toward interest?
2. **Loan Terminology.** You borrowed $7500 at an APR of 9%, which you are paying off with monthly payments of $95 for 10 years.
 a. Clearly identify each of the following: the starting loan principal, the annual interest rate, the number of payments per year, the loan term, and the payment amount.
 b. How many payments will you make in total? What total amount will you pay over the full term of the loan?
 c. Of the total amount you pay, how much will go toward principal and how much toward interest?

Loan Payments. For the loans described in Exercises 3–12, do the following:
 a. Calculate the monthly payment.
 b. Determine the total payment over the term of the loan.
 c. Determine how much of the total payment over the loan term goes to principal and how much to interest.

3. A student loan of $25,000 at a fixed APR of 10% for 20 years
4. A student loan of $12,000 at a fixed APR of 8% for 10 years
5. A home mortgage of $150,000 with a fixed APR of 7.5% for 30 years
6. A home mortgage of $150,000 with a fixed APR of 7.5% for 15 years
7. A home mortgage of $100,000 with a fixed APR of 9% for 15 years
8. A home mortgage of $100,000 with a fixed APR of 8.5% for 15 years
9. You borrow $5000 over a period of 3 years at a fixed APR of 12%.
10. You borrow $10,000 over a period of 5 years at a fixed APR of 10%.
11. You borrow $50,000 over a period of 15 years at a fixed APR of 8%.
12. You borrow $100,000 over a period of 30 years at a fixed APR of 7%.

Principal and Interest Payments. For the loans described in Exercises 13–14, calculate the monthly payment and the portions of the payments that go to principal and to interest during the first 3 months. (Hint: Use a table as in Example 2.)

13. A home mortgage of $100,000 with a fixed APR of 8.5% for 30 years

14. A student loan of $12,000 at a fixed APR of 8% for 15 years

15. **Choosing an Auto Loan.** You need to borrow $10,000 to buy a car and you determine that you can afford monthly payments of $220. The bank offers three choices: a 3-year loan at 7% APR, a 4-year loan at 7.5% APR, or a 5-year loan at 8% APR. Which loan best meets your needs? Explain your reasoning.

16. **Choosing a Personal Loan.** You need to borrow $4000 to pay off your credit cards and you can afford monthly payments of $150. The bank offers three choices: a 2-year loan at 8% APR, a 3-year loan at 9% APR, or a 4-year loan at 10% APR. Which loan best meets your needs? Explain your reasoning.

Credit Card Debt. For Exercises 17–20, assume you have a balance of $2500 on your credit card that you want to pay off. Calculate your monthly payment and total payment under the conditions listed. Assume you make no additional charges to the card.

17. The credit card APR is 18% and you want to pay off the balance in 1 year.

18. The credit card APR is 20% and you want to pay off the balance in 2 years.

19. The credit card APR is 21% and you want to pay off the balance in 3 years.

20. The credit card APR is 22% and you want to pay off the balance in 1 year.

21. **Credit Card Debt.** Assume you have a balance of $1200 on a credit card with an APR of 18%, or 1.5% per month. You start making monthly payments of $200, but at the same time you charge an additional $75 per month to the credit card. Assume that interest for a given month is based on the balance for the previous month. The following table shows how you can calculate your monthly balance.

Month	Payment	Expenses	Interest	New Balance
0	—	—	—	$1200
1	$200	$75	1.5% × $1200 = $18	$1200 − $200 + $75 + 18 = $1093
2	$200	$75		
3	$200	$75		

Complete and extend the table to show your balance at the end of each month until the debt is paid off. How long does it take to pay off the credit card debt?

22. **Credit Card Debt.** Repeat the table of Exercise 21, but this time assume that you make monthly payments of $300. Extend the table as long as necessary until your debt is paid off. How long does it take to pay off your debt?

23. **Credit Card Woes.** The following table shows the expenses and payments for 8 months on a credit card account with an initial balance of $300. Assume that the interest rate is 1.5% per month (18% APR) and that interest for a given month is charged on the balance for the previous month. Complete the table. After 8 months, what is the balance on the credit card? Comment on the effect of the interest and the initial balance, in light of the fact that for 7 of the 8 months expenses never exceeded payments.

Month	Payment	Expenses	Interest	Balance
0	—	—	—	$300
1	$300	$175	1.5% × $300 = $4.50	$179.50
2	$150	$150		
3	$400	$350		
4	$500	$450		
5	0	$100		
6	$100	$100		
7	$200	$150		
8	$100	$80		

24. **Teaser Rate.** You have a total credit card debt of $4000. You receive an offer to transfer this debt to a new card with an introductory APR of 6% for the first 6 months. After that, the rate becomes 24%.

 a. What is the monthly interest payment on $4000 during the first 6 months? (Assume you pay nothing toward principal and don't charge any further debts.)

 b. What is the monthly interest payment on $4000 *after* the first 6 months? Comment on the change from the teaser rate.

Fixed Rate Options. Compare your monthly payments and total loan cost under the two options listed in each of Exercises 25–28. Assume that the loans are fixed rate and that closing costs are the same in both cases. Briefly discuss the pros and cons of each option.

25. You need a $100,000 loan.

 Option 1: a 30-year loan at an APR of 8%

 Option 2: a 15-year loan at 7.5%

26. You need a $75,000 loan.

 Option 1: a 30-year loan at an APR of 8%

 Option 2: a 15-year loan at 7%

27. You need a $120,000 loan.

 Option 1: a 30-year loan at an APR of 7.15%

 Option 2: a 15-year loan at 6.75%

28. You need a $180,000 loan.

 Option 1: a 30-year loan at an APR of 7.25%

 Option 2: a 15-year loan at 6.8%

Closing Costs. You need a loan of $80,000 to buy a home. Each of Exercises 29–32 offers two choices. Calculate your monthly payments and total closing costs in each case. Briefly discuss how you would decide between the two choices.

29. Choice 1: 30-year fixed rate at 8% with closing costs of $1200 and no points

 Choice 2: 30-year fixed rate at 7.5% with closing costs of $1200 and 2 points

30. Choice 1: 30-year fixed rate at 8.5% with no closing costs and no points

 Choice 2: 30-year fixed rate at 7.5% with closing costs of $1200 and 4 points

31. Choice 1: 30-year fixed rate at 7.25% with closing costs of $1200 and 1 point

 Choice 2: 30-year fixed rate at 6.75% with closing costs of $1200 and 3 points

32. Choice 1: 30-year fixed rate at 7.5% with closing costs of $1000 and no points

 Choice 2: 30-year fixed rate at 6.5% with closing costs of $1500 and 4 points

33. **Accelerated Loan Payment.** Suppose you have a student loan of $25,000 with an APR of 9% for 20 years.

 a. What are your required monthly payments?

 b. Suppose you would like to pay the loan off in 10 years instead of 20. What monthly payments will you need to make?

 c. Compare the total amounts you'll pay over the loan term if you pay the loan off in 20 years versus 10 years.

34. **Accelerated Loan Payment.** Suppose you have a student loan of $60,000 with an APR of 8% for 25 years.

 a. What are your required monthly payments?

 b. Suppose you would like to pay the loan off in 15 years instead of 25. What monthly payments will you need to make?

 c. Compare the total amounts you'll pay over the loan term if you pay the loan off in 25 years versus 15 years.

35. **ARM Rate Approximations.** You have a choice between a 30-year fixed rate loan at 7% and an ARM with a first-year rate of 5%. Neglecting compounding and changes in principal, estimate your monthly savings with the ARM during the first year on a $150,000 loan. Suppose that the ARM rate rises to 8.5% at the start of the third year. Approximately how much extra will you then be paying over what you would have paid if you had taken the fixed rate loan?

36. **ARM Rate Approximations.** You have a choice between a 30-year fixed rate loan at 8.5% and an ARM with a first-year rate of 5.5%. Neglecting compounding and changes in principal, estimate your monthly savings with the ARM during the first year on a $125,000 loan. Suppose that the ARM rate rises to 10% at the start of the second year. Approximately how much extra will you then be paying over what you would have paid if you had taken the fixed rate loan?

37. **How Much House Can You Afford?** You can afford monthly payments of $500. If current mortgage rates are 9% for a 30-year fixed rate loan, what loan principal can you afford? If you are required to make a 20% down payment and you have the cash on hand to do it, what price home can you afford? (Hint: You will need to solve the loan payment formula for P.)

38. **How Much House Can You Afford?** You can afford monthly payments of $1200. If current mortgage rates are 7.5% for a 30-year fixed rate loan, what loan principal can you afford? If you are required to make a 20% down pay-

ment and you have the cash on hand to do it, what price home can you afford? (Hint: You will need to solve the loan payment formula for P.)

39. **Student Loan Consolidation.** Suppose you have the following three student loans: $10,000 with an APR of 8% for 15 years, $15,000 with an APR of 8.5% for 20 years, and $12,500 with an APR of 9% for 10 years.

 a. Calculate the monthly payment for each loan individually.

 b. Calculate the total you'll pay in payments during the life of all three loans.

 c. A bank offers to consolidate your three loans into a single loan with an APR of 8.5% and a loan term of 20 years. What will your monthly payments be in that case? What will your total payments be over the 20 years? Discuss the pros and cons of accepting this loan consolidation.

40. **Bad Deals: Car-Title Lenders.** Some "car-title lenders" offer quick cash loans in exchange for being allowed to hold the title to your car as collateral (you lose your car if you fail to pay off the loan). In many states, these lenders operate under pawnbroker laws that allow them to charge fees as a percentage of the unpaid balance. Suppose you need $2000 in cash, and a car-title company offers you a loan at an interest rate of 2% per month *plus* a monthly fee of 20% of the unpaid balance.

 a. How much will you owe in interest and fees on your $2000 loan at the end of the first month?

 b. Suppose that you pay only the interest and fees each month. How much will you pay over the course of a full year?

 c. Suppose instead that you obtain a loan from a bank with a term of 3 years and an APR of 10%. What are your monthly payments in that case? Compare these to the payments to the car-title lender.

41. **Project: Choosing a Mortgage.** Imagine that you work for an accounting firm and a client has told you that he is buying a house and needs a loan of $120,000. His monthly income is $4000 and he is single with no children. He has $14,000 in savings that can be used for a down payment. Find the current rates available from local banks for both fixed rate mortgages and adjustable rate mortgages (ARMs). Analyze the offerings and summarize orally or in writing the best options for your client, along with the pros and cons of each option.

Web Projects

Find useful links for Web Projects on the text Web site: www.aw.com/bennett-briggs

1. **Credit Card Comparisons.** Visit a Web site that gives comparisons between credit cards. Briefly explain the factors that are considered in the comparisons. How does your own credit card compare to other credit cards? Based on this comparison, do you think you would be better off with a different credit card?

2. **Home Financing.** Visit a Web site that offers online home financing. Describe the terms of a particular home mortgage. Discuss the advantages and disadvantages of financing a home online rather than at a local bank.

3. **Online Car Purchase.** Find a car online that you might want to buy. Find a loan that you would qualify for, and calculate your monthly payments and total payments over the life of the loan. Next, suppose that you started a savings plan instead of buying the car, depositing the same amounts that would have gone to car payments. Estimate how much you would have in your savings plan by the time you graduate from college. Explain your assumptions.

4. **Student Financial Aid.** There are many Web sites that offer student loans. Visit a Web site that offers student loans and describe the terms of a particular loan. Discuss the advantages and disadvantages of financing a student loan online rather than through a bank or through your university or college.

5. **Scholarship Scams.** The Federal Trade Commission keeps track of many financial scams related to college scholarships. Read about two different types of scams, and report on how they work and how they hurt people who are taken in by them.

6. **Financial Scams.** Many Web sites keep track of current financial scams. Visit some of these sites and report on one scam that has already hurt a lot of people. Describe how the scam works and how it hurts those who are taken in by it.

CHAPTER 4 SUMMARY

UNIT	KEY TERMS	KEY IDEAS AND SKILLS
4A	principal simple interest compound interest annual percentage rate (APR) annual percentage yield (APY) variable definitions: A, P, i, N, n, Y	**General form of the compound interest formula:** $A = P \times (1 + i)^N$ **Compound interest formula for interest paid once a year:** $A = P \times (1 + \text{APR})^Y$ **Compound interest formula for interest paid n times a year:** $A = P\left(1 + \dfrac{\text{APR}}{n}\right)^{(nY)}$ **Compound interest formula for continuous compounding:** $A = P \times e^{(\text{APR} \times Y)}$
4B	savings plan total return annual return	**Savings plan formula:** $A = \text{PMT} \times \dfrac{\left[\left(1 + \dfrac{\text{APR}}{n}\right)^{(nY)} - 1\right]}{\left(\dfrac{\text{APR}}{n}\right)}$ **Return on investments:** $\text{total return} = \dfrac{(A - P)}{P}$ $\text{annual return} = \left(\dfrac{A}{P}\right)^{(1/Y)} - 1$
4C	installment loan mortgages down payment closing costs points fixed rate mortgage adjustable rate mortgage	**Loan payment formula:** $\text{PMT} = \dfrac{P \times \left(\dfrac{\text{APR}}{n}\right)}{\left[1 - \left(1 + \dfrac{\text{APR}}{n}\right)^{(-nY)}\right]}$ **Understand** the uses and dangers of credit cards. **Understand** strategies for early payment of loans. **Understand** considerations in choosing a mortgage.

5
STATISTICAL REASONING

Statistical thinking will one day be as necessary for efficient citizenship as the ability to read and write.
—H. G. WELLS

Is your drinking water safe? Do most people approve of the President's tax plan? How much is the cost of health care rising? These questions and thousands more like them can be answered only through statistical studies. Indeed, statistical information appears in the news every day, making the ability to understand and reason with statistics crucial to modern life. In this chapter, you will study several key concepts necessary to understanding the statistics that appear daily in the news.

UNIT 5A Fundamentals of Statistics: We'll discuss how statistical studies are conducted, with emphasis on the importance of sampling.

UNIT 5B Should You Believe a Statistical Study? We'll develop eight useful guidelines for evaluating statistical claims.

UNIT 5C Statistical Tables and Graphs: We'll investigate basic frequency tables and graphs, including bar graphs, pie charts, and histograms.

UNIT 5D Graphics in the Media: We'll interpret common types of media graphics and explore a few important cautions about graphics.

UNIT 5E Correlation and Causality: We'll investigate how to interpret correlations and how to decide whether a correlation is the result of causality.

UNIT 5F Characterizing Data: We often describe data by their average and their distribution. We'll study such descriptions and discover the multiple meanings of "average."

UNIT 5A

FUNDAMENTALS OF STATISTICS

The subject of statistics plays a major role in modern society. It's used to determine whether a new drug is effective in treating cancer. It's involved when agricultural inspectors check the safety of the food supply. It's used in every opinion poll and survey. In business, it's used for market research. Sports statistics are part of daily conversation for millions of people. Indeed, you'll be hard-pressed to think of a topic that is not linked in some way to statistics.

But what *is* (or *are*) statistics? There are two answers, because the term *statistics* can be either singular or plural. When it is singular, *statistics* refers to the *science* of statistics. The science of statistics helps us collect, organize, and interpret *data*, which are numbers or other pieces of information about some topic. When it is plural, the word *statistics* refers to the data themselves, especially those that describe or summarize something. For example, if there are 30 students in your class and they range in age from 17 to 64, the numbers "30 students," "17 years," and "64 years" are statistics that describe your class.

by the way...
You'll sometimes hear the word *data* used as a singular synonym for *information*, but technically *data* is plural; one piece of information is called a *datum*, and two or more pieces are called *data*.

TWO DEFINITIONS OF STATISTICS

- Statistics is the *science* of collecting, organizing, and interpreting data.
- Statistics are the *data* that describe or summarize something.

HOW STATISTICS WORKS

Although statistical studies are conducted in many different ways and for many different purposes, they all share a few characteristics. To get the basic ideas, consider the *Nielsen ratings*, which are used to estimate the number of people watching various television shows. These ratings are used, for example, to determine the most popular television show of the week.

Suppose the Nielsen ratings tell you that *Millionaire* was last week's most popular show, with 22 million viewers. You probably know that no one actually counted all 22 million people. But you may be surprised to learn that the Nielsen ratings are based on the television-viewing habits of people in only 5000 homes. To understand how Nielsen can draw a conclusion about millions of Americans from only 5000 homes, we need to investigate the principles behind statistical research.

Nielsen's goal is to draw conclusions about the viewing habits of all Americans. In the language of statistics, we say that Nielsen is interested in the **population** of all Americans. The characteristics of this population that Nielsen seeks to learn—such as the number of people watching each television show—are called **population parameters.** Note that, although we usually think of a population as a group of people, in statistics a population can be any kind of group—people, animals, or things. For example, in a study of college costs, the population might be *all colleges and*

historical note
Statistics originated with the collection of census and tax data, which are affairs of state. That is why the word *state* is at the root of the word *statistics*.

universities, and the population parameters might include prices for tuition, fees, and housing.

Nielsen seeks to learn about the population of all Americans by studying a much smaller **sample** of Americans in depth. More specifically, Nielsen has devices (called "people meters") attached to televisions in 5000 homes, so the people who live in these homes make up the sample of Americans that Nielsen studies. The individual measurements that Nielsen collects from the sample, such as who is watching each show at each time, constitute the *raw data.* Nielsen then consolidates these raw data into a set of numbers that characterize the sample, such as the percentage of young male viewers watching *Millionaire;* these numbers are called **sample statistics.**

> **DEFINITIONS**
>
> The **population** in a statistical study is the *complete* set of people or things being studied. The **sample** is the subset of the population from which the raw data are actually obtained.
>
> **Population parameters** are specific characteristics of the population that a statistical study is designed to estimate. **Sample statistics** are numbers or observations that summarize the raw data.

EXAMPLE 1 *Population and Sample*

For each of the following cases, describe the population, sample, population parameters, and sample statistics.

a. Agricultural inspectors for Jefferson County measure the levels of residue from three common pesticides on 25 ears of corn from each of the 104 corn-producing farms in the county.

b. Anthropologists determine the average brain size of early Neanderthals in Europe by studying skulls found at three sites in southern Europe.

Solution

a. The inspectors seek to learn about the population of all ears of corn grown in the county. They do this by studying a sample that consists of 25 ears from each farm. The population parameters are the average levels of residue from the three pesticides on all corn grown in the county. The sample statistics describe the average levels of residue that are actually measured on the corn in the sample.

b. The anthropologists seek to learn about the population of all early Neanderthals in Europe. Specifically, they seek to determine the average brain size of all Neanderthals, which is the population parameter in this case. The sample consists of the relatively few individual Neanderthals whose skulls are found at the three sites. The sample statistic is the average brain size (skull size) of the individuals in the sample.

historical note

Arthur C. Nielsen founded his company and invented market research in 1923. He began ratings for radio programs in 1942 and added television ratings in the 1960s. Nielsen's people meters, attached to all the televisions in 5000 homes, tell the company when each television is on and what show is being watched. People in the homes are supposed to push buttons that tell Nielsen who is watching each television. Nielsen can thereby determine the breakdown of viewership by age, sex, and ethnicity, as well as total viewing numbers.

> **by the way...**
> Statisticians often divide their subject into two major branches. *Descriptive statistics* is the branch that deals with *describing* data in the form of tables, graphs, or sample statistics. *Inferential statistics* is the branch that deals with *inferring* (or estimating) population characteristics from sample data.

The Process of a Statistical Study

Because Nielsen does not study the entire population of all Americans, it cannot actually measure any population parameters. Instead, the company tries to infer reasonable values for population parameters from the sample statistics (which it *did* measure).

The process of inference is simple in principle, though it must be carried out with great care. For example, suppose Nielsen finds that 7% of the people in its sample watched *Millionaire*. If this sample accurately represents the entire population of all Americans, then Nielsen can infer that *approximately* 7% of all Americans watched the show. In other words, the sample statistic of 7% is used as an estimate for the population parameter. Once Nielsen has estimates of the population parameters, it can draw general conclusions about what Americans were watching. The process used by Nielsen Media Research is similar to that used in many statistical studies. Figure 5.1 summarizes the general relationships among a population, a sample, the sample statistics, and the population parameters.

Basic Steps in a Statistical Study

1. State the goal of your study precisely; that is, determine the population you want to study and exactly what you'd like to learn about it.
2. Choose a representative sample from the population.
3. Collect raw data from the sample and summarize these data by finding sample statistics of interest.
4. Use the sample statistics to infer the population parameters.
5. Draw conclusions; determine what you learned and whether you achieved your goal.

FIGURE 5.1 Elements of a statistical study.

EXAMPLE 2 Unemployment Survey

Each month, the U.S. Labor Department surveys 60,000 households to determine characteristics of the U.S. work force. One population parameter of interest is the U.S. unemployment rate, defined as the percentage of people who are unemployed among all those who are either employed or actively seeking employment. Describe how the five basic steps of a statistical study apply to this research.

Solution The steps apply as follows.

Step 1: The goal of the research is to learn about the employment (or unemployment) within the population of all Americans who are either employed or actively seeking employment.

Step 2: The Labor Department chooses a sample consisting of people employed or seeking employment in 60,000 households.

Step 3: The Labor Department asks questions of the people in the sample; the responses constitute the raw data for the research. The Department then consolidates these data into sample statistics, such as the percentage of people in the sample who are unemployed.

Step 4: Based on the sample statistics, the Labor Department makes estimates of the corresponding population parameters, such as the unemployment rate for the entire United States.

Step 5: The Labor Department draws conclusions based on the population parameters and other information. For example, it might use the current and past unemployment rates to draw conclusions about whether jobs have been created or lost.

> *by the way . . .*
>
> According to the Labor Department, someone who is not working is not necessarily unemployed. For example, stay-at-home moms and dads are not counted among the unemployed unless they are actively trying to find a job, and people who had been trying to find work but gave up in frustration are not counted as unemployed.

Choosing a Sample

Choosing a sample may well be the most important step in any statistical study. If the sample fairly represents the population as a whole, then it's reasonable to make inferences from the sample to the population. But if the sample is *not* representative, then there's little hope of drawing accurate conclusions about the population.

Suppose you want to determine the average height and weight of students at a large university by measuring the heights and weights of a sample of 100 students. A sample consisting only of members of the football and basketball teams would not be reliable, because these athletes tend to be larger than most students. In contrast, suppose you select your sample with a computer program that randomly draws student numbers from the entire university population. In this case, the 100 students in your sample are likely to be representative of the entire student body. You can therefore expect that the average height and weight of students in the sample are reasonable estimates of the averages for all students.

> **DEFINITION**
>
> A **representative sample** is a sample in which the relevant characteristics of the sample members match those of the population.

A sample drawn with a computer program that selects students at random is an example of a **simple random sample.** More technically, simple random sampling

Simple Random Sampling:
Every sample of the same size has an equal chance of being selected. Computers are often used to generate random telephone numbers.

Convenience Sampling:
Use results that are readily available.

Systematic Sampling:
Select every *k*th member.

Stratified Sampling:
Partition the population into at least two strata, then draw a sample from each.

FIGURE 5.2 Common sampling techniques.

means that every sample of a particular size has the same chance of being selected. In the case of the student sample, every set of 100 students has an equal chance of being selected by the computer program.

Simple random sampling is usually the best way to choose a representative sample. However, it is not always practical or necessary, so other sampling techniques are sometimes used. The following box summarizes four of the most common sampling techniques, and Figure 5.2 illustrates the ideas.

Common Sampling Methods

Simple random sampling: We choose a sample of items in such a way that every sample of a given size has an equal chance of being selected.

Systematic sampling: We use a simple system to choose the sample, such as selecting every 10th or every 50th member of the population.

Convenience sampling: We use a sample that is convenient to select, such as people who happen to be in the same classroom.

Stratified sampling: We use this method when we are concerned about differences among subgroups, or *strata*, within a population. We first identify the subgroups and then draw a simple random sample within each subgroup. The total sample consists of all the samples from the individual subgroups.

Regardless of what type of sampling is used, always keep the following two key ideas in mind:

- No matter how a sample is chosen, the study can be successful only if the sample is representative of the population.
- Even if a sample is chosen in the best possible way, it is still just a sample (as opposed to the entire population). Thus, we can never be *sure* that a sample is representative of the population. In general, a larger sample is more likely to be representative of the population, as long as it is chosen well.

EXAMPLE 3 *Sampling Methods*

Identify the type of sampling used in each of the following cases, and comment on whether the sample is likely to be representative of the population.

a. You are conducting a survey of students in a dormitory. You choose your sample by knocking on the door of every 10th room.
b. To survey opinions on a possible property tax increase, a research firm randomly draws the addresses of 150 homeowners from a public list of all homeowners.
c. Agricultural inspectors for Jefferson County check the levels of residue from three common pesticides on 25 ears of corn from each of the 104 corn-producing farms in the county.
d. Anthropologists determine the average brain size of early Neanderthals in Europe by studying skulls found at three sites in southern Europe.

Solution

a. Choosing every 10th room makes this a systematic sample. The sample may be representative, as long as students were randomly assigned to rooms.
b. The records presumably list all homeowners, so drawing randomly from this list produces a simple random sample. It has a good chance of being representative of the population.
c. Each farm may have different pesticide use, so the inspectors are considering corn from each farm as a subgroup (stratum) of the full population. By checking 25 ears of corn from each of the 104 farms, the inspectors are using stratified sampling. If the ears are collected randomly on each farm, each set of 25 is likely to be representative of its farm.
d. By studying skulls found at selected sites, the anthropologists are using a convenience sample. They have little choice, because only a few skulls remain from the many Neanderthals who once lived in Europe. However, it seems reasonable to assume that these skulls are representative of the larger population.

Watching Out for Bias

Consider a study designed to estimate the average weight of all men at a college. As we discussed earlier, a sample consisting only of football players would *not* be representative of the population with respect to weight. We say that this sample is *biased* because the men in the sample differ in a critical way from "typical" men at the college. More generally, the term **bias** refers to any problem in the design or conduct of a statistical study that tends to favor certain results.

Besides occurring in a poorly chosen sample, bias can arise in many other ways. For example, a researcher may be biased if he or she has a personal stake in the outcome of the study. In that case, the researcher might distort (intentionally or unintentionally) the true meaning of the data. You should always be on the lookout for any type of bias that may affect the results or interpretation of a statistical study. We'll discuss sources of bias further in Unit 5B.

> **DEFINITION**
>
> A statistical study suffers from **bias** if its design or conduct tends to favor certain results.

Time out to think

Thinking about issues of bias, explain why television networks use Nielsen to measure ratings rather than doing it themselves.

TYPES OF STATISTICAL STUDY

Broadly speaking, most statistical studies fall into one of two categories: observational studies and experiments. Nielsen's studies of television viewing are **observational** because they are designed to *observe* the television-viewing behavior of the people in its 5000 sample homes. Note that observational studies may still involve some interaction. For example, an opinion poll is observational, even though researchers may conduct in-depth interviews, because the poll's goal is to learn (observe) people's opinions, not to change them. Similarly, a study in which individuals in the sample are weighed is also observational, because the measurement process records (observes) but does not change a person's weight.

In contrast, consider a medical study designed to test whether large doses of vitamin C can help prevent colds. To conduct this study, the researchers must ask some people in the sample to take large doses of vitamin C. This type of statistical study is called an **experiment,** because some participants receive a treatment (in this case, vitamin C) that they would not otherwise receive.

It is difficult to determine whether an experimental treatment works unless you compare groups that receive the treatment to groups that don't. In the vitamin C study, for example, researchers might create two groups of people: a **treatment group** that takes large doses of vitamin C and a **control group** that does not take vitamin C. The researchers can then look for differences in the numbers of colds among people in the two groups. Having a control group is crucial to interpreting the results of experiments.

In an experiment, it is very important for the treatment and control groups to be alike in all respects except for the treatment. For example, if the treatment group consisted of active people with good diets and the control group consisted of sedentary people with poor diets, we could not attribute any differences in colds to vitamin C alone. To avoid this type of problem, assignments to the control and treatment groups must be done randomly.

With proper treatment, a cold can be cured in a week. Left to itself, it may linger for seven days.

—A medical folk saying

> **Two Basic Types of Statistical Study**
>
> 1. In an **observational study,** researchers observe or measure characteristics of the sample members, but do not attempt to influence or modify these characteristics.
> 2. In an **experiment,** researchers divide the sample into two (or more) groups, each of which is treated differently. The **treatment group** receives the experimental treatment, while the **control group** does not.

The Placebo Effect and Blinding

For experiments involving people, using a treatment and a control group might not be enough to get reliable results. The problem is that people can be affected by their beliefs as well as by real treatments. For example, stress and other psychological factors have been shown to affect resistance to colds. If people taking vitamin C get fewer colds than people who don't, we can't conclude that the vitamin C was responsible; it might be that people stayed healthier because they believed that vitamin C works. Therefore, people in the control group should be given a **placebo**—in this case, pills that look like vitamin C pills but don't actually contain vitamin C. As long as the participants don't know whether they are in the treatment or control group (that is, whether they got the real pills or the placebo), any effect arising from psychological factors—known as a **placebo effect**—should affect both groups equally. Then, if people in the vitamin C group get fewer colds than people in the control group, we have evidence that vitamin C really works.

> **DEFINITIONS**
>
> A **placebo** lacks the active ingredients of a treatment being tested in a study, but is identical in appearance to the treatment. Thus, study participants cannot distinguish the placebo from the real treatment.
>
> The **placebo effect** refers to the situation in which patients improve simply because they believe they are receiving a useful treatment.

In statistical terminology, the practice of keeping people in the dark about who is in the treatment group and who is in the control group is called **blinding.** A **single-blind** experiment is one in which the participants don't know which group they belong to, but the experimenters (the people administering the treatment) do know. Using a placebo is one way to create a single-blind experiment. Sometimes, a single-blind experiment can still be unreliable if the experimenters can subtly influence outcomes. For example, in an experiment that involves interviews, the experimenters might speak differently to people who received the real treatment than to those who received the placebo. This type of problem can be avoided by making the experiment **double-blind,** which means neither the participants nor the

by the way...
The placebo effect can be surprisingly powerful. Consider a drug now used to combat balding, which was tested on balding men. The drug maker was pleased to learn that 86% of the men receiving the drug either stopped balding or grew new hair. But remarkably, so did 42% of the men who received the placebo! In other studies, as many as 75% of participants receiving a placebo have actually improved.

experimenters know who belongs to each group. (Of course, someone must keep track of the two groups in order to evaluate the results at the end; in typical double-blind experiments, researchers hire experimenters to make any necessary contact with the participants.)

> **Blinding in Experiments**
>
> An experiment is **single-blind** if the participants do not know whether they are members of the treatment group or members of the control group, but the experimenters do know.
>
> An experiment is **double-blind** if neither the participants nor the experimenters (people administering the treatment) know who belongs to the treatment group and who belongs to the control group.

EXAMPLE 4 What's Wrong with This Experiment?

For each of the experiments described below, identify any problems and explain how the problems could have been avoided.

a. A chiropractor wants to know if his adjustments relieve back pain. He performs adjustments on 25 patients with back pain. Afterward, 18 of the patients say they feel better. He concludes that the adjustments are an effective treatment.

b. A new drug for attention deficit disorder (ADD) is supposed to make the affected children more polite. Randomly selected children suffering from ADD are divided into treatment and control groups; those in the control group receive a placebo that looks just like the real drug. The experiment is single-blind. Experimenters interview the children one on one to decide whether they became more polite.

DILBERT reprinted by permission of United Feature Syndicate, Inc.

Solution

a. The 25 patients who receive adjustments represent a treatment group, but this study lacks a control group. The patients may be feeling better because of a placebo effect rather than any real effect of the adjustments. The chiropractor might have improved his study by hiring an actor to do a fake adjustment (one that feels like a real manipulation, but doesn't actually conform to chiropractic guidelines) on a control group. Then he could have compared the results in the two groups to see whether a placebo effect was involved.

b. Because the experimenters know which children received the real drug, during the interviews they may inadvertently speak differently or interpret behavior differently with these children. In that case, their conclusions might not be valid. The experiment should have been double-blind, so that experimenters conducted interviews without knowing which children received the real drug and which children received the placebo.

Case-Control Studies

Sometimes it may be impractical or unethical to conduct an experiment. For example, suppose we want to study how alcohol consumed during pregnancy affects newborn babies. Because it is already known that alcohol can be harmful during pregnancy, it would be unethical to divide a sample of pregnant mothers randomly into two groups and then force the members of one group to consume alcohol. However, we may be able to conduct a **case-control study,** in which the participants naturally form groups by choice. In this example, the **cases** consist of mothers who consume alcohol during pregnancy *by choice,* and the **controls** consist of mothers who choose *not* to consume alcohol.

A case control study is observational because the researchers do not change the behavior of the participants. But it also resembles an experiment because the cases effectively represent a treatment group and the controls represent a control group.

> **DEFINITIONS**
>
> A **case-control study** is an observational study that resembles an experiment because the sample naturally divides into two (or more) groups. The participants who engage in the behavior under study form the **cases,** which makes them like a treatment group in an experiment. The participants who do not engage in the behavior are the **controls,** making them like a control group in an experiment.

EXAMPLE 5 *Which Type of Study?*

For each of the following questions, what type of statistical study is most likely to lead to an answer? Why?

a. What is the average income of stock brokers?

b. Do seat belts save lives?
 c. Can lifting weights improve runners' times in a 10-kilometer race?
 d. Can a new herbal remedy reduce the severity of colds?

Solution
 a. An *observational study* can tell us the average income of stock brokers. We need only survey (observe) the brokers.
 b. It would be unethical to do an experiment in which some people were told to wear seat belts and others were told *not* to wear them. Instead, we can conduct an observational *case-control study*. Some people choose to wear seat belts (the cases) and others choose not to wear them (the controls). By comparing the death rates in accidents between cases and controls, we can learn whether seat belts save lives. (They do.)
 c. We need an *experiment* to determine whether lifting weights can improve runners' 10K times. One group of runners will be put on a weight-lifting program, and a control group will be asked to stay away from weights. We must try to ensure that all other aspects of their training are similar. Then we can see whether the runners in the lifting group improve their times more than those in the control group. Note that we cannot use blinding in this experiment because there is no way to prevent participants from knowing whether they are lifting weights.
 d. We should use a *double-blind experiment,* in which some participants get the actual remedy while others get a placebo. We need double-blind conditions because the severity of a cold may be affected by mood or other factors that experimenters might inadvertently influence.

SURVEYS AND OPINION POLLS

Surveys and opinion polls may be the most common types of statistical study, and we must be very careful when we interpret them. Fortunately, survey and poll results usually include something called the **margin of error.**

Suppose a poll finds that 76% of the public supports the President, with a margin of error of 3 percentage points. The 76% is a sample statistic; that is, 76% of the people in a sample said they support the President. The margin of error helps us understand how well this sample statistic is likely to approximate the true population parameter (in this case, the percentage of all Americans who support the President). By adding and subtracting the margin of error from the sample statistic, we find a range of values, or a **confidence interval,** *likely* to contain the population parameter. In this case, we add and subtract 3 percentage points to find a confidence interval from 73% to 79%.

by the way...
Politicians and marketers often pretend they are trying to conduct a true opinion poll or survey when, in fact, they are deliberately trying to get particular results. These types of surveys are called *push polls* because they try to "push" people's opinions.

DEFINITION

The **margin of error** in a statistical study is used to describe a **confidence interval** that is likely to contain the true population parameter. We find this interval by subtracting and adding the margin of error from the sample statistic obtained in the study. That is, the confidence interval is

from (sample statistic − margin of error)
to (sample statistic + margin of error)

How confident can we be in a poll result? Unless we are told otherwise, we assume that the margin of error is defined to give us 95% confidence that the confidence interval contains the population parameter. You can think of "95% confidence" as follows: If the poll were repeated 20 times with 20 different samples, 19 of the 20 polls (that is, 95% of the polls) would have a confidence interval that contains the true population parameter.

EXAMPLE 6 Nielsen's Margin of Error

For the Nielsen ratings based on 5000 households, the margin of error is 1 percentage point. Suppose that 7% of the people in the sample watched *Millionaire*. What can we conclude about the percentage of all Americans who watched the show?

Solution We subtract and add the 1 percentage point margin of error from the sample statistic of 7%, which gives a confidence interval from 6% to 8%. Thus, we can be 95% confident that the actual percentage of Americans watching *Millionaire* was somewhere between 6% and 8%.

> **TECHNICAL NOTE**
> Because a household can have more than 1 person, Nielsen's survey of 5000 households includes more than 5000 people. The margin of error of 1 percentage point comes from Nielsen's own research.

EXAMPLE 7 Close Election

An election eve poll finds that 52% of surveyed voters plan to vote for Smith, and she needs a majority (more than 50%) to win without a runoff. The margin of error in the poll is 3 percentage points. Will she win?

Solution We subtract and add the margin of error of 3 percentage points to find a confidence interval

$$\text{from } 52\% - 3\% = 49\% \text{ to } 52\% + 3\% = 55\%$$

We can be 95% confident that the actual percentage of people planning to vote for her is between 49% and 55%. Because this confidence interval leaves open the possibility of both a majority and less than a majority, this election is too close to call.

Time out to think

In Example 7, suppose the poll found the candidate had 55% of the vote. Should she be confident of a win?

Review Questions

1. Why do we say that the term *statistics* has two meanings? Describe both meanings.
2. Define the terms *population, sample, population parameter,* and *sample statistics* as they apply to statistical studies.
3. Describe the five basic steps in a statistical study, and give an example of their application.
4. Why is it so important that a statistical study use a representative sample? Briefly describe four common sampling methods.
5. What is *bias*? How can it affect a statistical study? Give examples of several forms of bias.
6. Describe and contrast observational studies and experiments. What do we mean by the treatment group and control group in an experiment? What do we mean by the cases and controls in an observational case-control study?
7. What is a placebo? Describe the placebo effect and how it can make experiments difficult to interpret. How can making an experiment single-blind or double-blind help?
8. What is meant by the margin of error in a survey or opinion poll? How is it used to identify a confidence interval?

In the News

1. **Statistics in the News.** Select three news stories from the past week that involve statistics in some way. In each case, write one or two paragraphs describing the role of statistics in the story.

2. **Statistics in Your Major.** Write two to three paragraphs describing the ways in which you think the science of statistics is important in your major field of study. (If you have not chosen a major, answer this question for a major that you are considering.)

3. **Statistics in Sports.** Choose a sport and describe three different statistics commonly tracked by participants in or spectators of the sport. In each case, briefly describe the importance of the statistic to the sport.

4. **Sample and Population.** Find a report in today's news concerning any type of statistical study. What is the population being studied? What is the sample? Why do you think the sample was chosen as it was?

5. **Poor Sampling.** In a recent newspaper or magazine, find an article about a study that attempts to describe some characteristic of a population, but that you believe involved poor sampling (for example, a sample that was too small or unrepresentative of the population under study). Describe the population, the sample, and what you think was wrong with the sample. Briefly discuss how you think the poor sampling affected the study results.

6. **Good Sampling.** In a recent newspaper or magazine, find an article that describes a statistical study in which the sample was well chosen. Describe the population, the sample, and why you think the sample was a good one.

7. **Margin of Error.** Find a report of a recent survey or poll. Interpret the sample statistic and margin of error quoted for the survey or poll.

Exercises

Population and Sample. For the studies described in Exercises 1–6, describe the population, sample, population parameters, and sample statistics.

1. In order to gauge public opinion about the candidates in a recent election for Governor of California, CBS News conducted telephone interviews with 1026 people who were registered to vote in California.

2. Astronomers typically determine the distance to a galaxy (a galaxy is a huge collection of billions of stars) by measuring the distances to just a few stars within it and taking the mean (average) of these distance measurements.

3. To rate the speed of new computers, *Byte* magazine editors purchased five new computers and measured the time required by each to run a standard program.

4. The Gallup Organization conducted a poll of 1200 adults to determine how Americans respond to the question "Do you think college coaches use physical force with their athletes?"

5. A nationwide Harris Poll of 1010 adults surveyed by telephone revealed that the professions viewed by the public as most prestigious are doctor, scientist, teacher, and minister/clergy.

6. The American Institute of Education conducts an annual study of attitudes of incoming college students by surveying approximately 261,000 first-year students at 462 colleges and universities. There are approximately 1.6 million first-year college students in this country.

Steps in a Study. Describe how you would apply the five basic steps of a statistical study to the issues in Exercises 7–12.

7. You want to determine the average amount of pizza consumed each year by students at your school.

8. As an airline marketing executive, you want to know if there has been an increase in frustration with air travel among business travelers.

9. You want to know the mean (average) height of adult American women.

10. You want to know the typical percentage of the bill that is left as a tip in restaurants.

11. You want to know how long alkaline batteries last in Kodak cameras.

12. You want to know the percentage of high school students who are vegetarians.

13. **Representative Sample?** You want to determine the mean (average) number of calories eaten daily by girls on high school track teams. Which of the following samples is most likely to be representative, and why? Also explain why each of the other choices is *not* likely to make a representative sample for this study.
 - The seniors on your high school's girls' track team
 - The shot putters and javelin throwers on your high school's girls' track team
 - The sprinters on your high school's girls' track team
 - The entire girls' track team at your high school

14. **Representative Sample?** You want to determine the typical dietary habits of students at a college. Which of the following would make the best sample, and why? Also explain why each of the other choices would *not* make a good sample for this study.
 - Students in a single dormitory
 - Students majoring in public health
 - Students who participate in intercollegiate sports
 - Students enrolled in a required mathematics class

Identify the Sampling Method. Exercises 15–20 each describe a sample. Identify the sampling method as simple random sampling, systematic sampling, convenience sampling, or stratified sampling. Briefly explain why you think this sampling method was chosen.

15. An IRS (Internal Revenue Service) auditor randomly selects for audits 30 taxpayers whose gross income is less than $30,000 and 30 taxpayers whose gross income is more than $30,000.

16. *People* magazine chooses its "best dressed celebrities" by looking at responses from readers who choose to mail in a survey printed in the magazine.

17. A study of sleep habits uses 50 participants whose ages are between 20 and 29, 50 participants whose ages are between 30 and 39, and 50 participants whose ages are between 40 and 49.

18. Every 100th roll of rope that is produced is given a strength test.

19. A computer randomly selects 400 names from a list of all registered voters; those selected are surveyed to predict who will win the election for Mayor.

20. At a football game, a computer randomly selects 50 ticket numbers; those whose tickets are selected are invited to participate in a contest.

Type of Study. For Exercises 21–26, state whether the study is an observational study or an experiment. If it is an experiment, describe the treatment and control groups and discuss whether single- or double-blinding is needed. If it is observational, state whether it is a case-control study and, if it is, distinguish between the cases and the controls.

21. A 1999 Michigan State University study of starting salaries for college graduates found that computer science majors had the highest salaries ($42,500), while communications majors had the lowest ($25,600).

22. A National Cancer Institute study of 716 melanoma patients and 1014 cancer-free patients matched by age, sex, and race found that those having a single large mole had twice the risk of melanoma; having 10 or more moles was associated with a 12 times greater risk of melanoma (*Journal of the American Medical Association*, September 1997).

23. A Gallup survey done for CNN/*USA Today* found that 80% of Americans think we're less civil than we were ten years ago and 67% think that we are more likely to use vulgar language than we were ten years ago.

24. A breast cancer study began by asking 25,624 women questions about how they spent their leisure time. The health of these women was tracked over the next 15 years. Those women who said they exercise regularly were found to have a lower incidence of breast cancer (*New England Journal of Medicine*, May 1, 1997).

25. A (hypothetical) study found that tomato plants treated with a new fertilizer had a 23% greater yield than plants with no fertilizer.

26. A survey of 275,811 first-year college students revealed that 32.4% of these students had an A average in high school (Higher Education Research Institute, 1998).

Which Type of Study? For each of the questions in Exercises 27–32, what type of statistical study is most likely to lead to an answer? Why?

27. What is the average income of public school teachers?

28. Can listening to jazz music while studying improve students' grades?

29. Are high levels of artificial flavoring harmful to humans when eaten in foods?

30. Does playing soccer help swimmers improve their times?

31. Does an aspirin a day reduce the incidence of heart attacks?

32. Does a self-proclaimed mind-reader really have supernatural abilities?

Experiment Results. Consider an experiment designed to determine the effectiveness of a new drug. The drug is given to participants in the treatment group, while participants in the control group receive a placebo. For each set of results described in Exercises 33–36, discuss whether there appears to be evidence that the treatment is effective.

33. 60% of those in the treatment group showed improvement; 20% of those in the placebo group showed improvement.

34. 40% of those in the treatment group showed improvement; 40% of those in the placebo group showed improvement.

35. 90% of those in the treatment group showed improvement; 50% of those in the placebo group showed improvement.

36. 30% of those in the treatment group showed improvement; 40% of those in the placebo group showed improvement.

Margin of Error. Each of Exercises 37–40 states both a sample statistic and a margin of error. Find the confidence interval in each case, and answer any additional questions asked. Be sure to explain your answers clearly.

37. A poll is conducted the day before a state election for Senator. There are only two candidates running. The poll shows that 53% of the voters surveyed favor the Republican candidate, with a margin of error of 2.5 percentage points. Should the Republican plan a victory party? Why or why not?

38. A poll is conducted the day before an election for U.S. Representative. There are only two candidates running. The poll shows that 48.5% of the voters surveyed favor the Democratic candidate, with a margin of error of 2.0 percentage points. Based on this poll, should the Democratic candidate expect to lose the election? Why or why not?

39. The U.S. Labor Department reports that the January unemployment rate was 4.2% and the February unemployment rate was 4.3%. The margin of error for each report is 0.2 percentage point. Can we conclude that unemployment increased from January to February?

40. In a survey of 1002 people, 701 (which is 70%) said that they voted in the most recent presidential election (based on data from ICR Research Group). The margin of error for the survey was 3 percentage points. However, actual voting records show that only 61% of all eligible voters actually did vote. Does this necessarily imply that people lied when they answered the survey?

Interpreting Real Studies. For each of Exercises 41–46, do the following:

a. Based on the given information, determine the goal of the study. Identify the population and the population parameter of interest.

b. Briefly describe the sample and sample statistic for the study.

c. Find the confidence interval likely to contain the population parameter of interest.

41. In a TIME/CNN poll, 748 adults were asked whether they believed their children would have a higher standard of living than they have; 63% of those polled said "yes." The margin of error was 3.6 percentage points.

42. A Yankelovich poll determined that 70% of 4000 people surveyed agreed with the statement "People have to realize that they can only count on their own skills and abilities if they're going to win in this world." The margin of error in the poll was 1.6 percentage points.

43. Based on its survey of 60,000 households (see Example 2), the U.S. Labor Department reported an unemployment rate of 4.2% in September 1999. The margin of error for the report was 0.2 percentage point.

44. The Pew Research Center asked 1546 adult Americans whether humans would land on Mars within the next 50 years; 76% of these people said either definitely yes or

probably yes. The margin of error for the poll was 3 percentage points.

45. The Pew Research Center asked 1546 adult Americans whether they believed that humans would be cloned within the next 50 years; 51% answered either definitely or probably yes. The margin of error was 3 percentage points.

46. A Roper Organization survey of 2000 adults revealed that 64% of those surveyed kept money in a regular savings account. The margin of error for the survey was 2.0 percentage points.

Web Projects

Find useful links for Web Projects on the text Web site: www.aw.com/bennett-briggs

1. **Current Nielsen Ratings.** Find the Nielsen ratings for the past week. What were the three most popular television shows? Explain both the "rating" and the "share" for each show.

2. **Nielsen Sample.** Use information available on the Nielsen Media Research Web site to answer each of the following questions.

 a. How does Nielsen select the sample of homes to be included in a viewer survey?
 b. Describe a few ways by which Nielsen attempts to check that the results from its people meter surveys are accurate.
 c. Based on what you have learned, do you think the Nielsen ratings are reliable? If so, why? If not, why not?

3. **Attitude Update.** The Pew Research Center for the People and the Press studies public attitudes toward the press, politics, and public policy issues. Go to its Web site and find the latest survey about attitudes. Write a one-page summary of what Pew surveyed, how it conducted the survey, and what it found.

4. **Labor Statistics.** Use the Bureau of Labor Statistics Web page to learn about its monthly survey. Choose one aspect of the survey, such as how the sample is chosen or how it is used to compare unemployment rates over time. Write a short summary of what you learn.

5. **Professional Polling.** Visit the Web site of a national polling organization and report on a recent poll. Write a short description of the poll and its results, commenting on features such as sampling technique, sample size, and margin of error.

UNIT 5B

SHOULD YOU BELIEVE A STATISTICAL STUDY?

Most statistical research is carried out with integrity and care. Nevertheless, statistical research is sufficiently complex that bias can arise in many different ways. Thus, we must always examine reports of statistical research carefully, looking for anything that might make us question the results. In this unit, we discuss eight guidelines that can help you answer the question "Should I believe a statistical study?"

GUIDELINE 1: IDENTIFY THE GOAL, POPULATION, AND TYPE OF STUDY

Before evaluating the details of a statistical study, we must know what it is about. Based on what you hear or read, try to answer basic questions such as these:

- What was the goal of the study?
- What was the population under study? Was the population clearly and appropriately defined?
- What type of study was used? Was the type appropriate for the goal?

> **by the way ...**
> Surveys show that nearly half of Americans believe their horoscopes. However, in controlled experiments, the predictions of horoscopes come true no more often than would be expected by chance.

If you can't find reasonable answers to these questions, it will be difficult to evaluate other aspects of the study.

EXAMPLE 1 *Appropriate Type of Study?*

A newspaper reports: "Researchers gave each of the 100 participants their astrological horoscopes, and asked them whether the horoscopes appeared to be accurate. Eighty-five percent of the participants reported that the horoscopes were accurate. The researchers concluded that horoscopes are valid most of the time." Analyze this study according to Guideline 1.

Solution The goal of the study was to determine the validity of horoscopes. Based on the news report, it appears that the study was *observational:* The researchers simply asked the participants about the accuracy of the horoscopes. However, because the accuracy of a horoscope is somewhat subjective, this study should have been a controlled experiment in which some people were given their actual horoscope and others were given a fake horoscope. Then the researchers could have looked for differences between the two groups. Moreover, because researchers could easily influence the results by how they questioned the participants, the experiment should have been double-blind. In summary, the type of study was inappropriate to the goal and its results are meaningless. ■

EXAMPLE 2 *Appropriate Population?*

A news report states: "Researchers tracked 1000 men for 15 years. They found that the men who exercised regularly had lower mortality rates than those who did not. They concluded that regular exercise leads to longer lives." Analyze this headline according to Guideline 1.

Solution This study is an observational case-control study, because the men formed two groups: a case group that exercised regularly and a control group that did not. This type of study seems appropriate, which should give us some confidence in its conclusions. However, the population is not clearly defined—at least in the news report. Does the population include *all* men? If so, several factors may complicate interpretation of the results. For example, it may not have been a good idea to include smokers in the population, because exercise might affect smokers and nonsmokers differently. Because the news report does not clearly state how the population was defined, we should withhold final judgment on the conclusions until we learn more. ■

GUIDELINE 2: CONSIDER THE SOURCE

Statistical studies are supposed to be objective, but the people who carry them out and fund them may be biased. Thus, it is important to consider the source of a study and evaluate the potential for biases that might invalidate its conclusions.

EXAMPLE 3 *Is Smoking Healthy?*

By 1963, enough research on the health dangers of smoking had accumulated that the Surgeon General of the United States publicly announced that smoking is bad for health. Research done since that time has built further support for this claim.

> **by the way ...**
> After decades of arguing to the contrary, in October 1999 the Philip Morris Company—the world's largest seller of tobacco products—publicly acknowledged that smoking causes lung cancer, heart disease, emphysema, and other serious diseases.

However, while the vast majority of studies show that smoking is unhealthy, a few studies have consistently shown no dangers from smoking, and perhaps even health *benefits*. These studies generally were carried out by the Tobacco Research Institute, funded by the tobacco companies. Analyze the Tobacco Research Institute studies according to Guideline 2.

Solution Tobacco companies had a financial interest in minimizing the dangers of smoking. Because the studies carried out at the Tobacco Research Institute were funded by the tobacco companies, there may have been pressure on the researchers to produce results to the companies' liking. This *potential* for bias does not mean their research *was* biased, but the fact that it contradicts virtually all other research on the subject should be cause for concern.

GUIDELINE 3: LOOK FOR BIAS IN THE SAMPLE

Look for bias that may prevent the sample from being representative of the population. There are two particularly common forms of bias that can affect sample selection.

Copyright © 1998 by Sidney Harris.

> **Bias in Choosing a Sample**
>
> **Selection bias** occurs whenever researchers *select* their sample in a way that tends to make it unrepresentative of the population. For example, a pre-election poll that surveys only registered Republicans has selection bias because it is unlikely to reflect the opinions of all voters.
>
> **Participation bias** occurs primarily with surveys and polls; it arises whenever people *choose* whether to participate. Because people who feel strongly about an issue are more likely to participate, their opinions may not represent the larger population that is less emotionally attached to the issue.

by the way...
Surveys or polls in which people choose whether to participate are often called *self-selected* or *voluntary response* surveys.

CASE STUDY *The 1936 Literary Digest Poll*

The *Literary Digest*, a popular magazine of the 1930s, successfully predicted the outcomes of several elections using large polls. In 1936, editors of the *Literary Digest* conducted a particularly large poll in advance of the presidential election. They randomly chose a sample of 10 million people from various lists, including names in telephone books and rosters of country clubs. They mailed a postcard "ballot" to each of these 10 million people. About 2.4 million people returned the postcard ballots. Based on the returned ballots, the editors of the *Literary Digest* predicted that Alf Landon would win the presidency by a margin of 57% to 43% over Franklin Roosevelt. Instead, Roosevelt won with 62% of the popular vote. How did such a large survey go so wrong?

historical note

A young pollster named George Gallup conducted his own survey prior to the 1936 election. Sending postcards to only 3000 randomly selected people, he correctly predicted not only the outcome of the election, but also the outcome of the *Literary Digest* poll to within 1%. Gallup went on to establish a very successful polling organization.

The sample suffered from both selection bias and participation bias. The selection bias arose because the *Literary Digest* chose its 10 million names in ways that favored affluent people. For example, selecting names from telephone books meant choosing only from those who could afford telephones back in 1936. Similarly, country club members are usually quite wealthy. The selection bias favored Landon because he was the Republican, and affluent voters of the 1930s tended to vote for Republican candidates.

The participation bias arose because return of the postcard ballots was voluntary. People who felt most strongly about the election were more likely to be among those who returned their postcard ballots. This bias also tended to favor Landon because he was the challenger—people who did not like President Roosevelt could express their desire for change by returning the postcards. Together, the two forms of bias made the sample results useless, despite the sample's unusually large size of 2.4 million people.

The Franklin Delano Roosevelt Memorial in Washington D.C.

EXAMPLE 4 *Self-Selected Poll*

The television show *Nightline* conducted a poll in which viewers were asked whether the United Nations headquarters should be kept in the United States. Viewers could respond to the poll by paying 50 cents to call a "900" phone number with their opinions. The poll drew 186,000 responses, of which 67% favored moving the United Nations out of the United States. Around the same time, a poll using simple random sampling of 500 people found that 72% wanted the United Nations to *stay* in the United States. Which poll is more likely to be representative of the general opinions of Americans?

Solution The *Nightline* sample suffered from severe participation bias. Not only did viewers choose whether to call in for the survey, but they had to *pay* to participate. This cost made it even more likely that respondents would be those who felt a need for change. Thus, despite its large number of respondents, the *Nightline* survey was

too biased to be trusted. In contrast, a simple random sample of 500 people is quite likely to be representative, so the finding of this small survey has a better chance of representing the true opinions of all Americans.

GUIDELINE 4: LOOK FOR PROBLEMS IN DEFINING OR MEASURING THE VARIABLES OF INTEREST

Statistical studies usually attempt to measure *something*, and we call the things being measured the **variables** of interest in the study; the term *variable* simply refers to an item or quantity that can vary or take on different values. For example, variables in the Nielsen ratings include *show being watched* and *number of viewers*.

> **DEFINITION**
>
> A **variable** is any item or quantity that can vary or take on different values. The *variables of interest* in a statistical study are the items or quantities that the study seeks to measure.

by the way...
More than a third of all Americans routinely shut the door or hang up the phone when contacted for a survey, thereby making self-selection a problem for legitimate pollsters. One reason people hang up may be the proliferation of selling under the guise of market research (often called "sugging"), in which a telemarketer pretends you are part of a survey in order to get you to buy something.

Results of a statistical study may be especially difficult to interpret if the variables under study are difficult to define or measure. For example, imagine trying to conduct a study of how exercise affects resting heart rates. The variables of interest would be *amount of exercise* and *resting heart rate*. However, both variables are difficult to define and measure. In the case of *amount of exercise*, it's not clear what the definition covers: Does it include walking to class? Even if we specify the definition, how can we measure *amount of exercise* given that some forms of exercise are more vigorous than others? The following two examples describe real cases in which defining or measuring variables caused problems in statistical studies.

Time out to think

How would you measure your resting heart rate? Describe some difficulties in defining and measuring *resting heart rate*.

EXAMPLE 5 Can Money Buy Love?

A Roper poll reported in *USA Today* involved a survey of the wealthiest 1% of Americans. The survey found that these people would pay an average of $487,000 for "true love," $407,000 for "great intellect," $285,000 for "talent," and $259,000 for "eternal youth." Analyze this result according to Guideline 4.

Solution The variables in this study are very difficult to define. How, for example, do you define "true love"? And does it mean true love for a day, a lifetime, or something else? Similarly, does the ability to balance a spoon on your nose constitute "talent"? Because the variables are so poorly defined, it's likely that different people interpreted them differently, making the results very difficult to interpret.

EXAMPLE 6 *Illegal Drug Supply*

Law enforcement authorities try to stop illegal drugs from entering the country. A commonly quoted statistic is that they succeed in stopping only about 10% to 20% of the drugs entering the United States. Should you believe this statistic?

Solution There are essentially two variables in the study: *quantity of illegal drugs intercepted* and *quantity of illegal drugs NOT intercepted*. It should be relatively easy to measure the quantity of illegal drugs that law enforcement officials intercept. However, because the drugs are illegal, it's unlikely that anyone is reporting the quantity of drugs that are *not* intercepted. How, then, can anyone know that the intercepted drugs are 10% to 20% of the total? In a *New York Times* analysis, a police officer was quoted as saying that his colleagues refer to this type of statistic as "P.F.A.," for "pulled from the air."

GUIDELINE 5: WATCH OUT FOR CONFOUNDING VARIABLES

Variables that are *not intended* to be part of the study can sometimes make it difficult to interpret results properly. Such variables are often called *confounding* variables, because they confound (confuse) a study's results.

It's not always easy to discover confounding variables. Sometimes they are discovered years after a study was completed, and sometimes they are not discovered at all. Fortunately, confounding variables are sometimes more obvious and can be discovered simply by thinking hard about factors that may have influenced a study's results.

EXAMPLE 7 *Radon and Lung Cancer*

Radon is a radioactive gas produced by natural processes (the decay of uranium) in the ground. The gas can leach into buildings through the foundation and can accumulate to relatively high concentrations if doors and windows are closed. Imagine a study that seeks to determine whether radon gas causes lung cancer by comparing the lung cancer rate in Colorado, where radon gas is fairly common, with the lung cancer rate in Hong Kong, where radon gas is less common. Suppose the study finds that the lung cancer rates are nearly the same. Is it fair to conclude that radon is *not* a significant cause of lung cancer?

Solution The variables under study are *amount of radon* and *lung cancer rate*. However, because smoking can also cause lung cancer, *smoking rate* may be a confounding variable in this study. In particular, the smoking rate in Hong Kong is much higher than the smoking rate in Colorado, so any conclusions about radon and lung cancer must take the smoking rate into account. In fact, careful studies have shown that radon gas *can* cause lung cancer, and the U.S. Environmental Protection Agency (EPA) recommends taking steps to prevent radon from building up indoors.

by the way...
Many hardware stores sell simple kits that you can use to test whether radon gas is accumulating in your home. If it is, the problem can be eliminated by installing an appropriate "radon mitigation" system, which usually consists of a fan that blows the radon out from under the house before it can get in.

GUIDELINE 6: CONSIDER THE SETTING AND WORDING IN SURVEYS

Even when a survey is conducted with proper sampling and with clearly defined terms and questions, it's important to watch out for problems in the setting or wording that might produce inaccurate or dishonest responses. Dishonest responses are

particularly likely when the survey concerns sensitive subjects, such as personal habits or income. For example, the question "Do you cheat on your income taxes?" is unlikely to elicit honest answers from those who cheat, especially if the setting does not guarantee complete confidentiality. In other cases, even honest answers may not be accurate if the wording of questions invites bias.

EXAMPLE 8 *Word Order Matters*

A poll conducted in Germany asked the following two questions:

- *Would you say that traffic contributes more or less to air pollution than industry?*
- *Would you say that industry contributes more or less to air pollution than traffic?*

With the first question, 45% answered traffic and 32% answered industry. With the second question, only 24% answered traffic while 57% answered industry. What happened?

Solution Note that the order of the words *traffic* and *industry* was reversed in the two questions. Apparently, this change in order dramatically changed the survey results. In fact, this effect is quite common, because people tend to focus more on the first thing they hear or see.

EXAMPLE 9 *Do You Want a Tax Cut?*

When Republicans in Congress proposed a 10% across-the-board cut in income taxes, the Republican National Committee commissioned a poll to find out whether Americans supported the proposal. Asked whether they favored the tax cut, 67% of respondents answered yes. Should we conclude that Americans supported the proposal?

Solution A question like "Do you favor a tax cut?" is biased because it does not give other options (much like the fallacy of *limited choice* discussed in Unit 1A). In fact, an independent poll conducted at the same time gave respondents a list of options for using surplus revenues. This poll found that 31% wanted the money devoted to Social Security, 26% wanted it used to reduce the national debt, and only 18% favored using it for a tax cut. (The remaining 25% of respondents chose a variety of other options.)

> **by the way . . .**
> People are more likely to choose the item that comes first in a survey because of what psychologists call the *availability error*—the tendency to make judgments based on what is *available* in the mind. Professional polling organizations must be very careful to avoid this problem, sometimes by posing the question to some people in one order and to others in the opposite order.

GUIDELINE 7: CHECK THAT RESULTS ARE PRESENTED FAIRLY

Even when a statistical study is done well, it may be misrepresented in graphs or concluding statements. Researchers may occasionally misinterpret the results of their own studies or jump to conclusions that are not supported by the results, particularly when they have personal biases toward certain interpretations. In other cases, news reporters or others may misinterpret a survey or jump to unwarranted conclusions that make a story seem more spectacular. Misleading graphs are an

especially common problem (see Unit 5D). In general, you should look for inconsistencies between the interpretation of a study (in pictures and words) and any actual data given with it.

EXAMPLE 10 Does the School Board Need a Statistics Lesson?

The school board in Boulder, Colorado created a hubbub when it announced that 28% of Boulder school children were reading "below grade level," and hence concluded that methods of teaching reading needed to be changed. The announcement was based on reading tests on which 28% of Boulder school children scored below the national average for their grade. Do these data support the board's conclusion?

Solution The fact that 28% of Boulder children scored below the national average for their grade implies that 72% scored at or above the national average. Thus, the school board's ominous statement about students reading "below grade level" makes sense only if "grade level" means the national average score for a particular grade. This interpretation of "grade level" is curious because it means that half the students in the nation are always below grade level—no matter how high the scores. The conclusion that teaching methods needed to be changed was not justified by these data.

GUIDELINE 8: STAND BACK AND CONSIDER THE CONCLUSIONS

Finally, even if a study seems reasonable according to all the previous guidelines, you should stand back and consider the conclusions. Ask yourself questions such as these:

- Did the study achieve its goals?

- Do the conclusions make sense?

- Can you rule out alternative explanations for the results?

- If the conclusions do make sense, do they have any practical significance?

Extraordinary claims require extraordinary evidence.
—CARL SAGAN (1934–1996)

EXAMPLE 11 Practical Significance

An experiment is conducted in which the weight losses of people who try a new "Fast Diet Supplement" are compared to the weight losses of a control group of people who try to lose weight in other ways. After eight weeks, the results show that the treatment group lost an average of $\frac{1}{2}$ pound more than the control group. Assuming that it has no dangerous side effects, does this study suggest that the Fast Diet Supplement is a good treatment for people wanting to lose weight?

Solution Compared to the average person's body weight, the difference of $\frac{1}{2}$ pound hardly matters at all. Thus, while the statistics in this case may be interesting, they don't seem to have much practical significance.

> **SUMMARY** Eight Guidelines for Evaluating a Statistical Study
>
> 1. Identify the goal of the study, the population being considered, and the type of study used.
> 2. Consider the source, particularly with regard to whether the researchers may be biased.
> 3. Look for bias that may prevent the sample from being representative of the population.
> 4. Look for problems in defining or measuring the variables of interest, which can make it difficult to interpret results.
> 5. Watch out for confounding variables that can invalidate the conclusions of a study.
> 6. Consider the setting and the wording of questions in any survey, looking for anything that might tend to produce inaccurate or dishonest responses.
> 7. Check that results are presented fairly in graphs and concluding statements, since both researchers and media often create misleading graphics or jump to conclusions that the results do not support.
> 8. Stand back and consider the conclusions. Did the study achieve its goals? Do the conclusions make sense? Do the results have any practical significance?

Review Questions

1. Briefly describe each of the eight guidelines for evaluating statistical studies. Give an example to which each guideline applies.
2. Describe and contrast selection bias and participation bias in sampling. Give an example of each.
3. What do we mean by the variables of interest in a study?
4. What are confounding variables, and what problems can they cause?

In the News

1. **Applying the Guidelines.** Find a recent newspaper article or television report about a statistical study on a topic that you find interesting. Write a short report applying each of the eight guidelines given in this section. (Some of the guidelines may not apply to the particular study you are analyzing; in that case, explain why the guideline is not applicable.)
2. **Believable Results.** Find a recent news report about a statistical study whose results you believe are meaningful and important. In one page or less, summarize the study and explain why you find it believable.
3. **Unbelievable Results.** Find a recent news report about a statistical study whose results you *don't* believe are meaningful or important. In one page or less, summarize the study and why you don't believe its claims.
4. **Legal Experts.** Find a news report concerning a major ongoing trial. Find out whether any of the "expert witnesses" are being paid by either side. Based on what you learn, describe whether you think the experts are giving biased testimony.
5. **Biased Questioning?** Find a recent news report of responses to a single question in an opinion poll. State the exact words of the question and the results of the poll. Analyze the question and the reported results for potential biases. At the end of your analysis, state whether you believe the results, and defend your opinion.

Exercises

Bias. Exercises 1–8 present situations in which bias may be an issue. Describe one potential source of bias in the situation, and briefly discuss whether the bias should affect your view of the situation.

1. A film critic on ABC gives her opinion of the latest movie from Disney, which happens to own ABC.

2. Market researchers conduct a survey at a supermarket on a weekday between 10:00 a.m. and noon to determine what fraction of customers uses coupons.

3. An exit poll designed to predict the winner of a local election uses interviews with everyone who votes between 7:00 and 7:30 a.m.

4. An exit poll designed to predict the winner of a national election uses interviews with randomly selected voters in New York.

5. In order to determine the opinions of people in the 18- to 24-year age group on capital punishment, researchers survey a random sample of 500 Marines in this age group.

6. A college mails survey forms to all current seniors, asking for the students' choices of their all-time best and worst professors. Students are asked to return the survey in the campus mail.

7. Monsanto conducts a study to determine whether its new, genetically engineered soybean poses any threat to the environment.

8. Scientists working for Greenpeace (which opposes genetically engineered crops) conduct a study to determine whether Monsanto's new, genetically engineered soybean poses any threat to the environment.

9. **It's All in the Wording.** In 1998, Princeton Survey Research Associates did a study for *Newsweek* magazine illustrating the effects of wording in a survey. Two questions were asked:
 - Do you personally believe that abortion is wrong?
 - Whatever your own personal view of abortion, do you favor or oppose a woman in this country having the choice to have an abortion with the advice of her doctor?

 To the first question, 57% of the respondents replied yes, while 36% responded no. In response to the second question, 69% of the respondents favored the choice, while 24% opposed the choice. Discuss why the two questions produced seemingly contradictory results. How could the results of the questions be used selectively by various groups?

10. **Tax or Spend?** A Gallup poll in March 1999 asked the following two questions:
 - Do you favor a tax cut or "increased spending on other government programs"? *Result:* 75% for tax cut.
 - Do you favor a tax cut or "spending to fund new retirement savings accounts, as well as increased spending on education, defense, Medicare and other programs"? *Result:* 60% for the spending.

 Discuss why the two questions produced seemingly contradictory results. How could the results of the questions be used selectively by various groups?

Stat-Bytes. Politicians must make their political statements (often called sound-bytes) very short because the attention span of listeners is so short. A similar effect occurs in reporting statistical news. Major statistical studies are often reduced to one or two sentences. The summaries of statistical reports in Exercises 11–16 are taken from various news sources. Discuss what crucial information is missing and what more you would want to know before you acted on the report.

11. *USA Today* (April 28, 1999) reports on a Harris poll claiming that the percentage of adults with a "great deal of confidence" in military leaders stands at 54% (up from 37% in 1997).

12. CNN reports on a Zagat Survey of America's Top Restaurants which found that "only nine restaurants achieved a rare 29 out of a possible 30 rating and none of those restaurants are in the Big Apple."

13. A University of Michigan study concluded that drunken driving by high school seniors had declined. The report was based on two studies. In 1984, 31.2% of high school seniors reported driving after drinking sometime during the previous two weeks. By 1997, only 18.3% of high school seniors reported driving after drinking sometime during the previous two weeks.

14. An estimated 2% of Americans deceased last year paid estate taxes, while 60% of Americans favor repealing estate taxes.

15. The size of network TV audiences has decreased by 24% since 1991, while the annual advertising revenue of the networks increased by 42%.

16. Thirty percent of newborns in India would qualify for intensive care if they were born in the United States.

Accurate Headlines? Exercises 17–19 give a headline and a brief description of the statistical news story that accompanied the headline. In each case, discuss whether the headline accurately represents the story.

17. Headline: "Drugs shown in 98 percent of movies"
 Story summary: A "government study" claims that drug use, drinking, or smoking was depicted in 98% of the top movie rentals (Associated Press, April 29, 1999).

18. Headline: "Sex more important than jobs"
 Story summary: A survey found that 82% of 500 people interviewed by phone ranked a satisfying sex life as important or very important, while 79% ranked job satisfaction as important or very important (Associated Press, April 30, 1999).

19. Headline: "Grape juice may fight disease"
 Story summary: A study of 15 people, partially funded by Welch Foods, found that grape juice helps to expand blood vessels and increase the levels of LDL cholesterol; both constricted blood vessels and low LDL levels are risk factors for heart disease (*Milwaukee Journal Sentinel*, September 8, 1999).

Web Projects

Find useful links for Web Projects on the text Web site: www.aw.com/bennett-briggs

1. **Polling Organization.** Go to the Web site for a major professional polling organization. Study results from a recent poll, and evaluate the poll according to the guidelines in this section.

2. **Harper's Index.** Go to the Web site for the Harper's Index and study a few of the recently quoted statistics. Be sure to select the option on the page that allows you to see the sources for the statistics. Choose three statistics that you find particularly interesting, and discuss whether, in accord with the guidelines given in this section, you believe them.

UNIT 5C

STATISTICAL TABLES AND GRAPHS

Whether you look at a newspaper, a corporate annual report, or a government study, you are almost sure to see tables and graphs of statistical data. Some of these tables and graphs are simple; others can be quite complex. Some make it easy to understand the data; others may be confusing or even misleading. In this unit, we'll investigate some of the basic principles behind tables and graphs, preparing for more complex graphics in Unit 5D.

FREQUENCY TABLES

A teacher makes the following list of the grades she gave to her 25 students on an essay:

A C C B C D C C F D C C C B B A B D B A A B F C B

This list contains all the raw data, but it isn't easy to read. A better way to display these data is with a **frequency table**—a table showing the number of times, or **frequency**, that each grade appears (Table 5.1). The five possible grades are called the **categories** for the table.

There are two common variations on the idea of frequency. The **relative frequency** for a category expresses its frequency as a fraction or percentage of the total. For example, 4 of the 25 students received A grades, so the relative frequency for A grades is 4/25, or 16%. Note that the total relative frequency must always be 1, or 100%.

The **cumulative frequency** is the number of responses in a particular category and all preceding categories. For example, the cumulative frequency for grades of C and above is 20, because 20 students received grades of either A, B, or C.

TABLE 5.1

GRADE	FREQUENCY
A	4
B	7
C	9
D	3
F	2
Total	**25**

> **DEFINITION**
>
> A basic **frequency table** has two columns:
> - The first column lists all the **categories** of data.
> - The second column lists the **frequency** of each category, which is the number of times each category appears in the data set.
>
> Additional columns may include **relative frequency** (frequency expressed as a fraction or percentage of the total) or **cumulative frequency** (total of frequencies for the given category and all previous categories).

EXAMPLE 1 *Relative and Cumulative Frequency*

Add to Table 5.1 columns showing the relative and cumulative frequencies.

Solution Table 5.2 shows columns and calculations for relative and cumulative frequency.

TABLE 5.2

GRADE	FREQUENCY	RELATIVE FREQUENCY	CUMULATIVE FREQUENCY
A	4	4/25 = 16%	4
B	7	7/25 = 28%	7 + 4 = 11
C	9	9/25 = 36%	9 + 7 + 4 = 20
D	3	3/25 = 12%	3 + 9 + 7 + 4 = 23
F	2	2/25 = 8%	2 + 3 + 9 + 7 + 4 = 25
Total	25	1 = 100%	25

Time out to think

Briefly explain why the total relative frequency must always be 1, or 100%.

Data Types

Essay grades represent subjective ratings, not actual measurements or counts. We say that the grade categories are **qualitative,** because they represent *qualities* such as bad or good. In contrast, scores on a multiple-choice exam are **quantitative,** because they represent an actual count (or measurement) of the number of correct answers. As we'll see shortly, distinguishing between qualitative and quantitative data can be useful in creating tables or graphs.

> **Data Types**
>
> **Qualitative data** describe *qualities* or nonnumerical categories.
>
> **Quantitative data** represent counts or measurements.

EXAMPLE 2 Data Types

Classify each of the following types of data as either qualitative or quantitative.

 a. Brand names of shoes in a consumer survey
 b. Heights of students
 c. Audience ratings of a film on a scale of 1 to 5, where 5 means excellent

Solution

 a. Brand names are nonnumerical categories, so they are qualitative data.
 b. Heights are measurements, so they are quantitative data.
 c. Although the film rating categories involve numbers, the numbers represent subjective opinions about a film, not counts or measurements. Thus, they are qualitative data, despite being stated as numbers.

Time out to think

Give another example in which numbers are used to represent qualitative data rather than quantitative data.

Binning Data

When we deal with quantitative data categories, it's often useful to group, or *bin*, the data into categories that cover a range of possible values. For example, in a table of income levels, it might be useful to create bins of $0 to $20,000, $20,001 to $40,000, and so on. In this case, the frequency of each bin is simply the number of people with incomes in that bin.

EXAMPLE 3 Binned Exam Scores

Consider the following set of 20 scores from a 100-point exam:

76 80 78 76 94 75 98 77 84 88 81 72 91 72 74 86 79 88 72 75

Determine appropriate bins and make a frequency table. Include columns for relative and cumulative frequency, and interpret the cumulative frequency for this case.

Solution The scores range from 72 to 98. One way to group the data is with 5-point bins. The first bin represents scores from 95 to 99, the second bin represents scores from 90 to 94, and so on; note that there is no overlap between bins. We then count the frequency (the number of scores) in each bin. For example, only 1 score is in bin 95 to 99 (the high score of 98) and 2 scores are in bin 90 to 94 (the scores of 91 and 94). Table 5.3 on the next page shows the complete frequency table. In this case, we interpret the cumulative frequency of any bin to be the total number of scores in *or above* that bin. For example, the cumulative frequency of 6 for the bin 85 to 89 means that 6 scores are either between 85 to 89 or higher than 89.

TABLE 5.3	Frequency Table for Binned Exam Scores		
SCORES	FREQUENCY	RELATIVE FREQUENCY	CUMULATIVE FREQUENCY
95 to 99	1	0.05 = 5%	1
90 to 94	2	0.10 = 10%	3
85 to 89	3	0.15 = 15%	6
80 to 84	3	0.15 = 15%	9
75 to 79	7	0.35 = 35%	16
70 to 74	4	0.20 = 20%	20
Total	**20**	**1.00 = 100%**	**20**

BAR GRAPHS AND PIE CHARTS

Bar graphs and pie charts are commonly used to show data when the categories are qualitative. You are probably familiar with both, but let's review the basic ideas.

Consider the essay grade data in Table 5.1. A bar graph would show each category with a bar whose length corresponded to its frequency. If you make a bar graph by hand (as opposed to with a computer), you should measure the bar lengths carefully to make sure they correctly correspond to the frequencies. In Figure 5.3, for example, the vertical axis is marked with frequencies $\frac{1}{2}$ centimeter apart. Thus, the bar for A grades is 2 centimeters long, because the frequency of A grades is 4. Note that the left side of the bar graph in Figure 5.3 is marked with frequency, while the right side is marked with relative frequency. As you can see, bar graphs make it easy to display both frequencies simultaneously.

In contrast, pie charts are used primarily for relative frequencies, because the total pie must always represent the total relative frequency of 100%. The size of each wedge is proportional to the relative frequency of the category it represents. Figure 5.4 shows a pie chart for the essay grade data. To make comparisons easier, relative frequencies are often written on pie chart wedges.

FIGURE 5.3 Bar graph for the essay grade data in Table 5.1.

FIGURE 5.4 Pie chart for the essay grade data in Table 5.1.

Nowadays, most people make graphs with the aid of computers that measure bar lengths or wedge sizes automatically. However, you must still specify any labels or axis marks you want on a graph. This labeling is extremely important: Without proper labels, a graph is meaningless. The following summary lists the important labels for graphs; of course, not all labels are necessary in all cases. For example, pie charts do not require a vertical or horizontal scale. Notice how these rules were applied in Figure 5.3.

> **Important Labels for Graphs**
>
> **Title/caption:** The graph should have a title or caption (or both) that explains what is being shown and, if applicable, lists the source of the data.
>
> **Vertical scale and title:** Numbers along the vertical axis should clearly indicate the scale. The numbers should line up with the tick marks—the marks along the axis that precisely locate the numerical values. Include a label that describes the variable shown on the vertical axis.
>
> **Horizontal scale and title:** The categories should be clearly indicated along the horizontal axis. (Tick marks may not be necessary for qualitative data, but should be included for quantitative data.) Include a label that describes the variable shown on the horizontal axis.
>
> **Legend:** If multiple data sets are displayed on a single graph, include a legend or key to identify the individual data sets.

EXAMPLE 4 *Carbon Dioxide Emissions*

Carbon dioxide is released into the atmosphere by the combustion of fossil fuels (oil, coal, natural gas). Because carbon dioxide in the atmosphere can potentially lead to global warming, the U.S. Department of Energy tracks carbon dioxide emissions of countries around the world. Table 5.4 lists the eight countries that emit the most carbon dioxide each year. Make a bar graph for these data. Put the bars in descending order of size.

TABLE 5.4 The World's Eight Biggest Emitters of Carbon Dioxide, 1998

COUNTRY	TOTAL CARBON DIOXIDE EMISSIONS	COUNTRY	TOTAL CARBON DIOXIDE EMISSIONS
United States	1495	United Kingdom	147
China	740	Germany	227
Russia	405	India	253
Japan	288	Canada	138

NOTE: Emissions are given in millions of metric tons of carbon.

Solution The categories are the countries. Because country names are qualitative data, a bar graph is appropriate. The emissions are measurements that we use to determine the lengths of the bars. The values for total carbon dioxide emissions go

historical note

A bar graph with the bars in descending order is often called a *Pareto chart*, after Italian economist Vilfredo Pareto (1848–1923).

Total CO₂ Emissions

FIGURE 5.5 Bar graph for the data in Table 5.4.

Registered Voters in Rochester County

FIGURE 5.6 Party affiliations of registered voters in Rochester County.

from 138 to 1495, so a range of 0 to 1500 makes a good choice for the vertical scale. Each bar's height corresponds to its data value, and we label the category (country) under the bar. Figure 5.5 shows the bar graph, with bars in order of decreasing height.

Time out to think

Note that China is the second largest total emitter of carbon dioxide. But is it also second in per capita (per person) emissions? Why or why not? In general, how do you think a graph of per capita carbon dioxide emissions would look different from Figure 5.5?

EXAMPLE 5 *Simple Pie Chart*

Among the registered voters in Rochester County, 25% are Democrats, 25% are Republicans, and 50% are Independents. Make a pie chart showing the breakdown of party affiliations in Rochester County.

Solution The wedge sizes should correspond to the relative frequencies. Thus, the wedges for Republicans and Democrats each occupy one-fourth of the pie, while the wedge for Independents occupies the remaining half of the pie (Figure 5.6). Note the importance of clear labeling.

EXAMPLE 6 *Student Majors*

Figure 5.7 is a pie chart showing planned major areas for first-year college students. Make a bar graph showing the same data, with the bars in order of decreasing size. What are the three most popular major areas? Comment on the relative ease with which this question can be answered with the pie chart and the bar graph.

Solution Figure 5.8 shows the bar graph for the data. Note that, because we have only relative frequency data from the pie chart, we can show only relative frequencies on the bar graph. This bar graph makes it immediately obvious that the three most popular major areas are business (16.5%), professional (15.2%), and education (11.1%). ("Professional" includes fields with professional licensing, such as architecture, nursing, and pharmacy.) In contrast, it takes a fair amount of study of the pie chart before we can easily list the three most popular major areas.

Time out to think

Example 6 discussed an advantage of a bar graph over a pie chart for showing the data concerning major areas. Do you think the pie chart has any advantages over the bar graph? If so, what?

5C STATISTICAL TABLES AND GRAPHS

What Students Expect to Major In

- Other Fields 10.8%
- Undecided 7.5%
- Arts and Humanities 9.8%
- Technical 4.4%
- Biological Sciences 5.6%
- Social Sciences 8.4%
- Business 16.5%
- Professional 15.2%
- Physical Sciences 2.1%
- Engineering 8.2%
- Education 11.1%

FIGURE 5.7 Planned major areas for first-year college students, 1999. SOURCE: *Chronicle of Higher Education*, January 29, 1999, p. A48.

What Students Expect to Major In

(Bar graph: Business, Professional, Education, Other, Arts and Humanities, Social Sciences, Engineering, Undecided, Biology, Technical, Physical Sciences)

FIGURE 5.8 Bar graph for the data in Figure 5.7.

HISTOGRAMS AND LINE CHARTS

For quantitative data categories, the two most common types of graphics are *histograms* and *line charts*. Figure 5.9a shows a histogram for the binned exam data of Table 5.3; Figure 5.9b shows a line chart for the same data.

Note that a **histogram** is essentially a bar graph in which the data categories are quantitative. Thus, the bars on a histogram must follow the natural order of the numerical categories. In addition, the widths of histogram bars have a specific meaning; for example, the width of each bar in Figure 5.9a represents 5 points on the exam. Because there are no gaps between the categories, the bars on a histogram touch each other.

TECHNICAL NOTE
Different books define the terms *histogram* and *bar graph* differently. In this book, a bar graph is any graph that uses bars, and histograms are bar graphs used for quantitative data categories.

Exam Scores (a) Histogram

Exam Scores (b) Line chart

FIGURE 5.9 (a) Histogram for data in Table 5.3. (b) Line chart for the same data.

U.S. Homicide Rate

FIGURE 5.10 U.S. homicide rate per 100,000 people. SOURCE: U.S. Bureau of Justice Statistics.

A **line chart** serves the same basic purpose as a histogram, but instead of using bars, a line chart connects a series of dots. When data are binned, the dot is placed at the center of each bin. Histograms and line charts are often used to show how some variable changes with time. For example, the line chart in Figure 5.10 shows how the U.S. homicide rate has changed with time. The categories are time intervals; in this case, each bin represents a year in the data. Histograms and line charts with time on the horizontal axis are often called **time-series diagrams.**

> **DEFINITIONS**
>
> A **histogram** is a bar graph for quantitative data categories; the bars have a natural order and the bar widths have specific meaning.
>
> A **line chart** shows the data value for each category as a dot, and the dots are connected with lines. For each dot, the horizontal position is the *center* of the bin it represents and the vertical position is the data value for the bin.
>
> A **time-series diagram** is a histogram or line chart in which the horizontal axis represents time.

TABLE 5.5

AGE	NUMBER OF ACTRESSES
20–29	7
30–39	15
40–49	6
50–59	1
60–69	3
70–79	1
80–89	1

EXAMPLE 7 Oscar-Winning Actresses

Table 5.5 shows the ages of 34 recent Academy Award–winning actresses at the time when they won their award. Make a histogram and a line chart to display these data. Discuss the results.

Solution The fact that the categories are 10-year bins makes them quantitative data; thus, a histogram is appropriate. Figure 5.11a shows the histogram. Note that the bars touch one another because there are no gaps between the categories. Figure 5.11b shows the same data as a line chart; the histogram is overlaid with a line to show how it relates to the line chart. In looking at these data, we see that

Ages of 34 Academy Award–Winning Actresses

(a)

Ages of 34 Academy Award–Winning Actresses

(b)

FIGURE 5.11 (a) Histogram for ages of 34 recent Academy Award–winning actresses. (b) Line chart for the same data, with histogram overlaid for comparison.

actresses are most likely to win Oscars when they are fairly young. In contrast, male actors tend to win Oscars at older ages (see Exercise 22).

EXAMPLE 8 *Reading a Time-Series Diagram*

Figure 5.12 shows a time-series line chart of stock, bond, and gold prices over a 12-week period. Suppose that, on July 7, you invested $100 in a stock fund that tracks the S&P 500, $100 in a bond fund that follows the Lehman Index, and $100 in gold. If you sold all three funds on August 4, how much did you gain or lose?

Solution The graph shows that the $100 in the stock fund would have been worth about $101 on August 4. The $100 bond investment would have declined in value to about $96. The gold investment would have held its initial value of $100. Thus, on August 4, your complete portfolio would have been worth $101 + $96 + 100 = $297. You would have lost $3 on your total investment of $300.

MARKET GAUGE: COMPARING INVESTMENTS

How $100 invested 12 weeks ago in stocks (measured by the S.&P. 500), bonds (Lehman Treasury Bond Index) and gold would have fared through yesterday.

FIGURE 5.12

historical note

Gold was once considered to be a solid investment and an important part of any investment portfolio. However, gold prices have languished in recent decades. By 2000, gold was worth only about $300 per ounce—much less than half its peak price of $850 per ounce in 1980.

Review Questions

1. What is a frequency table? Explain what we mean by the categories and frequencies. What do we mean by relative frequency? What do we mean by cumulative frequency?
2. What is the distinction between qualitative data and quantitative data? Give a few examples of each.
3. What is the purpose of binning? Give an example in which binning is useful.
4. What two types of graphs are most common when the categories are qualitative data? Describe the construction of each.
5. Describe the importance of labeling on a graph, and briefly discuss the kinds of labels that should be included on graphs.
6. What two types of graphs are most common when the categories are quantitative data? Describe the construction of each.
7. What is a time-series diagram? Give an example.

In the News

1. **Frequency Tables.** Find a recent news article that includes some type of frequency table. Briefly describe the table and how it is useful to the news report. Do you think the table was constructed in the best possible way for the article? If so, why? If not, what would you have done differently?
2. **Bar Graph.** Find a recent news article that includes a bar graph with qualitative data categories. Briefly explain what the graph shows, and discuss whether it helps make the point of the news article.
3. **Pie Chart.** Find a recent news article that includes a pie chart. Briefly discuss the effectiveness of the pie chart. For example, would it be better if the data were displayed in a bar graph rather than a pie chart? Could the pie chart be improved in other ways?
4. **Histogram.** Find a recent news article that includes a histogram. Briefly explain what the histogram shows, and discuss whether it helps make the point of the news article. Are the labels clear? Is the histogram a time-series diagram? Explain.
5. **Line Chart.** Find a recent news article that includes a line chart. Briefly explain what the line chart shows, and discuss whether it helps make the point of the news article. Are the labels clear? Is the line chart a time-series diagram? Explain.

Exercises

Qualitative vs. Quantitative. In Exercises 1–8, determine whether the variable described is qualitative or quantitative, and explain why.

1. The blood types of individuals
2. The average waiting time in check-out lines at a grocery store
3. The responses of people to potato chips in a taste test where $0 =$ inedible up to $5 =$ outstanding
4. The high temperature in Anchorage, Alaska on each day of the year
5. The responses of people to the question "Will you vote to reelect your mayor?"
6. The federal income tax paid by individuals
7. The choice of TV shows to watch on a given night
8. The number of people watching various TV shows on a given night

Frequency Tables. Make a frequency table for the data in each of Exercises 9–10. Include columns for relative frequency and cumulative frequency. Briefly explain the meaning of each column.

9. Final grades of 24 students in a math class:

 A A A A B B B B B B B C C C
 C C C C C D D D F F

10. A guide book for New York City lists 5 five-star restaurants (the highest rating), 10 four-star restaurants, 20 three-star restaurants, 15 two-star restaurants, and 5 one-star restaurants.

Binned Frequency Tables. In Exercises 11–12, use the indicated bin size to make a frequency table for the following set of exam scores:

67 54 89 67 98 68 88 78 87 82 91
80 73 59 85 86 76 99 38 75 84 87
77 69 90

Include columns for relative frequency and cumulative frequency. Briefly explain the meaning of each column.

11. Use 5-point bins (95 to 99, 90 to 94, etc.).

12. Use 10-point bins (90 to 99, 80 to 89, etc.).

Statistical Graphs. Each of Exercises 13–21 gives a table of data. For each exercise, do the following:

 a. Explain whether the data categories are qualitative or quantitative.

 b. If the data categories are qualitative, draw either a bar graph or pie chart for the data. If the data categories are quantitative, draw either a histogram or a line chart for the data.

 c. Write a one-paragraph summary of any interesting information revealed by the graphic.

13. The following table gives colleges and universities that have won more than one NCAA men's basketball championship since 1960.

Team	Number of championships
UCLA	11
Kentucky	3
Indiana	3
Cincinnati	2
Duke	2
Louisville	2
North Carolina	2
North Carolina State	2
Michigan State	2

14. The following table lists the top eight retail companies in the United States in 1997.

Company	Sales (billions of dollars)
Wal-Mart	118.0
Sears, Roebuck	41.5
Kmart	32.1
JC Penney	30.5
Dayton Hudson	27.8
Kroger	26.6
Home Depot	24.1
Safeway	22.4

SOURCE: Wall Street Journal Almanac.

15. The following table shows high school graduation rates for various ethnic groups in the United States in 1997.

Ethnic group	Graduation rate
white	83.0%
black	74.9%
Asian/Pacific Islander	84.9%
Hispanic	54.7%

16. The following table gives world population by continent in 1998.

Continent	Population (billions)
Asia	3.6
Africa	0.8
North America	0.3
Latin America	0.5
Europe	0.7
Oceania	0.03

17. The following table lists areas of the world's major land masses.

Land mass	Area (millions of sq. miles)
Asia	17.2
Africa	11.6
North America	9.3
South America	6.9
Australia	3.0
Europe	3.8
Antarctica	5.1
All others	2.1

18. The following table gives the ages of first-year college students on December 31, 1998. (Hint: For the "16 or less" and "21 or more" categories, position the bars or dots at precisely 16 and 21, respectively, but label the bins clearly with their actual meanings.)

Age	Percent of sample
16 or less	0.1
17	1.9
18	66.9
19	26.3
20	2.0
21 or more	2.8

SOURCE: UCLA Higher Education Research Institute.

19. The following table gives the stated religions of first-year college students. (Note: The "other religions" category consists of religions that were stated by less than 1% of the students in the sample.)

Religion	Percent of sample
Baptist	16.8
Episcopal	1.8
Jewish	1.6
Lutheran	4.3
Mormon	1.1
Methodist	8.1
Presbyterian	3.9
Catholic	28.6
United Church of Christ	1.9
Other religions	17.4
No religion	14.5

SOURCE: UCLA Higher Education Research Institute.

20. The following table gives the number of "other colleges" (besides the one they attend) to which first-year college students applied. (Hint: For the "11 or more" category, position the bar or dot at precisely 11, but label the bin clearly.)

Other colleges	Percent of sample
0	34.0
1	14.2
2	15.2
3	14.5
4	8.9
5	5.4
6	3.3
7–10	3.8
11 or more	0.7

SOURCE: UCLA Higher Education Research Institute.

21. The following table gives average family size in the United States since 1940.

Year	Family size	Year	Family size
1940	3.76	1975	3.42
1950	3.54	1980	3.29
1960	3.67	1985	3.23
1965	3.70	1990	3.17
1970	3.58	1995	3.19

SOURCE: U.S. Bureau of Census.

22. **Oscar-Winning Actors.** The following data show the ages of 34 recent Academy Award–winning actors at the time they won their award. Make a frequency table and a histogram for these data, using bins of 20–29, 30–39, and so on. Do actors appear to be more likely to win Oscars when they are younger, older, or neither? Contrast your results with those found for actresses in Example 7.

32 37 36 32 51 53 33 61 35 45 55 39
76 37 42 40 32 60 38 56 48 48 40 43
62 43 42 44 41 56 39 46 31 47

Web Projects

Find useful links for Web Projects on the text Web site: www.aw.com/bennett-briggs

1. **CO₂ Emissions.** Look for updated data concerning international carbon dioxide emissions at the Web site for the *International Energy Annual,* published by the U.S. Energy Information Administration (EIA). Create an updated or expanded version of Figure 5.5. Discuss any new features of your updated graph.

2. **Energy Table.** Explore some of the many energy tables at the U.S. Energy Information Administration (EIA) Web site. Choose a table that you find interesting, and make a graph of its data. You may choose any of the graph types discussed in this section. Explain how you made your graph, and briefly discuss what can be learned from it.

3. **Statistical Abstract.** Go to the Web site for the *Statistical Abstract of the United States.* Explore the selection of "frequently requested tables." Choose one table of interest to you, and make a graph from its data. You may choose any of the graph types discussed in this section. Explain how you made your graph, and briefly discuss what can be learned from it.

UNIT 5D

GRAPHICS IN THE MEDIA

Now that you understand basic types of statistical graphs, you are ready to explore some of the fancier graphics that appear daily in the news. Equally important are several cautions to keep in mind when interpreting media graphics.

GRAPHICS BEYOND THE BASICS

Many graphical displays of data go beyond the basic types discussed in Unit 5C. Here, we explore a few of the types that are most common in the news media.

Multiple Bar Graphs

A **multiple bar graph** is a simple extension of a regular bar graph: It has two or more sets of bars that allow comparison between two or more data sets. All the data sets must involve the same categories so that they can be displayed on the same graph. For example, Figure 5.13 is a multiple bar graph showing trends in home computing. The categories are years. The two sets of bars represent two different measures of home computing: ownership of personal computers and connection to the Internet. Note that a legend clearly identifies the two sets of bars.

EXAMPLE 1 *Computing Trends*

Summarize two major trends shown in Figure 5.13.

Solution The most obvious trend is that both data sets show an increase with time. That is, the number of homes with computers and the number of online homes both increased with time. We see a second trend by comparing the bars within each year. In 1995, the number of online homes (about 10 million) was less than one-third the number of homes with computers (about 33 million). By 2000, the number of online homes (about 36 million) was nearly two thirds the number of homes with computers (about 55 million). This suggests that as more people get computers, most of them are going online. If we project the trends into the future, it seems likely that the number of online households will approach the number of households with computers, even while both numbers continue to grow larger.

FIGURE 5.13 Trends in home computing. SOURCE: *Wall Street Journal Almanac,* 1999.

Source: Jupiter Communications

Death Rates for Various Diseases, 1900–2000

FIGURE 5.14 Trends in death rates from four diseases.

Stack Plots

Another common type of graph, called a **stack plot,** shows different data sets in a vertical stack. Figure 5.14 uses a stack plot to show trends in death rates (deaths per 100,000 people) for four diseases over the period from 1900 to 2000. Each disease has its own color-coded region, or wedge; note the importance of the legend. The *thickness* of a wedge at a particular time tells you its value at that time: When a wedge is thick it has a large value, and when it is thin it has a small value.

EXAMPLE 2 *Stacked Line Chart*

Based on Figure 5.14, what was the death rate for pneumonia in 1980? Discuss the general trends visible on this graph.

Solution For 1980, the pneumonia wedge extends from about 620 to 640 on the vertical axis. Thus, the thickness of this wedge is 20, meaning that the death rate in 1980 for pneumonia was 20 deaths per 100,000 people. The graph shows several important trends. First, the downward slope of the top wedge shows that the overall death rate from these four diseases decreased substantially, from nearly 800 deaths per 100,000 in 1900 to about 600 in 2000. The drastic decline in the thickness of the tuberculosis wedge shows that this disease was once a major killer, but has been nearly wiped out since 1950. Meanwhile, the thickening of the cancer wedge shows that the death rate from cancer has gradually increased.

Graphs of Geographical Data

We are often interested in geographical patterns in data. Figure 5.15 shows one common way of displaying geographical data. In this case, the data on per capita (per person) income are shown state by state. The legend explains that different colors represent different income levels. Similar colors are used for similar income levels. Thus, it is easy to see that income levels tend to be highest in the northeast and lowest in the south.

by the way ...

Since the mid-1980's, there has been a small but noticeable resurgence of tuberculosis in the United States. Part of the resurgence is due to new strains of the disease that are resistant to most common drug treatments.

State Per Capita Income

Key:
- $15,000–$19,999
- $20,000–$24,999
- $25,000–$29,999
- $30,000–$34,999
- $35,000–$39,999

FIGURE 5.15 Per capita income in the 50 states (1998). SOURCE: *Statistical Abstract of the United States.*

The display in Figure 5.15 works well because each state is associated with a unique income level. For data that vary continuously across geographical areas, a **contour map** is more convenient. For example, Figure 5.16 shows a contour map of temperature over the United States at a particular time. Each of the *contours* connects locations with the same temperature. For example, the temperature is 50°F everywhere along the contour labeled 50° and 60°F everywhere along the contour labeled 60°. Between these two contours, the temperature is between 50°F and 60°F. Note that in regions where contours are tightly spaced, there is a greater difference in temperature. For example, the closely packed curves in the northeast indicate that the temperature varies substantially over small distances. To make the graph easier to read, the regions between adjacent contours are color-coded.

Time out to think

Look for a weather map in today's news. How are the temperature contours shown? Interpret the temperature data.

EXAMPLE 3 *Melanoma Mortality*

Figure 5.17 shows how mortality from *melanoma* (a form of skin cancer) varies on a county-by-county basis across the United States. The legend explains that the darker the shading is in a county, the higher the mortality rate. Summarize the major geographical pattern revealed in the figure, and suggest a possible reason for this pattern. Also explain how this graph helps you identify counties that don't follow the general pattern; why might these regions warrant special study?

Solution The map shows a general pattern in which southern and western states have higher melanoma mortality than northern or eastern states. This pattern might result from the fact that people in southern and western states spend more time outdoors.

The greatest value of a picture is when it forces us to notice what we never expected to see.

—JOHN TUKEY

FIGURE 5.16 A contour map of temperature.

FIGURE 5.17 Female melanoma mortality rates by county. SOURCE: Professor Karen Kafadar, Mathematics Department, University of Colorado at Denver.

Counties that don't follow the general pattern stand out—either darker or lighter in color than surrounding counties. These counties might warrant special study to learn why they differ from the general pattern.

Three-Dimensional Graphics

Today, computer software makes it easy to give almost any graph a three-dimensional appearance. For example, Figure 5.18 shows the bar graph of Figure 5.3, but "dressed up" with a three-dimensional look. It may look nice, but the three-dimensional effects are purely cosmetic; that is, they don't provide any information that wasn't already in the two-dimensional graph in Figure 5.3. As this example shows, many "three-dimensional" graphics really depict only two-dimensional data, but make it look a little fancier. In contrast, Figure 5.19, in

FIGURE 5.18 This graph has a three-dimensional appearance, but shows only two-dimensional data.

SONIC MAPPING TRACES BIRD MIGRATION

Sensors across New York State counted each occurrence of the nocturnal flight call of the bobolink to trace the fall migration on the night of Aug. 28–29, 1993. Computerized, the data showed the heaviest swath passing over the eastern part of the state.

Source: Bill Evans/Cornell Laboratory of Ornithology

FIGURE 5.19 This graph shows true three-dimensional data. SOURCE: *New York Times*, October 3, 1995.

which each of the three axes carries specific information, shows true three-dimensional data. Researchers studying migration patterns of a bird species (the *Bobolink*) counted the number of birds flying over seven New York cities throughout the night. As shown on the inset map, the cities were aligned east-west so that the researchers would learn what parts of the state the birds flew over, and at what times of night, as they headed south for the winter. Thus, the three axes measure *number of birds*, *time of night*, and *east-west location*.

EXAMPLE 4 *Three-Dimensional Bird Migration*

Based on Figure 5.19, at about what time was the largest number of birds flying over the east-west line marked by the seven cities? Over what part of New York did most of the birds fly? More specifically, approximately how many birds passed over Oneonta at about 12:30 a.m.?

Solution The number of birds detected in all the cities peaked between 4 and 6 hours after 8:30 p.m., or between about 12:30 and 2:30 a.m. More birds flew over the two easternmost cities of Oneonta and Jefferson than over cities farther west. Thus, most of the birds were flying over the eastern part of the state. To answer the specific question about Oneonta, note that 12:30 a.m. is 4 hours after 8:30 p.m. On the graph, this time appears to align with the lower peak on the line at Oneonta. Looking across to the *number of birds* axis, we see that about 30 to 40 birds were flying over Oneonta at that time.

Combination Graphics

All of the graphic types we have studied so far are common and fairly easy to create. But the media today are often filled with many varieties of even more complex graphics. For example, Figure 5.20 shows a graphic concerning the participation of women in the Olympics. This single graphic combines a line chart, many pie charts, and numerical data. It is certainly a case of a picture being worth far more than a thousand words.

EXAMPLE 5 *Olympic Women*

Describe three trends shown in Figure 5.20.

Solution The line chart shows that the total number of women competing in the Olympics has risen fairly steadily, especially since the 1960s, reaching nearly 4500 in the 2000 games. The pie charts show that the percentage of women among all competitors has also increased, reaching 42% in the 2000 games. The bold numbers at the bottom show that the number of events for women has also increased dramatically, reaching 121 in the 2000 games.

Time out to think

Do you think the upward trend of the pie charts in Figure 5.20 will continue over the next few Olympic games? Why or why not?

The Ever-Growing Presence of Women in Summer Olympics

Figure shows percentages of women participants and total number of women participating from 1900 to 2000:

Year	% Women	Events for women
1900	1.6%	3
'04	0.9	2
'08	1.8	3
'12	2.2	6
'16	no Games	—
'20	2.9	6
'24	4.4	11
'28	9.6	14
'32	9.0	14
'36	8.1	15
'40–'44	no Games	—
'48	9.4	19
'52	10.5	25
'56	16.1	26
'60	11.4	29
'64	13.3	33
'68	14.2	39
'72	14.8	43
'76	20.7	49
'80	21.5	50
'84	23.0	62
'88	25.8	86
'92	28.8	98
'96	34.2	108
'00	42.0	121

Number of events for women shown on x-axis. Women participating scale reaches 4,500.

Source: International Olympic Committee

FIGURE 5.20 SOURCE: *New York Times*, August 20, 2000.

A FEW CAUTIONS ABOUT GRAPHICS

As we have seen, graphics can offer clear and meaningful summaries of statistical data. However, even well-made graphics can be misleading if we are not careful in interpreting them, and poorly made graphics are almost always misleading. Moreover, some people use graphics in deliberately misleading ways. Here, we discuss a few of the more common ways in which graphics can lead us astray.

Perceptual Distortions

Many graphics are drawn in a way that distorts our perception of them. Figure 5.21 shows one of the most common types of distortion. The dollar-shaped bars are used to represent the declining value of the dollar over time. Although it is the *length* of each bar that represents the data, our eyes tend to focus on the *areas* of the bars. For example, the bottom bar is supposed to show that a dollar in 2000 was worth only 48% as much as a dollar in 1980. Its length is indeed 48% that of the top bar, but its area is much smaller in comparison (about 23% of the area of the top bar). This gives the perception that the value of the dollar shrank even more than it really did. (German researchers in the latter part of the 19th century studied many types of graphics. The type of distortion shown in Figure 5.21 was so common that they gave it its own name, which translates roughly as "the old goosing up the effect by squaring the eyeball trick.")

1980 = $1.00

1990 = $0.63

2000 = $0.48

FIGURE 5.21 The lengths of the dollars are proportional to their spending power, but our eyes are drawn to the areas, which decline more than the lengths.

Women as a Percentage of All College Students

(a)

(b)

FIGURE 5.22 SOURCE: National Center for Education Statistics.

Computer Speed

(a)

(b)

FIGURE 5.23

Watch the Scales

Figure 5.22a shows the percentage of college students between 1910 and 1990 who were women. At first glance, it appears that there was a huge increase in this percentage between 1950 and 1990. But note that the vertical axis scale does not begin at zero and does not end at 100%. The increase looks far less dramatic if we redraw the graph with the vertical axis covering the full range of 0 to 100% (Figure 5.22b). From a mathematical point of view, leaving out the zero point on a scale is perfectly honest and can make it easier to see small-scale trends in the data. Nevertheless, as this example shows, it can be visually deceptive if you don't study the scale carefully.

Sometimes the scale may not be deceptive, but still requires care to avoid misinterpretation. Consider Figure 5.23a, which shows the change in computer speed from 1950 to 2000. At first glance, it appears that speeds have been increasing linearly; for example, it might look as if the speed increased by the same amount from 1990 to 2000 as it did from 1950 to 1960. However, if we look closely, we see that each tick mark on the vertical scale represents a *tenfold* increase in speed. Now we see that computer speed grew from about 1 to 100 calculations per second between 1950 and 1960, and from about 100 million to 10 billion calculations per second between 1990 and 2000. This type of scale is called an *exponential scale* (or *logarithmic scale*), because each unit corresponds to a power of 10. In general, exponential scales are useful for displaying data that vary over a huge range of values. You can see this usefulness by looking at Figure 5.23b, where the computer data have been recast with an ordinary scale. Because the speeds have grown so rapidly, the

ordinary scale makes it impossible to see any detail in the early years shown on the graph.

Time out to think

Based on Figure 5.23a, can you predict the speed of the fastest computers in 2010? Could you make the same prediction with Figure 5.23b? Explain.

Percentage Change Graphs

Is college getting more or less expensive? A quick look at the downward-sloping lines in the lower graph in Figure 5.24 might give the impression that college costs fell steadily in the 1990s. But on closer examination, we find that this is not the case at all. The vertical axis on the lower graph represents the *percentage increase* in costs. Thus, the downward-sloping lines show only that the rate of increase in college costs decreased, *not* that the actual costs decreased. In fact, the actual costs have risen substantially, as shown in the upper graph. Moreover, because the rate of inflation has been less than the rate of increase in college costs, the *real cost* of public colleges has steadily risen. Graphs that show percentage change are very common, particularly with economic data. Although they are perfectly honest, you can be easily misled unless you interpret them with great care.

> **historical note**
>
> In 1965, Intel founder Gordon E. Moore predicted that advances in technology would allow computer chips to double in power roughly every two years. This idea is now called *Moore's law*, and it has held fairly true ever since Moore first stated it.

FIGURE 5.24 SOURCE: *New York Times*, October 6, 1999.

FIGURE 5.25 SOURCE: James Abundis and Bob Laird, *USA Today,* October 18, 1999.

Pictographs

Pictographs are graphs embellished with additional artwork. The artwork may make the graph more appealing, but it can also distract or mislead. Figure 5.25 is a pictograph showing the rise in world population from 1804 to 2054 (numbers for future years are based on United Nations projections). The lengths of the bars correspond correctly to world population for the different years listed. However, the artistic embellishments of this graph can be deceptive in several ways. For example, your eye may be drawn to the figures of people lining the globe. Because this line of people rises from the left side of the pictograph to the center and then falls, it might give the impression that future world population will be declining. In fact, the line of people is purely decorative and carries no information.

Perhaps the most serious problem with this pictograph is that it makes it appear that world population has been rising linearly. However, notice that the time intervals on the horizontal axis are not uniform in size. For example, the interval between the bars for 1 billion and 2 billion people is 123 years (from 1804 to 1927), but the interval between the bars for 5 billion and 6 billion people is only 12 years (from 1987 to 1999). Pictographs are very common, but as this example shows, you have to study them carefully to extract the essential information and not be distracted by the cosmetic effects.

by the way...

If world population continued to double at the same rate as in the late 20th century, it would reach 34 billion by 2100 and 192 billion by 2200. By about 2650, human population would be so large that it would not fit on the Earth, even if everyone stood elbow-to-elbow everywhere.

Review Questions

1. Briefly describe the construction and use of multiple bar charts and stack plots.
2. What are geographical data? Briefly describe at least two ways to display geographical data. Be sure to explain the meaning of contours on a contour map.
3. What are three-dimensional graphics? Explain the difference between graphics that only appear three-dimensional and those that show truly three-dimensional data.
4. Describe how perceptual distortions can arise in graphics and how they can be misleading.

5. How can graphics be misleading when the scales do not go all the way to zero? Why are such graphics sometimes useful?
6. What is an exponential scale? When is it useful to use an exponential scale on a graphic?
7. Explain how a graph that shows percentage change can show descending bars (or a descending line) even when the variable of interest is increasing.
8. What is a pictograph? How can a pictograph enhance a graph? How can it make a graph misleading?

In the News

1. **News Graphics.** Find a recent news report that shows a multiple bar chart or stack plot. Comment on the effectiveness of the display. Could another display have been used to depict the same data?
2. **Geographical Data.** Find an example of a graph of geographical data in a recent news report. Comment on the effectiveness of the display. Could another display have been used to depict the same data?
3. **Three-Dimensional Effects.** Find an example of a three-dimensional display in a recent news report. Are the data three-dimensional or are the three-dimensional effects cosmetic? Comment on the effectiveness of the display. Could another display have been used to depict the same data?
4. **Graphic Confusion.** Find an example in a recent news report of a graph that is misleading in one of the ways discussed in this unit. Explain what makes the graph misleading, and describe how it could have been drawn more honestly.
5. **Outstanding News Graph.** Find a graph from a recent news report that, in your opinion, is truly outstanding in displaying data visually. Discuss what the graph shows, and explain why you think it is so outstanding.

Exercises

1. **Net Grain Production.** Net grain production is the difference between the amount of grain a country produces and the amount of grain its citizens consume. It is positive if the country produces more than it consumes, and negative if the country consumes more than it produces. Figure 5.26 shows the net grain production of four countries in 1990 and projected for 2030.

FIGURE 5.26

a. Which of the four countries had to import grain to meet its needs in 1990?
b. Which of the four countries are expected to need to import grain to meet needs in 2030?
c. Given that India and China are the world's two most populous countries, what does this graph tell you about how world agriculture will have to change between now and 2030?

2. **Education and Earnings.** Figure 5.27 shows mean (average) earnings in three different years according to level of education.

Source: Census Bureau, Current Population Survey

FIGURE 5.27 SOURCE: *TIME* Almanac, 1999, p. 886.

a. Briefly explain the meaning of each of the three sets of bars on the graph.

b. Compare the change in earnings between 1975 and 1995 for people with bachelor's degrees to the change for people who did not graduate from high school. What do these data say about the value of a college education?

c. The graph has a three-dimensional appearance. Is it showing true three-dimensional data, or is the appearance purely cosmetic? Do you think the three-dimensional appearance helps or hinders the display?

3. **Stack Plot.** Answer the following based on Figure 5.14.

 a. State whether the death rate for each the four diseases individually decreased or increased between 1900 and 2000.

 b. When was the death rate due to cardiovascular diseases the greatest, and what was it?

 c. What was the death rate due to cancer in 2000?

 d. Based on the trends in the graph, which of these four diseases will be responsible for the most deaths in 2050? Explain.

4. **College Degrees.** Figure 5.28 shows the numbers of college degrees awarded to men and women over time.

College Degrees Awarded

FIGURE 5.28

a. Estimate the numbers of college degrees awarded to men and to women (separately) in 1930 and in 2000.

b. Compare the numbers of degrees awarded to men and to women (separately) in 1980 and 2000.

c. During what decade did the *total* number of degrees awarded increase the most?

d. Compare the *total* numbers of degrees awarded in 1950 to 2000.

e. Do you think the stack plot is an effective way to display these data? Briefly discuss other ways that might have been used instead.

5. **Federal Spending.** Figure 5.29 shows the changes in major spending categories of the federal budget. (Payments to individuals include Social Security and Medicare; *net interest* represents interest payments on the national debt; *all other* represents non-defense discretionary spending.) Interpret the stack plot and discuss some of the trends it reveals.

Percentage of Composition of Federal Government Outlays

FIGURE 5.29 SOURCE: Budget for Fiscal Year 2001.

a. Find the percentage of the budget that went to net interest in 1980, 1990, 1995, and 2000.

b. Find the percentage of the budget that went to defense in 1960, 1980, and 2000.

c. Find the percentage of the budget that went to payments to individuals in 1960, 1980, and 2000.

6. **Federal Trends.** Consider Figure 5.29. Summarize at least three trends shown in the figure.

Probability That a Black Student Would Have White Classmates

■ Less than 10%
■ 20% – 40%
□ 40% – 60%
■ 60% – 80%
■ More than 80%
□ Counties with no data or no black students

FIGURE 5.30 SOURCE: *New York Times*, April 2, 2000.

7. **School Segregation.** One way of measuring segregation is to determine the likelihood that a black student will have white classmates. A *New York Times* study found that, by this measure, segregation increased significantly in the 1990s. Figure 5.30 shows the probability that a black student had white classmates, by county, during the 1997–1998 academic year. Do there appear to be any significant regional differences? Can you pick out any differences between urban and rural areas? Discuss possible explanations for a few of the trends that you see in the figure.

8. **Contour Elevations.** Contour maps are often used to show geographical elevations. Figure 5.31 shows elevation contours around Boulder, Colorado. Discuss a few key features shown on the map.

FIGURE 5.31

U.S. Age Distribution. Exercises 9–10 refer to Figure 5.32.

U.S. Age Groups

■ <5
■ 5–19
□ 20–44
□ 45–65
■ >65

FIGURE 5.32

9. a. Briefly describe the meaning of each bar.
 b. Based on this graph, how has the percentage of the youngest Americans changed since 1860?
 c. Does this graph show true three-dimensional data, or is the three-dimensional look purely cosmetic? Explain.

10. a. Estimate the percentages of 5- to 19-year-olds in 1860 and 1990.
 b. Estimate the percentages of 20- to 44-year-olds in 1860 and 1990.
 c. Describe and explain the change in the percentage of under-5-year-olds from 1950 to 1960.

Creating Graphics. Exercises 11–14 give tables of real data. For each table, make a graphical display of the data. You may choose any graphic type that you feel is appropriate to the data set. In addition to making the display, write a few sentences explaining why you chose this type of display and a few sentences describing interesting patterns in the data.

11. **Percent Never Married.** The following table shows the percentages, for 1970 and 1996, of men and women in various age categories who were never married.

Women	1970	1996	Men	1970	1996
20–24	35.8	68.5	20–24	54.7	81.0
25–29	10.5	37.6	25–29	19.1	52.0
30–34	6.2	20.5	30–34	9.4	29.6
35–39	5.4	13.1	35–39	7.2	20.8
40–44	4.9	9.7	40–44	6.3	14.2

Source: U.S. Census Bureau.

12. **Alcohol on the Road.** The following table gives the number of automobile fatalities in which the (i) no alcohol was involved, (ii) moderate alcohol was involved (blood-alcohol content between 0.01 and 0.09), and (iii) a high level of alcohol was involved (blood-alcohol content above 0.1). All figures are in thousands of deaths.

Year	No alcohol	Moderate alcohol	High alcohol
1982	18.7	4.8	20.4
1984	20.5	4.8	19.0
1986	22.0	5.1	18.9
1988	23.5	4.9	18.7
1990	22.5	4.4	17.7
1992	21.4	3.6	14.2
1994	24.1	3.5	13.1
1996	24.8	3.8	13.4

Source: National Highway Traffic Safety Administration.

13. **Daily Newspapers.** The following table gives the number of daily newspapers and their total circulation (in millions) for selected years since 1920.

Year	Number of daily newspapers	Circulation (millions)
1920	2042	27.8
1930	1942	39.6
1940	1878	41.1
1950	1772	53.9
1960	1763	58.8
1970	1748	62.1
1980	1747	62.2
1990	1611	62.3
2000 (est)	1485	56.09

Source: Editor & Publisher.

14. **Firearm Fatalities.** The following table summarizes death due to firearms in different nations.

Country	Total	Homicides	Suicides	Fatal accidents
U. S. (1995)	35,563	15,835	18,503	1225
Germany (1995)	1197	168	1004	25
Canada (1994)	1189	176	975	38
Australia (1994)	536	96	420	20
Spain (1994)	396	76	219	101
U. K. (1994)	277	72	193	12
Sweden (1992)	200	27	169	4
Vietnam (1995)	131	85	16	30
Japan (1995)	93	34	49	10

Source: Coalition to Stop Gun Violence.

15. **Volume Distortion.** Figure 5.33 uses television sets to represent the numbers of homes with cable in 1980 and

Homes with Cable TV

18 million homes 72 million homes

FIGURE 5.33

1997. Note that the heights of the TVs represent the numbers of homes. Briefly explain how the graph creates a perceptual distortion that exaggerates the true change in the number of homes with cable.

16. **Three-Dimensional Pies.** The pie charts in Figure 5.34 represent the percentage of Americans in three age categories in 1990 and 2050 (projected). The 1990 data are 87.5%, 11.4%, and 1.2%, and the 2050 data are 80.0%, 15.4%, and 4.6%. Briefly explain how the three-dimensional effects create a perceptual distortion in this case. Why would flat pies (without the three-dimensional) effects give a more accurate representation of the data?

FIGURE 5.34 SOURCE: U.S. Census Bureau.

17. **Braking Distances.** Figure 5.35 shows the braking distance for four different cars. Discuss the ways in which it might be deceptive. How much greater is the braking distance of Lincolns than the braking distance of Oldsmobiles? Draw the display in a fairer way.

FIGURE 5.35 SOURCE: *Car and Driver.*

18. **Comparing Earnings.** Figure 5.36 compares the average weekly earnings of men and women. Identify any misleading aspects of the display. Draw the display in a fairer way.

FIGURE 5.36 Source: U.S. Census Bureau.

19. **Computer Sales.** The following table shows the number of worldwide shipments of personal computers since 1980. Draw the graph both with an ordinary vertical scale and with an exponential vertical scale. (Hint: For the exponential scale, use ticks at 1 million, 10 million, and 100 million.) Which graph is more useful? Why?

Year	Worldwide shipments	Year	Worldwide shipments
1981	1,630,000	1991	26,970,000
1983	11,120,000	1993	38,850,000
1985	14,700,000	1995	60,170,000
1987	16,680,000	1997	82,400,000
1989	21,330,000	1999	112,000,000

20. **World Population.** Recast Figure 5.25 with a proper horizontal axis. What trends are clear in your new graph that are not clear in the original? Explain.

Web Projects

Find useful links for Web Projects on the text Web site: www.aw.com/bennett-briggs

1. **Weather Maps.** Many Web sites offer contour maps with current weather data. For example, you can use the Yahoo Weather site to generate many different contour weather maps. Generate at least two contour weather maps and discuss what they show.

2. **Cancer Cure.** As shown in Figure 5.14, cancer is one of the leading causes of death today. Nevertheless, scientists have made great progress in treating many forms of cancer. Go to the American Cancer Society Web site and investigate research into cancer cures. Read about one or two recent studies, and write a short report on what you learn. Be sure to include graphics in your report.

3. **USA Snapshot.** The *USA Today* Web site offers a daily pictograph for its "USA Snapshot." Study today's snapshot. Briefly discuss its purpose and effectiveness.

UNIT 5E

CORRELATION AND CAUSALITY

A major goal of many statistical studies is to determine whether one factor *causes* another. For example, does smoking cause lung cancer? In this unit, we will discuss how statistics can be used to search for *correlations* that might suggest a cause-and-effect relationship. Then we'll explore the more difficult task of establishing causality.

Smoking is one of the leading causes of statistics.
— FLETCHER KNEBEL

SEEKING CORRELATION

What does it mean when we say that smoking *causes* lung cancer? It certainly does *not* mean that you'll get lung cancer if you smoke a single cigarette. It does not even mean that you'll definitely get lung cancer if you smoke regularly and heavily for many years, since some heavy smokers do not get lung cancer. Rather, it is a *statistical* statement meaning that you are *much more likely* to get lung cancer if you smoke than if you don't smoke.

Let's try to understand how researchers learned that smoking causes lung cancer. Before they could investigate *cause*, researchers first needed to establish correlations between smoking and cancer. The process of establishing correlations began with observations. The early observations were informal, as doctors noticed that, of their patients with lung cancer, a surprisingly high proportion were smokers. This suggestion of a linkage led to carefully conducted studies in which researchers compared lung cancer rates among smokers and nonsmokers. These studies showed clearly that heavier smokers were more likely to get lung cancer. In more formal terms, we say that there is a correlation between the variables *amount of smoking* and *incidence of lung cancer*. A correlation is a special type of relationship between variables, in which a rise or fall in one goes along with a corresponding rise or fall in the other.

> **DEFINITION**
>
> A **correlation** exists between two variables when higher values of one variable consistently go with higher values of another or when higher values of one variable consistently go with lower values of another.

Here are a few other examples of correlations:

- There is a correlation between the variables *height* and *weight* for people; that is, taller people tend to weigh more than shorter people.

- There is a correlation between the variables *demand for apples* and *price of apples;* that is, demand tends to decrease as prices increase.
- There is a correlation between *practice time* and *skill* among piano players; that is, those who practice more tend to be more skilled.

It's important to realize that establishing a correlation between two variables does *not* mean that a change in one variable *causes* a change in the other. Thus, finding the correlation between smoking and lung cancer did not by itself prove that smoking causes lung cancer. We could imagine, for example, that some gene predisposes a person both to smoking and to lung cancer. Nevertheless, identifying the correlation was the crucial first step in learning that smoking causes lung cancer.

> **by the way...**
> Smoking is linked to many serious diseases besides lung cancer, including heart disease and emphysema. Smoking is also linked with less lethal health conditions such as premature skin wrinkling and sexual impotence.

Time out to think

Suppose there really were a gene that made people prone to both smoking and lung cancer. Explain why we would still find a strong correlation between smoking and lung cancer in that case, but would not be able to say that smoking caused lung cancer.

Scatter Diagrams

Table 5.6 shows the production cost and gross receipts (total revenue from ticket sales) for 12 recent movies. Movie executives presumably hope there is a favorable correlation between the production budget and the receipts; that is, they hope that spending more to produce a movie will result in higher box office receipts. But is there such a correlation? We can look for a correlation by making a **scatter diagram** showing the relationship between the variables *production cost* and *gross receipts*.

> **DEFINITION**
>
> A **scatter diagram** is a graph in which each point represents the values of two variables.

The following procedure describes how we make the scatter diagram, which is shown in Figure 5.37:

1. We assign one variable to each axis, and we label each axis with values that comfortably fit the data. Here, we assign *production cost* to the horizontal axis and *gross receipts* to the vertical axis. We choose a range of 0 to $150 million for the production cost axis and 0 to $250 million for the gross receipts axis.
2. For each movie in Table 5.6, we plot a *single point* at the horizontal position corresponding to its production cost and the vertical position corresponding to its gross receipts. For example, the point for the movie *Hard Rain* goes at a position of $70 million on the horizontal axis and $20 million on the vertical axis. The dashed lines on Figure 5.37 show how we locate this point.
3. (Optional) If we wish, we can label each data point, as is done in Figure 5.37.

> **by the way...**
> We often have some reason to think that one variable depends at least in part on the other. In the case of Figure 5.37, we might guess that gross receipts should depend on the production cost. We therefore call production cost the *explanatory variable* and gross receipts the *response variable,* because the production cost might help explain the gross receipts. The explanatory variable is usually plotted on the horizontal axis and the response variable on the vertical axis.

TABLE 5.6 — Movie Data (receipts as of 1999)

MOVIE	PRODUCTION COST	GROSS RECEIPTS
Armageddon	140	194
Contact	90	101
Dante's Peak	104	67
Deep Impact	75	140
Good Will Hunting	10	138
Men in Black	90	90
Hard Rain	70	20
I Know What You Did Last Summer	17	72
Saving Private Ryan	70	178
Speed 2	110	48
The Fifth Element	90	64
Volcano	90	47

NOTE: Amounts are in millions of dollars.

FIGURE 5.37 Scatter diagram for data in Table 5.6.

EXAMPLE 1 *Inflation and Unemployment*

Table 5.7 gives data for the rates of inflation (based on the Consumer Price Index) and unemployment during the 1990s. Make a scatter diagram for these data.

Solution We plot the unemployment rate on the horizontal axis and the inflation rate on the vertical axis. To make the graph easy to read, we use values ranging from 4% to 8% for the unemployment rate and 0 to 6% for the inflation rate. Figure 5.38 shows the result, with each point identified by its year.

TABLE 5.7 — U.S. Inflation and Unemployment

YEAR	UNEMPLOYMENT RATE (%)	INFLATION RATE (%)
1990	5.6	5.4
1991	6.8	4.2
1992	7.5	3.0
1993	6.9	3.0
1994	6.1	2.6
1995	5.6	2.8
1996	5.4	3.0
1997	4.9	2.3
1998	4.6	2.3
1999	4.3	2.2

SOURCE: U.S. Bureau of Labor Statistics.

FIGURE 5.38 Scatter diagram for data in Table 5.7.

FIGURE 5.39 A scatter diagram for diamond weights and prices.

FIGURE 5.40 A scatter diagram for life expectancy and infant mortality.

Types of Correlation

Look carefully at the scatter diagram for movies in Figure 5.37. The dots seem to be scattered about with no apparent pattern. In other words, at least for the movies shown, there appears to be little or no correlation between the amount of money spent producing the movie and the amount of money it earned in gross receipts.

Now consider the scatter diagram in Figure 5.39, which shows the weights (in carats) and retail prices of 23 diamonds. Here, the dots show a clear upward trend, indicating that larger diamonds generally cost more. The correlation is not perfect; for example, the heaviest diamond is not the most expensive. But the overall trend seems fairly clear. Because the prices tend to increase with the weights, we say that Figure 5.39 shows a **positive correlation.**

In contrast, Figure 5.40 shows a scatter diagram for the variables *life expectancy* and *infant mortality* in 16 countries. We again see a clear trend, but this time it is a **negative correlation:** Countries with *higher* life expectancy tend to have *lower* infant mortality.

Besides stating whether a correlation exists, we can also discuss its strength: The more closely the data follow the general trend, the stronger is the correlation.

Relationships Between Two Data Variables

No correlation: There is no apparent relationship between the two variables.

Positive correlation: Both variables tend to increase (or decrease) together.

Negative correlation: The two variables tend to change in opposite directions, with one increasing while the other decreases.

Strength of a correlation: The more closely two variables follow the general trend, the stronger the correlation (which may be either positive or negative). In a perfect correlation, all data points lie on a straight line.

by the way...

In statistics, the *correlation coefficient* provides a quantitative measure of the strength of a correlation. It is defined to be 1 for a perfect positive correlation (i.e., all data points lie on a single, straight line), −1 for a perfect negative correlation, and 0 for no correlation.

EXAMPLE 2 *Are Inflation and Unemployment Correlated?*

Until the 1990s, most economists assumed that unemployment rates were negatively correlated with inflation rates. That is, when unemployment goes down, inflation goes up, and vice versa. Do the 1990s data in Figure 5.38 support this assumption?

Solution A close look at the scatter diagram shows a very weak trend in which inflation goes *up* as unemployment goes up. In other words, the 1990s data show a weak *positive* correlation between unemployment and inflation. This contradicts the long-standing belief among economists that inflation and unemployment are negatively correlated. ■

EXAMPLE 3 *Accuracy of Weather Forecasts*

The scatter diagrams in Figure 5.41 show two weeks of data comparing the actual high temperature for the day with the same-day forecast (left diagram) and the three-day forecast (right diagram). Discuss the types of correlation on each diagram.

FIGURE 5.41 Comparison of actual high temperatures with same-day and three-day forecasts.

Solution Both scatter diagrams show a general trend in which higher predicted temperatures mean higher actual temperatures. Thus, both show positive correlations. However, the points in the left diagram lie more nearly on a straight line, indicating a stronger correlation than in the right diagram. This makes sense, because we expect weather forecasts to be more accurate on the same day than three days in advance. ■

Summary of Possible Explanations for a Correlation

We began by stating that correlations can help us search for cause-and-effect relationships. But we've already seen that causality is not the only possible explanation for a correlation. For example, the predicted temperatures on the horizontal axis of Figure 5.41 certainly do not *cause* the actual temperatures on the vertical axis. The following box summarizes three possible explanations for a correlation.

> **Possible Explanations for a Correlation**
>
> 1. The correlation may be a *coincidence*.
> 2. Both variables might be directly influenced by some *common underlying cause*.
> 3. One of the correlated variables may actually be a *cause* of the other. Note that, even in this case, we may have identified only one of several causes.

EXAMPLE 4 Explanation for a Correlation

Consider the correlation between infant mortality and life expectancy in Figure 5.40. Which of the three possible explanations for a correlation applies? Explain.

Solution The negative correlation between infant mortality and life expectancy is probably an example of common underlying cause: Both variables respond to an underlying variable that we might call *quality of health care*. In countries where health care is better in general, infant mortality is lower and life expectancy is higher.

EXAMPLE 5 How to Get Rich in the Stock Market (Maybe)

Every financial advisor has a strategy for predicting the direction of the stock market. Most focus on fundamental economic data, such as interest rates and corporate profits. But an alternate strategy relies on a remarkable correlation between the Super Bowl winner in January and the direction of the stock market for the rest of the year: The stock market tends to rise when a team from the old, pre-1970 NFL wins the Super Bowl, and tends to fall otherwise. This correlation successfully matched 28 of the first 32 Super Bowls to the stock market. Suppose that the Super Bowl just ended and the winner was the Detroit Lions, an old NFL team. Should you invest all your spare cash (and maybe even some that you borrow) in the stock market?

Solution Based on the reported correlation, you might be tempted to invest, since the old-NFL winner suggests a rising stock market over the rest of the year. However, this investment would make sense only if you believed that the Super Bowl result actually *causes* the stock market to move in a particular direction. This belief is clearly preposterous, and the correlation is undoubtedly a coincidence. If you are going to invest, don't base your investment on this correlation.

by the way...
Another famous (or infamous) indicator of the stock market comes from the world of fashion: When women's skirt hemlines rise, the stock market supposedly is also about to rise.

THE SEARCH FOR CAUSALITY

Suppose you have discovered a correlation and suspect causality. How can you decide whether there really is a cause-and-effect relationship? While no general procedure applies to all cases, we can get an idea about how to proceed by returning to the issue of smoking and lung cancer.

First, researchers looked for and found correlations between smoking and lung cancer among many groups of people: women, men, and people of different races

The truth is rarely pure and never simple.
—OSCAR WILDE

and cultures. Second, among groups of people who seem otherwise identical, lung cancer was found to be rarer for the nonsmokers. Third, people who smoked more and for longer periods of time were found to have higher rates of lung cancer. Fourth, when researchers accounted for other potential causes of lung cancer (such as exposure to radon gas or asbestos), they found that almost all the remaining lung cancer cases occurred among smokers.

These four lines of evidence made a strong case, but still did not rule out the possibility that some other factor, such as genetics, predisposes people both to smoking and to lung cancer. But two additional lines of evidence made this possibility highly unlikely. First, animal experiments with randomly chosen treatment and control groups also showed a correlation between inhalation of cigarette smoke and lung cancer, which seems to rule out a genetic factor, at least in the animals. Second, biologists studying cell cultures (that is, small samples of human lung tissue) discovered the basic process by which ingredients in cigarette smoke can create cancer-causing mutations. This process does not appear to depend in any way on specific genetic factors, making it all but certain that lung cancer is caused by smoking and not by any preexisting genetic factor.

The following box summarizes these ideas about establishing causality. Generally speaking, the case for causality is stronger when more of these guidelines are met.

historical note

The first four guidelines for establishing causality are called *Mill's methods*. Individually, they are called, respectively, the methods of agreement, difference, concomitant variation, and residues. The methods are named after the English philosopher and economist John Stuart Mill (1806–1873). Mill was a leading scholar of his time and a prolific writer on a wide range of subjects. He was a leading proponent of a philosophy called utilitarianism and an early advocate of women's right to vote.

Guidelines for Establishing Causality

To investigate whether a suspected cause actually causes an effect:

1. Look for situations in which the effect is correlated with the suspected cause even while other factors vary.
2. Among groups that differ only in the presence or absence of the suspected cause, check that the effect is similarly present or absent.
3. Look for evidence that larger amounts of the suspected cause produce larger amounts of the effect.
4. If the effect might be produced by other potential causes (besides the suspected cause), make sure that the effect still remains after accounting for these other potential causes.
5. If possible, test the suspected cause with an experiment. If the experiment cannot be performed with humans for ethical reasons, consider doing the experiment with animals, cell cultures, or computer models.
6. Try to determine the physical mechanism by which the suspected cause produces the effect.

Time out to think

There's a great deal of controversy concerning whether animal experiments are ethical. What is your opinion of animal experiments? Defend your opinion.

CASE STUDY *Air Bags and Children*

By the mid-1990s, passenger-side air bags had become commonplace in cars. Statistical studies showed that the air bags saved many lives in moderate- to high-speed collisions. But a disturbing pattern also appeared. In at least some cases, young children, especially infants and toddlers in child car seats, were killed by air bags in low-speed collisions.

At first, many safety advocates found it difficult to believe that air bags could be the cause of the deaths. But the observational evidence became stronger, meeting the first four guidelines for establishing causality. For example, the greater risk to infants in child car seats fit Guideline 3, because it indicated that being closer to the air bags increased the risk of death. (A child car seat sits on top of the built-in seat, thereby putting a child closer to the air bags than the child would be otherwise.)

To seal the case, safety experts undertook experiments using dummies. They found that children, because of their small size, often sit where they could be easily hurt by the explosive opening of an air bag. The experiments also showed that an air bag could impact a child car seat hard enough to cause death, thereby revealing the physical mechanism by which the deaths occurred.

by the way . . .
Based on these studies, the government now recommends that child car seats *never* be used on the front seat, and that children under age 12 sit in the back seat if possible. Other research has shown that, even in the back seat, children between ages 4 and 8 are prone to more severe injury than adults in crashes because shoulder belts do not fit them properly. The solution is to make sure that such children sit in booster seats that allow the shoulder belts to fit them better.

CASE STUDY *What Causes Global Warming?*

Statistical measurements suggest that the average temperature of the entire Earth, or the *global average temperature,* has risen slightly (by about 1°F) in the past 50 years. But what is causing this so-called *global warming*?

Most scientists attribute the rise to the increasing concentration of carbon dioxide and other greenhouse gases in our atmosphere (Figure 5.42). Comparative studies of the Earth and other planets, particularly Venus and Mars, show that greenhouse gases meet the first three guidelines for establishing causality. Indeed, there is virtually no doubt that the average temperature of each planet is strongly influenced by its atmospheric concentration of greenhouse gases. The case for greenhouse gases also meets Guideline 6, as scientists have clearly established the physical mechanism by which these gases can cause global warming. (Basically, greenhouse gases slow the escape of heat from a planet's surface, thereby causing the surface to warm up.) Thus, greenhouse gases, the suspected cause of global warming, satisfy at least four of the six guidelines for establishing causality.

Nevertheless, we cannot be certain that the observed global warming of the past 50 years has been caused by greenhouse gases. The trouble comes mainly from Guideline 4, which tells us to account for other potential causes. One such potential cause is an increase in heating of the Earth by the Sun. Some evidence suggests that the Sun has become slightly warmer during the past century, thereby increasing the amount of heat the Earth receives by a few hundredths of 1%. If true, this could account for some or all of the observed global warming.

The best way to sort out the causes would be with an experiment (Guideline 5), but it is impossible to perform experiments with entire planets. Instead, researchers run experiments with *computer models* of planets that account for the way weather and climate work. In this way, researchers can compare results among models with varying amounts of heat from the Sun and varying amounts

by the way . . .
Some greenhouse gases are present naturally in the Earth's atmosphere. Without them, the global average temperature would be a frigid −10°F; with them, the global average temperature is about 59°F. Thus, the greenhouse effect is a good thing for life on Earth. On Venus, where the temperature would be pleasantly warm with an Earth-like atmosphere, a very strong greenhouse effect produces a global average temperature of about 870°F— demonstrating that it's possible to have too much of a good thing.

by the way ...
The North Pole has been covered by ice for most or all of the past 50 million years. But in August 2000, the North Pole was temporarily free of ice for the first time in recorded history. This alarming finding is consistent with climate models, which suggest that global warming should be more extreme near the North Pole than at other latitudes. While it still does not prove that human activity is the cause of global warming, it makes it clear that warming *is* occurring.

FIGURE 5.42 This graph shows data for carbon dioxide concentration collected on Mauna Loa in Hawaii. The yearly wiggles show seasonal changes in the concentration, but the long-term trend is clearly upward.
SOURCE: Dave Keeling and Tim Whorf, Scripps Institution of Oceanography.

of greenhouse gases. Unfortunately, the models made by different research groups tend to give different results. Unless everyone can agree on what constitutes a valid computer model, the controversy about the cause of global warming will persist.

by the way ...
Substantial global warming (say, a few degrees Fahrenheit) would almost certainly cause consequences such as a rise in sea level, an increase in the frequency of severe storms, changes in weather patterns that would reduce agricultural production on many existing farms, and extinctions of plant and animal species that cannot adapt to the changing climate.

CASE STUDY *Ozone and CFCs*

While the human connection to global warming remains controversial, evidence is much more definitive for another environmental issue: the depletion of the Earth's atmospheric ozone. Ozone is a gas that exists naturally at high altitudes, where it absorbs harmful ultraviolet radiation from the Sun. Ozone depletion (a reduction in the total amount of ozone) could pose a serious problem by allowing more of this ultra-violet radiation to reach the ground, where it can cause mutations, cancers, and deaths of both plants and animals.

By the early 1970s, experiments had shown that human-made chemicals called *chlorofluorocarbons* (CFCs) could destroy ozone. CFCs were invented in the 1930s and have been widely used in air conditioners and refrigerators, in the manufacture of packaging foam, as propellants in spray cans, in industrial solvents used in the computer industry, and for many other purposes. Because of their many everyday uses, it seemed unwise to ban CFCs on the basis of a few experiments, especially since there was no evidence at the time that any ozone depletion was actually occurring.

FIGURE 5.43 This map shows ozone distribution on October 1, 2000, as measured from a satellite. The key explains how colors represent ozone concentrations. The regions of gray, pink, and purple around Antarctica represent the ozone hole, where ozone concentrations are far below normal. The white strips on the map are regions where the satellite did not collect data.
SOURCE: NASA/NOAA Earth Probe TOMS data.

The situation changed in the mid-1980s with the discovery of the so-called *ozone hole* over Antarctica (Figure 5.43). Researchers found that the formation of this hole (generally lasting from August through November) was strongly correlated with the presence of chlorine and fluorine—both components of CFCs—in the atmosphere. They soon identified the physical mechanism that led to the fairly extreme ozone depletion over Antarctica compared to other places. (The mechanism involves chemical reactions that occur on the surfaces of ice crystals that form in the Antarctic clouds.)

With CFCs confirmed as a cause of damage to the Earth's ozone, negotiations soon began on international treaties that now ban the manufacture of CFCs. Unfortunately, these treaties have not by themselves put the problem of ozone depletion behind us. Most of the CFCs made during the 20th century are either still in use or in discarded appliances and materials in landfills, and probably will leak into the atmosphere over the next few decades. In addition, because substitute chemicals tend to be more expensive than CFCs, there is now a large international black market for CFCs manufactured in violation of the international treaties.

by the way . . .
Paul Crutzen, Mario Molina, and F. Sherwood Rowland received the 1995 Nobel Prize in chemistry for their work in discovering the linkage between CFCs and ozone destruction.

Time out to think

Suppose your car air conditioner is broken and will cost $200 to fix. You hear about a shop that can fix it for only $100—undoubtedly by using black market CFCs. Given that your single air conditioner won't make much difference to the ozone layer, is it worth the $100 difference to avoid the black market CFCs? Defend your opinion.

CONFIDENCE IN CAUSALITY

If human activity is *causing* global warming, we'd be wise to change our activities so as to stop it. But we cannot yet be sure that this is the case. Moreover, the changes needed to slow global warming might be very expensive. Given the huge cost and the uncertainty as to whether the changes would even make a difference, how can we set sensible policies?

In an ideal world, we would continue to study the issue until we could establish whether human activity is the cause of global warming. However, we have seen that it is difficult to establish causality and often impossible to *prove* causality beyond all doubt. We are therefore forced to make decisions about global warming, and many other important issues, despite great uncertainty about cause and effect.

In other areas of mathematics, accepted techniques help us deal with uncertainty by allowing us to calculate numerical measures of possible errors. But there are no accepted ways to assign such numbers to the uncertainty that comes with questions of causality. Fortunately, another area of study has dealt with practical problems of causality for hundreds of years: our legal system. You may be familiar with the following three broad ways of expressing a legal level of confidence.

> **Broad Levels of Confidence in Causality**
>
> **Possible cause:** We have discovered a correlation, but cannot yet determine whether the correlation implies causality. In the legal system, possible cause (such as thinking that a particular suspect possibly caused a particular crime) is often the reason for starting an investigation.
>
> **Probable cause:** We have good reason to suspect that the correlation involves cause, perhaps because some of the guidelines for establishing causality are satisfied. In the legal system, probable cause is the general standard for getting a judge to grant a warrant for a search or wiretap.
>
> **Cause beyond reasonable doubt:** We have found a physical model that is so successful in explaining how one thing causes another that it seems unreasonable to doubt the causality. In the legal system, cause beyond reasonable doubt is the usual standard for conviction and generally demands that the prosecution has shown how and why (essentially the physical model) the suspect committed the crime. Note that beyond *reasonable* doubt does *not* mean beyond *all* doubt.

While these broad levels remain fairly vague, they give us at least some common language for discussing confidence in causality. If you study law, you will learn much more about the subtleties of interpreting these terms. However, because statistics has little to say about them, we will not discuss them much further in this book.

Time out to think

Given what you know about global warming, do you think that human activity is a possible cause, probable cause, or cause beyond reasonable doubt? Defend your opinion. Based on your level of confidence in the causality, how would you recommend setting policies with regard to global warming?

by the way...

As you might guess, it is very difficult to define *reasonable* doubt. One legal definition, endorsed by Supreme Court Justice Ruth Bader Ginsburg, reads: "Proof beyond a reasonable doubt is proof that leaves you firmly convinced of the defendant's guilt. There are very few things in this world that we know with absolute certainty, and in criminal cases the law does not require proof that overcomes every possible doubt. If, based on your consideration of the evidence, you are firmly convinced that the defendant is guilty of the crime charged, you must find him guilty. If on the other hand, you think there is a real possibility that he is not guilty, you must give him the benefit of the doubt and find him not guilty."

Review Questions

1. What is a *correlation*? Give three examples of pairs of variables that are correlated.
2. What is a *scatter diagram*, and how do you make one? How can we use a scatter diagram to look for a correlation?
3. Define and distinguish among *positive correlation, negative correlation,* and *no correlation.* How do we determine the strength of a correlation?
4. Describe the three general categories of explanation for a correlation. Give an example of each.
5. Briefly describe each of the six guidelines presented in this section for establishing causality. Give an example of the application of each guideline.
6. Briefly describe three levels of confidence in causality and how they can be useful when we do not have absolute proof of causality.

In the News

1. **Correlations in the News.** Find a recent news report that describes some type of correlation. Describe the correlation. Does the article give any sense of the strength of the correlation? Does it suggest that the correlation reflects any underlying causality? Briefly discuss whether you believe the implications the article makes with respect to the correlation.
2. **Causation in the News.** Find a recent news report in which a statistical study has led to a conclusion of causation. Describe the study and the claimed causation. Do you think the claim of causation is legitimate? Explain.
3. **Legal Causation.** Find a news report concerning an ongoing legal case, either civil or criminal, in which establishing causality is important to the outcome. Briefly describe the issue of causation in the case and how the ability to establish or refute causality will influence the outcome of the case.

Exercises

Types of Correlation. Exercises 1–8 list pairs of variables. State the units you would use to measure each of the two variables (for example, pounds, years, or miles per hour). Then state whether you believe the two variables are correlated. If you believe they are correlated, state whether the correlation is positive or negative and strong or weak. Explain your reasoning.

1. Body weight and intake of calories per day
2. Height and shoe size
3. Rate of pedaling and speed of the bicycle
4. Altitude on a mountain hike and surrounding air temperature
5. Blood-alcohol level and reaction time (for an individual)
6. Weight of car and gas mileage
7. Weight of car and price of car
8. Retail price of an item and the number of items that can be sold in a day

Making Scatter Diagrams. Exercises 9–14 each give a table of data. In each case, do the following:

a. Make a scatter diagram for the data.

b. State whether the two variables appear to be correlated and, if so, whether the correlation is positive or negative and strong or weak.

c. Suggest a reason for the correlation (or lack of correlation). If you suspect causality, briefly discuss what further evidence you would need to establish it.

9. The following table gives speed limits and death rates for automobile accidents in selected countries in the 1980s.

Country	Death rate (per 100 million vehicle miles)	Speed limit (miles per hour)
Norway	3.0	55
United States	3.3	55
Finland	3.4	55
Britain	3.5	70
Denmark	4.1	55
Canada	4.3	60
Japan	4.7	55
Australia	4.9	65
Netherlands	5.1	60
Italy	6.1	75

SOURCE: *New York Times,* November 25, 1986.

10. The following table gives number of home runs and batting averages for baseball's Most Valuable Players, 1995–1999 (NL = National League and AL = American League).

Player	Home runs	Batting average
Barry Larkin (1995 NL)	15	.319
Mo Vaughn (1995 AL)	39	.300
Ken Caminiti (1996 NL)	40	.326
Juan Gonzalez (1996 AL)	47	.314
Larry Walker (1997 NL)	49	.366
Ken Griffey Jr. (1997 AL)	56	.304
Sammy Sosa (1998 NL)	66	.308
Juan Gonzalez (1998 AL)	45	.318
Chipper Jones (1999 NL)	45	.319
Ivan Rodriguez (1999 AL)	35	.332

11. The following table gives percent change in population and birth rate (per 1000 of population) for ten states between 1980 and 1990.

State	Percent change in population	Birth rate
Nevada	50.1%	16.3
California	25.7%	16.9
New Hampshire	20.5%	12.5
Utah	17.9%	21.0
Colorado	14.0%	14.6
Minnesota	7.3%	13.7
Montana	1.6%	12.3
Illinois	0%	15.5
Iowa	−4.7%	13.0
West Virginia	−8.0%	11.4

SOURCE: U.S. Bureau of Census.

12. The following table gives the average hours of television watched in households in five categories of annual income. (Hint: For the first and last categories of the household income data, place the dot at the position corresponding to $25,000 and $65,000, respectively; for other categories, place the dot at the center of each bin.)

Household income	Weekly TV hours
Less than $30,000	56.3
$30,000–$40,000	51.0
$40,000–$50,000	50.5
$50,000–$60,000	49.7
More than $60,000	48.7

SOURCE: Nielsen Media Research, 1997.

13. The following table gives total sales (revenue) and total earnings (profits) for the ten largest retailers in the United States.

Company	Total sales (billions of dollars)	Earnings (billions of dollars)
Wal-Mart	118	3.53
Sears, Roebuck	41	1.19
Kmart	32	0.25
JCPenney	31	0.57
Dayton Hudson	28	0.75
Kroger	27	0.41
Home Depot	24	1.16
Safeway	22	0.62
Costco	22	0.31
American Stores	19	0.28

SOURCE: National Retail Federation, 1997.

14. The following table gives mean daily caloric intake (all residents) and infant mortality rate (per 1000 births) for ten countries.

Country	Mean daily Calories	Infant mortality rate (per 1000 births)
Afghanistan	1523	154
Austria	3495	6
Burundi	1941	114
Columbia	2678	24
Ethiopia	1610	107
Germany	3443	6
Liberia	1640	153
New Zealand	3362	7
Turkey	3429	44
United States	3671	7

Correlation and Causality. Exercises 15–20 make statements about a correlation. In each case, state the correlation clearly (for example, there is a positive correlation between variable A and variable B). Then state whether the correlation is most likely due to coincidence, a common underlying cause, or a direct cause. Explain your answer.

15. In one U.S. city, the crime rate increased at the same time that the number of people in prison increased.

16. Over the past three decades, the number of miles of freeways in Los Angeles has grown, and traffic congestion has worsened.

17. When gasoline prices rise, the number of airline passengers increases.

18. Sales of ice tea in a local restaurant are positively correlated with ticket sales at the local swimming pool.

19. When deer hunting permit costs were gradually increased in one region, the incidence of deer sightings increased.

20. Over a period of twenty years, the number of ministers and priests increased, as did attendance at movies.

21. **Identifying Causes: Headaches.** You are trying to identify the cause of late-afternoon headaches that plague you several days each week. For each of the following tests and observations, explain which of the six guidelines for establishing causality you used and what you concluded.

 a. The headaches occur only on days that you go to work.

 b. If you stop drinking Coke at lunch, the headaches persist.

 c. In the summer, the headaches occur less frequently if you open the windows of your office slightly. They occur even less often if you open the windows of your office fully.

 d. Having made all these observations, what reasonable conclusion can you reach about the cause of the headaches?

22. **Smoking and Lung Cancer.** There is a strong correlation between tobacco smoking and incidence of lung cancer, and most physicians believe that tobacco smoking causes lung cancer. Yet, not everyone who smokes gets lung cancer. Briefly describe how smoking could cause cancer when not all smokers get cancer.

23. **Longevity of Orchestra Conductors.** A famous study in *Forum on Medicine* (1978) concluded that the mean lifetime of conductors of major orchestras was 73.4 years, about 5 years longer than that of all American males at the time. The author claimed that a life of music *causes* a longer life. Evaluate the claim of causality and propose other explanations for the longer life expectancy of conductors.

24. **High-Voltage Power Lines.** Suppose that people living near a particular high-voltage power line have a higher incidence of cancer than people living farther from the power line. Can you conclude that the high-voltage power line is the cause of the elevated cancer rate? If not, what other explanations might there be for it? What other types of research would you like to see before you conclude that high-voltage power lines cause cancer?

Web Projects

Find useful links for Web Projects on the text Web site:
www.aw.com/bennett-briggs

1. **Success in the NFL.** Use the Web to find last season's NFL team statistics. Make a table showing the following for each team: number of wins, average yards gained on offense per game, and average yards allowed on defense per game. Make scatter diagrams to explore the correlations between offense and wins and between defense and wins. Discuss your findings. Do you think that there are other team statistics that would yield stronger correlations with the number of wins?

2. **Statistical Abstract.** Explore the "frequently requested tables" at the Web site for the *Statistical Abstract of the United States*. Choose data that are of interest to you and explore at least two correlations. Briefly discuss what you learn from the correlations.

3. **Air Bags and Children.** Starting from the Web site of the National Highway Traffic Safety Administration, research the latest studies on the safety of air bags, especially with regard to children. Write a short report summarizing your findings and offering recommendations for improving child safety in cars.

4. **Global Warming.** Use the Web to find recent information about global warming and its potential consequences. Discuss the evidence linking human activity to global warming. In light of your findings, suggest how we should deal with the issue of global warming.

5. **Ozone Depletion.** Starting with information available at the Environmental Protection Administration (EPA) Web site, research the latest information about ozone depletion and efforts to combat it. Summarize your findings and suggest what should be done in the future to help prevent ozone depletion.

6. **Tobacco Lawsuits.** Tobacco companies have been the subject of many lawsuits relating to the dangers of smoking. Research one recent lawsuit. What were the plaintiffs trying to prove? What statistical evidence did they use? How well do you think they established causality? Did they win? Summarize your findings in one to two pages.

UNIT 5F

CHARACTERIZING A DATA DISTRIBUTION

Frequency tables and graphs (see Units 5C and 5D) show us how data are distributed over various categories. More formally, we say that a table or graph gives the **distribution** of a data set. For example, a distribution of essay grades shows how many students received grades of A, B, C, D, and F; in this case, the categories are the grades.

> **DEFINITION**
>
> The **distribution** of a variable (or data set) describes the values taken on by the variable and the frequency (or relative frequency) of these values.

In many cases, however, we are less interested in the complete distribution than in a few descriptive terms that summarize it. In this unit, we'll study how we can characterize a data distribution by its center, or average, and by its shape.

WHAT IS AVERAGE?

The term *average* comes up frequently in the news, but you may be surprised to learn that it does not always have the same meaning. Let's begin by studying three common ways of characterizing the center of a data distribution, all of which are sometimes called the "average."

Mean, Median, and Mode

Table 5.8 shows the number of movies (original and sequels or prequels) in each of five popular science fiction series.

TABLE 5.8 Five Science Fiction Movie Series

TITLE	NUMBER OF MOVIES IN SERIES (AS OF 2000)
Alien	4
Back to the Future	3
Planet of the Apes	5
Star Trek	9
Star Wars	4

What is the average number of films in these series? One way to answer this question is to compute the **mean**. We find the mean by dividing the total number of movies by five (because there are five series):

$$\text{mean} = \frac{4 + 3 + 5 + 9 + 4}{5} = \frac{25}{5} = 5$$

In other words, these five series have a mean of 5 movies. More generally, we find the mean of any data set by summing all the data values and then dividing by the number of data values. The mean is what most people think of as the average. In essence, it represents the balance point for a data distribution, as shown in Figure 5.44.

We could also describe the average number of films in these five series by computing the **median**, or middle value, of the data set. To find a median, we arrange the data values in ascending (or descending) order, repeating data values that appear more than once. If the number of values is odd, there is exactly one

by the way...
In terms of total box office receipts, the four *Star Wars* films rank as four of the top ten movies of all time (with ranks of 2nd, 3rd, 8th, and 10th through 2000).

value in the middle of the list, and this value is the median. If the number of values is even, there are two values in the middle of the list, and the median is the number that lies halfway between them. Putting the data in Table 5.8 in ascending order gives the list 3, 4, 4, 5, 9. The median number of movies is 4 because 4 is the middle number in the list.

Another value sometimes used as a measure of average is the **mode,** which is the most common value (or group of values) in a data set. In the case of the movies, the mode is 4 because this value occurs twice in the data set, while the other values occur once. A data set may have one mode, more than one mode, or no mode. Sometimes the mode refers to a group of closely spaced values rather than a single value.

FIGURE 5.44 A histogram made from blocks would balance at the position of its mean.

Measures of Center in a Distribution

The **mean** is what we most commonly call the average value. It is defined as follows:

$$\text{mean} = \frac{\text{sum of all values}}{\text{total number of values}}$$

The **median** is the middle value in the sorted data set (or halfway between the two middle values if the number of values is even).

The **mode** is the most common value (or group of values) in a distribution.

EXAMPLE 1 *Price Data*

Eight grocery stores sell the PR energy bar for the following prices:

$1.09 $1.29 $1.29 $1.35 $1.39 $1.49 $1.59 $1.79

Find the mean, median, and mode for these prices.

Solution The mean price is $1.41:

$$\text{mean} = \frac{\begin{pmatrix} \$1.09 + \$1.29 + \$1.29 + \$1.35 \\ + \$1.39 + \$1.49 + \$1.59 + \$1.79 \end{pmatrix}}{8} = \$1.41$$

To find the median, we first sort the data in ascending order:

$1.09 $1.29 $1.29 | $1.35 $1.39 | $1.49 $1.59 $1.79
 3 values below 2 middle values 3 values above

by the way...
A 1999 study at the University of Chicago discovered that the median allowance for teenagers in this country is $50 per week. The study estimates that teenagers receive over $1 billion per week in allowances.

Because there are eight prices (an even number), there are two values in the middle of the list: $1.35 and $1.39. Therefore the *median* lies halfway between these two values, which we calculate by adding them and dividing by 2:

$$\text{median} = \frac{\$1.35 + \$1.39}{2} = \$1.37$$

The mode is $1.29, because this is the only price that occurs more than once.

Effects of Outliers

To explore the differences among the mean, median, and mode, imagine that the five graduating seniors on a college basketball team receive the following first-year contract offers to play in the National Basketball Association (zero indicates that the player did not receive a contract offer):

$$0 \quad 0 \quad 0 \quad 0 \quad \$3,500,000$$

The mean contract offer is

$$\text{mean} = \frac{0 + 0 + 0 + 0 + \$3,500,000}{5} = \$700,000$$

Is it therefore fair to say that the *average* contract offer was $700,000?

Not really. The problem is that the single player receiving the large offer makes the mean much larger than it would be otherwise. In fact, if we ignore this one player and look only at the other four, the mean contract offer is zero! Because this one value of $3,500,000 is so extreme compared to the others, we call it an **outlier** (or *outlying value*). As our example shows, an outlier can pull the mean significantly upward (or downward), thereby making the mean unrepresentative of the data set as a whole.

> **DEFINITION**
>
> An **outlier** is a data value that is much higher or much lower than almost all other values.

While the outlier pulls the mean contract offer upward, it has no effect on the median contract offer, which remains zero for the five players. In fact, the value of an outlier never has an effect on the median, because outliers don't lie in the middle of a data set. Outliers do not affect the mode either.

Time out to think

Do you think it is fair to use the median as the *average* contract offer for the five players? Why or why not?

EXAMPLE 2 *Mistake?*

A track coach wants to determine an appropriate heart rate for her athletes during their workouts. She chooses five of her best runners and asks them to wear heart rate monitors during a workout. In the middle of the workout, she reads the following

by the way . . .
A survey once found that geography majors at the University of North Carolina had a far higher mean starting salary than geography majors from other schools. The reason for the high mean turned out to be a single outlier—the basketball superstar and geography major named Michael Jordan.

heart rates for the five athletes: 130, 135, 140, 145, 325. Which is a better measure of the average in this case: the mean or the median? Why?

Solution Four of the five values are fairly close together and seem reasonable for mid-workout heart rates. The high value of 325 is an outlier. This outlier seems likely to be a mistake (perhaps caused by a faulty heart monitor), because anyone with such a high heart rate should be in cardiac arrest. If the coach uses the mean as the average, she will be including this outlier—which means she will be including any mistake made when it was recorded. If she uses the median as the average, she'll have a more reasonable and defensible value. ■

Confusion About "Average"

The different meanings of "average" can lead to confusion. Sometimes this confusion arises because we are not told whether the "average" is the mean or the median, and other times because we are not given enough information about how the average was computed. The following two examples illustrate such situations.

EXAMPLE 3 Wage Dispute

A newspaper surveys wages for assembly workers in regional high-tech companies and reports an average of $22 per hour. The workers at one large firm immediately request a pay raise, claiming that they work as hard as employees at other companies but their average wage is only $19. The management rejects their request, telling them that they are *overpaid* because their average wage, in fact, is $23. Can both sides be right? Explain.

Figures won't lie, but liars will figure.

—Charles H. Grosvenor

Solution Both sides can be right if they are using different definitions of average. In this case, the workers may be using the median while the management uses the mean. For example, imagine that there are only five workers at the company and their wages are $19, $19, $19, $19, and $39. The median of these five wages is $19 (as the workers claimed), but the mean is $23 (as management claimed). ■

EXAMPLE 4 What Mean?

All 100 first-year students at a small college take three courses in the Core Studies Program. The first two courses are taught in large lectures, with all 100 students in a single class. The third course is taught in ten classes of 10 students each. Students and administrators get into an argument about whether classes are too large. The students claim that the mean size of their Core Studies classes is 70. The administrators claim that the mean class size is only 25 students. Can both sides be right? Explain.

Solution Both sides are right, but they are using different means. The students have calculated the mean size of the classes in which each student is personally enrolled. Each student is taking two classes with enrollments of 100 each and one class with an enrollment of 10, so the mean size of each student's classes is

$$\frac{\text{total enrollment in student's classes}}{\text{number of classes student is taking}} = \frac{100 + 100 + 10}{3} = 70$$

The administrators have calculated the mean enrollment in all classes. There are two classes with 100 students each and ten classes with 10 students each, making a total enrollment of 300 students in 12 classes. Thus, the mean enrollment per class is

$$\frac{\text{total enrollment}}{\text{number of classes}} = \frac{300}{12} = 25$$

The two claims about the mean are both correct, but very different, because the students and administrators used different means. The students calculated the mean *class size per student,* while the administrators calculated the mean number of *students per class.* ■

Time out to think

In Example 4, could the administrators redistribute faculty assignments so that all classes have 25 students each? How? Discuss the advantages and disadvantages of such a change.

SHAPES OF DISTRIBUTIONS

We next turn our attention to describing the overall shape of a distribution. We can *see* the complete shape of a distribution on a graph. In essence, our goal is to describe the general shape in words. Although such descriptions carry less information than a table or graph, they are still useful. Here, we will focus on three characteristics of a distribution: its number of peaks, its symmetry or skewness, and its variation.

Number of Peaks

Because we are interested primarily in the *general* shapes of distributions, it's often easier to work with smooth curves than actual data. Figure 5.45 shows three examples of this idea, two in which the actual distributions are shown as histograms and one in which the actual distribution is shown as a line chart. The smooth curves approximate the actual distributions, but do not show all of their details.

The first way we might characterize the shape is by its number of peaks. The distributions in Figures 5.45a and 5.45c are single-peaked (or *unimodal*); note that the peak represents the mode of each distribution. The distribution in Figure 5.45b has two

FIGURE 5.45 The smooth curves approximate the actual shapes of the distributions. Note that (a) and (c) are single-peaked, while (b) is double-peaked (or bimodal).

peaks; we say that it is double-peaked (or *bimodal*). Other distributions may have no peaks (in which case they are called *uniform distributions*) or more than two peaks.

EXAMPLE 5 *Number of Peaks*

How many peaks would you expect for each of the following distributions? Why?

a. Heights of all women at a college
b. Heights of all students at a college
c. The number of people with particular last digits (0 through 9) in their Social Security numbers

Solution

a. The distribution of heights of women is single-peaked because many women are at or near the mean height, with fewer and fewer women at heights much taller or shorter than the mean.
b. The distribution of heights of all students (men and women) will have two peaks, one corresponding to the mode of men's heights and one corresponding to the mode of women's heights.
c. The last digits of Social Security numbers are essentially random, so the frequencies of the different digits (0 through 9) should be nearly equal. That is, about 10% of all Social Security numbers end in 0, 10% end in 1, and so on. The result is a uniform distribution with no peak.

Symmetry or Skewness

A second way to describe shape is in terms of symmetry or skewness. A distribution is **symmetric** if its left half is a mirror image of its right half. The distributions in Figure 5.46 are all symmetric.

FIGURE 5.46 These distributions are all symmetric because their left halves are mirror images of their right halves. Note that (a) and (b) are single-peaked while (c) is triple-peaked (or trimodal).

A distribution that is not symmetric must have values that tend to be more spread out on one side than the other. In this case, we say that the distribution is **skewed.** Figure 5.47a shows a distribution in which the values are more spread out on the left, making it **left-skewed.** Sometimes it is helpful to think of such a distribution as having a tail that has been pulled toward the left. Figure 5.47b shows a distribution in which the values are more spread out on the right, making it **right-skewed;** it looks as if it has a tail pulled toward the right.

FIGURE 5.47 (a) Skewed to the left (left-skewed): The mean and median are less than the mode. (b) Skewed to the right (right-skewed): The mean and median are greater than the mode. (c) Symmetric distribution: The mean, median, and mode are the same.

Figure 5.47 also shows what the skewness of a distribution tells us about the relative positions of the mean, median, and mode. By definition, the mode is the peak in a single-peaked distribution. In a left-skewed distribution, the tail pulls the mean and median to the left of the mode. Moreover, the low values in the data set tend to make the mean less than the median. Similarly, a right-skewed distribution pulls the mean and median to the right—that is, to values greater than the mode. In such cases, the large values in the data set tend to make the mean greater than the median. When the distribution is symmetric and single-peaked, both the mean and the median are equal to the mode.

> **Symmetry and Skewness**
>
> A distribution is **symmetric** if its left half is a mirror image of its right half.
>
> A distribution is **left-skewed** if its values are more spread out on the left side.
>
> A distribution is **right-skewed** if its values are more spread out on the right side.

Time out to think

In a skewed distribution, the median is a better measure of the center of the distribution than the mean. Why do you think this is true?

EXAMPLE 6 Skewness

For each of the following situations, state whether you expect the distribution to be symmetric, left-skewed, or right-skewed. Explain.

 a. Heights of a sample of 100 women
 b. Family income in the United States
 c. Speeds of cars on a road where a visible patrol car is using radar to detect speeders

Solution

 a. The distribution of heights of women is symmetric because roughly equal numbers of women are shorter and taller than the mean, and extremes of height are rare on either side of the mean.

b. The distribution of family income is right-skewed. Most families are middle-class, so the mode of this distribution is a middle-class income (somewhere around $40,000 in the United States). But a few very high-income families pull the mean to a considerably higher value, stretching the distribution to the right (high-income) side.

c. Drivers usually slow down when they are aware of a patrol car looking for speeders. Few if any drivers will be exceeding the speed limit, but some drivers tend to slow to well below the speed limit. Thus, the distribution of speeds will be left-skewed, with a mode near the speed limit but a few cars going well below the speed limit. ■

> **by the way...**
> Speed kills. On average in the United States, someone is killed in an auto accident about every 12 minutes. About one-third of these fatalities involve a speeding driver.

Time out to think

In ordinary English, the term *skewed* is often used to mean something that is distorted or depicted in an unfair way. How is this use of *skew* related to its meaning in statistics?

Variation

A third way to describe a distribution is by its **variation,** which is a measure of how much the data values are spread out. A distribution in which most data values are clustered together has a low variation. As shown in Figure 5.48a, such a distribution has a fairly sharp peak. The variation is higher when the data are distributed more widely around the center, which makes the peak broader. Figure 5.48b shows a distribution with moderate variation and Figure 5.48c shows a distribution with high variation. Variation is often described quantitatively by the *standard deviation,* which you may study if you do further course work in statistics.

(a) Low variation (b) Moderate variation (c) High variation

FIGURE 5.48 From left to right, these three distributions have increasing variation.

> **DEFINITION**
>
> **Variation** describes how widely data values are spread out about the center of a distribution.

EXAMPLE 7 *Variation in Marathon Times*

How would you expect the variation to differ between times in the Olympic marathon and times in the New York marathon? Explain.

Solution The Olympic marathon invites only elite runners, whose times are likely to be clustered relatively near world record times. The New York marathon allows runners of all abilities, whose times are spread over a very wide range (from near the world record to many hours). Therefore, the variation among the times should be greater in the New York marathon than in the Olympic marathon.

Review Questions

1. Define and distinguish among *mean, median,* and *mode.*
2. What are outliers? Describe the effects of outliers on the mean, median, and mode.
3. Briefly describe at least two possible sources of confusion about the "average."
4. What do we mean by a data distribution? Briefly explain how we can use smooth curves to approximate a distribution in a way that helps us see its general features.
5. Give simple examples of a single-peaked distribution and a double-peaked distribution.
6. What do we mean when we say that a distribution is symmetric? Give simple examples of a symmetric distribution, a left-skewed distribution, and a right-skewed distribution.
7. What do we mean by the variation of a distribution? Give simple examples of distributions with different amounts of variation.

In the News

1. **Averages in the News.** Find three recent news articles that refer to some type of average. In each case, explain whether the average is a mean, a median, or some other type of average.
2. **Daily Averages.** Cite three examples of averages that you deal with in your own life (such as grade point average or batting average). In each case, explain whether the average is a mean, a median, or some other type of average. Briefly describe how the average is useful to you.
3. **Distributions in the News.** Find three recent examples in the news of distributions shown as histograms or line charts. Over each distribution, draw a smooth curve that captures its general features. Then classify the distribution according to its number of peaks, symmetry or skewness, and variation.

Exercises

Mean, Median, and Mode. Exercises 1–6 each list a set of numbers. In each case, find the mean, median, and mode of the listed numbers.

1. Weights (in milligrams) of Bufferin aspirin tablets:

 672.2 679.2 669.8 672.6 672.2 662.2

2. Body temperatures (in degrees Fahrenheit) of randomly selected normal and healthy adults:

 98.6 98.6 98.0 98.0 99.0
 98.4 98.4 98.4 98.4 98.6

3. Blood-alcohol concentrations of drivers involved in fatal crashes and then convicted and given jail sentences (based on data from the U.S. Department of Justice):

 0.27 0.17 0.17 0.16 0.13 0.24
 0.29 0.24 0.14 0.16 0.12 0.16

4. Golf scores from a recent WPGA tournament:

 69 70 70 71 73 73 75 76 76 78
 80 81 81 82 83 84 85 86 87 88

5. High temperatures (in °F) during a 15-day period in Alaska in March:

 15 11 10 9 0 2 4 5
 5 7 10 12 15 18 19

6. Weights (in grams) of randomly selected M&M plain candies:

 0.957 0.912 0.842 0.925 0.939 0.886
 0.914 0.913 0.958 0.947 0.920

Appropriate Average. Exercises 7–12 list "averages" that someone might want to know. In each case, state whether the mean, median, or some other measure would give a better description of the "average." Explain your reasoning.

7. The average income of all adults in a large city
8. The average weight of the oranges in a large box
9. The average number of times that people change jobs during their careers
10. The average number of pieces of lost luggage per flight for each airline company
11. The average daily high temperature over a period of a month
12. The average waiting time in the checkout lines of a supermarket
13. **Outlier Coke.** Cans of Coca-Cola vary slightly in weight. Here are the measured weights of seven cans, in pounds:

 0.8161 0.8194 0.8165 0.8176
 0.7901 0.8143 0.8126

 Find the mean and median of these weights. Which, if any, of these weights would you consider to be an outlier? What are the mean and median weights if the outlier is excluded?

14. **Alphabetic States.** The following table gives the total area in square miles (land and water) of the seven states whose names begin with the letters A through C.

State	Area
Alabama	52,200
Alaska	615,200
Arizona	114,000
Arkansas	53,200
California	158,900
Colorado	104,100
Connecticut	5,500

 Find the mean area and median area for these states. Which state is an outlier on the high end? If you eliminate this state, what are the new mean and median areas for this data set? Which state is an outlier on the low end? If you eliminate this state, what are the new mean and median areas for this data set?

Smooth Distributions. For each of the histograms in Exercises 15–17, draw a smooth curve that captures its general features. Then classify the distribution according to its number of peaks, symmetry or skewness, and variation. In words, summarize the meaning of your results.

15. Times between 300 eruptions of Old Faithful geyser in Yellowstone National Park, shown in Figure 5.49

Times Between Eruptions of Old Faithful

FIGURE 5.49 SOURCE: Hard et al., *Handbook of Small Data Sets*, 1994.

16. Time until failure for a sample of 108 computer chips that failed, shown in Figure 5.50

Failure Time of Computer Chips

FIGURE 5.50 SOURCE: Hard et al., *Handbook of Small Data Sets*, 1994.

17. Weights of a sample of 391 rugby players, shown in Figure 5.51

Weight of Rugby Players

[Histogram showing frequency of rugby player weights from about 65 to 130 kilograms, roughly symmetric and bell-shaped, peaking around 90–95 kg]

FIGURE 5.51 SOURCE: Hard et al., *Handbook of Small Data Sets*, 1994.

18. **Family Income.** Suppose you study family income in a random sample of 300 families. You find that the mean family income is $41,000; the median is $35,000; and the highest and lowest incomes are $250,000 and $2400, respectively.

 a. Draw a rough sketch of the income distribution, with clearly labeled axes. Describe the distribution as symmetric, left-skewed, or right-skewed.

 b. How many families in the sample earned less than $35,000? Explain how you know.

 c. Based on the given information, can you determine how many families earned more than $41,000? Why or why not?

Describing Distributions. For each distribution described in Exercises 19–26, answer the following:

 a. How many peaks would you expect for the distribution? Explain.

 b. Would you expect the distribution to be symmetric, left-skewed, or right-skewed? Explain.

 c. Would you expect the variation of the distribution to be small or large? Explain.

19. The exam scores on an exam where 50 students got an A, 20 got a B, and 5 got a C

20. The exam scores on an exam where 5 students got an A, 20 got a B, 40 got a C, 20 got a D, and 5 got an F

21. The weights of cars in the new car lot at a car dealership in which about half the inventory consists of compact cars and the other half consists of sport utility vehicles

22. The numbers of people with particular last digits (0 through 9) in their phone numbers

23. The monthly sales of swimming suits over a one-year period at a store in San Diego

24. The monthly sales of milk over a one-year period at a large grocery store

25. The salaries of major league baseball players

26. The batting averages of major league baseball players

Web Projects

Find useful links for Web Projects on the text Web site: www.aw.com/bennett-briggs

1. **Salary Data.** Many Web sites offer data on salaries in different careers. Find salary data for a career you are considering. What are the mean and median salaries for this career? How do these salaries compare to those of other careers that interest you?

2. **New York Marathon.** The Web site for the New York marathon gives frequency data for finish times in the most recent marathon. Study the data, make a rough sketch of the distribution, and describe the shape of the distribution in words.

3. **Tax Statistics.** The IRS Web site provides statistics collected from tax returns on income, refunds, and much more. Choose a set of statistics from this Web site and study the distribution. Describe the distribution in words, and discuss anything you learn that is relevant to national tax policies.

4. **Education Statistics.** Visit the Web site of the National Center for Education Statistics and view the most recent copy of the *Digest of Educational Statistics*. Choose one of the many tables or graphs in the report that display a distribution. Describe the variables that are presented, discuss the effectiveness of the presentation, and interpret the results.

CHAPTER 5 SUMMARY

UNIT	KEY TERMS	KEY IDEAS AND SKILLS
5A	statistics *is* a science *are* data population, sample population parameters, sample statistics bias observational study case-control study experiment placebo, placebo effect blinding single-blind double-blind margin of error confidence interval	**Understand and interpret** the five basic steps in a statistical study. **Understand** the importance of a representative sample. **Be familiar with** four common sampling methods: simple random sampling systematic sampling convenience sampling stratified sampling **Distinguish between** observational studies and experiments; also recognize observational case-control studies. **Understand** the placebo effect and the importance of blinding in experiments. **Find** a confidence interval from a margin of error: from (sample statistic $-$ margin of error) to (sample statistic $+$ margin of error)
5B	selection bias participation bias variable (in a statistical study)	**Understand and apply** eight guidelines for evaluating a statistical study.
5C	frequency table categories frequencies data types qualitative quantitative bar chart pie chart histogram line chart time-series diagram	**Interpret and create** frequency tables. **Interpret and create** bar graphs and pie charts. **Interpret and create** histograms and line charts.
5D	multiple bar graph stack plot contour map	**Interpret** multiple bar graphs, stack plots, contour maps, and other media graphs. **Distinguish between** true three-dimensional data and graphs that have a three-dimensional look for cosmetic reasons only. **Be aware of** common cautions about graphs.

(continued)

5E	correlation cause scatter diagram	**Distinguish between** correlation and causality. **Create and interpret** scatter diagrams and use them to identify correlations: positive, negative, or no correlation strength of correlation **Know** three possible explanations for a correlation: coincidence common underlying cause true cause **Understand and apply** six guidelines for establishing causality.
5F	distribution mean median mode outlier symmetric distribution skewed (left or right) distribution variation	**Measures of center in a distribution:** $\text{Mean} = \dfrac{\text{sum of all values}}{\text{total number of values}}$ Median is the middle value. Mode is the most common value. **Understand** effects of outliers. **Be aware of** potential confusion with the term *average*. **Characterize** the shape of a distribution by number of peaks symmetric, left-skewed, or right-skewed variation

6

You can take it as understood
That your luck changes only if it's good.
—Ogden Nash, *Roulette Us Be Gay*

PROBABILITY: LIVING WITH THE ODDS

Probability is involved in virtually every decision we make. Sometimes the role of probability is clear, as in deciding whether to buy a lottery ticket or whether to plan a picnic based on the probability of rain. In other cases, probability guides decisions on a deeper level. For example, you might choose a particular college because you believe that it is most likely to meet your personal needs. In this chapter, we will look at just how practical and powerful probability can be in our everyday lives.

UNIT 6A Fundamentals of Probability: We will examine basic concepts of probability and three ways to determine probabilities: theoretically, empirically, and subjectively.

UNIT 6B Combining Probabilities: Using basic rules for adding and multiplying probabilities, we will calculate probabilities for more complex and interesting events.

UNIT 6C The Law of Large Numbers: We cannot predict a single coin toss, but the law of large numbers allows us to make surprisingly accurate statements about thousands of coin tosses. We will study this idea and its application to lotteries, insurance, and more.

UNIT 6D Counting and Probability: We will study permutations and combinations and apply them to probability and to exploring coincidences.

UNIT 6A

FUNDAMENTALS OF PROBABILITY

Probability plays such a huge role in our lives that it is essential to understand how it works. While you probably understand the general meaning of *probability*, you may never have studied how probabilities are formally calculated. In this unit, we'll discuss fundamental concepts required for probability calculations. Then, in subsequent units, we'll build on these ideas to see some of the remarkable uses of probability.

Let's begin by considering a toss of two coins. As shown in Figure 6.1, there are four different ways the coins could fall. We say that each of these four ways is a different **outcome** of the coin toss; that is, the outcomes are the most basic possible results of the coin toss. But suppose we are interested only in the number of heads. Because the two middle outcomes in Figure 6.1 each have 1 head, we say that these two outcomes represent the same **event.** Similarly, the right outcome represents the event of tossing 2 heads and the left outcome represents the event of tossing 0 heads. We see that an event describes one or more possible outcomes that all have the same property of interest—in this case, the same number of heads. Figure 6.1 shows that there are four possible outcomes for the two-coin toss, but only three possible events: 0 heads, 1 head, and 2 heads.

FIGURE 6.1 The four possible outcomes for a toss of two coins. The two middle outcomes both represent the same event of 1 head.

> **DEFINITIONS**
>
> In probability, **outcomes** are the most basic possible results of observations or experiments. An **event** consists of one or more outcomes that share a property of interest.

Mathematically, we express probabilities with numbers between 0 and 1. For example, the probability of a coin landing on heads is one-half, or 0.5. If an event is impossible, such as the event of meeting a married bachelor, we assign it a probability of 0. At the other extreme, an event that is certain to occur has a probability of 1. The probability of the Sun rising tomorrow is essentially 1. We use the notation $P(\text{event})$ to mean the probability of an event. We often denote events by letters or symbols. For example, using H to represent the event of heads on a coin toss, we write $P(H) = 0.5$.

The false ideas prevalent among all classes of the community respecting chance and luck illustrate the truth that common consent argues almost of necessity of error.

—Richard Proctor, *Chance and Luck* (1887 textbook)

> **Expressing Probability**
> The probability of an event, expressed as $P(\text{event})$, is always between 0 and 1 (inclusive). A probability of 0 means the event is impossible and a probability of 1 means the event is certain.

Time out to think

Suppose you are considering families of three children. List the different outcomes (birth orders) that result in the *event* of two boys and one girl. How might you express this probability with the notation introduced above?

With these basic definitions and notation, we are ready to discuss the three basic techniques for finding probabilities. They go by the names *theoretical method*, *empirical method*, and *subjective method*. We'll explore each in turn.

THEORETICAL PROBABILITIES

When we say that the probability of heads on a coin toss is 1/2, we are *assuming* that the coin is fair, so it is equally likely to land heads or tails. In essence, the probability is based on a *theory* of how the coin behaves, so we say that the probability of 1/2 comes from the **theoretical method**. As another example, consider rolling a single die. Because there a six equally likely outcomes for a rolled die (Figure 6.2), the theoretical probability for each outcome is 1/6. For example, the theoretical probability of rolling a 3 is 1/6. As long as all outcomes are equally likely, we can use the following procedure to calculate theoretical probabilities.

TECHNICAL NOTE
Except when stated otherwise, we assume that dice are fair, six-sided cubes with face values of 1 through 6.

FIGURE 6.2 The six possible outcomes for a roll of one die.

> **Theoretical Method for Equally Likely Outcomes**
>
> **Step 1:** Count the total number of possible outcomes.
>
> **Step 2:** Among all the possible outcomes, count the number of ways the event of interest, *A*, can occur.
>
> **Step 3:** Determine the probability, $P(A)$, by
>
> $$P(A) = \frac{\text{number of ways } A \text{ can occur}}{\text{total number of outcomes}}$$

by the way...
Theoretical methods are also called *a priori* methods. The words *a priori* are Latin for "before the fact" or "before experience."

> **by the way...**
> In fact, births are not evenly distributed over the twelve months. Even when the different lengths of months are accounted for, the lowest birth rate is in January and the highest rates are in June and July.

EXAMPLE 1 Guessing Birthdays

Suppose you select a person at random from a large conference group. What is the probability that the person has a birthday in July? Assume 365 days in a year.

Solution If we assume that births are equally likely to occur on any day of the year, then we can use the three-step theoretical method.

Step 1: Each possible birthday represents an outcome, so there are 365 possible outcomes.

Step 2: July has 31 days, so 31 of the 365 total outcomes represent the event of a July birthday.

Step 3: The probability that a randomly selected person has a July birthday is

$$P(\text{July birthday}) = \frac{31}{365} \approx 0.085$$

which is slightly more than 1 in 12.

Time out to think

Continuing Example 1, find the probability of selecting a person with a July 4 birthday.

EXAMPLE 2 Two Boys and a Girl

What is the probability that a family with three children has two boys (and one girl)? Assume boys and girls are equally likely.

Solution We apply the theoretical method for equally likely outcomes.

Step 1: There are two possible outcomes for each birth: boy or girl. For a family with three children, the total number of possible outcomes (birth orders) is $2 \times 2 \times 2 = 8$ (see the Brief Review of the multiplication principle on p. 307). The possible birth orders are BBB, BBG, BGB, BGG, GBB, GBG, GGB, and GGG.

Step 2: Of these eight possible outcomes, three have two boys; they are the three birth orders BBG, BGB, and GBB.

Step 3: The probability that a family with three children has two boys (and one girl) is

$$P(2 \text{ boys}) = \frac{3}{8} = 0.375$$

In words, the probability that a family with three children has two boys is 3/8, or 0.375.

EMPIRICAL PROBABILITIES

Suppose that, from historical and geological records, we discover that a river has risen above flood level 9 times in the past 1000 years. We might reasonably claim that the probability of such a flood in any given year is about 9 in 1000, or 0.009. Note that this probability is not based on any theoretical ideas; instead, it is based on observations. When we determine a probability based on observations or experiments, we

> *He that leaves nothing to chance will do few things ill, but will do very few things.*
> —GEORGE SAVILE HALIFAX

A brief review...

THE MULTIPLICATION PRINCIPLE

Let's review an important method for counting outcomes. Suppose we toss two coins and want to count the total number of possible outcomes. Individually, each coin has two possible outcomes: heads (H) and tails (T). Together, the two coins have $2 \times 2 = 4$ possible outcomes: HH, HT, TH, and TT (see Figure 6.1).

We can extend this thinking to additional coins. For example, if we toss three coins, the total number of possible outcomes is $2 \times 2 \times 2 = 8$, as shown in Figure 6.3.

We can generalize these examples with the following rule, often called the **multiplication principle.**

The Multiplication Principle

Suppose there are M possible outcomes for one process and N possible outcomes for a second process. The total number of possible outcomes for the two processes combined is $M \times N$. This idea extends to any number of processes. For example, if a third process has R possible outcomes, the total number of possible outcomes for the three processes combined is $M \times N \times R$.

Example: How many outcomes are possible when you roll two dice?
Solution: A single die has six possible outcomes: the numbers 1 through 6 (see Figure 6.2). A second die also has six possible outcomes. The total number of possible outcomes for two dice is $6 \times 6 = 36$.

Example: A restaurant menu offers two choices for an appetizer, five choices for a main course, and three choices for a dessert. How many different three-course (appetizer, main course, dessert) meals are possible?
Solution: There are two possible outcomes when selecting an appetizer, five possible outcomes when selecting a main course, and three possible outcomes when selecting dessert. The total number of possible outcomes is $2 \times 5 \times 3 = 30$; that is, 30 different three-course meals are possible.

Example: A college offers 12 natural science classes, 15 social science classes, 10 English classes, and 8 fine arts classes. You must take one class from each category to meet your core requirements. How many possible ways are there to meet the core requirements?
Solution: The selection of a course from a category represents a distinct process. Thus, we find the total number of ways to meet the requirements by multiplying the number of choices in each category: $12 \times 15 \times 10 \times 8 = 14,400$. There are 14,400 different ways to choose your courses to meet the requirements.

Coin 1	Coin 2	Coin 3	Outcome
H	H	H	HHH
H	H	T	HHT
H	T	H	HTH
H	T	T	HTT
T	H	H	THH
T	H	T	THT
T	T	H	TTH
T	T	T	TTT

FIGURE 6.3 The first column shows the two possible outcomes for the first coin. Each of these two outcomes branches into two possible outcomes for the second coin (second column), which in turn branch into two possible outcomes for the third coin (third column). The last column shows the eight possible outcomes.

by the way...

The term "100-year flood" is often used to describe a flood of a magnitude that occurs only once every 100 years, on average. Because this is a probability, it is possible (though rare) to have 100-year floods in two or more consecutive years.

use the **empirical method.** Because we could also say that the relative frequency of years with floods is 9 in 1000, the empirical method is sometimes called the *relative frequency method*.

Besides being useful when there is no theoretical basis for finding a probability, the empirical method is useful for checking the validity of theoretical probabilities. For example, suppose you toss a coin 100 times and get 75 heads. In this case, the empirical probability for heads is 75/100, or 0.75, which is significantly higher than the theoretical probability of 0.5 for a fair coin. You therefore have reason to suspect that the theoretical probability does not apply and that the coin is unfair.

EXAMPLE 3 500-Year Flood

Geological records indicate that a river has crested above flood level just four times in the past 2000 years. What is the empirical probability that the river will crest above flood level next year?

Solution Based on the observation of 4 floods in 2000 years, the empirical probability is

$$\frac{4}{2000} = \frac{1}{500} = 0.002$$

The empirical probability of a flood next year is 1 in 500, or 0.002. Because a flood of this magnitude occurs on average once every 500 years, it could be called a "500-year flood."

EXAMPLE 4 Empirical Coin Tossing

If you are interested only in the number of heads in coin tosses, then the possible events when tossing two coins are 0 heads, 1 head, and 2 heads (see Figure 6.1). Suppose you repeat a two-coin toss 100 times and your results are as follows:

- 0 heads occurs 22 times.
- 1 head occurs 51 times.
- 2 heads occurs 27 times.

Compare the empirical probabilities to the theoretical probabilities. Do you have reason to suspect that the coins are unfair?

Solution Let's begin by finding the theoretical probabilities for two tossed coins. Figure 6.1 shows that there are four possible outcomes for the two coins. Only 1 of these 4 outcomes has 0 heads, so the theoretical probability for the event of 0 heads is 1/4, or 0.25. Similarly, only 1 of the 4 outcomes has 2 heads, so the theoretical probability for the event of 2 heads is also 0.25. Finally, 2 of the 4 outcomes have 1 head, so this event has a probability of 2/4, or 0.5.

We find the empirical probabilities by dividing the number of times each event occurred by the total of 100 tosses. The event of 0 heads occurred 22 times, so its empirical probability is 22/100, or 0.22. Similarly, the empirical probabilities for the events of 1 head and 2 heads are 0.51 and 0.27, respectively. Because each empirical probability is quite close to the expected theoretical probability, we have no reason to suspect that the coins are unfair.

The principal means for ascertaining truth—induction and analogy—are based on probabilities; so that the entire system of human knowledge is connected with the theory of probability.

—PIERRE SIMON, Marquis de Laplace (1819)

SUBJECTIVE PROBABILITIES

The third method for determining probabilities is to make a **subjective estimate** using experience or intuition. For example, you could make a subjective estimate of the probability that a friend will be married in the next year. Weather forecasts are often subjective probabilities based on the experience of the forecaster combined with current weather data. There is not much we can say mathematically about subjective probabilities because their validity depends on personal experience and judgment.

Three Types of Probability

A **theoretical probability** is based on a model in which all outcomes are equally likely. It is determined by dividing the number of ways an event can occur by the total number of possible outcomes.

An **empirical probability** is based on observations or experiments; it is the relative frequency of the event of interest.

A **subjective probability** is an estimate based on experience or intuition.

by the way...
Another approach to finding probabilities, called the *Monte Carlo method,* uses computer simulations. This technique essentially finds empirical probabilities, but the experiments are simulations in a computer rather than actual experiments in a lab.

EXAMPLE 5 Which Type of Probability?

Identify the method that resulted in the following statements.

 a. I'm 100% certain that you'll be happy with this car.
 b. Based on data from the Department of Transportation, the chance of dying in an automobile accident is 1 in 7000 (per year).
 c. The chance of rolling a 7 with a twelve-sided die is 1/12.

Solution

 a. This is a subjective probability based on the opinion of the speaker.
 b. This is an empirical probability because it is based on observations recorded in the data set.
 c. This is a theoretical probability because it is based on a model of a fair twelve-sided die that is equally likely to land on any of its twelve sides.

PROBABILITY OF AN EVENT *NOT* OCCURRING

Suppose we are interested in the probability that a particular event or outcome does *not* occur. For example, suppose we want to know the probability that a 4 does *not* appear on one roll of a die. We can start by finding the probability of getting a 4: There are 6 possible outcomes for the roll of the die (the numbers 1 through 6), so the probability of rolling a 4 is $P(4) = 1/6$. Now, note that it is certain (which means that the probability is 1) that either we roll a 4 or we do not roll a 4. We can write this fact as

$$P(4) + P(\text{not } 4) = 1$$

historical note
As early as 3600 B.C., rounded bones called *astragali* were used much like dice for games of chance in the Middle East. Our familiar cubical dice appeared about 2000 B.C. in Egypt and in China. Playing cards were invented in China in the 10th century and brought to Europe in the 14th century.

We can use this fact to find the probability that a 4 does not occur:

$$P(\text{not } 4) = 1 - P(4) = 1 - \frac{1}{6} = \frac{5}{6}$$

This answer makes sense because there are five possible events besides 4 (rolls of 1, 2, 3, 5, and 6). We can generalize this observation for any event.

> **Probability of an Event *Not* Occurring**
>
> Suppose the probability of an event *A* is *P(A)*. Then the probability that event *A* does *not* occur is 1 − *P(A)*.

EXAMPLE 6 Not Two Boys

What is the probability that a randomly chosen family with three children does *not* have two boys (and one girl)? Assume boys and girls are equally likely.

Solution In Example 2, we found that the probability of two boys in a family with three children is 3/8, or 0.375. Therefore, the probability that a family with three children does *not* have two boys is 1−3/8 = 5/8, or 0.625.

PROBABILITY DISTRIBUTIONS

Consider again the possible outcomes and events for tossing two coins. As shown in Figure 6.1, there are four possible outcomes, each with a probability of 1/4. But if we are interested only in the total number of heads, two of the four outcomes—the outcomes HT and TH—represent the same event of 1 head. Therefore, the event of 1 head has a probability of 2/4, or 1/2. Table 6.1 summarizes the probabilities of the three possible events of 0 head, 1 head, and 2 heads. This type of table shows a **probability distribution,** which is a display of the probability for *every possible event*. Note that the sum of all the probabilities must be 1, because it is certain that one of the events will occur when two coins are tossed.

TABLE 6.1 Tossing Two Coins

EVENT	PROBABILITY
2 heads, 0 tails	0.25
1 head, 1 tail	0.50
0 heads, 2 tails	0.25
Total	**1**

> **Making a Probability Distribution**
>
> A **probability distribution** represents the probabilities of all possible events. To make a table of a probability distribution:
>
> **Step 1:** List all possible *outcomes;* use a table or figure if it is helpful.
>
> **Step 2:** Identify outcomes that represent the same *event*. Find the probability of each event (using the theoretical method).
>
> **Step 3:** Make a table in which one column lists each event and another column lists each probability. The sum of all the probabilities must be 1.

EXAMPLE 7 Tossing Three Coins

Make a table of the probability distribution for the number of heads that occur when three coins are tossed simultaneously.

Solution We apply the three-step process.

Step 1: The diagram in Figure 6.3 (in the Brief Review on p. 307) shows the eight possible outcomes when three coins are tossed. Each outcome is equally likely and has a probability of 1/8.

Step 2: Several of the outcomes represent the same event. For example, the outcomes HTT, THT, and TTH all represent the event of 1 head and 2 tails. This event therefore has a probability of 3/8. Similarly, the outcomes HHT, HTH, and THH all represent the event of 2 heads and 1 tail. This event also has a probability of 3/8.

Step 3: There are four possible events: 0 heads, 1 head, 2 heads, and 3 heads. We have their probabilities from steps 1 and 2. Table 6.2 shows the complete probability distribution.

TABLE 6.2 Tossing Three Coins

RESULT	PROBABILITY
3 heads, 0 tails	1/8
2 heads, 1 tail	3/8
1 head, 2 tails	3/8
0 heads, 3 tails	1/8
Total	**1**

Time out to think

When four coins are tossed, how many different outcomes are there? If you record the number of heads, how many different events are possible?

EXAMPLE 8 Two Dice Distribution

Make a probability distribution for the sum of the dice when two dice are rolled. What is the most probable sum?

Solution Because there are six ways for each die to land, there are 6 × 6 = 36 outcomes of rolling two dice. We can enumerate all 36 outcomes by making a table in which we list one die along the rows and one die along the columns. Each of the 36 outcomes is equally likely, so each has a probability of 1/36. For each outcome, the table also shows the sum of the numbers on the two dice.

	1	2	3	4	5	6
1	1 + 1 = 2	1 + 2 = 3	1 + 3 = 4	1 + 4 = 5	1 + 5 = 6	1 + 6 = 7
2	2 + 1 = 3	2 + 2 = 4	2 + 3 = 5	2 + 4 = 6	2 + 5 = 7	2 + 6 = 8
3	3 + 1 = 4	3 + 2 = 5	3 + 3 = 6	3 + 4 = 7	3 + 5 = 8	3 + 6 = 9
4	4 + 1 = 5	4 + 2 = 6	4 + 3 = 7	4 + 4 = 8	4 + 5 = 9	4 + 6 = 10
5	5 + 1 = 6	5 + 2 = 7	5 + 3 = 8	5 + 4 = 9	5 + 5 = 10	5 + 6 = 11
6	6 + 1 = 7	6 + 2 = 8	6 + 3 = 9	6 + 4 = 10	6 + 5 = 11	6 + 6 = 12

The possible *events* are the sums from 2 to 12. We find the probability of each event by counting all the outcomes for each sum and then dividing the number of outcomes by 36. For example, the five highlighted outcomes in the table have a sum of 8, so

TABLE 6.3	Probability Distribution for the Sum of Two Dice											
EVENT (SUM)	2	3	4	5	6	7	8	9	10	11	12	Total
PROBABILITY	1/36	2/36	3/36	4/36	5/36	6/36	5/36	4/36	3/36	2/36	1/26	1

the probability of a sum of 8 is 5/36. Table 6.3 shows the complete probability distribution. The most likely event is a sum of 7, which has a probability of 6/36, or 1/6.

STATING THE ODDS

You've probably noticed that there are several ways to express the idea of *probability*. For example, saying that the *chance* of getting heads on a coin toss is 1/2 is equivalent to saying that the *probability* is 1/2. Another common term for probability is *odds*, but strictly speaking the two terms have different meanings. Odds are often stated as the *ratio* of the probability that a particular event occurs to the probability that it does not occur.

DEFINITION

The **odds** for an event A are

$$\text{odds for event } A = \frac{P(A)}{P(\text{not } A)}$$

For example, the probability of a coin landing heads is 1/2 and the probability of a coin *not* landing heads also is 1/2. Thus, by this definition, the odds for heads are

$$\text{odds for heads} = \frac{P(\text{heads})}{P(\text{not heads})} = \frac{1}{1}$$

This result is usually phrased as "The odds for a coin landing heads are 1 to 1, or even."

In gambling, the term *odds* usually expresses how much a bet will pay if you win. For example, suppose the odds on a particular horse at a horse race are 3 to 1. Then for each $1 you bet on this horse, you will gain $3 if the horse wins. (If you place a $1 bet and the horse wins, you will receive $3 plus your $1 bet.) If you place a $2 bet and the horse wins, you will gain 3 × $2 = $6.

EXAMPLE 9 *Two Coin Odds*

What are the odds for getting two heads in tossing two coins?

Solution As we found earlier, the probability of getting two heads in tossing two coins is 1/4; that is, $P(2 \text{ heads}) = 1/4$. The probability of not getting two heads in tossing two coins is $P(\text{not } 2 \text{ heads}) = 1 - 1/4 = 3/4$. Thus, the odds for getting 2 heads are

$$\text{odds for 2 heads} = \frac{P(2 \text{ heads})}{P(\text{not } 2 \text{ heads})} = \frac{1/4}{3/4} = \frac{1}{3}$$

The odds for getting 2 heads in two coin tosses are 1 to 3. We could also express the odds by turning them around: The odds *against* getting 2 heads in tossing two coins are 3 to 1.

EXAMPLE 10 Horse Race Payoff

At a horse race, the odds for Blue Moon are given as 7 to 2. If you bet $10 and Blue Moon wins, how much will you gain?

Solution The odds of 7 to 2 mean that, for each $2 you bet on Blue Moon, you gain $7 if you win. A $10 bet is equivalent to five $2 bets, so you gain 5 × $7 = $35. You also get your $10 back, so you will receive $45 when you collect on your winning ticket.

by the way ...
What's a blue moon? Many people now say that a blue moon is the second full moon in a calendar month. But the editors of *Sky and Telescope* recently discovered that this definition came from a mistake in a 1946 issue of their magazine. No one knows the original definition of blue moon, but the phrase "once in a blue moon" has come to mean an extremely low probability of occurrence.

Review Questions

1. Distinguish between an outcome and an event in probability. Give an example in which the same event can occur through two or more outcomes.
2. Explain the notation $P(\text{event})$. Why must the probability always be between 0 and 1 (inclusive)?
3. Briefly describe the differences among theoretical, empirical, and subjective techniques for finding probabilities.
4. Give an example that illustrates the three-step process for calculating theoretical probabilities.
5. How is the probability of an event *not* occurring related to the probability that it *does* occur? Why?
6. What is a probability distribution? Explain how to make a table of a probability distribution.
7. Explain the common usage of the term *odds* and its usage in gambling.

In the News

1. **Probability in the News.** Find a news article or research report that makes use of a probability. Interpret the probability and discuss whether it is theoretical, empirical, or subjective.
2. **Probability in Your Life.** Describe a recent instance from your own life when you used probability to make a decision. What type of probability did you use? How did it help you make the decision?
3. **Gambling Odds.** Find an advertisement from a gambling establishment or lottery in which the term *odds* is used, and interpret its meaning.

Exercises

Review of the Multiplication Principle. Exercises 1–2 use skills covered in the Brief Review on p. 307.

1. a. How many different choices of car do you have if a particular model comes in eight colors and three styles (sedan, station wagon, or hatchback)?
 b. Your car radio has five preset buttons, each of which can be tuned to either an AM or an FM station. How many different stations can you preset?
 c. The local paint store offers wallpaper in eight colors, each of which comes in four different patterns. How many different styles of wallpaper are available?
 d. You have a choice of six TV sets that can be matched with any of five VCR players. How many different TV/VCR systems can you assemble?

2. a. A local ski sale features eight types of skis, six types of bindings, and seven types of boots. How many different ski/binding/boot packages are available?
 b. Three districts elect representatives to the County Commission. There are two candidates in District 1, three candidates in District 2, and 4 candidates in District 3. How many different Commissions are possible?

c. You are required to take five courses, one each in humanities, sociology, science, math, and music. You have a choice of four humanities courses, three sociology courses, five science courses, two math courses, and three music courses. How many different sets of five courses are possible?

d. The car model you are considering comes with or without air conditioning, with or without a sun roof, with or without a CD-ROM player, and in eight stock colors. How many different versions of the car are available?

Theoretical Probabilities. In Exercises 3–10, use the theoretical method to determine the probability of the given outcomes and events. State any assumptions that you need to make.

3. Meeting someone with a phone number that ends in 0
4. Meeting someone born between midnight and 1:00 a.m.
5. Randomly selecting a two-child family with two girl children
6. Randomly selecting a three-child family with exactly two girl children
7. Rolling two dice and getting a sum of 5
8. Selecting a blue M&M from a bag containing 10 red M&Ms, 15 blue M&Ms, and 20 yellow M&Ms
9. Selecting a pair of blue socks from a drawer that holds 4 pairs of black socks, 6 pairs of blue socks, and 12 pairs of white socks
10. Guessing a correct answer on a multiple-choice question with five possible answers

Empirical Probabilities. Use the empirical method to estimate the probabilities in Exercises 11–16.

11. After recording the forecasts of your local weatherman for 30 days, you conclude that he gave a correct forecast 12 times. What is the probability that his next forecast will be correct?
12. What is the probability of a 100-year flood this year?
13. Halfway through the season, a basketball player has hit 86% of her free throws. What is the probability that her next free throw will be successful?
14. You count 66 tails when you toss a coin 100 times. If you don't know whether the coin is fair, what is the probability that the next toss will be a head?
15. A doctor diagnosed pneumonia in 100 patients and 34 of them actually had pneumonia. What is the probability that the next patient diagnosed with pneumonia will actually have pneumonia?
16. What is the probability of a 200-year flood this year?

Event *Not* Occurring. Determine the probability of the events in Exercises 17–24. State any assumptions that you use.

17. What is the probability of not rolling a 4 with a fair die?
18. What is the probability of not tossing two heads with two fair coins?
19. What is the probability that a 76% free-throw shooter will miss his next free throw?
20. What is the probability of meeting a person not born in January, February, or March?
21. What is the probability of not rolling a sum of 7 with two fair dice?
22. What is the probability of a sunny day after a forecast of a 20% chance of rain?
23. What is the probability of not rolling an odd number with a fair die?
24. What is the probability of not rolling a double-6 with two fair dice?
25. **Gender Politics.** The following table gives the gender and political party of the 100 delegates at a political convention. Suppose you encounter a delegate at random. What is the probability that the delegate will *not* be a man? What is the probability that the delegate will *not* be a Republican?

	Women	Men
Republicans	21	28
Democrats	25	16
Independents	6	4

26. **Senior Citizen.** In 2000, there were 34.7 million people over 65 years of age out of a U.S. population of 274 million. By 2050, it is estimated that there will be 78.9 million people over 65 years of age out of a U.S. population of 394 million. Would your chances of meeting a person over 65 at random be greater in 2000 or 2050? Explain.

Probability Distributions. In Exercises 27–28, make a probability distribution for the given set of events.

27. The number of heads when four fair coins are tossed
28. The sums that appear when two fair four-sided dice (tetrahedrons) are tossed

29. **Fair Coins?** You toss three coins 1000 times; the following table shows the results. Compute the empirical probability for each outcome. Compare the empirical probabilities to the theoretical probabilities for three tossed coins. Do you have reason to believe the coins are unfair? Explain.

Result	Number of Occurrences
0 heads	260
1 head	495
2 heads	245
3 heads	0
Total	1000

30. **Marriage Age.** The following table gives percentages of women and men married for the first time in several age categories. What is the probability that a randomly encountered married woman was married, for the first time, between the ages of 35 and 44? Make a probability distribution for ages of first marriage for men and women.

	Women	Men
Under 20	16.6	6.6
20–24	40.8	36.0
25–29	27.2	34.3
30–34	10.1	14.8
35–44	4.5	7.1
45–64	0.7	1.1
65 and over	0.1	0.1

The Odds. Use the definition given in the text to find the *odds* of the events in Exercises 31–34.

31. Rolling a fair die and getting a 1 or a 2
32. Flipping two fair coins and getting two tails
33. Rolling a fair die and getting a 5 or a 6
34. Flipping two fair coins and getting a head and a tail

Gambling Odds. Use the definition of odds in betting for Exercises 35–36.

35. Suppose that you are given odds of 3 to 4 on a bet. If you bet $20 and win, how much will you gain?
36. Suppose that you are given odds of 5 to 4 on a bet. If you bet $20 and win, how much will you gain?
37. **Deceptive Odds.** Suppose event *A* has a 0.99 probability of occurring and event *B* has a 0.96 probability of occurring—both high probabilities. Compute the odds for event *A* and the odds for event *B*. Comment on the relative difference between the odds for the two events compared to the relative difference between the probabilities. How are the odds deceptive in this case?

38. **Project: Thumb Tack Probabilities.** Find a standard thumb tack and practice tossing it onto a flat surface. Notice that there are two different outcomes: The tack can land with its point facing down or with its point facing up.

 a. Toss the tack 50 times and record the outcomes.
 b. Give the empirical probabilities of the two outcomes based on these results.
 c. If possible, ask several other people to repeat the process. How well do your probabilities agree?

39. **Project: Three-Coin Experiment.** Toss three coins (at once) 50 times and record the outcomes in terms of the number of heads. Based on your observations, give the empirical probability of each result. Do the empirical results agree with the theoretical probabilities? Explain.

40. **Project: Randomizing a Survey.** Suppose you want to conduct a survey involving a sensitive question that not all participants may choose to answer honestly (for example, a question involving cheating on taxes or drug use). Here is a way to conduct the survey and protect the identity of respondents. We will assume that the sensitive question requires a YES or NO answer. First ask all respondents to toss a fair coin. Then give the following instructions:

 • If you toss a head, then answer the decoy question: Were you born on an even day of the month? (YES/NO).
 • If you toss a tail, then answer the real question of the survey (YES/NO).

 Ask all participants to answer YES or NO to the question they were assigned; then count the total numbers of YES and NO responses.

 a. Choose a particular question that may not produce totally honest responses. Conduct a survey in your class using this technique.
 b. Given only the total numbers of YES and NO responses to both questions, explain how you can estimate the numbers of people who answered YES and NO to the real question.
 c. Will the results computed in part b be exact? Explain.
 d. Suppose the decoy question was replaced by these instructions: If you toss a head, then answer YES. Can you still determine the number of people who answered YES and NO to the real question?

Web Projects

Find useful links for Web Projects on the text Web site: www.aw.com/bennett-briggs

1. **Blood Groups.** The four major blood groups are designated A, B, AB, and O. Within each group there are two Rh types: positive and negative. Using library resources or the World Wide Web, find data on the relative frequency of blood groups, including the Rh types. Make a table showing the probability of meeting someone in each of the eight blood groups.

2. **Age and Gender.** The proportions of men and women in the population change with age. Using current data from a Web site, construct a table showing the probability of meeting a male or a female in each of these age categories: 0–5, 6–10, 11–20, 21–30, 31–40, 41–50, 51–60, 61–70, 71–80, and over 80.

3. **Accidents.** Use the Web to obtain data to estimate the empirical probabilities of accidents (such as being killed by lightning, or by a shark bite.) Are the empirical probabilities consistent with your intuition?

UNIT 6B

COMBINING PROBABILITIES

historical note

Chevalier was a title of low-ranking nobility in France. The Chevalier de Mère was a man by the name of Antoine Gombaud.

In 1654, a French con man known as the Chevalier de Mère bet other gamblers even money that he could roll at least one 6 in four rolls of a standard six-sided die. The Chevalier made quite a profit on this bet, despite having calculated his chance of winning incorrectly!

Because the probability of rolling a 6 on one roll of a die is 1/6, the Chevalier falsely reasoned that his chances would be four times as great on four rolls, or 4/6 = 0.67. In fact, the chance of rolling a 6 within four tries is only 0.52 (as we will show shortly). Nevertheless, because a probability of 0.52 means winning more often than losing, the Chevalier profited over the long run.

As gamblers caught on to this "con game," they were no longer willing to play. The Chevalier therefore proposed a new game in which he rolled *two* dice and bet that he could throw a *double-6* within 24 rolls. He knew that, on any single roll of the two dice, the probability of a double-6 is 1/36 (see Unit 6A, Example 8). He then used his faulty reasoning to guess that, if he rolled the pair of dice 24 times, his odds of rolling at least one double-6 would be 24/36 = 0.67.

To his dismay, the Chevalier began losing money. To understand his losses, he turned to the mathematician Blaise Pascal, who, in turn, began a correspondence about probability with the mathematician Pierre de Fermat. Pascal and Fermat quickly recognized the error in the Chevalier's reasoning and developed methods for properly combining probabilities. In this unit, we will study some of these methods.

Time out to think

The Chevalier multiplied the single-roll probability of 1/6 by 4 rolls to get his guess that he had a 4/6 chance of rolling a 6 within 4 rolls. Based on this reasoning, what would he have guessed for the probability of rolling a 6 within 6 rolls? How does this show the error in his reasoning?

AND PROBABILITIES

Suppose you toss two dice and want to know the probability that *both* will come up 4. One way to find the probability is to consider the tossing of the two dice as a *single* toss of two dice. Then we can follow the three-step procedure for theoretical proba-

bilities from Unit 6A. Because a double-4 is one of 36 possible outcomes for two dice, its probability is 1/36.

Alternatively, we can consider the two dice individually. For each die, the probability of a 4 is 1/6. Rolling a double-4 means rolling a 4 on the first die *and* a 4 on the second die. We find the probability of a double-4 by multiplying the two individual probabilities:

$$P(\text{double 4}) = P(4) \times P(4) = \frac{1}{6} \times \frac{1}{6} = \frac{1}{36}$$

By either method, the probability of rolling a double-4 is 1/36. In general, we call the probability of event *A and* event *B* occurring an **and probability** (or *joint probability*).

The advantage of the multiplication technique is that it can easily be extended to other situations involving two or more events. For example, we can use it to find the probability of getting ten heads in a row on a coin toss or of having a baby *and* getting a pay raise in the same year. However, we must distinguish between events that are *independent* and events that are *dependent* upon each other. Let's investigate each case.

Independent Events

The repeated roll of a single die produces **independent events** because the outcome of one roll does not affect other rolls. Similarly, coin tosses are independent events because one coin toss does not affect others. For independent events, we calculate the *and* probability by multiplication.

> **And Probability: Independent Events**
>
> Two events are **independent** if the outcome of one does not affect the probability of the other event. Consider two independent events *A* and *B* with individual probabilities $P(A)$ and $P(B)$. The probability that *A* and *B* occur together is
>
> $$P(A \text{ and } B) = P(A) \times P(B)$$
>
> This principle can be extended to any number of independent events. For example, the *and* probability of three independent events *A*, *B*, and *C* is
>
> $$P(A \text{ and } B \text{ and } C) = P(A) \times P(B) \times P(C)$$

EXAMPLE 1 *Consecutive Floods*

Find the probability that a 100-year flood (a flood with a 0.01 probability of striking in a given year) will strike a city in two consecutive years. Assume that a flood in one year does not affect the likelihood of a flood in the next year.

Solution We are told to assume that the events of floods in different years are independent. Therefore, we find the probability of floods in two consecutive years by multiplying the probability for each individual year:

$$P(\text{2 consecutive flood years}) = \underbrace{P(\text{flood})}_{\text{year 1}} \times \underbrace{P(\text{flood})}_{\text{year 2}} = 0.01 \times 0.01 = 0.0001$$

historical note

Blaise Pascal was a precocious Parisian who made fundamental contributions to mathematics and physics. He was deeply religious and his work *Pensées* is an important treatise of western philosophy. Pascal died at the age of 39. Pierre de Fermat was a lawyer, but his passion was mathematics. He spent his life corresponding with mathematicians and preoccupied with difficult problems. One of his most famous problems, known as Fermat's last theorem, was solved only recently.

> **by the way ...**
> Whether events are independent is not always obvious. Consider a set of basketball free throws by a single player. Some people argue that the probability of a basket is the same for each free throw. Others suggest that the psychological impact of success or failure on one free throw affects the player's chances of making the next one. Despite many statistical studies of basketball free throws, no one has clearly established which view is correct.

The probability of a 100-year flood in two consecutive years is 1 in 10,000. While events with such low probability can certainly occur, remember that this calculation *assumes* that the chance of a flood is 0.01 in any year. If a 100-year flood were to occur in two consecutive years, we might wonder whether the 0.01 probability of a single flood was correct.

EXAMPLE 2 Three Coins

Suppose you toss three coins. What is the probability of getting three tails?

Solution Because coin tosses are independent, we multiply the probability of tails on each individual coin:

$$P(3 \text{ tails}) = \underbrace{P(\text{tails})}_{\text{coin 1}} \times \underbrace{P(\text{tails})}_{\text{coin 2}} \times \underbrace{P(\text{tails})}_{\text{coin 3}} = \frac{1}{2} \times \frac{1}{2} \times \frac{1}{2} = \frac{1}{8}$$

The probability of three tails on three tossed coins is 1/8.

Dependent Events

Suppose you draw candy randomly from a box that initially contains five chocolates and five caramels. Clearly, the probability of getting a piece of chocolate on your first try is 5/10, or 1/2. Now, suppose you get a chocolate on the first try and eat it. What is the probability of getting another piece of chocolate on your second draw?

Because you've already eaten one chocolate, the box now contains only nine pieces of candy, of which four are chocolate. Thus, the probability of getting a piece of chocolate on the second try is 4 out of 9, or 4/9. Note that this is *not* the same as the 1/2 probability on the first draw. Because the outcome of the first event affects the probability of the second event, we say that the events are **dependent events**.

Calculating the probability for dependent events still involves multiplying the individual probabilities, but we must take into account how prior events affect subsequent events. In the case of the candy box, we find the probability of getting two chocolates in a row by multiplying the 1/2 probability on the first draw by the 4/9 probability on the second draw:

$$P(2 \text{ chocolates}) = \underbrace{P(\text{chocolate})}_{\text{first draw}} \times \underbrace{P(\text{chocolate})}_{\text{second draw}} = \frac{1}{2} \times \frac{4}{9} = \frac{2}{9}$$

The probability of drawing two chocolates in a row is 2/9. We can generalize this example to any set of dependent events.

> **And Probability: Dependent Events**
>
> Two events are **dependent** if the outcome of one event affects the probability of the other event. The probability that dependent events A and B occur together is
>
> $$P(A \text{ and } B) = P(A) \times P(B \text{ given } A)$$
>
> where $P(B \text{ given } A)$ means "the probability of event B given the occurrence of event A." This principle can be extended to any number of dependent events. For example, the *and* probability of three dependent events A, B, and C is
>
> $$P(A \text{ and } B \text{ and } C) = P(A) \times P(B \text{ given } A) \times P(C \text{ given } A \text{ and } B)$$

TECHNICAL NOTE
Probabilities for dependent events are often called *conditional probabilities*. Some books use the notation $P(B|A)$ instead of $P(B \text{ given } A)$.

Time out to think

Suppose that instead of eating the first candy you draw, you put it back in the box. Without doing any calculations, compare the probability of drawing two chocolates in a row when you replace the candy in the box to the probability of drawing two chocolates in a row when you eat the first candy. Which probability is larger and why?

EXAMPLE 3 BINGO

The game of BINGO involves drawing labeled buttons, at random and without replacement, from a bin. There are 75 buttons, 15 for each of the letters B, I, N, G, and O. What is the probability of drawing two B buttons in the first two selections?

Solution BINGO involves dependent events because removing a button from the bin changes the contents of the bin. The probability of drawing a B on the first draw is 15/75 = 1/5. If this occurs, the bin has 74 buttons remaining, of which 14 are the letter B. Therefore, the probability of drawing a B button on the second draw is 14/74 = 7/37. The probability of drawing two B buttons in the first two selections is

$$P(\text{B and B}) = \underbrace{P(\text{B})}_{\text{first draw}} \times \underbrace{P(\text{B})}_{\text{second draw}} = \frac{15}{75} \times \frac{14}{74} \approx 0.0378$$

The probability of drawing the letter B on the first two selections is 0.0378, or just under 4 in 100.

EXAMPLE 4 Jury Selection

A three-person jury must be selected at random from a pool of 12 people that has 6 men and 6 women. What is the probability of selecting an all-male jury?

Solution Because there are 6 men in the pool of 12 people, the probability that the first juror is male is 6/12. If the first juror is male, the remaining pool has 5 men among 11 people. Therefore, the probability that the second juror is also male is 5/11.

> **by the way...**
> In the late 1960s, the famed baby doctor Benjamin Spock was convicted by an all-male jury of encouraging draft resistance during the Vietnam War. His defense argued that a jury with women would have been more sympathetic.

If both the first two jurors are male, the pool is left with 4 men among 10 people, making the probability of a third male juror 4/10. The combined probability is

$$P(3 \text{ men}) = \underbrace{P(\text{man})}_{\text{juror 1}} \times \underbrace{P(\text{man given male juror 1})}_{\text{juror 2}} \times \underbrace{P(\text{man given male jurors 1 and 2})}_{\text{juror 3}}$$

$$= \frac{6}{12} \times \frac{5}{11} \times \frac{4}{10} = \frac{120}{1320} \approx 0.091$$

The probability of selecting an all-male jury is roughly 0.09, or 9%.

EITHER/OR PROBABILITIES

Suppose we want to know the probability that *either* of two events occurs, rather than the probability that both events occur. In that case, we are looking for an **either/or probability,** such as the probability that either a #2 bus or a #9 bus will be the next to arrive at your stop. As with *and* probabilities, there are two cases to consider; we call them *overlapping* and *non-overlapping* events.

Non-Overlapping Events

A coin can land either heads *or* tails, but it can't land on both heads *and* tails at the same time. When two events cannot occur at the same time, they are said to be **non-overlapping** (or *mutually exclusive*) **events.** If we make a Venn diagram (see Unit 1C) in which the circles represent different events, non-overlapping events are those for which there is no overlap between the circles. For example, the circles for heads and tails in a coin toss do not overlap because a coin cannot land both heads and tails at the same time (Figure 6.4). We can calculate *either/or* probabilities for non–overlapping events with the following rule.

FIGURE 6.4 Venn diagram for non-overlapping events.

Either/Or Probability: Non-Overlapping Events

Two events are **non-overlapping** if they cannot occur together. If A and B are non-overlapping events, the probability that either A or B occurs is

$$P(A \text{ or } B) = P(A) + P(B)$$

This principle can be extended to any number of non-overlapping events. For example, the probability that either event A, event B, or event C occurs is

$$P(A \text{ or } B \text{ or } C) = P(A) + P(B) + P(C)$$

EXAMPLE 5 *Either/Or Dice*

Suppose you roll a single die. What is the probability of rolling either a 2 or a 3?

Solution The events of 2 and 3 are non-overlapping because a single die can land only one way. Each probability is 1/6 (because there are 6 ways for the die to land), so the combined probability is

$$P(2 \text{ or } 3) = P(2) + P(3) = \frac{1}{6} + \frac{1}{6} = \frac{2}{6}$$

The probability of rolling either a 2 or a 3 is 2/6, or 1/3.

Overlapping Events

Suppose you have a standard deck of 52 cards and want to know the probability of drawing either a queen or a club. Because there are 4 queens in the deck, the probability of drawing a queen is 4/52. Similarly, because there are 13 clubs in the deck, the probability of drawing a club is 13/52. The sum of the two probabilities is

$$\frac{4}{52} + \frac{13}{52} = \frac{17}{52}$$

However, this is *not* the probability of drawing a queen *or* a club. The Venn diagram in Figure 6.5 shows why. One circle represents the 4 queens in the deck and the other represents the 13 clubs in the deck. The circles overlap because one card—the queen of clubs—is both a queen *and* a club. That is, the events of drawing a queen and drawing a club are **overlapping events** because both can occur together. If we simply add the individual probabilities, as above, the queen of clubs gets counted twice (once as a queen and once as a club). Thus, we can correct the calculation by subtracting the double-counted probability of drawing the queen of clubs, which is 1/52 (because the queen of clubs is just 1 of the 52 cards in the deck), which gives us

FIGURE 6.5 Venn diagram for overlapping events. One circle represents queens in a deck of cards and the other represents clubs. The overlap region contains the queen of clubs, which belongs to both circles.

$$P(\text{queen or club}) = P(\text{queen}) + P(\text{club}) - P(\text{queen and club})$$
$$= \frac{4}{52} + \frac{13}{52} - \frac{1}{52} = \frac{16}{52}$$

The probability of drawing a queen or a club is 16/52. Note that you can also see this result from the Venn diagram, which shows 16 cards (out of 52 total) that are either a queen or a club (or both).

> *Either/Or* **Probability: Overlapping Events**
>
> Two events are **overlapping** if they *can* occur together. If *A* and *B* are overlapping events, the probability that either *A* or *B* occurs is
>
> $$P(A \text{ or } B) = P(A) + P(B) - P(A \text{ and } B)$$

EXAMPLE 6 *Democrats and Women*

You select one person at random from a room with eight people: two Democratic men, two Republican men, two Democratic women, and two Republican women. What is the probability that you will select *either* a woman or a Democrat?

Solution The probability of selecting a Democrat is 1/2, because half of the people in the room are Democrats. Similarly, half of the people are women, so the probability of selecting a woman is also 1/2. We cannot simply add these probabilities, because two of the eight people in the room are both Democrats and women. The probability

of choosing one of these two people is 2/8, or 1/4. Thus, the overall probability of selecting either a woman or a Democrat is

$$P(\text{woman or Democrat}) = P(\text{woman}) + P(\text{Democrat}) - P(\text{woman and Democract})$$
$$= \frac{1}{2} + \frac{1}{2} - \frac{1}{4} = \frac{3}{4}$$

The probability that you will select either a woman or a Democrat is 3/4, or 0.75.

THE *AT LEAST ONCE* RULE

Suppose you toss a coin four times. What is the probability of getting *at least one* head? There are four different ways to get at least one head: by getting 1 head, 2 heads, 3 heads, or 4 heads. Thus, one way to solve this problem is to add the individual probabilities:

$$P(\text{at least one H in 4 tosses}) = P(1\,H) + P(2\,H) + P(3\,H) + P(4\,H)$$

To complete this calculation, we'd need to make a probability distribution that showed the probability for each of the four ways of getting at least one head.

Fortunately, there is an easier way. Because you either *do* or *do not* get at least one head in the four coin tosses, the sum of these two probabilities is 1:

$$P(\text{at least one H in 4 tosses}) + P(\text{no heads in 4 tosses}) = 1$$

(Getting "no heads" means *not* getting at least one head.) Subtracting the probability of no heads from both sides, we find

$$P(\text{at least one H in 4 tosses}) = 1 - P(\text{no heads in 4 tosses})$$

It's easy to find the probability of no heads. The probability of not getting heads on one coin toss is $P(\text{not H}) = 1/2$, so the probability of no heads in four tosses is

$$P(\text{no H in 4 tosses}) = \underbrace{P(\text{not H})}_{\text{toss 1}} \times \underbrace{P(\text{not H})}_{\text{toss 2}} \times \underbrace{P(\text{not H})}_{\text{toss 3}} \times \underbrace{P(\text{not H})}_{\text{toss 4}} = [P(\text{not H})]^4$$

(The last step follows because $P(\text{not H})$ is the same for all four tosses.) By substituting $P(\text{not H}) = 1/2$, we find that the probability of at least one head in 4 coin tosses is

$$P(\text{at least one H in 4 tosses}) = 1 - P(\text{no H in 4 tosses})$$
$$= 1 - [P(\text{not H})]^4$$
$$= 1 - \left[\frac{1}{2}\right]^4 = 1 - \frac{1}{16} = \frac{15}{16}$$

We can generalize this example to any question of the form "What is the probability of *at least one* particular outcome in a series of independent events?"

> ### The *At Least Once* Rule (for Independent Events)
> Suppose the probability of an event A occurring in one trial is $P(A)$. If all trials are independent, the probability that event A occurs *at least once* in n trials is
> $$P(\text{at least one event } A \text{ in } n \text{ trials}) = 1 - P(\text{no events } A \text{ in } n \text{ trials})$$
> $$= 1 - [P(\text{not } A \text{ in one trial})]^n$$

EXAMPLE 7 100-Year Flood

What is the probability that a region will experience at least one 100-year flood (a flood that has a 0.01 chance of occurring in any given year) during the next 100 years? Assume that 100-year floods in consecutive years are independent events.

Solution Because there's a 0.01 probability of a flood in any one year, there is a 0.99 chance that a flood will *not* occur in any one year. The *at least once* rule gives us the probability of at least one flood in 100 years:

$$P(\text{at least one flood in 100 years}) = 1 - [P(\text{not flood in one year})]^{100}$$
$$= 1 - [0.99]^{100} \approx 0.634$$

There is a 0.634, or almost 2 out of 3, chance that a 100-year flood will occur at least once in the next 100 years.

Time out to think

Suppose a region has not experienced a 100-year flood in 100 years. Should you be surprised? Would this fact give you reason to suspect that 100-year floods don't really occur in this region?

EXAMPLE 8 Lottery Chances

You purchase 10 lottery tickets, for which the probability of winning on a single ticket is 1 in 10. What is the probability that you will have at least one winning ticket among the 10 tickets?

Solution Because the probability of winning with any one ticket is 0.1, the probability of not winning with one ticket is 0.9. Therefore, the probability of winning at least once with ten tickets is

$$P(\text{at least one winner in 10 tickets}) = 1 - [P(\text{not winning on one ticket})]^{10}$$
$$= 1 - [0.9]^{10} = 0.651$$

The probability of at least one winning ticket among 10 is 0.651. This also means that the probability of all ten tickets being losers is $1 - 0.651 = 0.349$. In other words, you stand a better than 1 in 3 chance of losing on all ten tickets that you buy. Again, this result surprises many people.

Lottery: a tax on people who are bad at math.

—Message circulated on the Internet

RETURN TO THE CHEVALIER

We are now ready to return to the story of the Chevalier de Mère. Recall that, in his first game, he bet that he could get at least one 6 within four rolls of a die. We can use the *at least once* rule to calculate his probability of winning. The probability of a 6 on any single roll is 1/6, so the probability of *not* 6 is 5/6. Thus, the probability of getting at least one 6 in four rolls is

$$P(\text{at least one 6 in 4 rolls}) = 1 - [P(\text{not 6 on one roll})]^4$$
$$= 1 - \left[\frac{5}{6}\right]^4 \approx 0.518$$

The Chevalier's probability of winning in the first game was 0.518. Although this is considerably lower than the 4/6 (0.667) probability he had guessed, it is still better than even odds. He was likely to come out ahead if he played the game many times.

In his second game, he bet that he could roll a double-6 within 24 tries. The probability of a double-6 on any single roll is 1/36, so the probability of not getting a double-6 on a single roll is 35/36. Therefore, the probability of getting at least one double-6 in 24 rolls is

$$P(\text{at least one double-6 in 24 rolls}) = 1 - [P(\text{not double-6 on one roll})]^{24}$$
$$= 1 - \left[\frac{35}{36}\right]^{24} \approx 0.491$$

The Chevalier's probability of winning in his second game was 0.491, which is slightly less than even odds. Therefore, he was likely to lose more games than he won.

Time out to think

Comment on how the difference between a 52% and a 49% chance of winning can be the difference between rags and riches in gambling. How can casinos exploit this idea?

Review Questions

1. How did the gambling habits of the Chevalier de Mère help launch the mathematical study of probability?
2. Give an example in which we would be interested in an *and* probability. How do we determine whether the events are independent or dependent? Give examples of each case, and explain how we calculate the probabilities.
3. Give an example in which we would be interested in an *either/or* probability. How do we determine whether the events are overlapping or non-overlapping? Give examples of each case, and explain how we calculate the probabilities.
4. What is the *at least once* rule? Give an example of its use.
5. Explain how the *at least once* rule can be used to find the correct probabilities in the games of the Chevalier de Mère.

In the News

1. **Combined Probability in the News.** Find a recent news story in which two or more probabilities were combined in some way. Describe how and why they were combined.
2. **Combined Probability in Your Life.** Cite a recent situation or decision in your life that involved an *and* probability or an *either/or* probability. How did the probability affect the situation or decision?

Exercises

And **Probabilities.** Determine whether the events given in Exercises 1–8 are independent or dependent. Then find the probability of the event.

1. Purchasing three winning lottery tickets in a row when one ticket has a 1 in 10 chance of being a winner
2. Rolling four successive 6s with four rolls of a fair die
3. Selecting all women for a six-person jury from a pool of 12 men and 12 women
4. Discovering that your five best friends all have telephone numbers ending in 1
5. Selecting two defective computer chips from a large batch in which the defect rate is 1.5%
6. The next five births at a hospital being all girls
7. Tossing four heads, then four tails on the first eight tosses of a fair coin
8. Randomly selecting 3 Americans from a pool of 12 British people and 9 Americans

Either/Or **Probabilities.** Determine whether the events in Exercises 9–16 are overlapping or non-overlapping. Then find the probability of the event.

9. Rolling either a 1 or a 2 on a single die
10. Rolling either a 3, a 4, or a 5 on a single die

11. Getting a sum of either 7 or 8 on a roll of two dice
12. Getting a sum of either 2, 3, or 4 on a roll of two dice
13. Selecting either a woman or a Democrat when you select a person randomly from a group composed of 25 Democratic men, 25 Republican men, 25 Democratic women, and 25 Republican women
14. Selecting either a woman or an American when you select a person randomly from a group composed of 20 English women, 25 English men, 25 American women, and 20 American men
15. Drawing either a king or a heart from a standard deck of 52 cards
16. Drawing either an ace or a king from a standard deck of 52 cards

At Least Once Problems. Use the *at least once* rule to find the probabilities of the events described in Exercises 17–24.

17. Getting at least one 6 in five rolls of a single die
18. Getting at least one 1 in four rolls of a single die
19. Getting at least one head when tossing three coins
20. Getting at least one tail when tossing ten coins
21. Getting rain at least once in five days if the probability of rain on each single day is 0.2
22. Getting rain at least once in ten days if the probability of rain on each single day is 0.1
23. Drawing at least one king when you draw cards from a standard deck 20 times; assume you replace the card each time you draw, so there are always 52 cards to draw from.
24. Drawing at least one ace when you draw cards from a standard deck 52 times; assume you replace the card each time you draw, so there are always 52 cards to draw from.
25. **Probability and Court.** The data in the following table show the outcome of guilty and not-guilty pleas in 1028 criminal court cases.
 a. What is the probability that a randomly selected defendant either pled guilty or was sent to prison?
 b. What is the probability that a randomly selected defendant either pled not guilty or was not sent to prison?

	Guilty plea	Not-guilty plea
Sent to prison	392	58
Not sent to prison	564	14

Source: Brereton and Casper, *Law and Society Review*, Vol. 16, No. 7.

26. **Testing a Drug.** An allergy drug was tested by giving 120 people the drug and 100 people a placebo. A control group consisted of 80 people who were given no treatment. The number of people in each group who showed improvement is shown in the table below.

	Allergy drug	Placebo	Control	Total
Improvement	65	42	31	138
No improvement	55	58	49	162
Total	120	100	80	300

a. What is the probability that a randomly selected person in the study was given either the drug or a placebo?
b. What is the probability that a randomly selected person either improved or did not improve?
c. What is the probability that a randomly selected person either was given the drug or improved?
d. What is the probability that a randomly selected person was given the drug and improved?

27. **Polling Calls.** A telephone pollster has a list of names and telephone numbers for 45 voters, 20 of whom are listed as registered Democrats and 25 of whom are listed as registered Republicans. Calls are made in random order. Suppose you want to find the probability that the first two calls are to Republicans.
 a. Are these independent or dependent events? Explain.
 b. If you treat them as dependent events, what is the probability that the first two calls are to Republicans?
 c. If you treat them as independent events, what is the probability that the first two calls are to Republicans?
 d. Compare the results of parts b and c.

28. **Dominant and Recessive Genes.** Many traits are controlled by a dominant gene, **A**, and a recessive gene, **a**. A child gets two genes, one from each parent. Suppose a child's parents each have the gene combination **Aa**; that is, each parent is equally likely to pass either the **A** or the **a** gene to the child. Make a table showing the probability distribution for the child's possible gene combinations, which are **AA, Aa,** and **aa**. If the combinations **AA** and **Aa** both result in the same dominant trait (say, brown hair) and **aa** results in the reccessive trait (say, blond hair), what is the probability that a child will have the dominant trait? What is the probability that the child will have the recessive trait?

29. **Better Bet for the Chevalier.** Suppose that the Chevalier de Mère had bet that he could roll a double-6

within 25 rolls, rather than 24. In that case, what would have been his probability of winning? Had he made this bet, would he still have lost over time? Explain.

30. **Lottery Odds.** The probability of a $2 winner in a particular state lottery is 1 in 10, the probability of a $5 winner is 1 in 50, and the probability of a $10 winner is 1 in 500.

 a. What is the probability of getting either a $2, $5, or $10 winner? Compare to the probability of getting only a $2 winner.

 b. If you buy 50 lottery tickets, what is the probability that you will get at least one $5 winner?

 c. If you buy 500 lottery tickets, what is the probability that you will get at least one $10 winner?

31. **Miami Hurricanes.** Studies of the Florida everglades show that, historically, the Miami region is hit by a hurricane about every 40 years.

 a. Based on the historical record, what is the empirical probability that Miami will be hit by a hurricane next year?

 b. What is the probability that Miami will be hit by hurricanes in two consecutive years?

 c. What is the probability that Miami will be hit by at least one hurricane in the next 10 years?

32. **HIV Among College Students.** Suppose that 2% of the students at a particular college are infected with HIV.

 a. If a student has ten sexual partners over a period of time, what is the probability that at least one of these partners is infected with HIV?

 b. If a student has 20 different sexual partners over a period of time, what is the probability that at least one partner is infected with HIV?

Web Projects

Find useful links for Web Projects on the text Web site: www.aw.com/bennett-briggs

1. **Lottery Chances.** Find a lottery Web site and study the probabilities of winning. Based on your findings, what is the probability of winning at least once if you play ten times? Explain any assumptions you make.

2. **HIV Probabilities.** Research the empirical probabilities of HIV among various groups in the population (for example, groups based on ethnicity, nationality, or sexual orientation). Based on your findings, discuss the relative dangers of sexual promiscuity within various populations.

UNIT 6C

THE LAW OF LARGE NUMBERS

*I*f you toss a coin once, you cannot predict exactly how it will land; you can only state that the probability of a head is 0.5. If you toss the coin 100 times, you still cannot predict precisely how many heads will occur. However, you can reasonably expect to get heads *close to* 50% of the time. If you toss the coin 1000 times, you can expect the proportion of heads to be even closer to 50%. In general, the more times you toss the coin, the closer the percentage of heads will be to exactly 50%. While individual events may be unpredictable, large collections of events should show some pattern or regularity. This principle is called the **law of large numbers** (or the *law of averages*).

> **The Law of Large Numbers**
>
> The **law of large numbers** applies to a process for which the probability of an event A is $P(A)$ and the results of repeated trials are independent. It states: *If the process is repeated through many trials, the proportion of the trials in which event A occurs will be close to the probability $P(A)$. The larger the number of trials, the closer the proportion should be to $P(A)$.*

6C THE LAW OF LARGE NUMBERS 327

We can illustrate the law of large numbers with a die rolling experiment. The probability of a 1 on a single roll is $P(1) = 1/6 = 0.167$. To avoid the tedium of rolling the die many times, we can let a computer *simulate* random rolls of the die. Figure 6.6 shows a computer simulation of rolling a single die 5000 times. The horizontal axis gives the number of rolls, and the height of the curve gives the proportion of 1s. Although the curve bounces around when the number of rolls is small, the proportion of 1s approaches the probability of 0.167, in agreement with the law of large numbers.

FIGURE 6.6 Results of computer simulation of rolling a die.

EXAMPLE 1 Roulette

A roulette wheel has 38 numbers: 18 are black, 18 are red, and the numbers 0 and 00 are green.

a. What is the probability of getting a red number on any spin?
b. If patrons in a casino spin the wheel 100,000 times, how many times should the casino expect a red number?

Solution

a. There are 18 red numbers among 38 total, so the probability of red on a single spin is 18/38, or 0.474.
b. The law of large numbers tells us that as the game is played more and more times, the proportion of times that the wheel comes up red should get closer to the 0.474 probability. Thus, in 100,000 tries, the wheel should come up red close to 47.4% of the time, or about 47,400 times.

EXPECTED VALUE

Suppose the InsureAll Company sells a special type of insurance in which it promises to pay you $100,000 in the event that you must quit your job because of serious illness. Based on data from past claims, the empirical probability that a policyholder will make a claim for loss of job is 1 in 500. Should the insurance company expect to earn a profit if it sells the policies for $250 each?

If InsureAll sells only a few policies, we cannot predict the profit or loss. For example, selling 100 policies for $250 each generates revenue of 100 × $250 = $25,000. If none of the 100 policyholders file a claim, InsureAll makes a tidy profit. But if even one policyholder makes a $100,000 claim, InsureAll faces a huge loss.

In contrast, if InsureAll sells a large number of policies, the law of large numbers tells us that the proportion of policies for which claims must be paid should be very close to the 1 in 500 probability for a single policy. For example, if InsureAll sells

> **historical note**
>
> The first known use of life insurance was in the Roman empire in the 3rd century A.D. The idea reappeared in England in the late 16th century. The first careful analysis of birth and death data was done by John Graunt in London in the 1660s. His life expectancy tables were improved by the astronomer Edmund Halley (the discoverer of Halley's comet). These tables led to a thriving insurance industry in London. In 1687, Edward Lloyd opened a coffee house in London that specialized in insuring almost all risks, including maritime travel and trade; Lloyd's of London remains in the insurance business today.

1 million policies, it should expect that the number policyholders making the $100,000 claim will be close to

$$\underbrace{1{,}000{,}000}_{\text{number of policies}} \times \underbrace{\frac{1}{500}}_{\substack{\text{probability of}\\ \$100{,}000 \text{ claim}}} = 2000$$

Paying these 2000 claims will cost

$$2000 \times \$100{,}000 = \$200 \text{ million}$$

This cost of $200 million represents an *average* of $200 for each of the 1 million policies. Thus, if the policies sell for $250 each, InsureAll should expect to earn an average of $250 − $200 = $50 per policy. This amounts to a profit of $50 million on sales of 1 million policies.

We can find this same answer with a more formal procedure. Note that the InsureAll example involves two distinct events, each with a particular *probability* and *value* for the company:

1. In the event that a person buys a policy, the value to the company is the $250 price of the policy. The probability of this event is 1 because everyone who buys a policy pays the $250.
2. In the event that a person files a claim, the value to the company is −$100,000; it is negative because the company loses $100,000 in this case. The probability of this event is 1/500.

If we multiply the value of each event by its probability and then add the results, we find the average or **expected value** of each insurance policy:

$$\text{expected value} = \underbrace{\$250}_{\substack{\text{value of policy sale}}} \times \underbrace{1}_{\substack{\text{probability of}\\ \text{earning \$250 on sale}}} + \underbrace{(-\$100{,}000)}_{\substack{\text{value of}\\ \text{claim}}} \times \underbrace{\frac{1}{500}}_{\substack{\text{probability of}\\ \text{paying claim}}}$$

$$= \$250 - \$200 = \$50$$

This expected profit of $50 per policy is the same answer we found earlier. Keep in mind that this value is expected only if InsureAll sells enough policies for the law of large numbers to come into play.

Time out to think

Should InsureAll expect to see a profit of $50 on each individual policy? Should it expect a profit of $50,000 on 1000 policies? Explain.

Expected Value

Consider two events, each with its own value and probability. The **expected value** is

$$\text{expected value} = \binom{\text{event 1}}{\text{value}} \times \binom{\text{event 1}}{\text{probability}} + \binom{\text{event 2}}{\text{value}} \times \binom{\text{event 2}}{\text{probability}}$$

The expected value formula can be extended to any number of events by including more terms in the sum.

EXAMPLE 2 *Lottery Expectations*

Suppose that $1 lottery tickets have the following probabilities: 1 in 5 to win a free ticket (worth $1), 1 in 100 to win $5, 1 in 100,000 to win $1000, and 1 in 10 million to win $1 million. What is the expected value of a lottery ticket? Discuss the implications.

Solution The easiest way to proceed is to make a table of all the relevant events with their values and probabilities. We are calculating the expected value of a lottery ticket to *you*, so the ticket price has a negative value (it costs you money), while the values of winnings are positive.

Event	Value	Probability	Value × probability
Ticket purchase	−$1	1	−$1 × 1 = −$1
Win free ticket	$1	$\frac{1}{5}$	$1 × $\frac{1}{5}$ = $0.20
Win $5	$5	$\frac{1}{100}$	$5 × $\frac{1}{100}$ = $0.05
Win $1000	$1000	$\frac{1}{100,000}$	$1000 × $\frac{1}{100,000}$ = $0.01
Win $1 million	$1,000,000	$\frac{1}{10,000,000}$	$1,000,000 × $\frac{1}{10,000,000}$ = $0.10
			Sum of last column: −$0.64

The expected value is the sum of all the products' *value × probability*, which the table shows to be −$0.64. Thus, averaged over many tickets, you should expect to lose 64¢ for each lottery ticket that you buy. If you buy, say, 1000 tickets, you should expect to *lose* about 1000 × $0.64 = $640. ■

Time out to think

Many states use lotteries to finance worthy causes such as parks, recreation, and education. Lotteries also tend to keep state taxes at lower levels. On the other hand, research shows that lotteries are played primarily by people with low incomes. Do you think lotteries are good social policy? Do you think lotteries are good economic policy?

> **historical note**
>
> The French mathematician and philosopher Blaise Pascal used expected value to argue for believing in God. He used *p* to represent the probability that God exists, even if it is very small. He then assigned values for the consequences of believing and not believing in God. In particular, he claimed that the consequence of not believing in God in the case that God actually exists is infinitely bad (negative). Thus, regardless of how small *p* might be, the expected value of believing in God is positive and the expected value of not believing in God is infinitely negative. He concluded that one should believe in God.

DILBERT reprinted by permission of United Features Syndicate, Inc.

by the way ...

On November 9, 1999, a Pablo Picasso oil painting, *Nu au fauteuil noir*, sold for $45,102,500 at Christie's Auction House in New York.

EXAMPLE 3 Art Auction

You are at an art auction, trying to decide whether to bid $50,000 for a particular painting. You believe that you have a 1/2 probability of reselling the painting to a client in New York for $70,000 and a 1/4 probability of reselling the painting to a client in San Francisco for $80,000. Otherwise, you will have to keep the painting—an option that has benefits, but not monetary ones. What is the expected value of a $50,000 bid?

Solution There are three relevant events to consider:

- With probability 1, you will pay $50,000 to buy the painting. This option represents a monetary loss to you, so it has a negative value.
- With probability 1/2, you will sell the painting for $70,000. This option is a gain, so it has a positive value.
- With probability 1/4, you will sell the painting for $80,000. This option is also a gain, so it has a positive value.

Thus, the expected value of a $50,000 bid is

$$\text{expected value} = \binom{\text{value of}}{\text{event 1}} \times \binom{\text{prob of}}{\text{event 1}} + \binom{\text{value of}}{\text{event 2}} \times \binom{\text{prob of}}{\text{event 2}} + \binom{\text{value of}}{\text{event 3}} \times \binom{\text{prob of}}{\text{event 3}}$$

$$= (-\$50{,}000 \times 1) + \left(\$70{,}000 \times \frac{1}{2}\right) + \left(\$80{,}000 \times \frac{1}{4}\right)$$

$$= -\$50{,}000 + \$35{,}000 + \$20{,}000 = \$5000$$

With a $50,000 bid, you could expect to gain $5000 by the purchase. It's important to remember that expected value refers to expected gain or loss over *many* such purchases. If this is your only purchase, it might be better to reconsider or to assign a high value to admiring the painting in your own house! ■

The value of our expectations always signifies something in the middle between the best we can hope for and the worst we can fear.

—JACOB BERNOULLI,
17th-century
mathematician

THE GAMBLER'S FALLACY

Consider a simple game involving a coin toss: You win $1 if the coin lands heads and you lose $1 if it lands tails. Now, suppose you toss the coin 100 times and get 45 heads and 55 tails, putting you $10 in the hole. Are you "due" for a streak of better luck?

You probably recognize that the answer is *no*; each coin toss is independent of previous tosses, so your past bad luck has no bearing on your future chances. However, many gamblers—especially compulsive gamblers—guess just the opposite. They believe that when their luck has been bad, it's due for a change. This mistaken belief is often called the **gambler's fallacy** (or the *gambler's ruin*).

DEFINITION

The **gambler's fallacy** is the mistaken belief that a streak of bad luck makes a person "due" for a streak of good luck.

Time out to think

A farmer knows that at this time of year in his part of the country, the probability of rain on a given day is 0.5. It hasn't rained in 10 days and he needs to decide whether to start irrigating. Is he justified in postponing irrigation because he is due for a rainy day? Explain.

One reason people succumb to the gambler's fallacy is a misunderstanding of the law of large numbers. In our coin toss game, the law of large numbers tells us that the proportion of heads tends to be closer to 50% for larger numbers of games. But this does *not* mean that you are likely to recover early losses. The following example shows why.

EXAMPLE 4 Continued Losses

You are playing the coin toss game, in which you win $1 for heads and lose $1 for tails. After 100 tosses, you are $10 in the hole because you have 45 heads and 55 tails. You continue playing until you've tossed the coin 1000 times, at which point you've gotten 480 heads and 520 tails. Does the result agree with what we expect from the law of large numbers? Have you gained back any of your losses? Explain.

Solution The proportion of heads in your first 100 tosses was 45%. After 1000 tosses, the proportion of heads has increased to 480 out of 1000, or 48%. This agrees with the law of large numbers, because the proportion grew closer to 50%. However, you've now won $480 for the 480 heads while losing $520 for the 520 tails, for a net loss of $40. Thus, your losses have actually *increased*, despite the fact that the proportion of heads grew closer to 50%.

Streaks

Another common misunderstanding that contributes to the gambler's fallacy involves expectations about streaks. Suppose you toss a coin six times and see the outcomes HHHHHH (all heads). Then you toss it six more times and see the outcomes HTTHTH. Most people would say that the latter outcome is "natural" while the streak of all heads is surprising. But, in fact, all possible sets of six outcomes are equally likely: The probability of any *single* particular outcome is 1/2, so all sets of six particular outcomes have the same probability of $(1/2)^6 = 1/64$.

Time out to think

Consider a lottery game in which six balls are drawn randomly from a set of balls numbered 1 through 42. Should you be surprised if the winning combination consists of the numbers 21, 4, 17, 33, 40, 9? Should you be surprised if the winning combination consists of the numbers 1, 2, 3, 4, 5, 6? Discuss.

EXAMPLE 5 Hot Hand at the Craps Table

The popular casino game of craps involves rolling dice. Suppose you are playing craps and suddenly find yourself with a "hot hand": You roll a winner on ten consecutive bets. Is your hand really "hot"? Should you increase your bet because you are on a hot streak? Assume that you are

historical note

One of the most famous compulsive gamblers in history was the Russian writer Fyodor Dostoyevsky. He played at roulette throughout Europe in the 19th century. He once reported that he "won 10,000 francs at first, . . . but I got carried away and dropped half my winnings." His experience with gambling formed the basis of his work *The Gambler*.

making bets with a 0.486 probability of winning on a single play (the best odds available in craps).

Solution First, let's consider the chance of winning ten times in a row. With a probability of 0.486 on a single play, the chance of ten straight wins is

$$(0.486)^{10} \approx 0.000735$$

Thus, your streak of ten straight wins has a probability of only about 7 in 10,000, which might make you think that your hand really is "hot." However, look around the casino: If it is a large casino, several hundred people may be playing at the craps tables, and there may be *tens of thousands* of individual rolls each night. Thus, it is almost inevitable that someone will have a "hot streak" of ten straight wins; in fact, even longer hot streaks are likely during any given night. Your apparent "hot hand" is a mere coincidence and your probability of winning on your next bet is still only 0.486. Increasing your bet based on a mere coincidence would be foolish.

> Luck is a mighty queer thing. All you know about it for certain is that it's bound to change.
>
> —BRET HARTE, *The Outcasts of Poker Flat*

The House Edge

A casino makes money because games are set up so that the expected earnings of patrons are negative (that is, losses). Because the casino earns whatever patrons lose, the casino's earnings are positive. The amount that the casino, or *house*, can expect to earn per dollar bet is called the **house edge.** That is, the house edge is the expected value *to the casino* of a particular bet.

The house edge varies from game to game. It tends to be greatest in games where big winnings are possible, such as slot machines. It tends to be least in games where strategy can improve your odds, such as blackjack. But the house edge is always present, and gamblers who believe they can beat the house regularly are sadly succumbing to the gambler's fallacy.

EXAMPLE 6 *The House Edge in Roulette*

The game of roulette is usually set up so that betting on red is a 1 to 1 bet; that is, you win the same amount of money as you bet if red comes up. Betting on a single number is a 35 to 1 bet; that is, you win 35 times as much as you bet if your number comes up. What is the house edge in each of these two cases? If patrons wager $1 million on such bets, how much should the casino expect to earn?

Solution For each case, the house edge is the casino's expected earnings on the bet. The probability of red is 18/38 (see Example 1), so the probability of not red is 20/38. Let's call the *value* of red for the casino −$1, indicating that the casino pays the amount of the patron's bet when red comes up. The value of not red for the casino is $1, because it wins the patron's bet if red does not come up. The house edge when patrons bet on red is

$$\underbrace{\$1}_{\substack{\text{value to casino}\\\text{of not red}}} \times \underbrace{\frac{20}{38}}_{\substack{\text{probability}\\\text{of not red}}} + \underbrace{-\$1}_{\substack{\text{value to casino}\\\text{of red}}} \times \underbrace{\frac{18}{38}}_{\substack{\text{probability}\\\text{of red}}} \approx \$0.053$$

That is, the casino can expect to gain $0.053 per dollar gambled. Thus, if patrons bet $1 million on red, the casino can expect to earn about $1 million × 0.053 = $53,000.

The probability of winning a bet on a single number is 1/38 because there are 38 numbers on the roulette wheel. The 35 to 1 payoff offered by the casino on this bet means the casino pays the patron 35 times the amount of the bet; thus, its value to the casino is −$35. The probability that the bet does not win is 37/38; the casino gains the patron's bet in this case, so its value is $1. Thus, the house edge when patrons bet on a single number is

$$\underbrace{\$1}_{\text{value to casino of not hitting the number}} \times \underbrace{\frac{37}{38}}_{\text{probability of not hitting the number}} + \underbrace{-\$35}_{\text{value to casino of hitting the number}} \times \underbrace{\frac{1}{38}}_{\text{probability of hitting the number}} \approx \$0.053$$

This is the same house edge as for the red bet. Again, if patrons bet $1 million on single numbers, the house can expect to win $53,000.

Time out to think

Given the house edge, is it possible for a casino to lose money on its gambling results alone? Explain.

Review Questions

1. Explain the meaning of the law of large numbers. Does this law say anything about what will happen in a single observation or experiment? Why or why not?

2. In 10 tosses of a fair coin, should you be surprised to see 6 heads? In 1000 tosses of a fair coin, should you be surprised to see 600 heads? Explain in terms of the law of large numbers.

3. What is an expected value, and how is it computed? Should we always expect to get the expected value? Why or why not?

4. What is the gambler's fallacy? Give an example.

5. Explain why the probability is the same for any particular set of ten coin toss outcomes. How does this idea affect our thinking about streaks?

6. What is the house edge in gambling? Explain how it virtually guarantees that a casino will win more money than it loses.

In the News

1. **Law of Large Numbers.** Describe a news story in which the law of large numbers is mentioned. Is the term used in an accurate way? Explain.

2. **Personal Law of Large Numbers.** Describe a situation in which you personally have made use of the law of large numbers, either correctly or incorrectly. Why did you use the law of large numbers in this situation? Was it helpful?

3. **The Gambler's Fallacy.** Describe a situation in which you or someone you know has fallen victim to the gambler's fallacy. How should the situation have been dealt with?

Exercises

1. **Understanding the Law of Large Numbers.** Suppose you toss a fair coin 10,000 times. Should you expect to get exactly 5000 heads? Why or why not? What does the law of large numbers tell you about the results you are likely to get?

2. **Speedy Driver.** Suppose a person who has a habit of driving fast has never had an accident or traffic citation. What does it mean to say that "the law of large numbers will catch up with him"? Is it true? Explain.

Expected Value in Games. In Exercises 3–6, find the expected value (to you) of the described game. Would you expect to win or lose money in 1 game? in 100 games? Explain.

3. You are given 5 to 1 odds against tossing three heads in three tosses of a fair coin, meaning you win $5 if you succeed and you lose $1 if you fail.

4. You are given 10 to 1 odds against tossing three heads in three tosses of a fair coin, meaning you win $10 if you succeed and you lose $1 if you fail.

5. You are given 5 to 1 odds against rolling two even numbers with the roll of two fair dice, meaning you win $5 if you succeed and you lose $1 if you fail.

6. You are given 10 to 1 odds against rolling a double number (for example, two 1s or two 2s) with the roll of two fair dice, meaning you win $10 if you succeed and you lose $1 if you fail.

Insurance Claims. In Exercises 7–8, find the expected value (to the company) per policy sold. If the company sells 100,000 policies, what is the expected profit or loss? Explain.

7. An insurance policy sells for $1000. Based on past data, an average of 1 in 100 policyholders will file a $20,000 claim, an average of 1 in 200 policyholders will file a $50,000 claim, and an average of 1 in 500 policyholders will file a $100,000 claim.

8. An insurance policy sells for $500. Based on past data, an average of 1 in 50 policyholders will file a $5000 claim, an average of 1 in 100 policyholders will file a $10,000 claim, and an average of 1 in 200 policyholders will file a $30,000 claim.

9. **Expected Wait.** Suppose you arrive at a bus stop randomly, so all arrival times are equally likely. The bus arrives regularly every 30 minutes without delay (say, on the hour and on the half hour). What is the expected value of your waiting time? Explain.

10. **Expected Wait.** A bus arrives at a bus stop at noon, 12:20, and 1:00. You arrive at the bus stop at random times between noon and 1:00 every day, so all arrival times are equally likely.

 a. What is the probability that you will arrive at the bus stop between noon and 12:20? What is your mean waiting time in that case? (Exercise 9 may be helpful.)

 b. What is the probability that you will arrive at the bus stop between 12:20 and 1:00? What is your mean waiting time in that case?

 c. Overall, what is your expected waiting time for the bus?

 d. Would your expected waiting time be longer or shorter if the bus arrived at equally spaced intervals (say, noon, 12:30, 1:00)? Explain.

Powerball. Exercises 11–12 state a jackpot value for the 17-state Powerball lottery, for which the advertised prizes and probabilities of winning (on a single $1 ticket) are shown in the following table. (The table has two entries each for $100 and $7 because these prizes can be won in two different ways.) For the given jackpot, find the expected value of the winnings for a single lottery ticket. Then state how much you can expect to win or lose if you spend $365 per year on the lottery. Briefly discuss your results.

Prize	Probability
Jackpot	1 in 80,089,128
$100,000	1 in 1,953,393
$5000	1 in 364,042
$100	1 in 8879
$100	1 in 8466
$7	1 in 207
$7	1 in 605
$4	1 in 188
$3	1 in 74

11. The jackpot is $30 million.

12. The jackpot is $100 million.

Big Game. Exercises 13–14 state a jackpot value for the Multi-State Big Game lottery, for which the advertised prizes and probabilities of winning (on a single $1 ticket) are shown in the following table. (The table has two entries for $5 because this prize can be won in two different ways.) For the given jackpot, find the expected value of the winnings for a single lottery ticket. Then state how much you can expect to win or lose if you spend $365 per year on the lottery. Briefly discuss your results.

Prize	Probability
Jackpot	1 in 76,275,360
$150,000	1 in 2,179,296
$5000	1 in 339,002
$150	1 in 9686
$100	1 in 7705
$5	1 in 220
$5	1 in 538
$2	1 in 102
$1	1 in 62

13. The jackpot is $3 million.

14. The jackpot is $25 million.

15. **Extra Points in Football.** Football teams have the option of trying to score either 1 or 2 extra points after a touchdown. They can get 1 point by kicking the ball through the goal posts or 2 points by running or passing the ball across the goal line. In the NFL in 1999, 1-point kicks were successful 94% of the time, while 2-point attempts were successful only 37% of the time. In either case, failure means zero points. Calculate the expected values of the 1-point and 2-point attempts. Based on these expected values, which option makes more sense in most cases? Explain.

16. **Average American Age.** The following table gives the distribution of Americans by age categories. Use the midpoint of each age category (for example, 6.5 for the 0–13 category) to find the expected value of American ages; use 75 for the 65 and over category. Explain why the expected value is also the mean age of the population.

Age category	Percentage
0–13	20.0%
14–24	15.3%
25–34	13.6%
35–44	16.3%
45–64	22.2%
65 and over	12.6%

17. **Household Size.** It is estimated that 57% of Americans live in households with 1 or 2 people, 32% live in households with 3 or 4 people, and 11% live in households with 5 or more people. Using 1.5 for the expected number of people in households with 1 or 2 people, 3.5 for households with 3 or 4 people, and 6 for households of 5 or more people, find the expected number of people in an American household. How is this related to the mean household size?

18. **Gambler's Fallacy and Coins.** Suppose you play a coin toss game in which you win $1 if a head appears and lose $1 if a tail appears. In the first 100 coin tosses, heads comes up 47 times and tails comes up 53 times.

 a. What percentage of times has heads come up in the first 100 tosses? What is your net gain or loss at this point?

 b. Suppose you toss the coin 200 more times (a total of 300 tosses), and at that point heads has come up 48% of the time. Is this increase in the percentage of heads consistent with the law of large numbers? Explain. What is your net gain or loss at this point?

 c. How many heads would you need in the next 100 tosses in order to break even after 400 tosses? Is this likely to occur?

 d. Suppose that, still behind after 400 tosses, you decide to keep playing because you are "due" for a winning streak. Explain how this belief would illustrate the gambler's fallacy.

19. **Gambler's Fallacy and Dice.** Suppose you roll a die with the following rules: You win $1 if the die comes up with an even number and you lose $1 if it comes up odd.

 a. Suppose you get 45 even numbers in your first 100 rolls. How much money have you won or lost?

 b. On the second 100 rolls, your luck improves and you roll 47 even numbers. How much money have you won or lost over 200 rolls?

 c. Your luck continues to improve, and you roll 148 even numbers in your next 300 rolls. How much money have you won or lost over your total of 500 rolls?

 d. How many even numbers would you have to roll in the next 100 rolls to break even? Is this likely? Explain.

 e. What were the percentages of even numbers after 100, 200, and 500 rolls? Explain how this illustrates the law of large numbers, even while your losses increased.

20. **Can You Catch Up?** Suppose you toss a fair coin 100 times, getting 42 heads and 58 tails, which is 16 more tails than heads.

 a. Explain why, on your next toss, the *difference* between the numbers of heads and tails is as likely to grow to 17 as it is to shrink to 15.

 b. Extend your explanation from part a to explain why, if you toss the coin 1000 more times, the final difference between the numbers of heads and tails is as likely to be larger than 16 as it is to be smaller than 16.

 c. Suppose you are betting on heads with each coin toss. After the first 100 tosses, you are well on the losing side. Explain why, if you continue to bet, you will most likely remain on the losing side. How is this answer related to the gambler's fallacy?

21. **Profitable Casino.** Averaged over all games and all bets being played, a particular casino has a house edge of 0.07, or 7¢ per dollar gambled. If a total of $100 million is wagered in the casino over the course of a year, what is the casino's total profit? Explain.

22. **House Edge in Blackjack.** In a large casino, the house wins on its blackjack tables with a probability of 50.7%. All bets at blackjack are 1 to 1: If you win, you gain the amount you bet; if you lose, you lose the amount you bet.

 a. What is the expected value to you of a single game? What is the house edge?

 b. If you played 100 games of blackjack in an evening, betting $1 on each hand, how much should you expect to win or lose? Explain.

 c. If you played 100 games of blackjack in an evening, betting $5 on each hand, how much should you expect to win or lose? Explain.

d. If patrons bet $1,000,000 on blackjack in one evening, how much should the casino expect to earn? Explain.

23. Mail Sweepstakes. You receive a notice in the mail saying that you will become eligible to win $1 million in a sweepstakes simply by filling out and returning the enclosed card. As it turns out, you are one of 10 million people who return the card, which makes your odds of winning 1 in 10 million.

 a. Assume you spend 34¢ to mail the card back to the sweepstakes. Calculate your expected earnings (or losses) in sending back the card.

 b. Suppose you decide to purchase three magazine subscriptions, which are offered to you as an option when you return the card. Later, you learn that you could have purchased the same subscriptions through a student discount service for $7 less than you paid. What are your expected earnings now?

24. Psychology of Expected Values. In 1953, French economist Maurice Allais studied how people assess risk. Here are two survey questions that he used:

 Decision 1

 Option A: 100% chance of gaining $1,000,000

 Option B: 10% chance of gaining $2,500,000; 89% chance of gaining $1,000,000; and 1% chance of gaining nothing

 Decision 2

 Option A: 11% chance of gaining $1,000,000 and 89% chance of gaining nothing

 Option B: 10% chance of gaining $2,500,000 and 90% chance of gaining nothing

Allais discovered that for decision 1, most people chose option A, while for decision 2, most people chose option B.

 a. For each decision, find the expected value of each option.

 b. Are the responses given in the surveys consistent with the expected values?

 c. Give a possible explanation for the responses in Allais's surveys.

Web Projects

Find useful links for Web Projects on the text Web site: www.aw.com/bennett-briggs

1. Lotteries. Go to the Web site for all U.S. lotteries and study the summary of state and multi-state lottery odds and prizes. Pick five lotteries and determine the expected value for winnings in each case. Discuss your results.

2. The Morality of Gambling. Republican Senator Richard Lugar attacked state-supported lotteries and the trend toward increasing legalization of gambling, saying "the spread of gambling is a measure of the moral erosion taking place in our country.... It says that if you play enough, you can hit the jackpot and be freed from the discipline of self-support through a job or the long commitment to ongoing education." Do a Web search on "gambling and morality" to find other opinions on this issue. Write a short summary of your findings and your own opinion, which you should defend clearly.

UNIT 6D

COUNTING AND PROBABILITY

As we have seen, computing theoretical probabilities requires counting outcomes. In previous units, we considered problems in which the counting was straightforward or the multiplication principle could be used (see the Brief Review on p. 307). Here, we'll discuss counting tools that will enable us to solve a much wider range of probability problems and to explore the nature of coincidences.

ARRANGEMENTS WITH REPETITION

Suppose that license plates in your state display seven numerals (Figure 6.7). How many different license plates are possible? There are 10 possibilities (the numerals 0 through 9) for the first position and the same 10 possibilities for the second position, so there are $10 \times 10 = 100$ possibilities for the first two positions combined. Similarly, because there are also 10 possibilities for the third position, there are $10 \times 10 \times 10 = 1000$ possibilities for the first three positions. For all seven positions, each with 10 possibilities, the total number of possibilities is

$$10 \times 10 \times 10 \times 10 \times 10 \times 10 \times 10 = 10^7$$

That is, this system allows for 10 million different license plates.

The license plate problem involves making selections from a *single* group of choices—in this case, the numerals 0–9. However, the same choices may be selected over and over again. This type of counting problem, in which we repeatedly select from the same group of choices, is called **arrangements with repetition.**

> **Arrangements with Repetition**
> If we make r selections from a group of n choices, a total of n^r different arrangements are possible.

FIGURE 6.7 For each of the seven positions on the license plate, there are ten choices, resulting in 10^7 possible license plates.

EXAMPLE 1 *Arrangements with Repetition*

a. How many seven-symbol license plates are possible if both numerals *and* letters can be used in any order?
b. How many six-character passwords can be made by combining lower-case letters, upper-case letters, numerals, and the characters @, $, and &?

Solution

a. With 10 numerals and 26 letters in the alphabet, there are 36 choices for each of the seven symbols. Thus, we are selecting $r = 7$ symbols with $n = 36$ choices for each symbol, so the number of seven-symbol license plates is

$$n^r = 36^7 = 78{,}364{,}164{,}096$$

There are more than 78 billion possible seven-symbol license plates.

b. Each password character can be selected from the set of 26 lower-case letters, 26 upper-case letters, 10 numerals, and the three symbols @, $, and &. Thus, there are 65 characters to choose from. For six-character passwords, we select $r = 6$ symbols with $n = 65$ choices for each symbol. The total number of possible passwords is

$$n^r = 65^6 = 75,418,890,625$$

More than 75 billion passwords are possible with these choices.

Time out to think

Compare the numbers of possibilities in Example 1 to the human population. Will a state using seven-symbol license plates ever run out of possibilities? Do you think a six-symbol password is secure against hackers?

PERMUTATIONS AND COMBINATIONS

Suppose you coach a team of four swimmers. How many different ways can you put together a four-person relay team?

You can choose any of the four swimmers for the first leg. Once you've chosen the first swimmer, you have three swimmers left to choose from for the second leg. Thus, the number of possible choices for the first two legs combined is $4 \times 3 = 12$. You then have two swimmers to choose from for the third leg. After the first three legs are filled, you have only one choice for the last leg. The total number of possible arrangements for the relay is

$$4 \times 3 \times 2 \times 1 = 24$$

by the way ...
The word *permute* (root of *permutation*) comes from a Latin term meaning "to change throughout."

Figure 6.8 shows all 24 possible relay orders with a tree diagram in which the four swimmers are labeled A, B, C, and D. The first row shows the four possibilities for the first leg. The second row shows the three possible second-leg choices for each of the first-leg choices. The third row shows each of the prior possibilities branching into two possibilities for the third leg, and the fourth row shows the single remaining choice for the last leg in each case. Counting across the bottom row, we see the total of 24 possibilities. Each of the 24 different relay orders is called a **permutation** of the four letters A, B, C, and D.

> **Permutations**
> Mathematically, we are dealing with **permutations** whenever all selections come from a single group of items, no item may be selected more than once, and the *order of arrangement matters* (for example, ABCD is considered different from DCBA). The total number of permutations possible with a group of n items is $n!$ ("n factorial"), where $n! = n \times (n - 1) \times \cdots \times 2 \times 1$.

EXAMPLE 2 Class Schedules

A middle school principal needs to schedule six different classes—algebra, English, history, Spanish, science, and gym—in six different time periods. How many different class schedules are possible?

Solution This is a case of permutations because all selections come from the same group of classes, each class is scheduled only once, and the order of the classes matters (because a schedule that begins with Spanish is different from one that begins with gym). The principal may choose any of the six classes for the first period, leaving five choices for the second period. After the first two periods are filled, there are four choices left for the third period. Similarly, there are three choices for the fourth period, two choices for the fifth period, and only one choice for the sixth period. Altogether, the number of possible schedules is

$$6! = 6 \times 5 \times 4 \times 3 \times 2 \times 1 = 720$$

The principal can schedule the six classes in any of 720 possible ways.

The Permutations Formula

Suppose you coach a team of ten swimmers, from which you must put together a four-person relay team. How many possibilities do you have in this case?

This time, you have ten swimmers from which to choose for the first leg of the relay. Once you make the first choice, you have nine swimmers from which to choose for the second leg, then eight choices for the third leg, and seven choices for the fourth leg (Figure 6.9). The total number of relay possibilities is

$$10 \times 9 \times 8 \times 7 = 5040$$

Using your CALCULATOR

Factorials

Many calculators have a special key for factorials, usually labeled *n*! For example, to calculate the factorial 6!, press 6 (n!) =

First leg	A			B			C			D		
Second leg	B	C	D	A	C	D	A	B	D	A	B	C
Third leg	C D	B D	B C	C D	A D	A C	B D	A D	A B	B C	A C	A B
Fourth leg	D C	D B	C B	D C	D A	C A	D B	D A	B A	C B	C A	B A

Each path from the first leg to the fourth leg gives a relay order:

```
A A A A A A   B B B B B B   C C C C C C   D D D D D D
B B C C D D   A A C C D D   A A B B D D   A A B B C C
C D B D B C   C D A D A C   B D A D A B   B C A C A B
D C D B C B   D C D A C A   D B D A B A   C B C A B A
```

FIGURE 6.8 Tree diagram for the relay possibilities with four swimmers A, B, C, and D. There are four choices for the first leg. Each possible choice for the first leg leads to three remaining choices for the second leg. After the swimmers have been chosen for the first two legs, two choices remain for the third leg. Once the first three legs have been chosen, only one swimmer remains for the fourth leg.

a brief review...

FACTORIALS

Products of the form $4 \times 3 \times 2 \times 1$ come up so frequently in counting problems that they have a special name. Whenever a positive integer n is multiplied by all the preceding positive integers, the result is called ***n* factorial** and is denoted $n!$ (the exclamation mark is read as "factorial"). For example:

$$1! = 1$$
$$2! = 2 \times 1 = 2$$
$$3! = 3 \times 2 \times 1 = 6$$
$$4! = 4 \times 3 \times 2 \times 1 = 24$$
$$5! = 5 \times 4 \times 3 \times 2 \times 1 = 120$$

In general,

$$n! = n \times (n-1) \times (n-2) \times \cdots \times 2 \times 1$$

Note that $n!$ grows rapidly with n. For example, $20! \approx 2.4 \times 10^{18}$, $40! \approx 8.2 \times 10^{47}$, and $60! \approx 8.3 \times 10^{81}$. In fact, factorials quickly become so large that many calculators cannot handle them above $n = 69$ (if the calculator limit is 10^{100}). Note also that, by definition, $0! = 1$. This definition might not seem to make a lot of sense, but it turns out to be important to making counting formulas work properly.

Example: Calculate each of the following *without* using the factorial key on your calculator.

a. $\dfrac{6!}{4!}$ b. $\dfrac{25!}{22!}$ c. $\dfrac{200!}{199!}$

Solution:

a. We can write out the entire calculation, but it is easier if we simplify it by canceling like terms in the numerator and denominator:

$$\frac{6!}{4!} = \frac{6 \times 5 \times \cancel{4} \times \cancel{3} \times \cancel{2} \times \cancel{1}}{\cancel{4} \times \cancel{3} \times \cancel{2} \times \cancel{1}} = \frac{6 \times 5 \times \cancel{4!}}{\cancel{4!}}$$
$$= 6 \times 5 = 30$$

b. This calculation is easier if we recognize that $25! = 25 \times 24 \times 23 \times 22!$, giving us

$$\frac{25!}{22!} = \frac{25 \times 24 \times 23 \times \cancel{22!}}{\cancel{22!}} = 25 \times 24 \times 23 = 13{,}800$$

c. If you tried 200! on your calculator, you would probably get an error because it is too large for most calculators to handle. But this calculation is easy if we recognize that $200! = 200 \times 199!$, so

$$\frac{200!}{199!} = \frac{200 \times \cancel{199!}}{\cancel{199!}} = 200$$

FIGURE 6.9 With ten swimmers to choose from for a four-person relay team, you have ten choices for the first position, nine for the second position, eight for the third position, and seven for the fourth position. This results in $10 \times 9 \times 8 \times 7 = 5040$ possible relay teams.

Each of the 5040 relays represents a different *permutation* because each relay swimmer is selected from the same group of ten swimmers, no swimmer can swim more than once, and the order of the swimmers in the relay is important. However, in this case the relays make use of only four of the total of ten swimmers. Therefore, the number of possible relays made from "ten swimmers selected four at a time" is 5040.

We can find a general formula for permutations by writing the product $10 \times 9 \times 8 \times 7$ in a slightly more complex way:

$$10 \times 9 \times 8 \times 7 = \frac{10 \times 9 \times 8 \times 7 \times \cancel{6!}}{\cancel{6!}} = \frac{10!}{(10-4)!}$$

Note that, in the final formula on the right,

- the number 10 in the numerator is the number of swimmers from which the coach may choose the four relay participants, and
- the number $10 - 4 = 6$ in the denominator is the number of swimmers who do *not* swim in the relay.

Permutations have their own special notation: We read $_{10}P_4$ as "the number of permutations of ten items selected four at a time." Using this compact notation, we have

$$_{10}P_4 = \frac{10!}{(10-4)!} = 5040$$

Generalizing, we have a formula for calculating the number of permutations.

> **The Permutations Formula**
> If we make r selections from a group of n items, the number of permutations is
> $$_nP_r = \frac{n!}{(n-r)!}$$
> where $_nP_r$ is read as "the number of permutations of n items taken r at a time."

EXAMPLE 3 Leadership Election

A city has 12 candidates running for three leadership positions. The top vote-getter will become the mayor, the second vote-getter will become the deputy mayor, and the third vote-getter will become the treasurer. How many outcomes are possible for the three leadership positions?

Solution The voters are choosing $r = 3$ leaders from a group of $n = 12$ candidates. The order in which leaders are chosen matters because each of the three leadership positions is different. Thus, we are looking for the number of permutations of the 12 candidates taken 3 at a time:

$$_{12}P_3 = \frac{12!}{(12-3)!} = \frac{12!}{9!} = \frac{12 \times 11 \times 10 \times 9!}{9!} = 12 \times 11 \times 10 = 1320$$

There are 1320 possible outcomes for the three leadership positions. ∎

EXAMPLE 4 Batting Orders

How many ways can the manager of a baseball team form a (9-player) batting order from a roster of 15 players?

Solution The manager forms a batting order by selecting $r = 9$ players from the roster of $n = 15$ players. Thus, the number of possible batting orders is the number of permutations of 15 players taken 9 at a time:

$$_{15}P_9 = \frac{15!}{(15-9)!} = \frac{15!}{6!} = 1,816,214,400$$

Nearly 2 *billion* batting orders are possible for a baseball team with a roster of 15 players! ∎

Time out to think

Major league baseball teams carry a roster of 25 players. Without actually calculating, describe how the calculation of the number of possible batting orders would be different for a major league team than for the team in Example 4.

Using your **CALCULATOR**

Permutations

Many calculators have a special key for permutations, usually labeled $_nP_r$. For example, to find the number of permutations of $n = 10$ items taken $r = 4$ at a time, press 10 [$_nP_r$] 4. You should get the result 5040.

Combinations

Suppose that the five members of a city council decide they need a three-person committee to study the impact of a new shopping center. The council members are Ursula, Vern, Wendy, Yolanda, and Zeke (U, V, W, Y, and Z, for short). How many committees could be formed from the five council members?

First, note that this problem is *not* a permutation problem because we are interested only in the makeup of the committee, *not* the order in which the names are listed. For example, if we were dealing with permutations, we would consider ZWU to be different from WZU. In this case, however, both arrangements are equivalent because they both represent the same committee (with members Zeke, Wendy, and Ursula). Problems of this type are called **combinations** problems.

One way to count the possible committees is to list them all. Going in alphabetical order, we begin by listing all three-person committees that include Ursula:

UVW UVY UVZ UWY UWZ UYZ

Next, we list the committees that include Vern but not Ursula:

VWY VWZ VYZ

This leaves only one committee that includes neither Ursula nor Vern:

WYZ

We've found a total of ten three-person committees that can be made from the five council members.

This method of listing all the possible committees works, but it can become quite tedious when the numbers are large. Fortunately, we can get the same answer with an alternative strategy. First, we find the number of *permutations* of the $n = 5$ council members selected $r = 3$ at a time:

$$_5P_3 = \frac{5!}{(5-3)!} = \frac{5!}{2!} = 5 \times 4 \times 3 = 60$$

That is, there are 60 possible permutations when we choose three people from a group of five. However, because order matters for permutations but does not matter for the committees, the number of permutations is an *overcount* of the actual number of different committees. More specifically, any three-person committee can be listed in $3! = 3 \times 2 \times 1 = 6$ different orders. For example, the committee consisting of Zeke, Yolanda, and Wendy has six different permutations:

ZYW ZWY YZW YWZ WZY WYZ

Because each three-person committee is counted $3! = 6$ times by the permutations formula, this formula gives us six times the actual number of committees. Thus, we must divide the number of permutations by 3! to find the number of committees:

$$\frac{_5P_3}{3!} = \frac{60}{3!} = \frac{60}{3 \times 2 \times 1} = 10$$

This is the same result we obtained by listing the three-person committees. Note that the permutations part of the equation is $_nP_r$, where $n = 5$ and $r = 3$. We then divided this term by $r! = 3!$ to correct for the overcounting. Thus, the general formula for combinations is the permutations formula divided by $r!$.

> *A committee is a group of important individuals who singly can do nothing, but who together can agree that nothing can be done.*
>
> —FRED ALLEN, comedian

> **Combinations**
>
> Mathematically, we are dealing with **combinations** whenever all selections come from a single group of items, no item may be selected more than once, and the order of arrangement does *not* matter (for example, ABCD is considered to be the same as DCBA). If we make *r* selections from a group of *n* items, the number of possible combinations is
>
> $$_nC_r = \frac{_nP_r}{r!} = \frac{n!}{(n-r)! \times r!}$$
>
> where $_nC_r$ is read as "the number of combinations of *n* items taken *r* at a time."

EXAMPLE 5 Ice Cream Combinations

Suppose that you select 3 different flavors of ice cream in a shop that carries 12 flavors. How many flavor combinations are possible?

Solution We are looking for the number of combinations of $n = 12$ flavors selected $r = 3$ at a time. From the combinations formula, the number of flavor combinations is

$$_{12}C_3 = \frac{12!}{(12-3)! \times 3!} = \frac{12!}{9! \times 3!} = \frac{12 \times 11 \times 10 \times 9!}{9! \times 3!} = \frac{12 \times 11 \times 10}{3 \times 2 \times 1} = \frac{1320}{6} = 220$$

There are 220 different three-flavor combinations possible from the 12 flavors.

EXAMPLE 6 Poker Hands

How many different five-card poker hands can be dealt from a standard deck of 52 cards? What is the probability of a particular hand, such as a *royal flush* of hearts (a hand consisting of the ace, king, queen, jack, and 10 of hearts)?

Solution We are dealing with a combinations problem, because the order of the cards in a hand does *not* matter. The number of combinations of $n = 52$ cards drawn $r = 5$ at a time is

$$_{52}C_5 = \frac{52!}{(52-5)! \times 5!} = \frac{52!}{47! \times 5!} = \frac{52 \times 51 \times 50 \times 49 \times 48}{5 \times 4 \times 3 \times 2 \times 1} = 2,598,960$$

More than 2.5 million hands are possible in five-card poker. Any particular hand, including the royal flush of hearts, has a probability of about 1 in 2.5 million.

Time out to think

Consider the following unimpressive poker hand: 5 of clubs, 7 of hearts, 3 of diamonds, 2 of spades, 9 of hearts. How does the probability of this hand, which has no value in poker, compare to the probability of a royal flush of spades, which is the highest-value hand in poker? Explain.

EXAMPLE 7 (Not) Winning the Lottery

Suppose you play a lottery in which the winner is chosen by drawing 6 balls at random from a drum containing 52 numbered balls (numbered 1 through 52). What is the probability that your 6 numbers will match the 6 winning numbers?

Using your
CALCULATOR

Combinations

Many calculators have a special key for combinations, usually labeled $_nC_r$. For example, to find the number of combinations of $n = 10$ items taken $r = 4$ at a time, press 10 $\boxed{nC_r}$ 4. You should get the result 210.

historical note

New Hampshire began the first legalized lottery in the United States in 1964. Since then, 37 states have adopted lotteries as a source of revenue. Many states also participate in multi-state lottery games (such as Powerball game or the Multi-State Big Game). Lotteries have existed in various forms and places for centuries. For example, lotteries have been traced to the Roman Empire (1st century) and to Italy in the Middle Ages.

Solution The order in which the balls are drawn does not matter, so finding the total number of possible outcomes is a combinations problem. The number of combinations for $n = 52$ balls selected $r = 6$ at a time is

$$_{52}C_6 = \frac{52!}{(52-6)! \times 6!} = \frac{52!}{46! \times 6!} = 20{,}358{,}520$$

There are more than 20 million possible combinations of 6 numbers, so the chance that your 6 numbers will match the 6 winning numbers is less than 1 in 20 million.

PROBABILITY AND COINCIDENCE

Probability has important applications to the idea of coincidences. For example, across the United States in any given year, only a few dozen of the many millions of people who play lotteries win multimillion-dollar prizes. Because the probability of any particular person winning is so low, winners often attribute their fortune to "good luck" or to some special "system" they used in choosing lottery numbers. In fact, *someone* was bound to win and the particular winners actually won by mere coincidence. Said differently, it is virtually certain that *someone* will win a big lottery prize during the next year. Yet the probability that *you* will win the big prize is exceedingly small.

Coincidences Are Bound to Happen

Although a *particular* outcome may be highly unlikely, *some* similar outcome may be extremely likely or even certain to occur. In general, this means that coincidences will inevitably be experienced by someone, even if these coincidences have a low probability for any particular person.

Coincidences come in many different forms. You may be surprised to find that two people at your dinner party have the same birthday or that a friend's mother met your mother on a trip to China. You might flip a coin and get heads ten times in a row or

CALVIN AND HOBBES © 1990 Watterson. Reprinted with permission of UNIVERSAL PRESS SYNDICATE. All rights reserved.

have a dream that seems to predict an event that later occurs. It is tempting to chalk these cases up to something more mysterious than mere coincidence. But the laws of probability dictate that many coincidences are bound to happen, even though the particular form of a coincidence is unpredictable. The following two examples about birthday coincidences demonstrate one of the more surprising aspects of this idea.

EXAMPLE 8 A Particular Birthday Coincidence

Suppose there are 25 students in your class. What is the probability that at least one person in the class has the same birthday as you? Assume 365 days in a year.

Solution This is an *at least once* question (see Unit 6B). We begin by recognizing that, with 365 days in a year, the probability that any particular student has your birthday is 1/365. Therefore, the probability that a particular student does not have your birthday is 364/365. From the *at least once* rule, the probability that at least one of the other 24 other students has your birthday is

$$P(\text{at least one with your birthday}) = 1 - [P(\text{not your birthday})]^{24}$$
$$= 1 - \left[\frac{364}{365}\right]^{24}$$
$$= 1 - 0.936 = 0.064$$

The probability that at least one other student shares your birthday is only 0.064, which is about 1 in 16.

EXAMPLE 9 Some Birthday Coincidence

What is the probability that *some* pair of students in the class of 25 share the *same* birthday? Compare this result to the result from Example 8.

Solution This question differs from the question in Example 8 because the pair sharing the same birthday does not have to include you; it can be *any* pair of two students. But this is still an *at least once* question, because we are looking for the probability of a shared birthday among at least one pair of students. Thus, the answer will take the following form:

$$P(\text{at least one pair of shared birthdays}) = 1 - P(\text{no shared birthdays})$$

The easiest way to find the probability of no shared birthdays is by calculating the probability that all 25 students have *different* birthdays. We begin by considering just two students from the class. The first student has a birthday on 1 of the 365 days of the year, so the probability that the second student has a different birthday is 364/365. Now consider a third student. If the first two students have different birthdays, then 2 of the 365 days are already "taken." Thus, the probability that the third student's birthday falls on a third different day is 363/365, since 363 days are not yet "taken" for birthdays. Putting the probabilities together, we find that the chance that the first three students all have different birthdays is

$$\underbrace{\frac{364}{365}}_{\substack{\text{probability that first two students} \\ \text{have different birthdays}}} \times \underbrace{\frac{363}{365}}_{\substack{\text{probability that third student} \\ \text{also has different birthday}}}$$

Similarly, if the first three students all have different birthdays, the probability that a fourth student has a different birthday from any of the first three is 362/365. And so on. Finally, if the first 24 students all have different birthdays, then 24 of the 365 days are "taken," leaving 365 − 24 = 341 possible different birthdays for the 25th student. Thus, the probability that all 25 students have different birthdays is

$$\frac{364}{365} \times \frac{363}{365} \times \cdots \times \frac{341}{365} = \frac{364 \times 363 \times \cdots \times 341}{365^{24}}$$

Although we can write the product in the numerator more compactly as 364!/340!, the factorials are too large for most calculators to handle. Therefore, the better way to proceed is by computing the numerator and denominator separately. The result is

$$\frac{364 \times 363 \times \cdots \times 341}{365^{24}} \approx \frac{1.348 \times 10^{61}}{3.126 \times 10^{61}} \approx 0.431$$

This is the probability that there are no shared birthdays among the 25 students in the class. Thus, the probability that there is at least one pair of shared birthdays is

$$P(\text{at least one pair of shared birthdays}) = 1 - P(\text{no shared birthdays})$$
$$= 1 - 0.431 = 0.569$$

There's a 0.569 probability—which is about a 57% chance—that at least two of the 25 students have the same birthday! Most people are very surprised at how much higher this probability is than the 0.064 probability that one of the students shares *your* birthday (see Example 8), but it illustrates the fact that *some* coincidence is far more likely than a *particular* coincidence.

Review Questions

1. What are arrangements with repetition? Give an example of a situation in which the n^r formula gives the number of possible arrangements.
2. What do we mean by *permutations*? Give an example of a situation in which we would want to know the number of permutations of something.
3. Explain the meaning of each of the terms in the permutations formula. Give an example of its use.
4. What do we mean by *combinations*? Explain the meaning of each of the terms in the combinations formula. Give an example of its use.
5. Explain what we mean when we say that *some* outcome is much more likely than a particular outcome. How does this idea affect our perception of coincidences?

In the News

1. **Lottery Chances.** Find a newspaper article or advertisement about a lottery in which choosing a winner involves combinations. Calculate the probability of winning. Does your calculation agree with the advertised probability? Explain.
2. **Amazing Coincidence.** Discuss a seemingly amazing coincidence that you've read about or experienced yourself. Based on ideas of probability, decide whether the coincidence was bound to happen to someone. Was it really as amazing as it seemed? Explain.

Exercises

Review of Factorials. Exercises 1–2 use skills covered in the Brief Review on p. 340.

1. Find the values of the following quantities *without* using the factorial key on your calculator (you may use the multiplication key). Show your work.

 a. $7!$ b. $\dfrac{7!}{3!}$ c. $\dfrac{7!}{4!3!}$

 d. $\dfrac{12!}{5!}$ e. $\dfrac{12!}{8!(12-8)!}$ f. $\dfrac{21!}{20!}$

2. Find the values of the following quantities *without* using the factorial key on your calculator (you may use the multiplication key). Show your work.

 a. $6!$ b. $\dfrac{6!}{3!}$ c. $\dfrac{6!}{4!3!}$

 d. $\dfrac{11!}{5!}$ e. $\dfrac{11!}{4!(11-4)!}$ f. $\dfrac{16!}{15!}$

Counting Possibilities. Answer the questions in Exercises 3–18 using the appropriate counting technique, which may be either arrangements with repetition, permutations, or combinations. Be sure to explain why this counting technique applies to the problem.

3. How many different four-digit house addresses can be formed with the numerals 0–9?

4. How many different ten-note "tunes" can be created from the notes C, D, E, F, G, A, and B?

5. Supposedly, if a monkey types randomly on a typewriter long enough, it eventually will type the works of Shakespeare. How many different five-letter words could a monkey type from the 26 letters of the alphabet?

6. A lock consists of a dial that can be set to any of the numbers 1–30. How many different three-number combinations are possible?

7. How many different five-letter passwords can be formed from the letters A, B, C, D, E, and F if no repetition of letters is allowed?

8. Twelve acts at a comedy club all perform on one evening. How many different ways can their appearances be scheduled?

9. How many different three-letter "words" can be formed with the four letters of the genetic code (A, C, G, T)?

10. Ten finalists in a talent show must give final performances. Five contestants will perform on the first night of the show. How many different ways can the schedule for the first night be made?

11. How many different five-person subcommittees can be formed from a group of ten people?

12. How many different four-card hands can be dealt from a deck that has 52 cards?

13. You own ten compact disks, but plan to take only six with you on a road trip. How many different sets of six disks can you take?

14. You deal five cards from a standard deck of cards face up on a table. How many different sequences of cards (in which order matters) can you deal?

15. How many different four-person subcommittees can be formed from a group of 12 people?

16. How many different five-card hands can be dealt from a deck that has only face cards and aces (16 cards altogether)?

17. Suppose that license plates are made with three letters followed by three numerals (0–9). How many different license plates are possible?

18. The coach of the debate club must choose a team of 8 people to travel to the next meet. How many different teams can be chosen from the 20 members of the club?

19. **Ice Cream Shop.** Josh and John's Ice Cream shop offers 12 different flavors of ice cream and 6 different toppings. Answer the following questions by using the appropriate counting technique (multiplication principle, arrangements with repetitions, permutations, or combinations). Explain why you chose the particular counting technique.

 a. How many different sundaes can you create using one of the ice cream flavors and one of the toppings?

 b. How many different triple cones can you create from the 12 flavors if the same flavor may be used more than once? Assume that you specify which flavor goes on the bottom, middle, and top.

 c. Using the 12 flavors, how many different triple cones can you create with 3 *different* flavors if you specify which flavor goes on the bottom, middle, and top?

 d. Using the 12 flavors, how many different triple cones can you create with 3 *different* flavors if you don't care about the order of the flavors on the cone?

20. **Telephone Numbers.** A seven-digit phone number in the United States consists of a three-digit *exchange* followed by a four-digit number.

 a. The first digit of the exchange is not allowed to be 0 or 1. How many different seven-digit phone numbers can be formed? Can a city of 2 million people be served by a single area code? Explain.

 b. How many exchanges are needed to serve a city of 80,000 people? Explain.

21. **Pizza Hype.** Luigi's Pizza Parlor advertises 84 different three-topping pizzas. How many individual toppings does Luigi actually use? Ramona's Pizzeria advertises 45 different two-topping pizzas. How many toppings does Ramona actually use? (Hint: In these problems, you are given the total number of combinations, and you must find the number of toppings that are used.)

22. **ZIP Codes.** The U.S. Postal Service uses both five-digit and nine-digit ZIP codes.

 a. How many five-digit ZIP codes are available to the U.S. Postal Service?

 b. For a U.S. population of 270 million people, what is the average number of people per ZIP code, if all possible ZIP codes are used? Explain.

 c. How many nine-digit ZIP codes are available to the U.S. Postal Service? Could everyone in the United States have his or her own personal nine-digit ZIP code? Explain.

Counting and Probability. In Exercises 23–28, find the probability of the given event.

23. Choosing five numbers that match five randomly selected balls when the balls are numbered 1 through 32

24. Choosing six numbers that match six randomly selected balls when the balls are numbered 1 through 40

25. Being dealt an ace, 2, 3, 4, and 5 all of the same suit, from a standard 52-card deck of cards

26. Being dealt a 10, jack, queen, king, and ace, all of the same suit, from a standard 52-card deck of cards

27. Guessing the top 3 winners (in order) from a group of 10 finalists in a soccer tournament

28. Guessing the top 3 winners (in any order) from a group of 16 finalists in a spelling bee

Birthday Coincidences. Suppose you are part of a group of people at a dinner party. For the situations given in Exercises 29–30, find the probability that at least one of the other guests has *your* birthday and the probability that *some* pair of guests share the same birthday. Discuss your results.

29. You are one of 10 people at the party.

30. You are one of 15 people at the party.

31. **Hot Streaks.** Suppose that 2000 people are all playing a game for which the chance of winning is 48%.

 a. What is the probability of one person winning five games in a row?

 b. What is the probability of one person winning ten games in a row?

 c. On average, how many of the 2000 people could be expected to have a "hot streak" of five games?

 d. On average, how many of the 2000 people could be expected to have a "hot streak" of ten games?

32. **Coin Streaks.** Toss a coin 100 times and record your results (heads or tails) in order. What was your longest streak of consecutive heads or tails? Calculate the probability of this streak by itself. For example, if your longest streak was four heads, what is the probability of four heads in four tosses? Should you be surprised that you got such a streak? Explain.

33. **Joe DiMaggio's Record.** One of the longest standing records in sports is the 56-game hitting streak (in baseball) of Joe DiMaggio. Assume that a player on a long "hot streak" is batting about .400, which is about the best that anyone ever hits over a period of 50 or more games. (A batting average of .400 means the batter gets a hit 40% of the time. Typically, only a handful of players each year hit that well for a period as long as 56 games.)

a. What is the probability that a player batting .400 will get at least one hit in four at-bats?

b. Use the result in part a to calculate the probability of a .400 hitter getting a hit in 56 consecutive games.

c. Suppose that, instead of batting .400, a player has a more ordinary average of .300. In that case, what is the probability of the player getting a hit in 56 consecutive games, assuming four at-bats per game?

d. Considering the results in parts b and c and the fact that baseball has been played for about 100 years, are you surprised that *someone* set a record of hitting in 56 consecutive games? Explain clearly.

Web Project

Find useful links for Web Projects on the text Web site: www.aw.com/bennett-briggs

1. **Let's Make a Deal.** In September 1990, controversy about a probability question erupted over an item in the "Ask Marilyn" column in *Parade* magazine (carried by many newspapers across the country). Her column answered a question loosely based on a TV game show called "Let's Make a Deal." Here's the question:

Suppose that you're on a game show and you're given the choice of three doors: Behind one door is a car; behind the others, goats. You pick a door, say No. 1, and the host, who knows what's behind the doors, opens another door, say No. 3, which has a goat. He then says to you, "Do you want to change your pick to door No. 2?" Is it to your advantage to switch your choice?

Marilyn answered that the probability of winning was higher if the contestant switched. This answer generated a huge number of letters, including a few from mathematicians, claiming that she was wrong. Marilyn answered with the following logic. When you first pick door No. 1, the chance that you picked the one with the car is 1/3. The probability that you chose a door with a goat is 2/3. When the host opens door No. 3 to reveal a goat, it does not change the 1/3 probability that you picked the right door in the first place. Thus, as only one other door remains, the probability that it contains the car is 2/3. Visit a few of the many Web sites devoted to this problem to gain some understanding of its subtleties. Do you agree with Marilyn's logic? If so, try to explain it in your own words. If not, present an alternative approach.

CHAPTER 6 SUMMARY

UNIT	KEY TERMS	KEY IDEAS AND SKILLS
6A	outcome event theoretical probability empirical probability subjective probability probability distribution odds	**Distinguish between** theoretical, empirical, and subjective probabilities. **Determine** theoretical probabilities: $$P(A) = \frac{\text{number of ways } A \text{ can occur}}{\text{total number of outcomes}}$$ **Make** a probability distribution.
6B	*and* probability independent events dependent events *either/or* probability non-overlapping events overlapping events	***And* probability, independent events:** $P(A \text{ and } B) = P(A) \times P(B)$ ***And* probability, dependent events:** $P(A \text{ and } B) = P(A) \times P(B \text{ given } A)$ ***Either/or* probability, non-overlapping events:** $P(A \text{ or } B) = P(A) + P(B)$ ***Either/or* probability, overlapping events:** $P(A \text{ or } B) = P(A) + P(B) - P(A \text{ and } B)$ **The *at least once* rule:** $P(\text{at least one event } A \text{ in } n \text{ trials}) = 1 - P(\text{no events } A \text{ in } n \text{ trials})$ $= 1 - [P(\text{not } A \text{ in one trial})]^n$
6C	law of large numbers expected value gambler's fallacy house edge	**Understand and apply** the law of large numbers. **Calculate and interpret** expected values: $$\text{expected value} = \binom{\text{event 1}}{\text{value}} \times \binom{\text{event 1}}{\text{probability}} + \binom{\text{event 2}}{\text{value}} \times \binom{\text{event 2}}{\text{probability}}$$
6D	arrangements with repetition permutations combinations	**Arrangements with repetition:** For r selections from a group of n choices, the number of arrangements is n^r. **Permutations:** For r selections from a group of n items, $$_nP_r = \frac{n!}{(n-r)!}$$ **Combinations:** For r selections from a group of n items, $$_nC_r = \frac{_nP_r}{r!} = \frac{n!}{(n-r)! \times r!}$$ **Coincidences are bound to happen**—understand why and the implications for probability.

7

The greatest shortcoming of the human race is [its] inability to understand the exponential function.

—Albert A. Bartlett, Professor of Physics, University of Colorado

EXPONENTIAL ASTONISHMENT

World population is currently growing by some 80 million people per year—enough to fill an entire new United States in just over three years. A growing population means new challenges for our species, which we can meet only if we understand this growth. In this chapter, we will investigate the mathematical laws of growth—specifically, exponential growth. We will focus on what we call *exponential astonishment:* the intuition-defying reality of exponential growth. Although we will emphasize population growth, we will also study many other important topics, including the decay of waste from nuclear power plants and the depletion of natural resources.

UNIT 7A Growth: Linear versus Exponential: We will contrast the two basic forms of growth: *linear growth* by the same absolute amount in any time period and *exponential growth* by the same relative amount in any time period. We will also explore the remarkable effects of the repeated doublings that characterize exponential growth.

UNIT 7B Doubling Time and Half-Life: Exponential growth is characterized by a constant *doubling time,* while exponential decay is characterized by a constant *half-life.* We will explore both ideas and some of their many applications.

UNIT 7C Exponential Modeling: Writing functions to describe exponential growth and decay allows us to build on the ideas introduced in Units 7A and 7B. We will explore many applications of exponential functions, including models of population growth and techniques for radioactive dating.

UNIT 7A

GROWTH: LINEAR VERSUS EXPONENTIAL

FIGURE 7.1 Straightown grows linearly, while Powertown grows exponentially.

Imagine two communities, Straightown and Powertown, each with an initial population of 10,000 people (Figure 7.1). Straightown grows at a constant rate of 500 people per year, so its population reaches 10,500 after 1 year, 11,000 after 2 years, 11,500 after 3 years, and so on. Powertown grows at a constant rate of 5 *percent* per year. Because 5% of 10,000 is 500, Powertown's population also reaches 10,500 after 1 year. In the second year, however, Powertown's population increases by 5% of 10,500, which is 525, to 11,025. In the third year, Powertown's population increases by 5% of 11,025, or by 551 people. Figure 7.1 contrasts the populations of the two towns over a 45-year period. Note that Powertown's population rises ever more steeply and quickly outpaces Straightown's.

Straightown and Powertown illustrate two fundamentally different types of growth. Straightown grows by the same *absolute* amount—500 people—from one year to the next, which is a characteristic of **linear growth.** In contrast, Powertown grows by the same *relative* amount—5%—each year, which is a characteristic of **exponential growth.**

> **Two Basic Growth Patterns**
>
> **Linear growth** occurs when a quantity grows by the same *absolute* amount in each unit of time.
>
> **Exponential growth** occurs when a quantity grows by the same *relative* amount—that is, by the same *percentage*—in each unit of time.

The terms *linear* and *exponential* can also be applied to quantities that decrease with time. For example, *exponential decay* occurs when a quantity decreases by the same relative amount in each unit of time. In the rest of this unit, we will explore the surprising properties of exponential growth.

EXAMPLE 1 *Linear or Exponential?*

In each of the following situations, state whether the growth (or decay) is linear or exponential, and answer the associated questions.

a. The number of students at Wilson High School has increased by 50 in each of the past four years. If the student population was 750 four years ago, what is it today?

b. The price of milk has been rising with inflation at 3% per year. If the price of a gallon of milk was $2 a year ago, what is it now?

c. Tax law allows you to depreciate the value of your equipment by $200 per year. If you purchased the equipment three years ago for $1000, what is its depreciated value today?

d. The memory capacity of state-of-the-art computer hard drives is doubling approximately every two years. If a company's top-of-the-line drive holds 100 gigabytes today, what will it hold in six years?

e. The price of DVD players has been falling by about 25% per year. If the price is $200 today, what can you expect it to be in two years?

Solution

a. The number of students increased by the same absolute amount each year, so this is *linear growth*. Because the student population increased by 50 students per year, in four years it grew by $4 \times 50 = 200$ students, from 750 to 950.

b. The price rises by the same percent each year, so this is *exponential growth*. If the price was $2 a year ago, it increased by $0.03 \times \$2 = \0.06, making the price $2.06.

c. The equipment value depreciates by the same amount each year, so this is *linear decay*. In three years, the value decreases by $3 \times \$200 = \600, so the value decreases from $1000 to $400.

d. A doubling is the same as a 100% increase, so the two-year doubling time represents *exponential growth*. With a doubling every two years, the capacity will double three times in six years: from 100 to 200 gigabytes after two years, from 200 to 400 gigabytes after four years, and from 400 to 800 gigabytes after six years.

e. The price decreases by the same percentage each year, so this is *exponential decay*. From $200 today, the price will fall by 25%, or $0.25 \times \$200 = \50, in one year. Therefore, next year's price will be $\$200 - \$50 = \$150$. The following year, the price will again fall by 25%, or $0.25 \times \$150 = \37.50, so the price after two years will be $\$150 - \$37.50 = \$112.50$.

THE IMPACT OF DOUBLINGS

Look again at the graph of Powertown's population in Figure 7.1. After about 14 years, the original population has doubled to 20,000. In the next 14 years, it doubles again to 40,000. It then doubles again, to 80,000, 14 years after that. This type of repeated *doubling,* in which each doubling occurs in the same amount of time, is a hallmark of exponential growth.

The time it takes for each doubling depends on the rate of the exponential growth; in Unit 7B, we'll see how the doubling time depends on the percentage growth rate. Here, we'll explore three parables that show how doublings make exponential growth so very different from linear growth.

Parable 1: From Hero to Headless in 64 Easy Steps

Legend has it that, when chess was invented in ancient times, a king was so enchanted that he said to the inventor, "Name your reward."

"If you please, king, put one grain of wheat on the first square of my chessboard," said the inventor. "Then, place two grains on the second square, four grains on the third square, eight grains on the fourth square, and so on." The king gladly agreed, thinking the man a fool for asking for a few grains of wheat when he could have had gold or jewels. But let's see how it adds up for the 64 squares on a chessboard.

Table 7.1 shows the calculations. Each square gets twice as many grains as the previous square, so the number of grains on any square is a power of 2. The third column shows the total number of grains up to each point, and the last column gives a simple formula for the total number of grains.

TABLE 7.1

SQUARE	GRAINS ON THIS SQUARE	TOTAL GRAINS THUS FAR	FORMULA FOR TOTAL GRAINS
1	$1 = 2^0$	1	$2^1 - 1$
2	$2 = 2^1$	$1 + 2 = 3$	$2^2 - 1$
3	$4 = 2^2$	$3 + 4 = 7$	$2^3 - 1$
4	$8 = 2^3$	$7 + 8 = 15$	$2^4 - 1$
5	$16 = 2^4$	$15 + 16 = 31$	$2^5 - 1$
⋮	⋮	⋮	⋮
64	2^{63}		$2^{64} - 1$

From the pattern in the last column, we see that the grand total for all 64 squares is $2^{64} - 1$ grains. How much wheat is this? With a calculator, you can confirm that $2^{64} = 1.8 \times 10^{19}$, or about 18 *billion billion*. Not only would it be difficult to fit so many grains on a chessboard, but this number happens to be larger than the total number of grains of wheat harvested in all human history. The king never finished paying the inventor and, according to legend, instead had him beheaded.

Parable 2: The Magic Penny

One lucky day, you meet a leprechaun who promises to give you fantastic wealth, but hands you only a penny before disappearing. You head home and place the penny under your pillow. The next morning, to your surprise, you find two pennies under your pillow. The following morning, you find four pennies, and the fourth morning, eight pennies. Apparently, the leprechaun gave you a *magic* penny: While you sleep, each magic penny turns into *two* magic pennies. Table 7.2 shows your growing wealth. Note that "day 0" is the day you met the leprechaun. Generalizing from the table, we find that the amount under your pillow after t days is

$$\$0.01 \times 2^t$$

TABLE 7.2

DAY	AMOUNT UNDER PILLOW
0	$\$0.01 \times 2^0 = \0.01
1	$\$0.01 \times 2^1 = \0.02
2	$\$0.01 \times 2^2 = \0.04
3	$\$0.01 \times 2^3 = \0.08
4	$\$0.01 \times 2^4 = \0.16

We can use this formula to figure out how long it will be until you have fantastic wealth. After $t = 9$ days, you'll have $\$0.01 \times 2^9 = \5.12, which is barely enough to

buy lunch. But by the end of a month, or $t = 30$ days, you'll have $\$0.01 \times 2^{30} = \$10,737,418.24$. That is, you'll be a millionaire within a month, and you'll need a much larger pillow! In fact, if your magic pennies keep doubling, by the end of just 50 days you'll have $\$0.01 \times 2^{50} \approx \11.3 trillion, more than enough to pay off the national debt of the United States.

Parable 3: Bacteria in a Bottle

For our third parable, we turn to a story with much more relevance to the future of the human race. Suppose you place a single bacterium in a bottle at 11:00 a.m. It grows and at 11:01 divides into two bacteria. These two bacteria each grow and at 11:02 divide into four bacteria, which grow and at 11:03 divide into eight bacteria, and so on.

Now, suppose the bacteria continue to double every minute, and the bottle is full at 12:00. You may already realize that the number of bacteria at this point must be 2^{60} (because they doubled every minute for 60 minutes), but the important fact is that we have a bacterial disaster on our hands: Because the bacteria have filled the bottle, the entire bacterial colony is doomed. Let's examine this tragedy in greater detail, by asking a few questions about the demise of the colony.

- Question 1: If the tragedy occurred because the bottle was full at 12:00, when was the bottle *half*-full?

Because it took one hour to fill the bottle, many people guess that it was half-full after a half-hour, or at 11:30. However, because the bacteria double in number every minute, they must also have doubled during the last minute. Thus, the bottle went from being half-full to full during the final minute; that is, the bottle was half-full at 11:59, just 1 minute before the disaster.

- Question 2: Imagine that *you* are a mathematically sophisticated bacterium, and at 11:56 you recognize the impending disaster. You immediately jump on your soapbox and warn that unless your fellow bacteria slow their growth dramatically, the end is just 4 minutes away. Will anyone believe you?

You've done your mathematics correctly; the bottle will indeed be full in just 4 minutes. But how will it appear to others who have not done the calculations? As we've already seen, the bottle will be half-full at 11:59. Continuing to work backward through the doublings each minute, we find that it will be $\frac{1}{4}$ full at 11:58, $\frac{1}{8}$ full at 11:57, and $\frac{1}{16}$ full at 11:56. Thus, if your fellow bacteria look around the bottle at 11:56, they'll see that only $\frac{1}{16}$ of the bottle has been used. In other words, the amount of unused space is *15 times* the amount of used space. You are asking them to believe that, in just the next 4 minutes, they'll fill 15 times as much space as they did in their entire 56-minute history. Unless they do the mathematics for themselves, they are unlikely to take your warnings seriously. Figure 7.2 shows the situation graphically; note that the bottle remains nearly empty for most of the 60 minutes, but the continued doublings fill it rapidly in the final 4 minutes.

- Question 3: It's 11:59 and, with the bottle half-full, your fellow bacteria are finally taking your warnings seriously. They quickly start a space program, sending little bacterial spaceships out into the lab in search of new bottles. Thankfully, they discover three more bottles (making a total of four, including the one already occupied). Working quickly, they initiate a mass migration by packing bacteria onto spaceships and sending them to the new

by the way...
The bacteria in a bottle parable was developed by Professor Albert A. Bartlett, University of Colorado, who has given more than 1200 lectures on the lessons of exponential growth around the country over the past 30 years.

Facts do not cease to exist because they are ignored.
—ALDOUS HUXLEY

FIGURE 7.2 The population of the bacteria in the bottle.

bottles. They successfully distribute the population evenly among the four bottles, just in time to avert the disaster. Given that they now have four bottles rather than just one, how much time have they gained for their civilization?

Since it took one hour to fill one bottle, you might guess that it would take four hours to fill four bottles. But the bacterial population still *doubles* each minute. If there are enough bacteria to fill one bottle at 12:00, there will be enough to fill two bottles by 12:01 and four bottles by 12:02. The discovery of three new bottles gives them only 2 additional minutes.

- Question 4: Suppose the bacteria continue their space program, constantly looking for more bottles. Is there any hope that further discoveries will allow the colony to continue its exponential growth?

Let's do some calculations. Note that, after n minutes, the bacterial population is 2^n; for example, it is $2^0 = 1$ when the first bacterium starts the colony at 11:00, $2^1 = 2$ at 11:01, $2^2 = 4$ at 11:02, and so on. There are 2^{60} bacteria when the first bottle fills at 12:00, and 2^{62} bacteria when four bottles are full at 12:02. Suppose that, somehow, the bacteria managed to keep doubling every minute until 1:00. By that time, the number of bacteria would be 2^{120}, because it has been 120 minutes since the colony began. Now, we must figure out how much space they'd require for this population.

The smallest bacteria measure approximately 10^{-7} m (0.1 micrometer) across. If we assume that the bacteria are roughly cube-shaped, the volume of a single bacterium is

$$(10^{-7} \text{ m})^3 = 10^{-21} \text{ m}^3$$

Therefore, the colony of 2^{120} bacteria would occupy a total volume of

$$2^{120} \times 10^{-21} \text{ m}^3 = 1.3 \times 10^{15} \text{ m}^3$$

With this volume, the bacteria would cover the entire surface of the Earth in a layer more than 2 meters deep! (See Exercise 19 to calculate this result for yourself.)

In fact, if the doublings continued for just $5\frac{1}{2}$ hours, the volume of bacteria would exceed the volume of the entire universe (see Exercise 20). Needless to say, this cannot happen. The exponential growth of the colony cannot possibly continue for long, no matter what technological advances might be imagined.

Time out to think

Some people have suggested that we could find room for an ever-growing human population by colonizing other planets in our solar system. Is this possible?

DOUBLING LESSONS

The three parables reveal at least two key lessons about the repeated doublings that arise with exponential growth. First, if you look back at Table 7.1, you'll notice that the number of grains on each square is nearly equal to the total number of grains on all previous squares combined. For example, the 16 grains on the fifth square are 1 more than the total of 15 grains on the first four squares combined.

practical matters

HUMAN POPULATION GROWTH

From the time of our earliest ancestors until just a few centuries ago, human population grew very slowly, because the global birth rate was almost precisely balanced against the death rate. But more recently, advances in agriculture, industry, and medicine have dramatically lowered death rates. As a result, births now outpace deaths significantly (even though global birth rates are at historic lows), which means that human population is growing exponentially. Figure 7.3 shows estimated human population over the past 12,000 years.

The sheer numbers involved in current population growth are astonishing. For example, in recent years:

- Every three years, the world has added as many people as live in the entire United States.
- Each month, world population has increased by the equivalent of population of Switzerland.

Indeed, if recent trends continue, world population will increase by about 10,000 people during the time you are reading this chapter.

World Population, 10,000 B.C. to Present

FIGURE 7.3 World population.

Clearly, the recent rate of growth cannot continue because, if it did, humanity would be elbow-to-elbow on Earth in only a few centuries. Thus, like the growth of bacteria in a bottle, the exponential growth of human population *will* soon stop. The only questions are *when* and *how*.

First consider the question of *when*. The maximum number of people that the Earth can support, which we call the Earth's *carrying capacity*, is a subject of great debate. Nevertheless, the most optimistic estimates put the carrying capacity around 40 billion people, which we would reach within 150 years at the current growth rate. Thus, regardless of other assumptions, population growth must stop or slow considerably within the next 150 years—which is quite soon on the scale of human history.

As for *how*, there are only two basic ways to slow the growth of a population:

- a decrease in the birth rate or
- an increase in the death rate.

Most people are already choosing the first option, as birth rates now are at historic lows. Indeed, population actually is *decreasing* in a few European nations. Nevertheless, worldwide birth rates still are much higher than death rates, so exponential growth continues.

If a decrease in birth rates doesn't halt the exponential growth, an increase in the death rate will. If population significantly overshoots the carrying capacity by the time this process begins, the increase in the death rate will be dramatic—probably on a scale never before seen. This forecast is not a threat, a warning, or a prophecy of doom. It is simply a law of nature: *Exponential growth always stops.*

As human beings, we can choose to slow our population growth through intelligent and careful decisions. Or, we can choose to do nothing, leaving ourselves to the mercy of natural forces over which we have no more control than we do over hurricanes, tornadoes, earthquakes, or the explosions of distant stars. Either way, it's a choice that each and every one of us must make, and upon which our entire future depends.

Second, all three parables show quantities growing to impossible proportions. We cannot possibly fit all the wheat harvested in world history on a chessboard; $5 trillion worth of pennies is far more than the number of pennies that exist in the world; and a colony of bacteria can't possibly fill the universe.

> **Key Facts About Exponential Growth**
>
> - Exponential growth leads to repeated doublings. With each doubling, the amount of increase is approximately equal to the *sum* of all preceding doublings.
> - Exponential growth cannot continue indefinitely. After only a relatively small number of doublings, exponentially growing quantities reach impossible proportions.

Review Questions

1. Describe the basic differences between linear growth and exponential growth.
2. Briefly explain how repeated doublings characterize exponential growth. Describe the impact of doublings, using the chessboard or magic penny parable.
3. Briefly summarize the story of the bacteria in the bottle. Be sure to explain the answers to the four questions asked in the text, and describe why the answers are surprising.
4. Explain the meaning of the two key facts about exponential growth given at the end of this unit.
5. Create your own example of exponential growth and describe the influence and impact of repeated doubling.

In the News

1. **Linear Growth.** Identify at least two news stories that describe a quantity undergoing linear growth or decay. Describe the growth or decay process.
2. **Exponential Growth.** Identify at least two news stories that describe a quantity undergoing exponential growth or decay. Describe the growth or decay process.

Exercises

Linear or Exponential? For each of Exercises 1–8, state whether the growth (or decay) is linear or exponential, and answer the associated questions.

1. The population of Danbury is increasing at a rate of 505 people per year. If the population is 15,000 today, what will it be in three years?
2. The population of Winesburg is increasing at a rate of 2% per year. If the population is 100,000 today, what will it be in three years?
3. During the worst periods of hyperinflation in Brazil, the price of food increased at a rate of 30% per month. If your food bill was $100 one month during this period, what was it two months later?
4. The price of a gallon of gasoline is decreasing by 2¢ per week. If the price is $1.60 per gallon today, what will it be in ten weeks?
5. The price of computer memory is decreasing at a rate of 12% per year. If a memory chip costs $80 today, what will it cost in two years?
6. The value of your car is decreasing by 10% per year. If the car is worth $10,000 today, what will it be worth in two years?
7. The value of your house is rising by $10,000 per year. If it is worth $100,000 today, what will it be worth in three years?
8. The value of your house is rising by 10% per year. If it is worth $100,000 today, what will it be worth in three years?

Chessboard Parable. Use the chessboard parable to answer Exercises 9–12. Assume that each grain of wheat weighs 1/7000 pound (see Unit 2B for the definition of a grain as a unit of weight).

9. How many grains of wheat should be placed on square 20 of the chessboard? Find the total number of grains and their total weight (in pounds) at this point.

10. How many grains of wheat should be placed on square 30 of the chessboard? Find the total number of grains and their total weight (in pounds) at this point.

11. What is the total weight of all the wheat when the chessboard is full?

12. According to the U.S. Department of Agriculture, the current world harvest of all wheat, rice, and corn is somewhat less than 2 billion tons per year. Compare this total to the weight of the wheat on the chessboard. (Hint: 1 ton = 2000 pounds.)

Magic Penny Parable. Use the magic penny parable to answer Exercises 13–16.

13. How much money would you have after 15 days?

14. Remember that all your money is in pennies. Suppose that you stacked the pennies after 15 days. How high would the stack rise? (Hint: Find a few pennies and a ruler.)

15. How many days would elapse before you had a total of over $1 billion? (Hint: Proceed by trial and error.)

16. Suppose that you could keep making a single stack of the pennies. After how many days would the stack be long enough to reach the nearest star (beyond the Sun), which is about 4.3 light-years (4.0×10^{13} km) away? (Hint: Proceed by trial and error.)

Bacteria in a Bottle Parable. Use the bacteria parable to answer Exercises 17–18.

17. How many bacteria are in the bottle at 11:50? What fraction of the bottle is full at that time?

18. How many bacteria are in the bottle at 11:10? What fraction of the bottle is full at that time?

19. **Knee-Deep in Bacteria.** The total surface area of the Earth is about 5.1×10^{14} m². Assume that the bacteria continued their doublings for two hours (as discussed in the text), at which point they were distributed uniformly over the Earth's surface. How deep would the bacterial layer be? Would it be knee-deep, more than knee-deep, or less than knee-deep? (Hint: Remember that a volume divided by an area gives a depth.)

20. **Bacterial Universe.** Suppose the bacteria in the parable continued to double their population every minute. How long would it take until their volume exceeded the total volume of the observable universe, which is about 10^{79} m³? (Hint: Proceed by trial and error.)

21. **Human Doubling.** Human population in the year 2000 was about 6 billion. Suppose this population increases exponentially with a doubling time of 50 years.

 a. Continue the following table, showing the population at 50-year intervals under this scenario, until you reach the year 3000. Use scientific notation, as shown.

Year	Population
2000	6×10^9
2050	$12 \times 10^9 = 1.2 \times 10^{10}$
2100	2.4×10^{10}
⋮	⋮

 b. The total surface area of the Earth is about 5.1×10^{14} m². Assuming that people could occupy all this area (in reality, most of it is ocean), approximately when would people be so crowded that every person would have only 1 m² of space?

 c. Suppose that, when we take into account the area needed to grow food and to find other resources, each person actually requires about 10^4 m² of area to survive. About when would we reach this limit?

 d. Suppose that we learn to colonize other planets and moons in our solar system. The total surface area of the worlds in our solar system that could potentially be colonized (not counting gas planets such as Jupiter) is roughly five times the surface area of the Earth. Under the assumptions of this problem, could humanity fit in our solar system in the year 3000? Explain.

Web Projects

Find useful links for Web Projects on the text Web site: www.aw.com/bennett-briggs

1. **Computing Power.** Computing power is often said to increase exponentially. Choose an aspect of computing power (such as processor speed or memory chip capacity) and investigate its growth. Has the growth been exponential? How much longer is the exponential growth likely to continue? Explain.

2. **Web Growth.** Investigate the growth of the Web itself, in terms of both number of users and number of Web pages. Has the growth been linear or exponential? How do you think the growth will change in the future? Explain.

UNIT 7B

DOUBLING TIME AND HALF-LIFE

Exponential growth leads to repeated doublings and exponential decay leads to repeated *halvings*. However, in most cases of exponential growth or decay, we are given the rate of growth or decay—usually as a percentage—rather than the time required for doubling or halving. In this unit, we'll convert between percentage rates and doubling (or halving) times.

DOUBLING TIME

The time required for each doubling in exponential growth is called the **doubling time**. For example, the doubling time for the magic penny (see Unit 7A) was one day, because your wealth doubled each day. The doubling time for the bacteria in the bottle was one minute.

Given the doubling time, we can easily calculate the value of a quantity at any time. Consider an initial population of 10,000 that grows with a doubling time of 10 years:

- In 10 years, or one doubling time, the population increases by a factor of 2, to a new population of $2 \times 10,000 = 20,000$.
- In 20 years, or two doubling times, the population increases by a factor of $2^2 = 4$, to a new population of $4 \times 10,000 = 40,000$.
- In 30 years, or three doubling times, the population increases by a factor of $2^3 = 8$, to a new population of $8 \times 10,000 = 80,000$.

We can generalize this idea by letting t be the amount of time that has passed and T_{double} be the doubling time. Note that after $t = 30$ years with a doubling time of $T_{double} = 10$ years, there have been $t/T_{double} = 30/10 = 3$ doublings. Generalizing, we find that the number of doublings after a time t is t/T_{double}. That is, the size of the population after time t is the initial population times $2^{t/T_{double}}$.

> **Calculations with the Doubling Time**
>
> After a time t, an exponentially growing quantity with a doubling time of T_{double} increases in size by a factor of $2^{t/T_{double}}$. The new value of the growing quantity is related to its initial value (at $t = 0$) by
>
> $$\text{new value} = \text{initial value} \times 2^{t/T_{double}}$$

Time out to think

Consider an initial population of 10,000 that grows with a doubling time of 10 years. Confirm that the above formula gives a population of 80,000 after 30 years, as we found earlier. What does the formula predict for the population after 50 years?

EXAMPLE 1 Doubling with Compound Interest

Compound interest (Unit 4A) is a form of exponential growth because an interest-bearing account grows by the same percentage each year. Suppose your bank account has a doubling time of 13 years. By what factor does your balance increase in 50 years?

Solution The doubling time is $T_{double} = 13$ years, so after $t = 50$ years your balance increases by a factor of

$$2^{t/T_{double}} = 2^{50 \text{ yr}/13 \text{ yr}} = 2^{3.85} = 14.382$$

For example, if you start with a balance of $1000, in 50 years it will grow to $1000 × 14.382 = $14,382.

EXAMPLE 2 World Population Growth

World population doubled from 2.6 billion in 1950 to 5.2 billion in 1990. Suppose that, from 1990 onward, world population continues to grow with a doubling time of 40 years. What will the population be in 2020? in 2190?

Solution The doubling time is $T_{double} = 40$ years. If we let $t = 0$ represent 1990, the year 2020 is $t = 30$ years later. If the 1990 population of 5.2 billion is used as the initial value, the population in 2020 will be

$$\text{new value} = \text{initial value} \times 2^{t/T_{double}}$$
$$= 5.2 \text{ billion} \times 2^{30 \text{ yr}/40 \text{ yr}}$$
$$= 5.2 \text{ billion} \times 2^{0.75} = 8.7 \text{ billion}$$

By 2190, which is $t = 200$ years after 1990, the population will reach

$$\text{new value} = \text{initial value} \times 2^{t/T_{double}}$$
$$= 5.2 \text{ billion} \times 2^{200 \text{ yr}/40 \text{ yr}}$$
$$= 5.2 \text{ billion} \times 2^5 = 166.4 \text{ billion}$$

If world population continues to grow at the same rate it did between 1950 and 1990, it will reach 8.7 billion by 2020 and more than 166 billion by 2190.

Time out to think

Do you think that it's really possible for human population to reach 166 billion in 200 years? Why or why not?

THE APPROXIMATE DOUBLING TIME FORMULA

Consider an ecological study of a prairie dog community. The community contains 100 prairie dogs when the study begins, and researchers soon determine that the population is increasing at a rate of 10% per month. That is, each month the population grows to 110% of, or 1.1 times, its previous value (see the "of versus more than" rule in Unit 3A). Table 7.3 tracks the population growth.

Dilbert reprinted by permission of United Feature Syndicate, Inc.

TABLE 7.3 Growth of a Prairie Dog Community

MONTH	POPULATION	MONTH	POPULATION
0	100	8	$(1.1)^8 \times 100 = 214$
1	$(1.1)^1 \times 100 = 110$	9	$(1.1)^9 \times 100 = 236$
2	$(1.1)^2 \times 100 = 121$	10	$(1.1)^{10} \times 100 = 259$
3	$(1.1)^3 \times 100 = 133$	11	$(1.1)^{11} \times 100 = 285$
4	$(1.1)^4 \times 100 = 146$	12	$(1.1)^{12} \times 100 = 314$
5	$(1.1)^5 \times 100 = 161$	13	$(1.1)^{13} \times 100 = 345$
6	$(1.1)^6 \times 100 = 177$	14	$(1.1)^{14} \times 100 = 380$
7	$(1.1)^7 \times 100 = 195$	15	$(1.1)^{15} \times 100 = 418$

Note that the population nearly doubles (to 195) after 7 months, then nearly doubles again (to 380) after 14 months. This roughly seven-month doubling time is related to the 10% growth rate as follows:

$$\text{doubling time} \approx \frac{70}{\text{percentage growth rate}} = \frac{70}{10/\text{mo}} = 7 \text{ mo}$$

This formula, in which the doubling time is approximately 70 divided by the percentage growth rate, works whenever the growth rate is relatively small (less than about 15%). It is often called the **rule of 70.**

Approximate Doubling Time Formula (Rule of 70)

For a quantity growing exponentially at a rate of $P\%$ per time period, the doubling time is *approximately*

$$T_{\text{double}} \approx \frac{70}{P}$$

This approximation works best for small growth rates and breaks down for growth rates over about 15%.

EXAMPLE 3 *Population Doubling Time*

World population was about 6.0 billion in 2000 and was growing at a rate of about 1.4% per year. What is the approximate doubling time at this growth rate? If this growth rate continues, what will world population be in 2020? Compare to the result in Example 2.

Solution We can use the approximate doubling time formula because the growth rate is much less than 15%. The percentage growth rate of 1.4% per year means we set $P = 1.4/\text{yr}$:

$$T_{\text{double}} \approx \frac{70}{P} = \frac{70}{1.4/\text{yr}} = 50 \text{ yr}$$

The doubling time is approximately 50 years. The year 2020 is $t = 20$ years after 2000, so the population in 2020 will be

$$\text{new value} = \text{initial value} \times 2^{t/T_{\text{double}}}$$
$$= 6.0 \text{ billion} \times 2^{20 \text{ yr}/50 \text{ yr}}$$
$$= 6.0 \text{ billion} \times 2^{0.4} = 7.9 \text{ billion}$$

At a 1.4 % growth rate, world population will be nearly 8 billion in 2020. This is some 700 million fewer people than predicted in Example 2, in which the doubling time was assumed to be 40 years rather than 50 years.

EXAMPLE 4 *Solving the Doubling Time Formula*

World population doubled in the 40 years from 1950 to 1990. What was the approximate percentage growth rate during this period? Contrast this growth rate with the 2000 growth rate of 1.4% per year.

Solution We can answer the question by solving the approximate doubling time formula for P. We multiply both sides by P and divide both sides by T_{double}; you should confirm that this gives the following result:

$$P \approx \frac{70}{T_{\text{double}}}$$

Substituting $T_{\text{double}} = 40$ years, we find

$$P \approx \frac{70}{T_{\text{double}}} = \frac{70}{40 \text{ yr}} = 1.75/\text{yr}$$

The population growth rate between 1950 and 1990 was about $P\% = 1.75\%$ per year. This is 0.35 percentage point higher than the 2000 growth rate of 1.4% per year.

EXPONENTIAL DECAY AND HALF-LIFE

Exponential *decay* occurs whenever a quantity *decreases* by the same percentage in every fixed time period (for example, by 20% every year). In that case, the value of the quantity repeatedly decreases to half its value, with each halving occurring in a time called the **half-life.**

> **by the way...**
> Plutonium-239 is the chemical element plutonium in a form (or *isotope*) with atomic weight 239. Atomic weight is the total number of protons and neutrons in the nucleus. Because all plutonium nuclei have 94 protons, Pu-239 nuclei have 239 − 94 = 145 neutrons.

You may have heard half-lives described for radioactive materials such as uranium or plutonium. For example, radioactive plutonium-239 (Pu-239) has a half-life of about 24,000 years. To understand the meaning of the half-life, suppose that 100 pounds of Pu-239 is deposited at a nuclear waste site. The plutonium gradually decays into other substances as follows:

- In 24,000 years, or one half-life, the amount of Pu-239 declines to 1/2 its original value, or to (1/2) × 100 pounds = 50 pounds.
- In 48,000 years, or two half-lives, the amount of Pu-239 declines to $(1/2)^2 = 1/4$ its original value, or to (1/4) × 100 pounds = 25 pounds.
- In 72,000 years, or three half-lives, the amount of Pu-239 declines to $(1/2)^3 = 1/8$ its original value, or to (1/8) × 100 pounds = 12.5 pounds.

We can generalize this idea much as we did earlier for the doubling time. Note that a single halving reduces a quantity by a factor of 1/2, two halvings reduce it by a factor of $(1/2)^2$, and three halvings reduce it by a factor of $(1/2)^3$. If we let t be the amount of time that has passed and T_{half} be the half-life, then the number of halvings after a time t is t/T_{half}. Thus, the quantity after time t is the original quantity times this factor of $(1/2)^{t/T_{half}}$.

> **Calculations with the Half-Life**
> After a time t, an exponentially decaying quantity with a half-life of T_{half} decreases in size by a factor of $(1/2)^{t/T_{half}}$. The new value of the decaying quantity is related to its initial value (at $t = 0$) by
>
> $$\text{new value} = \text{initial value} \times \left(\frac{1}{2}\right)^{t/T_{half}}$$

EXAMPLE 5 *Carbon-14 Decay*

Radioactive carbon-14 has a half-life of about 5700 years. It collects in organisms only while they are alive; once they are dead, it only decays. What fraction of the carbon-14 in an animal bone still remains 1000 years after the animal has died?

Solution The half-life is T_{half} = 5700 years, so the fraction of the initial amount remaining after t = 1000 years is

$$\left(\frac{1}{2}\right)^{t/T_{half}} = \left(\frac{1}{2}\right)^{1000 \text{ yr}/5700 \text{ yr}} = 0.885$$

For example, if the bone originally contained 1 kilogram of carbon-14, the amount remaining after 1000 years is 0.885 kilogram. As you can probably see, we can use this idea to determine the age of bones found at archaeological sites. We'll explore this idea of radioactive dating further in Unit 7C. ■

> **by the way...**
> Ordinary carbon is carbon-12, which is stable (not radioactive). Carbon-14 is produced in the Earth's atmosphere by high-energy particles coming from the Sun. It mixes with ordinary carbon and therefore becomes incorporated into living tissue through respiration (breathing).

EXAMPLE 6 *Plutonium After 100,000 Years*

Suppose that 100 pounds of Pu-239 is deposited at a nuclear waste site. How much of it will still be present in 100,000 years?

Solution The half-life of Pu-239 is $T_{\text{half}} = 24{,}000$ years. Given an initial amount of 100 pounds, the amount remaining after $t = 100{,}000$ years is

$$\text{new value} = \text{initial value} \times \left(\frac{1}{2}\right)^{t/T_{\text{half}}} = 100 \text{ lb} \times \left(\frac{1}{2}\right)^{100{,}000 \text{ yr}/24{,}000 \text{ yr}} = 5.6 \text{ lb}$$

About 5.6 pounds of the original 100 pounds of Pu-239 will still be present in 100,000 years. ■

Time out to think

Plutonium, which is not found naturally on the Earth, is made in nuclear reactors for use both as fuel for nuclear power plants and for nuclear weapons. Based on its half-life, explain why the safe disposal of Pu-239 poses a significant challenge.

THE APPROXIMATE HALF-LIFE FORMULA

The approximate doubling time formula (the rule of 70) found earlier works equally well for exponential decay if we replace the doubling time with the half-life and the percentage growth rate with the percentage decay rate.

Approximate Half-Life Formula

For a quantity decaying exponentially at a rate of $P\%$ per time period, the half-life is *approximately*

$$T_{\text{half}} \approx \frac{70}{P}$$

This approximation works best for small decay rates and breaks down for decay rates over about 15%.

EXAMPLE 7 *Devaluation of Currency*

Suppose that inflation causes the value of the Russian ruble to fall at a rate of 12% per year (relative to the dollar). Approximately how long does it take for the ruble to lose half its value?

Solution We can use the approximate half-life formula because the growth rate is less than 15%. The 12% decay rate means we set $P = 12/\text{yr}$:

$$T_{\text{half}} \approx \frac{70}{P} = \frac{70}{12/\text{yr}} = 5.8 \text{ yr}$$

The half-life is a little less than six years, meaning that the ruble loses half its value (against the dollar) in six years. ■

EXACT FORMULAS FOR DOUBLING TIME AND HALF-LIFE

The approximate doubling time and half-life formulas are useful because they are easy to remember. However, for more precise work or for cases of larger rates where the approximate formulas break down, we need exact formulas. The formulas are given below; in Unit 7C, we will see how they are derived. Note that these formulas use the **fractional growth rate,** $r = P/100$, rather than the percentage growth rate,

historical note

The atomic bomb that devastated Nagasaki during World War II generated its destructive power by fission of plutonium-239. (The Hiroshima bomb used uranium-235.)

$P\%$. For example, if the percentage growth rate is 5% per year, the fractional growth rate is 0.05 per year.

> **Exact Doubling Time and Half-Life Formulas**
>
> For an exponentially growing quantity with a fractional growth rate r, the doubling time is
>
> $$T_{double} = \frac{\log_{10} 2}{\log_{10}(1 + r)}$$
>
> For an exponentially decaying quantity, we use a *negative* value for r (for example, if the decay rate is 5% per year, we set $r = -0.05$ per year); the half-life is
>
> $$T_{half} = -\frac{\log_{10} 2}{\log_{10}(1 + r)}$$
>
> Note that *the units of time used for T and r must be the same.* For example, if the fractional growth rate is 0.05 per *month*, then the doubling time will also be measured in months.

EXAMPLE 8 *Large Growth Rate*

A population of rats is growing at a rate of 80% per month. Find the exact doubling time for this growth rate and compare it to the doubling time found with the approximate doubling time formula.

Solution The growth rate of 80% per month means $P = 80/\text{mo}$ or $r = 0.8/\text{mo}$. The doubling time is

$$T_{double} = \frac{\log_{10} 2}{\log_{10}(1 + 0.8)} = \frac{0.301\,030}{\log_{10}(1.8)} = \frac{0.301\,030}{0.255\,273} = 1.18 \text{ mo}$$

The doubling time is almost 1.2 months. Note that this answer makes sense: With the population growing by 80% in a month, we expect it to take a little over a month to grow by 100% (which is a doubling). In contrast, the approximate doubling time formula predicts a doubling time of $70/P = 70/80 = 0.875$ month, which is less than one month. We see that the approximate formula does not work well for large growth rates. ■

EXAMPLE 9 *Ruble Revisited*

Suppose the Russian ruble is falling in value against the dollar at 12% per year. How long does it take the ruble to lose half its value? Compare your answer to the approximate answer found in Example 7.

Solution The percentage decay rate is 12%/yr, so the fractional decay rate is $r = -0.12/\text{yr}$. The half-life is

$$T_{half} = -\frac{\log_{10} 2}{\log_{10}(1 - 0.12)} = \frac{0.301\,030}{-0.055\,517} = 5.42 \text{ yr}$$

The ruble loses half its value against the dollar in 5.42 years. This exact result is about 0.4 year less than the 5.8 years obtained with the approximate formula. We see that the approximate formula is reasonably accurate for the 12% rate. ■

Using Your CALCULATOR

Your calculator should have a button for computing common (base 10) logarithms. It is usually labeled *log,* but be careful because some calculators use this label for natural (base *e*) logarithms. With the right button, you will find $\log_{10} 2 = 0.3010\ldots$.

a brief review...

LOGARITHMS

A **logarithm** (or **log,** for short) is a number that represents a power or exponent. In this book, we will focus on only base 10 logs, also called **common logs,** which are defined as follows:

> $\log_{10} x$ is *the power to which 10 must be raised to obtain x.*

You may find it easier to remember the meaning with a less technical definition:

> $\log_{10} x$ means "10 to what *power* equals *x*?"

For example:

$\log_{10} 1000 = 3$ because $10^3 = 1000$

$\log_{10} 10{,}000{,}000 = 7$ because $10^7 = 10{,}000{,}000$

$\log_{10} 0.1 = -1$ because $10^{-1} = 0.1$

$\log_{10} 30 = 1.477$ because $10^{1.477} = 30$

Four important rules follow directly from the definition of a logarithm.

1. Taking the logarithm of a power of 10 gives the power. That is,

$$\log_{10} 10^x = x$$

2. Raising 10 to a power that is the logarithm of a number gives back the number. That is,

$$10^{\log_{10} x} = x \quad (x > 0)$$

3. Because powers of 10 are multiplied by adding their exponents, we have the *addition rule* for logarithms:

$$\log_{10} xy = \log_{10} x + \log_{10} y \quad (x > 0 \text{ and } y > 0)$$

4. We can "bring down" an exponent within a logarithm by applying the *power rule* for logarithms:

$$\log_{10} a^x = x \times \log_{10} a \quad (a > 0)$$

Most calculators have a key to compute $\log_{10}$ of any positive number. You should find this key on your calculator and use it to verify that $\log_{10} 1000 = 3$ and $\log_{10} 2 = 0.301\,030$.

Example: Given that $\log_{10} 2 = 0.301\,030$, find each of the following:

a. $\log_{10} 8$ **b.** $10^{\log_{10} 2}$ **c.** $\log_{10} 200$

Solution:

a. We notice that $8 = 2^3$. Thus, from Rule 4,

$$\log_{10} 8 = \log_{10} 2^3$$
$$= 3 \times \log_{10} 2 = 3 \times 0.301\,030 = 0.903\,09$$

b. From Rule 2,

$$10^{\log_{10} 2} = 2$$

c. We notice that $200 = 2 \times 100 = 2 \times 10^2$. Thus, from Rule 3,

$$\log_{10} 200 = \log_{10}(2 \times 10^2) = \log_{10} 2 + \log_{10} 10^2$$

From Rule 1, we know that $\log_{10} 10^2 = 2$, so

$$\log_{10} 200 = \log_{10}(2 \times 10^2) = 0.301\,030 + 2 = 2.301\,030$$

Example: Someone tells you that $\log_{10} 600 = 5.778$. Should you believe it?

Solution: Because 600 is between 100 and 1000, $\log_{10} 600$ must be between $\log_{10} 100$ and $\log_{10} 1000$. From Rule 1, we find that $\log_{10} 100 = \log_{10} 10^2 = 2$ and $\log_{10} 1000 = \log_{10} 10^3 = 3$. Thus, $\log_{10} 600$ must be between 2 and 3, so the claimed answer of 5.778 must be wrong.

Review Questions

1. What is a doubling time? Suppose a population has a doubling time of 25 years. By what factor will it grow in 25 years? in 50 years? in 100 years?
2. Given a doubling time, explain how you can calculate the value of an exponentially growing quantity at any time *t*.
3. State the approximate doubling time formula and the conditions under which it works well. Give an example.
4. What is a half-life? Suppose a radioactive substance has a half-life of 1000 years. What fraction will be left after 1000 years? after 2000 years? after 4000 years?
5. Given a half-life, explain how you can calculate the value of an exponentially decaying quantity at any time *t*.
6. State the approximate half-life formula and the conditions under which it works well. Give an example.
7. Briefly describe exact doubling time and half-life formulas. Give examples using the formulas.

In the News

1. **Doubling Time.** Find a news story that gives an exponential growth rate. Find the approximate doubling time from the growth rate and discuss the implications of the growth.
2. **Radioactive Half-Life.** Find a news story that discusses some type of radioactive material. If it is not given, look up the half-life of the material. Discuss the implications for disposal of the material.

Exercises

Logarithms. Exercises 1–4 use skills covered in the Brief Review on p. 367. In Exercises 1–2, determine whether each statement is true or false *without* doing any calculations. Explain your reasoning.

1. a. $10^{0.928}$ is between 10 and 100.
 b. $10^{3.334}$ is between 1000 and 10,000.
 c. $10^{-5.2}$ is between 100,000 and 1,000,000.
 d. $10^{-2.67}$ is between 0.001 and 0.01.
 e. $\log_{10} \pi$ is between 0 and 1.
2. a. $\log_{10} 96$ is between 3 and 4.
 b. $\log_{10} 1{,}600{,}000$ is between 6 and 7.
 c. $\log_{10}(8 \times 10^9)$ is between 12 and 13.
 d. $\log_{10} \frac{1}{4}$ is between 0 and 1.
 e. $\log_{10} 0.00045$ is between 5 and 6.
3. Given that $\log_{10} 2 = 0.301$, find each of the following *without* a calculator.
 a. $\log_{10} 4$ b. $\log_{10} 20{,}000$
 c. $\log_{10} 0.5$ d. $\log_{10} 32$
 e. $\log_{10} 0.25$ f. $\log_{10} 0.2$
4. Given that $\log_{10} 5 = 0.699$, find each of the following *without* a calculator.
 a. $\log_{10} 50$ b. $\log_{10} 5000$
 c. $\log_{10} 0.05$ d. $\log_{10} 25$
 e. $\log_{10} 0.20$ f. $\log_{10} 0.04$

Doubling Time. Exercises 5–12 give a doubling time for an exponentially growing quantity.

5. The doubling time of a population of flies is 3 hours. By what factor does the population increase in 24 hours? in 1 week?
6. The doubling time of a bank account balance is 10 years. By what factor does it grow in 30 years? in 50 years?
7. The doubling time of a city's population is 22 years. How long does it take for the population to quadruple?
8. Prices are rising with a doubling time of 4 weeks. By what factor do prices increase in a year?
9. The initial population of a town is 15,600, and it grows with a doubling time of 20 years. What will the population be in 12 years? in 24 years?
10. The initial population of a town is 15,600, and it grows with a doubling time of 8 years. What will the population be in 12 years? in 24 years?
11. The number of cells in a tumor doubles every 1.5 months. If the tumor begins as a single cell, how many cells will there be after 3 years? after 4 years?
12. The number of cells in a tumor doubles every 6 months. If the tumor begins with a single cell, how many cells will there be after 3 years? after 6 years?

World Population. In 2000, world population was 6 billion. In Exercises 13–14, use the given doubling time to predict the population in 2010, 2060, and 2100.

13. Assume a doubling time of 45 years.
14. Assume a doubling time of 60 years.
15. **Rabbits.** A community of rabbits begins with an initial population of 100 and grows 7% per month. Make a table,

similar to Table 7.3, that shows the population for each of the next 15 months. Based on the table, find the doubling time of the population and briefly discuss how well the approximate doubling time formula works for this case.

16. **Mice.** A community of mice begins with an initial population of 1000 and grows 20% per month. Make a table, similar to Table 7.3, that shows the population for each of the next 15 months. Based on the table, find the doubling time of the population and briefly discuss how well the approximate doubling time formula works for this case.

Doubling Time Formula. Use the approximate doubling time formula (rule of 70) in Exercises 17–20. Be sure to discuss whether the formula is valid for the case described.

17. The Consumer Price Index is increasing at a rate of 7% per year. What is its doubling time? By what factor will prices increase in 3 years?

18. A city's population is growing at a rate of 2.0% per year. What is its doubling time? By what factor will the population increase in 75 years?

19. Prices are rising at a rate of 0.7% per month. What is their doubling time? By what factor will prices increase in 1 year? in 8 years?

20. Oil consumption is increasing at a rate of 1.9% per year. What is its doubling time? By what factor will oil consumption increase in a decade?

Half-Life. Exercises 21–28 give a half-life for an exponentially decaying quantity.

21. The half-life of a radioactive substance is 35 years. If you start with some amount of this substance, what fraction will remain in 70 years? in 140 years?

22. The half-life of a radioactive substance is 250 years. If you start with some amount of this substance, what fraction will remain in 70 years? in 1500 years?

23. The half-life of a drug in the bloodstream is 12 hours. By what factor does the concentration of the drug decrease in 24 hours? in 36 hours?

24. The half-life of a drug in the bloodstream is 4 hours. By what factor does the concentration of the drug decrease in 24 hours? in 36 hours?

25. The current population of a threatened animal species is 1 million, but is declining with a half-life of 10 years. How many animals will be left in 40 years? in 70 years?

26. The current population of a threatened animal species is 1 million, but is declining with a half-life of 25 years. How many animals will be left in 40 years? in 70 years?

27. Cobalt-56 has a half-life of 77 days. If you start with 1 kilogram of cobalt-56, how much will remain after 100 days? after 1 year?

28. Radium-226 is a metal with a half-life of 1600 years. If you start with 1 kilogram of radium-226, how much will remain after 1000 years? after 10,000 years?

29. **Plutonium on Earth.** Scientists believe the Earth once had naturally existing plutonium-239. Suppose the Earth had 10 trillion tons of Pu-239 when it formed. Given plutonium's half-life of 24,000 years and the Earth's current age of 4.6 billion years, how much would remain today? Use your answer to explain why plutonium is not found naturally on Earth today.

30. **Nuclear Weapons.** Thermonuclear weapons use tritium for their nuclear reactions. Tritium is a radioactive form of hydrogen (containing 1 proton and 2 neutrons) with a half-life of about 12 years. Suppose a nuclear weapon contains 1 kilogram of tritium. How much will remain in 50 years? Use your answer to explain why thermonuclear weapons require regular maintenance.

Half-Life Formula. Use the approximate half-life formula in Exercises 31–34. Be sure to discuss whether the formula is valid for the case described.

31. Urban encroachment is causing the area of a forest to decline at a rate of 10% per year. What is the half-life of the forest? What fraction of the forest will remain in 100 years?

32. A clean-up project is reducing the concentration of a pollutant in the water supply, with an 8% decrease per week. What is the half-life of the concentration of the pollutant?

What fraction of the original amount of the pollutant will remain when the project ends after 1 year (52 weeks)?

33. Poaching is causing a population of elephants to decline by 7% per year. What is the half-life for the population? If there are 10,000 elephants today, how many will remain in 50 years?

34. The production of a gold mine is declining by 12% per year. If its current annual production is 4000 kilograms, what will its production be in 10 years?

Exact Formulas. In Exercises 35–38, use the approximate and exact formulas to find the doubling time. Also, answer the given question.

35. Inflation is causing prices to rise at a rate of 10% per year. For an item that costs $1000 today, what will the price be in 3 years?

36. Hyperinflation is driving up prices at a rate of 80% per month. For an item that costs $1000 today, what will the price be in 1 year?

37. A nation of 100 million people is growing at a rate of 3% per year. What will its population be in 50 years?

38. A family of 100 termites invades your house and grows at a rate of 20% per week. How many termites will be in your house after 1 year?

Web Projects

Find useful links for Web Projects on the text Web site: www.aw.com/bennett-briggs

1. **National Growth Rates.** Visit the Web site of the Population Reference Bureau, where you will find growth rates, doubling times, and population projections tabulated for many different countries. Select several countries from several continents and record relevant growth data. Comment on whether the doubling times and growth rates are consistent. Discuss how these data are used to make population projections.

2. **World Population Growth.** The Web site for the U.S. Bureau of Census contains a wealth of data on world population. Visit this site and gather data on world population growth over the last 50 years. Estimate the population growth rate over each decade. Compute the associated doubling times. Write a two-paragraph statement on the trends that you observe.

UNIT 7C

EXPONENTIAL MODELING

*I*n the previous two units, we worked with applications of exponential growth and decay that required knowing only a doubling time or half-life. To work with the many applications of exponential growth and decay more generally, we need an equation that allows us to see how an exponentially growing (or decaying) quantity changes with time. In mathematical terms, we need to work with an exponential *function* that describes the growing quantity's dependence on time. In this unit, we investigate the exponential function and some of its many applications.

THE EXPONENTIAL FUNCTION

Our first task is to find a general form for an exponential function. Consider a town that begins with a population of 10,000 and grows at a rate of 20% per year. As we saw in Unit 7A, this population grows exponentially because it increases the same relative amount, 20%, each year (see Figure 7.1). We say that the *percentage growth rate* is $P\% = 20\%$ per year or, equivalently, the *fractional growth rate* is $r = 0.2$ per year.

The initial population is 10,000. With a 20% growth rate, the population at the end of the first year is 20% *more than* the initial value of the population, or $120\% = 1.2$ *times* the initial population (see the "of versus more than" rule in Unit 3A):

$$\text{population after 1 year} = 10{,}000 \times 1.2 = 12{,}000$$

During the second year, the population again grows by 20%, so we can find the population at the end of the second year by multiplying again by 1.2:

$$\begin{aligned}\text{population after 2 years} &= \text{population after 1 year} \times 1.2 \\ &= (10{,}000 \times 1.2) \times 1.2 \\ &= 10{,}000 \times 1.2^2 = 14{,}400\end{aligned}$$

With the 20% per year increase continuing in the third year, we express the population at the end of the third year by multiplying once more by 1.2:

$$\begin{aligned}\text{population after 3 years} &= \text{population after 2 years} \times 1.2 \\ &= (10{,}000 \times 1.2^2) \times 1.2 \\ &= 10{,}000 \times 1.2^3 = 17{,}280\end{aligned}$$

This growth is illustrated in Figure 7.4. The pattern may now be clear. If we let t represent the time in years, we find that

$$\text{population after } t \text{ years} = \text{initial population} \times 1.2^t$$

For example, after $t = 25$ years, the population is $10{,}000 \times 1.2^{25} = 953{,}962$.

FIGURE 7.4 Exponential growth for three years with a growth rate of 20% per year. Each year, the population increases by a factor of 1.2.

We can extend this idea to *any* exponentially growing quantity Q, to write a general exponential function.

> **by the way...**
> You may notice that this exponential function is identical to the compound interest formula (Unit 4A) if we identify Q as the accumulated balance (called A in the compound interest formula), Q_0 as the starting principal, r as the interest rate, and t as the number of times interest is paid. In other words, compound interest is a form of exponential growth.

Exponential Functions

An **exponential function** grows (or decays) by the same relative amount per unit time. For any quantity Q growing exponentially with a fractional growth rate r,

$$Q = Q_0 \times (1 + r)^t$$

where

t = time
Q = value of the exponentially growing quantity at time t
Q_0 = initial value of the quantity (at $t = 0$)
r = fractional growth rate for the quantity

Negative values of r correspond to exponential decay. Note that *the units of time used for t and r must be the same*. For example, if the fractional growth rate is 0.05 per *month*, then t must also be measured in months.

We can use the same function for exponential decay if we make r negative. For example, if a quantity decreases by 10% per month, we identify the fractional growth rate as $r = -0.1$ per month.

EXAMPLE 1 U.S. Population Growth

The 2000 census found a U.S. population of about 281 million. Write an equation for the U.S. population that assumes exponential growth at 0.7% per year. Use the equation to predict the U.S. population in 2100.

a brief review...

ALGEBRA WITH LOGARITHMS

The basic properties of logarithms (see the Brief Review, p. 367) lead to two very useful algebraic techniques. These techniques apply to logarithms with any base, but we will focus on common (base 10) logarithms.

First, we can solve for a variable that appears in an exponent by taking the logarithm of both sides of the equation (as long as both sides are positive) and applying the rule that $\log_{10} a^x = x \log_{10} a$.

Example: Solve for x in the equation $2^x = 50$.

Solution: Taking the logarithm of both sides, we have

$$\log_{10} 2^x = \log_{10} 50$$

By the rule that $\log_{10} a^x = x \log_{10} a$, this equation becomes

$$x \log_{10} 2 = \log_{10} 50$$

Now we divide both sides by $\log_{10} 2$ to get the answer:

$$x = \frac{\log_{10} 50}{\log_{10} 2} = \frac{1.69897}{0.30103} = 5.644$$

Second, we can solve for a variable that appears within a logarithm by making both sides of the equation into powers of 10 and applying the rule that $10^{\log_{10} x} = x$ (as long as $x > 0$).

Example: Solve for x in the equation $2 \log_{10} x = 15$.

Solution: First, we isolate the term containing x by dividing both sides by 2:

$$\log_{10} x = \frac{15}{2} = 7.5$$

Because x is within a logarithmic expression, we make both sides of the equation into powers of 10:

$$10^{\log_{10} x} = 10^{7.5}$$

Applying the rule $10^{\log_{10} x} = x$ (as long as $x > 0$), we find

$$x = 10^{7.5} = 3.1623 \times 10^7$$

Solution The quantity Q is the U.S. population. The initial value is the 2000 population, $Q_0 = 281$ million. The percentage growth rate is $P\% = 0.7\%$ per year, so the fractional growth rate is $r = P/100 = 0.007$ per year. Thus, the equation takes the form

$$Q = Q_0 \times (1 + r)^t = 281 \text{ million} \times (1 + 0.007)^t$$
$$= 281 \text{ million} \times (1.007)^t$$

Note that, because the units of r are *per year*, t must be measured in years. When we substitute $t = 100$ years, this exponential function predicts a 2100 population of

$$Q = 281 \text{ million} \times (1.007)^{100} = 564 \text{ million}$$

At its present growth rate of 0.7% per year, the U.S. population will swell to over 560 million by 2100.

EXAMPLE 2 *Declining Population*

China's one-child policy (see Unit 2C, Example 7) was implemented with the goal of reducing China's population to 700 million by 2050. China's 2000 population was about 1.2 billion. Suppose China's population declines at a rate of 0.5% per year. Write an equation for the exponential decay of the population. Will this rate of decline be sufficient to meet the original goal?

Solution The quantity Q is China's population. The population is *declining* at a rate of $P\% = 0.5\%$ per year, so the fractional growth rate r should be negative; it is $r = -P/100 = -0.005$ per year. The initial value is the 2000 population, $Q_0 = 1.2$ billion. Thus, the equation takes the form

$$Q = Q_0 \times (1 + r)^t = 1.2 \text{ billion} \times (1 - 0.005)^t$$
$$= 1.2 \text{ billion} \times (0.995)^t$$

Again, because the units of r are *per year*, t must be measured in years. When we substitute $t = 50$ years, this exponential function predicts a 2050 population of

$$Q = 1.2 \text{ billion} \times (0.995)^{50} = 0.934 \text{ billion}$$

A decline of 0.5% per year will reduce China's population to 0.934 billion, or 930 million, by 2050. In that case, the goal of 700 million will not be met. ■

Doubling Time and Half-Life

In Unit 7B, we used both approximate and exact formulas for the doubling time and half-life. Now that we have an equation for the exponential function, we can see where these formulas came from.

For exponential growth, the doubling time is the time required for a quantity to double in size. That is, after a time $t = T_{\text{double}}$, the quantity has grown to twice its initial value, or $Q = 2Q_0$. Substituting these values into the exponential equation gives

$$2Q_0 = Q_0 \times (1 + r)^{T_{\text{double}}}$$

Dividing both sides by Q_0 and interchanging the two sides yields

$$(1 + r)^{T_{\text{double}}} = 2$$

To solve this equation for the doubling time, we must "bring down" the exponent by taking the logarithm of both sides (see the Brief Review on p. 373):

$$\log_{10}[(1 + r)^{T_{\text{double}}}] = \log_{10} 2$$

By the rule that $\log_{10} a^x = x \log_{10} a$, this equation becomes

$$T_{\text{double}} \log_{10}(1 + r) = \log_{10} 2$$

Dividing both sides by $\log_{10}(1 + r)$, we find the exact formula for the doubling time, as given in Unit 7B:

$$T_{\text{double}} = \frac{\log_{10} 2}{\log_{10}(1 + r)}$$

We find the half-life for exponential decay similarly. For decay, the quantity decreases to half its original value, or $Q = (\frac{1}{2})Q_0$, in a time $t = T_{\text{half}}$. Thus, there are two changes in the algebra: We have T_{half} instead of T_{double}, and $\log_{10}(\frac{1}{2})$ instead of $\log_{10} 2$. The result is

$$T_{\text{half}} = \frac{\log_{10}(\frac{1}{2})}{\log_{10}(1 + r)}$$

We can simplify this result by recognizing that $(\frac{1}{2}) = 2^{-1}$. Thus, using the rule that

by the way...
Between 1950 and 2000, China's population growth *rate* decreased by more than a factor of 2. However, the actual population increased from about 550 million to 1.2 billion.

$\log_{10} a^x = x \log_{10} a$, we find that $\log_{10} 2^{-1} = -\log_{10} 2$. With this substitution, the formula takes the form given in Unit 7B:

$$T_{\text{half}} = -\frac{\log_{10} 2}{\log_{10}(1+r)}$$

Remember that r is negative for exponential decay, so $1 + r$ is less than 1 in this case. (For example, if $r = -0.05$, then $1 + r = 0.95$.) Because the logarithm of a number between 0 and 1 is negative, the formula gives a positive value for the half-life.

Graphing Exponential Functions

The easiest way to graph an exponential function is to use points corresponding to several doubling times (or half-lives, in the case of decay). We start at the point $(0, Q_0)$ that represents the initial value at $t = 0$. For an exponentially growing quantity, we know that the value of Q is $2Q_0$ (double its initial value) after one doubling time (T_{double}), $4Q_0$ after two doubling times ($2T_{\text{double}}$), $8Q_0$ after three doubling times ($3T_{\text{double}}$), and so on. We simply fit a steeply rising curve between these points, as shown in Figure 7.5a.

FIGURE 7.5 (a) We can graph exponential growth by plotting points for the repeated doublings. (b) We can graph exponential decay by plotting points for the repeated halvings.

For an exponentially decaying quantity, we know that the value of Q falls to $Q_0/2$ (half its initial value) after one half-life (T_{half}), $Q_0/4$ after two half-lives ($2T_{\text{half}}$), $Q_0/8$ after three half-lives ($3T_{\text{half}}$), and so on. Fitting a falling curve to these points gives the graph shown in Figure 7.5. Note that this curve gets closer and closer to the horizontal axis, but never reaches it.

EXAMPLE 3 Sensitivity to Growth Rate

The growth rate of the U.S. population has varied substantially during the past century. It depends on the immigration rate as well as birth and death rates. Starting from a 2000 population of 281 million, project the population in 2100

using growth rates that are just 0.2 percentage point lower and higher than the 0.7% used in Example 1. Make a graph showing the population through 2100 for each growth rate.

Solution A growth rate 0.2 percentage point lower than 0.7% is 0.5%, or $r = 0.005$. At this growth rate, the 2100 population would be

$$Q = 281 \text{ million} \times (1.005)^{100} = 462 \text{ million}$$

A rate 0.2 percentage point higher than 0.7% is 0.9%, or $r = 0.009$, which gives a 2100 population of

$$Q = 281 \text{ million} \times (1.009)^{100} = 688 \text{ million}$$

Within a range of less than half a percentage point, from 0.5% to 0.9%, the projected population for 2100 varies by more than 200 million people (from 462 million to 688 million). Clearly, population projections are very sensitive to changes in the growth rate.

To make the graphs, note that we already have two points for each growth rate: the initial population in 2000, $Q_0 = 281$ million, and the population calculated for 2100. We find a third point in each case from the doubling time, which is the time when the population reaches $2Q_0 = 562$ million. You should use the doubling time formula to confirm that the doubling times are 77 years for $r = 0.009$, 99 years for $r = 0.007$, and 139 years for $r = 0.005$. Figure 7.6 shows the curves for all three growth rates. ■

FIGURE 7.6 Future world population growth with three growth rates.

Alternative Forms of the Exponential Function

Our general equation for the exponential function, $Q = Q_0 \times (1 + r)^t$, contains the growth rate r, but not the doubling time or half-life. Because we are often given the doubling time or half-life, it's useful to rewrite the equation in forms that use T_{double} or T_{half} rather than r.

As before, we let Q represent the quantity at any time t and Q_0 represent the initial value. Then, in terms of the doubling time T_{double}, the exponential function takes the form

$$Q = Q_0 \times 2^{t/T_{\text{double}}}$$

This is the same equation we used with the doubling time in Unit 7B, except that now we are using Q for the "new value" and Q_0 for the initial value. As it should, this equation gives $2Q_0$ when $t = T_{\text{double}}$, $4Q_0$ when $t = 2T_{\text{double}}$, and so on.

For exponential decay, we get a similar function with the half-life:

$$Q = Q_0 \times \left(\frac{1}{2}\right)^{t/T_{\text{half}}}$$

Again, this is the same equation we used with the half-life in Unit 7B, except that now we are using Q for the "new value" and Q_0 for the initial value. You should confirm that this equation gives $Q_0/2$ when $t = T_{half}$, $Q_0/4$ when $t = 2T_{half}$, and so on.

> **Forms of the Exponential Function**
>
> - If given the growth or decay rate r, use the exponential function in the form
>
> $$Q = Q_0 \times (1 + r)^t$$
>
> Remember that r is positive for growth and negative for decay.
> - If given the doubling time T_{double}, use the exponential function in the form
>
> $$Q = Q_0 \times 2^{t/T_{double}}$$
>
> - If given the half-life T_{half}, use the exponential function in the form
>
> $$Q = Q_0 \times \left(\frac{1}{2}\right)^{t/T_{half}}$$

TECHNICAL NOTE
A slightly different form of the exponential function is $Q = Q_0 \times B^t$. If the base B is greater than 1, the function describes exponential growth. If the base B is between 0 and 1, the function describes exponential decay.

SELECTED APPLICATIONS

We have seen several examples of using exponential functions to model population growth and decline. We also used exponential equations in Chapter 4 for compound interest calculations. In this section, we'll explore a few more of the many applications in which exponential functions are used to model growth or decay processes.

Inflation

Because prices tend to change with time, price comparisons from one time to another are meaningful only if the prices are adjusted for the effects of inflation. We can model the effects of inflation with an exponential function in which r represents the rate of inflation.

EXAMPLE 1 *Monthly and Annual Inflation Rates*

The U.S. government reports the rate of inflation (as measured by the Consumer Price Index) both monthly and annually. Suppose that, for a particular month, the *monthly* rate of inflation is reported as 0.8%. What *annual* rate of inflation does this imply? Is the annual rate 12 times the monthly rate? Explain.

Solution We can write an exponential function for the inflation by letting Q_0 represent a price at time $t = 0$ and Q represent a price at any later time t. Using the monthly inflation rate $r = 0.8\% = 0.008$, we have

$$Q = Q_0 \times (1 + r)^t = Q_0 \times (1 + 0.008)^t = Q_0 \times 1.008^t$$

Note that, because *r* is given *per month*, *t* must be measured in months in this equation. Thus, to find the *annual* inflation rate, we substitute $t = 12$ months into the equation. We find that the price, Q, after 12 months is

$$Q = Q_0 \times 1.008^{12}$$
$$= Q_0 \times 1.1$$
$$= 1.1 Q_0$$

The price after one year is 1.1 times the initial price, which means that prices rose by 10%. Thus, the *annual* inflation rate is 10%. Note that this is somewhat more than 12 times the monthly rate, or $12 \times 0.8\% = 9.6\%$. The reason is that inflation compounds with each passing month, much like interest compounded monthly. In this case, the 9.6% found by multiplying the monthly rate by 12 is analogous to the APR quoted for an interest-bearing account (see Unit 4A); the actual annual inflation rate is higher because it is analogous to the annual yield, APY, which also depends on the number of compounding periods. ■

Environment and Resources

Among the most important applications of exponential models are those involving the environment and resource depletion. Global concentrations of many pollutants in the water and atmosphere have increased exponentially, as has consumption of nonrenewable resources such as oil and natural gas. Two basic factors can be responsible for such exponential growth.

1. The *per capita* demand for a resource often increases exponentially. For example, per capita energy consumption in the United States increased exponentially during most of the 20th century.
2. An exponentially increasing population will mean an exponentially increasing demand for a resource even if per capita demand remains constant.

In most cases, the growth rate is determined by a combination of both factors.

by the way ...
World oil consumption in 2000 was about 77 million barrels per day and increasing at a rate of about 2.5% per year. In Asian countries, consumption has increased by as much as 9% per year. In Europe and North America, the annual rate of increase has been between 1% and 2%. Russia is one of the few countries where oil consumption decreased in the 1990s, falling by up to 10% per year.

EXAMPLE 5 Carbon Emissions from China and the United States

The burning of fossil fuels releases greenhouse gases into the atmosphere. In 1995, the United States emitted about 1.4 billion (1.4×10^9) tons of carbon into the atmosphere, nearly one-fourth of the world total. China was the second largest contributor, emitting about 850 million (8.5×10^8) tons of carbon. However, emissions from China were rising at a rate of about 4% per year, while U.S. emissions were rising at about 1.3% per year. Using these growth rates, project greenhouse gas emissions from the United States and China in 2020. Graph the projected emissions for both countries.

Solution We use 1995 as $t = 0$, so 2020 is $t = 25$ years later. For the United States, we set the 1995 emissions as $Q_0 = 1.4 \times 10^9$ tons of carbon and $r = 0.013$ per year. The exponential function for the U.S. emissions is

$$Q = Q_0 \times (1 + r)^t$$
$$= (1.4 \times 10^9) \times (1.013)^t$$

Substituting $t = 25$ years, we predict that U.S. carbon emissions in 2020 will be

$$Q = (1.4 \times 10^9) \times (1.013)^{25}$$
$$= 1.9 \times 10^9 \text{ tons of carbon}$$

For China, we set the 1995 emissions as $Q_0 = 8.5 \times 10^8$ tons of carbon and $r = 0.04$ per year. The exponential function for the Chinese emissions is

$$Q = Q_0 \times (1 + r)^t$$
$$= (8.5 \times 10^8) \times (1.04)^t$$

Substituting $t = 25$ years, we predict that Chinese carbon emissions in 2020 will be

$$Q = (8.5 \times 10^8) \times (1.04)^{25}$$
$$= 2.3 \times 10^9 \text{ tons of carbon}$$

Thus, for the given growth scenarios, China will become the largest emitter of greenhouse gases before 2020. We can graph the two functions by finding the doubling time for each country; using the doubling time formula, we find that the doubling time is 54 years for the United States and 18 years for China. Figure 7.7 shows the graphs for both countries. From the graph, this model predicts that Chinese carbon emissions will surpass U.S. carbon emissions in 2014.

> **by the way...**
> As of 2000, the United States had less than 5% of the world's population but was responsible for about 32% of greenhouse gas emissions.

FIGURE 7.7 Exponential growth models for annual carbon dioxide emissions from the United States and China.

Physiological Processes

Many physiological processes are exponential. For example, a cancer tumor grows exponentially, at least in its early stages, and the concentration of many drugs in the bloodstream decays exponentially. (Alcohol is a notable exception, as its concentration decays linearly.)

EXAMPLE 6 Drug Concentration

Consider an antibiotic that has a half-life in the bloodstream of 12 hours. A 10-milligram injection of the antibiotic is given at 1:00 p.m. How much antibiotic remains in the blood at 9:00 p.m.? Draw a graph that shows the amount of antibiotic remaining as the drug is eliminated by the body.

Solution Because we are given a half-life rather than a growth rate r, we use the form of the exponential function that contains the half-life:

$$Q = Q_0 \times \left(\frac{1}{2}\right)^{t/T_{half}}$$

In this case, $Q_0 = 10$ milligrams is the initial dose given at $t = 0$ and Q is the amount of antibiotic in the blood at any time t later. The half-life is given as $T_{half} = 12$ hours, so the equation is

$$Q = 10 \times \left(\frac{1}{2}\right)^{t/12}$$

We must remember that Q is measured in milligrams and t in hours. At 9:00 p.m., which is $t = 8$ hours after the injection, the amount of antibiotic remaining is

$$Q = 10 \times \left(\frac{1}{2}\right)^{8/12} = 10 \times \left(\frac{1}{2}\right)^{2/3} = 6.3 \text{ mg}$$

Eight hours after the injection, 6.3 milligrams of the antibiotic remain in the bloodstream. Graphing this exponential decay function up to $t = 100$ hours, we see that the amount of antibiotic decreases steadily toward zero (Figure 7.8).

FIGURE 7.8 Exponential decay of an antibotic with a half-life of 12 hours in the bloodstream.

Radioactive Dating

As we discussed in Unit 7B, we can take advantage of the exponential nature of radioactive decay to measure the ages of rocks, bones, pottery, or other solid objects that contain radioactive elements. The process is called **radioactive dating.**

The basic idea of radioactive dating is simple: If we know both current and original amounts of the radioactive substance in the object, then we can use an exponential function to calculate the time since the object formed. Of course, we must know the half-life of the substance, but half-lives for nearly all radioactive substances have been carefully measured in laboratories. Thus, the only real difficulty in radioactive dating is determining how much of the radioactive substance was originally in the object. Fortunately, because radioactive materials decay in very specific ways, we can often determine the original amount by studying the overall chemical composition of the object today.

by the way ...

Most atomic nuclei are *stable,* meaning they do not spontaneously fall apart. By definition, radioactive substances have unstable nuclei that *do* fall apart spontaneously, leaving other nuclei as by-products.

EXAMPLE 7 The Allende Meteorite

The famous Allende meteorite lit up the skies of Mexico as it fell to Earth on February 8, 1969. Scientists melted and chemically analyzed small pieces of the meteorite and found traces of both radioactive potassium-40 and argon-40. From laboratory studies, we know that potassium-40 becomes argon-40 when it decays. Moreover, because argon-40 is a gas, it could not have been present when the meteorite first solidified (because a meteorite is made from solids, not gases). Thus, the scientists concluded that the argon-40 must have been trapped in the meteorite after it formed, in which case it must have come from decay of potassium-40. From the amounts of the two substances in the meteorite samples, the scientists determined that 8.5% of the potassium-40 originally present in the rock remains today. The half-life of potassium-40 is 1.3 billion (1.3×10^9) years. How old is the rock that makes up the Allende meteorite?

Solution We can model radioactive decay with an exponential function in which Q is the current amount of the radioactive substance and Q_0 is the original amount. Because we are given the half-life, we use the function in the form

$$Q = Q_0 \times \left(\frac{1}{2}\right)^{t/T_{\text{half}}}$$

For radioactive dating, we compare the current amount Q to the original amount Q_0, so the equation is more useful if we divide both sides by Q_0:

$$\frac{Q}{Q_0} = \left(\frac{1}{2}\right)^{t/T_{\text{half}}}$$

In this case, we are given that 8.5% of the original potassium-40 remains, which means that $Q/Q_0 = 8.5\% = 0.085$. If we substitute this value and the half-life of 1.3×10^9 years and interchange the two sides, the equation becomes

$$\left(\frac{1}{2}\right)^{t/(1.3 \times 10^9)} = 0.085$$

We can solve for t by taking the log of both sides:

$$\log_{10}\left(\frac{1}{2}\right)^{t/(1.3 \times 10^9)} = \log_{10} 0.085$$

Applying the rule that $\log_{10} a^x = x \log_{10} a$, we find

$$\left(\frac{t}{1.3 \times 10^9}\right) \times \log_{10}\left(\frac{1}{2}\right) = \log_{10} 0.085$$

Finally, we solve for t by multiplying both sides by 1.3×10^9 and dividing both sides by $\log_{10}(\frac{1}{2})$:

$$t = 1.3 \times 10^9 \times \frac{\log_{10} 0.085}{\log_{10}(\frac{1}{2})} = 4.6 \times 10^9 \text{ yr}$$

We conclude that the Allende meteorite is 4.6 billion years old.

historical note

American scientist Willard Libby won the Nobel Prize in Chemistry in 1960 for inventing the method of radioactive dating.

by the way...

Many meteorites date to 4.6 billion years ago, but none have ever been found that are older. Scientists therefore conclude that the solar system itself must have formed 4.6 billion years ago, and hence that the Allende meteorite is a remnant from the formation of the solar system. (The photo shows a meteorite at the Hayden Planetarium in New York City.)

Review Questions

1. Describe the meanings of all the variables in the exponential function. Explain how the function is used for exponential growth and decay.
2. Briefly explain how to find the doubling time and half-life from the exponential equation.
3. Describe how you can graph an exponential function with the help of the doubling time or half-life. What is the general shape of an exponential growth function? What is the general shape of an exponential decay function?
4. Describe the meaning of each of the three forms of the exponential function given in this chapter. Under what circumstance is each form useful?
5. Briefly describe how exponential functions are useful for modeling inflation, environmental and resource issues, physiological processes, and radioactive decay.
6. Briefly describe the process of radioactive dating. What makes it difficult? How can the difficulty be alleviated?

In the News

1. **Inflation Rate.** Find a news report that states both monthly and annual inflation rates. Using the methods in this unit, check whether the rates agree. Explain.
2. **Exponential Process.** Find a news account of a process that illustrates either exponential growth or decay. Is the term *exponential* used in the account? How do you know that there is an exponential process at work? Describe how you could use an exponential function to model the process.
3. **Radioactive Dating.** Find a news report that gives the age of an archaeological site, fossil, or rock based on radioactive dating. Briefly describe how the dating process worked and how accurate we can expect it to be.

Exercises

Review of Logarithms. Exercises 1–4 use the skills covered in the Brief Review on p. 373. In each case, solve the equation for the unknown quantity x.

1. a. $2^x = 23$ b. $10^x = 43$
 c. $3^{2x} = 12$ d. $2 \times 8^x = 24$
2. a. $3^x = 15$ b. $7^x = 100$
 c. $2^{x/3} = 20$ d. $4 \times 10^x = 120$
3. a. $\log_{10} x = 1.5$ b. $\log_{10} x = -2$
 c. $\log_{10} x = -1.5$ d. $\log_{10} 2x = 4$
4. a. $2 \times \log_{10} x = 8$ b. $\log_{10}(2x) = 2$
 c. $\log_{10}(3 + x) = 1$ d. $3 \log_{10}(4x) = 2$

Exponential Growth and Decay Laws. For Exercises 5–12, do the following:

 a. Create an exponential function of the form $Q = Q_0 \times (1 + r)^t$ (where $r > 0$ for growth and $r < 0$ for decay) to model the situation described. Be sure to clearly identify each of the variables in your function.
 b. Create a table showing the value of the quantity Q for the first 10 units of time (either years, months, weeks, or hours) of growth or decay.
 c. Make a graph of the exponential function.

5. The population of a town with a 1998 population of 85,000 grows at a rate of 2.4% per year.
6. The homicide rate in a city that had 800 homicides in 2001 is decreasing at a rate of 3% per year.
7. A privately owned forest that had 1 million acres of old growth is being clear cut at a rate of 7% per year.
8. A town with a population of 10,000 is losing residents at a rate of 0.3% per month because of a poor economy.
9. The average price of a home in a town was $125,000 in 1995, but home prices are rising by 7% per year.
10. A certain drug breaks down in the human body at a rate of 15% per hour. The initial amount of the drug in the bloodstream is 8 milligrams.
11. Your starting salary at a new job is $2000 per month, and you get annual raises of 5% per year.

12. You hid 100,000 rubles in a mattress at the end of 1991, when they had a value of $10,000. However, the value of the ruble against the dollar then fell 50% per year.

13. **Metropolitan Population Growth.** A small city had a population of 85,000 in 2000. Concerned about rapid growth, the residents passed a growth control ordinance limiting population growth to 3% each year. If the population grows at this 3% annual rate, what will the population be in 2010? in 2100? In one or two paragraphs, discuss the implications of your results for the region.

14. **Rising Costs.** Between 1995 and 2000, the average rate of inflation was about 3% per year (as measured by the Consumer Price Index). If a cart of groceries cost $100 in 1995, what did it cost in 2000?

Annual versus Monthly Inflation. Answer the questions in Exercises 15–16 about monthly and annual inflation rates.

15. If prices increase at a monthly rate of 0.6%, by what percentage do they increase in a year?

16. If prices of gold decrease at a monthly rate of 0.8%, by what percentage do they decrease in a year?

17. **Hyperinflation in Brazil.** During the 1970s and 1980s, Brazil underwent tremendous *hyperinflation*—periods of extraordinarily large inflation in prices. During the worst periods, prices rose as rapidly as 80% per month. At this rate, by what percentage would prices have risen in 1 year? in 1 day?

18. **Hyperinflation in Bosnia.** During the war in Bosnia, inflation sometimes drove prices up at a rate of 100% per month. At this rate, by what percentage would prices rise in 1 year? in 1 day?

19. **Extinction by Poaching.** Suppose that poaching reduces the population of an endangered animal by 10% per year. Further suppose that when the population of this animal falls below 30, its extinction is inevitable (owing to the lack of reproductive options without severe in-breeding). If the current population of the animal is 1000, when will it face extinction? Comment on the validity of the exponential model.

20. **World Oil Production.** World oil production was 518 million tons in 1950. Between 1950 and 1972, production increased at a rate of 7% per year. World oil production in 1998 was approximately 3300 million tons.

 a. What was world oil production in 1972?

 b. Using the result of part a, determine how much oil would have been produced in 1998 if growth in production had continued at a rate of 7% between 1972 and 1998. Compare this result to the actual figure given above.

 c. Using the result of part a, determine how much oil would have been produced in 1998 if growth in production had proceeded at a rate of 3% between 1972 and 1998. Compare this result to the actual figure given above.

 d. By trial and error, estimate the annual growth rate in world oil production between 1972 and 1998 with an exponential function.

21. **Valium Metabolism.** The drug valium is eliminated from the bloodstream exponentially with a half-life of 36 hours. Suppose that a patient receives an initial dose of 20 milligrams of valium at midnight.

 a. How much valium is in the patient's blood at noon the next day?

 b. Estimate when the valium concentration will reach 10% of its initial level.

22. **Aspirin Metabolism.** Assume that for the average individual, aspirin has a half-life of 12 hours in the bloodstream. At 12:00 noon, you take a 200-milligram dose of aspirin.

 a. How much aspirin will be in your blood at 6:00 p.m. the same day? at midnight? at 12:00 noon the next day?

 b. Estimate when the amount of aspirin will decay to 5% of its original amount.

Radioactive Dating. Use the radioactive dating formula to answer the questions in Exercises 23–24.

23. Uranium-238 has a half-life of 4.5 billion years.

 a. You find a rock containing a mixture of uranium-238 and lead. You determine that 85% of the original uranium-238 remains; the other 15% decayed into lead. How old is the rock?

b. Analysis of another rock shows that it contains 55% of its original uranium-238; the other 45% decayed into lead. How old is the rock?

24. The half-life of carbon-14 is about 5700 years.

 a. You find a piece of cloth painted with organic dyes. By analyzing the dye in the cloth, you find that only 77% of the carbon-14 originally in the dye remains. When was the cloth painted?

 b. A well-preserved piece of wood found at an archaeological site has 6.2% of the carbon-14 that it must have had when it was alive. Estimate when the wood was cut.

 c. Is carbon-14 useful for establishing the age of the Earth? Why or why not?

25. **Radioactive Waste.** A toxic radioactive substance with a density of 2 milligrams per square centimeter is detected in the ventilating ducts of a nuclear processing building that was used 45 years ago. If the half-life of the substance is 20 years, what was the density of the substance when it was deposited 45 years ago?

26. **Project: Periodic Drug Doses.** It is common to take a drug (such as aspirin or an antibiotic) repeatedly at fixed time intervals. Suppose that an antibiotic has a half-life of 8 hours and a 100-milligram dose is taken every 8 hours.

 a. Write an exponential function that represents the decay of the antibiotic from the moment of the first dose to just *prior* to the next dose (i.e., 8 hours after the first dose). How much antibiotic is in the bloodstream just *prior* to this next dose? How much antibiotic is in the bloodstream just *after* this next dose?

 b. Following a procedure similar to that in part a, calculate the amounts of antibiotic in the bloodstream just prior to and just after the doses at 16 hours, 24 hours, and 32 hours.

 c. Make a graph of the amount of antibiotic in the bloodstream for the first 32 hours after the first dose of the drug. What do you predict will happen to the amount of drug if the doses every 8 hours continue for several days or weeks? Explain.

 d. Consult a pharmacist (or read the fine print on the information sheet enclosed with many medicines) to find the half-lives of some common drugs. Create a model for the metabolism of one drug using the above procedure.

27. **Project: Increasing Atmospheric Carbon Dioxide.** Between 1860 and 1990, carbon dioxide (CO_2) concentration in the atmosphere rose from roughly 290 parts per million to 350 parts per million. Assume that this growth can be modeled with an exponential function of the form $Q = Q_0 \times (1 + r)^t$.

 a. By experimenting with various values of the fractional growth rate r, find an exponential function that fits the given data for 1860 and 1990.

 b. Use this exponential model to predict when the CO_2 concentration will be double its 1990 level.

 c. In 1998, the Kyoto treaty aimed to slow the growth in carbon dioxide concentration. Research the current status of this treaty and its effects.

Web Projects

Find useful links for Web Projects on the text Web site: www.aw.com/bennett-briggs

1. **Radioactive Dating.** Use the Web to conduct additional research on radioactive dating. Choose a specific application of radioactive dating, explain how it is used, and summarize the results.

2. **Cost of Living.** Gather data on the Consumer Price Index (CPI) between 1990 and 2000 from the Web. Use the exponential growth law, $Q = Q_0(1 + r)^t$, and the data to find the value of the growth rate r that best fits all of the data. Use the model to predict the CPI in 2010 and 2020. What do these predictions say about the cost of a movie in 2010 and 2020?

3. **Carbon Dioxide Emissions.** Visit the Web to gather data on U.S. emissions of the greenhouse gas carbon dioxide for the 1990s. Use the exponential growth law, $Q = Q_0(1 + r)^t$, together with the data to find the value of the growth rate r that best fits all of the data. Use the model to predict emissions in 2010 and 2020.

4. **Resource Consumption.** Choose a particular natural resource (such as natural gas or oil) and find data on the Web on the consumption of that resource either nationally or worldwide. Use an exponential growth model together with the data to make the case that consumption of the resource has been increasing exponentially.

5. **The Inflation Calculator.** Visit the inflation calculator on the Web. Experiment with various amounts of money and various beginning and ending years. Also use the Web to find the prices of several common services or products (such as a gallon of milk, a haircut, a loaf of bread, and admission to the movies). Determine whether the rate of increase in price for these items has been faster or slower than the rate of inflation. Use your experiments to write a three-paragraph explanation of inflation for a friend.

CHAPTER 7 SUMMARY

UNIT	KEY TERMS	KEY IDEAS AND SKILLS
7A	linear growth exponential growth doublings	In **linear growth**, a quantity grows by the same *absolute* amount in each unit of time. In **exponential growth**, a quantity grows by the same *relative* amount in each unit of time. **Understand** the impact of doublings and why exponential growth cannot continue indefinitely.
7B	doubling time half-life rule of 70	**Doubling time calculations**: After time t, for an exponentially growing quantity with doubling time T_{double}, $$\text{new value} = \text{initial value} \times 2^{t/T_{double}}$$ The **approximate doubling time formula** for a quantity growing exponentially at a rate of $P\%$ per time period: $$T_{double} \approx \frac{70}{P}$$ The **exact doubling time formula** for a quantity growing exponentially at a fractional rate r: $$T_{double} = \frac{\log_{10} 2}{\log_{10}(1+r)}$$ **Half-life calculations**: After time t, for an exponentially decaying quantity with half-life T_{half}, $$\text{new value} = \text{initial value} \times \left(\frac{1}{2}\right)^{t/T_{half}}$$ The **approximate half-life formula** for a quantity decaying exponentially at a rate of $P\%$ per time period: $$T_{half} \approx \frac{70}{P}$$ The **exact half-life formula** for a quantity decaying exponentially at a fractional rate (r must be negative): $$T_{half} = -\frac{\log_{10} 2}{\log_{10}(1+r)}$$
7C	exponential function initial value fractional growth rate r radioactive dating	**Equation for an exponential function**: $$Q = Q_0 \times (1+r)^t$$ **Equation for an exponential function given the doubling time**: $$Q = Q_0 \times 2^{t/T_{double}}$$ **Equation for an exponential function given the half-life**: $$Q = Q_0 \times \left(\frac{1}{2}\right)^{t/T_{half}}$$

8

MATHEMATICS AND THE ARTS

[Mathematics] seems to stand for all that is practical, poetry for all that is visionary, but in the kingdom of the imagination you will find them close akin, and they should go together as a precious heritage to every youth.

—FLORENCE MILNER, *School Review*, 1898

The connections between art and mathematics go deep into history. The ties are most evident in architecture: The Great Pyramids in Egypt, the Eiffel Tower in France, and modern-day skyscrapers all required mathematics in their design. But mathematics has contributed in equally profound ways to music, painting, and sculpture. In this chapter, we will explore a few of the many connections between mathematics and the arts.

UNIT 8A Mathematics and Music: We will explore the connections between mathematics and music, particularly as they apply to the ideas of musical tones and scales.

UNIT 8B Perspective and Symmetry: Perspective and symmetry are key ideas in art that are best understood through mathematics. We will explore the role these ideas have played both in classical art and in modern tilings.

UNIT 8C Proportion and the Golden Ratio: Proportion is another key idea in art. We will discuss the idea generally and then devote most of our study to the famous golden ratio.

UNIT 8A

MATHEMATICS AND MUSIC

Music is the universal language of mankind.
—Henry Wadsworth Longfellow

The roots of mathematics and music are entwined in antiquity. Pythagoras (c. 500 B.C.) claimed that "all nature consists of harmony arising out of number." He imagined that the planets circled the Earth on invisible heavenly spheres, obeying specific numeric laws and emitting the ethereal sounds known as the "music of the spheres." Thus, he saw a direct connection between geometry and music. Over a thousand years later, the standard curriculum in medieval universities was the *quadrivium* (Latin for "crossroads"), consisting of arithmetic, geometry, music, and astronomy.

SOUND AND MUSIC

Any vibrating object produces sound. The vibrations produce a **wave** (much like a water wave) that propagates through the surrounding air in all directions. When such a wave impinges on the miraculously designed organ called the ear, it is perceived as sound. Of course, some sounds, such as speech and screeching tires, do not qualify as music. Most musical sounds are made by vibrating strings (violins, cellos, guitars, and pianos), vibrating reeds (clarinets, oboes, and some organ pipes), or vibrating columns of air (other organ pipes, horns, and flutes).

One of the most basic qualities of sound is **pitch.** For example, a tuba has a "lower" pitch than a flute, and a violin has a "higher" pitch than a bass guitar. To understand pitch, find a taut string (a guitar string works best, but a stretched rubber band will do). When you pluck the string, it produces a sound with a certain pitch. Next, use your finger to hold the midpoint of the string in place, and pluck either half of the string. Note that a higher-pitched sound is produced, demonstrating an ancient musical principle discovered by the Greeks: *The shorter the string, the higher the pitch.*

But what *is* pitch? It took many centuries to arrive at a precise understanding of pitch; it is attributed to both Galileo (1564–1642) and the French monk Marin Mersenne (1588–1648). Plucking the fully taut string causes it to vibrate up and down along its length (Figure 8.1a). For example, if the string vibrates up and down 100 times per second, its **frequency** is said to be 100 **cycles per second (cps);** each cycle corresponds to one vibration up *and* down. When you pluck only half the string (by holding down its midpoint), the resulting wave is *half* as long as the first wave (Figure 8.1b), and its frequency therefore is twice as much, or 200 cps. Thus, pitch is related to the frequency of the vibrating string: *The higher the frequency, the higher the pitch.*

For any particular string, plucking its full length causes it to vibrate at its **fundamental frequency.** Plucking half its length generates a wave with twice the fundamental frequency, called the *first harmonic* (or *overtone*). Halving the length of the string again (Figure 8.1c) doubles the frequency again, so the resulting tone, called the *second harmonic,* has four times the fundamental frequency.

historical note

Evidence of music is found in nearly all ancient cultures. Indeed, based on evidence dating back 30,000 years, some archaeologists suspect that music may predate speech.

historical note

String and wind instruments were designed at least 5000 years ago in Mesopotamia, and Egyptians played in small ensembles of lyres and flutes by 2700 B.C.

FIGURE 8.1 A vibrating string at (a) its fundamental frequency, (b) its first harmonic, and (c) its second harmonic.

Now we come to another discovery of the ancient Greeks: Doubling the frequency of a tone raises the pitch by an **octave**—perhaps the most appealing and natural combination of notes in music. The piano keyboard (Figure 8.2) is helpful here. An octave is the interval between, say, middle C and the next higher C. For example, middle C on the piano has a frequency of 260 cps, the C above middle C has a frequency of 2 × 260 cps = 520 cps, and the next higher C has a frequency of 2 × 520 cps = 1040 cps. Similarly, the C below middle C has a frequency of about $\frac{1}{2}$ × 260 cps = 130 cps.

FIGURE 8.2 Piano keys.

> *by the way ...*
> The frequencies at which a particular string vibrates depend on its length, how tightly it is strung, and several other factors. Human speech consists of sounds with frequencies of 200 to 400 cycles per second. The range of a piano extends from about 27 to 4200 cycles per second. The human ear can detect frequencies between about 20 and 20,000 cycles per second.

Time out to think

The note middle A (above middle C) has a frequency of about 440 cps. What is the frequency of the A an octave higher? What is the frequency of the A an octave below middle A?

> Human speech is like a cracked kettle on which we tap crude rhythms for bears to dance to, while we long to make music that will melt the stars.
>
> —Gustave Flaubert

SCALES

The musical tones that span an octave comprise a **scale.** The Greeks invented the 7-note (or diatonic) scale that corresponds to the white keys on the piano. In the 17th century, Johann Sebastian Bach adopted a 12-tone scale for his keyboard music, which corresponds to both the white and the black keys on a modern piano. Through Bach's music, the 12-tone scale spread throughout Europe, becoming a foundation of Western music. Many other scales are possible. For example, 3-tone scales are common in African music, scales with more than 12 tones occur in Asian music, and 19-tone scales are used in contemporary music.

On the 12-tone scale, the frequency separating each tone is called a **half-step,** corresponding to consecutive notes on the piano keyboard. For example, E and F are separated by a half-step, as are F and F# (read "F sharp"). In each half-step, the frequency increases by some *multiplicative* factor; let's call it f. Thus, the frequency of the note C is the frequency of B times the factor f, the frequency of C# is the frequency of C times the factor f, and so on. The frequencies of the notes across the entire scale are related as follows:

$$C \xrightarrow{f} C\# \xrightarrow{f} D \xrightarrow{f} D\# \xrightarrow{f} E \xrightarrow{f} F \xrightarrow{f} F\# \xrightarrow{f} G \xrightarrow{f} G\# \xrightarrow{f} A \xrightarrow{f} A\# \xrightarrow{f} B \xrightarrow{f} C$$

Because an octave corresponds to an increase in frequency of a factor of 2, the factor f must have the property

$$\underbrace{f \times f \times f \times f \times f \times f \times f \times f \times f \times f \times f \times f}_{12 \text{ times}} = f^{12} = 2$$

Thus, f must be the *twelfth root* of two, or $f = \sqrt[12]{2} \approx 1.05946$.

We can now calculate the frequency of every note of a 12-tone scale. Starting from middle C, with its frequency of 260 cps, we multiply by $f = \sqrt[12]{2}$ to find that the frequency of C# is 260 cps $\times \sqrt[12]{2} \approx 275$ cps. Multiplying again by $f = \sqrt[12]{2}$ gives the frequency of D as 275 cps $\times \sqrt[12]{2} \approx 292$ cps. Continuing in this way generates Table 8.1.

TABLE 8.1 Frequencies of Notes in the Octave Above Middle C

NOTE	FREQUENCY (CPS)	RATIO TO FREQUENCY OF PRECEDING NOTE	RATIO TO FREQUENCY OF MIDDLE C
C	260	$\sqrt[12]{2} \approx 1.05946$	$1.00000 = 1$
C#	275	$\sqrt[12]{2} \approx 1.05946$	1.05946
D (second)	292	$\sqrt[12]{2} \approx 1.05946$	1.12246
D#	309	$\sqrt[12]{2} \approx 1.05946$	1.18921
E (third)	328	$\sqrt[12]{2} \approx 1.05946$	$1.25992 \approx \frac{5}{4}$
F (fourth)	347	$\sqrt[12]{2} \approx 1.05946$	$1.33484 \approx \frac{4}{3}$
F#	368	$\sqrt[12]{2} \approx 1.05946$	1.41421
G (fifth)	390	$\sqrt[12]{2} \approx 1.05946$	$1.49831 \approx \frac{3}{2}$
G#	413	$\sqrt[12]{2} \approx 1.05946$	1.58740
A (sixth)	437	$\sqrt[12]{2} \approx 1.05946$	$1.68179 \approx \frac{5}{3}$
A#	463	$\sqrt[12]{2} \approx 1.05946$	1.78180
B (seventh)	491	$\sqrt[12]{2} \approx 1.05946$	1.88775
C (octave)	520	$\sqrt[12]{2} \approx 1.05946$	$2.00000 = 2$

by the way ...
All entries in column 3 are the same because the same factor, $f \approx \sqrt[12]{2}$, separates every pair of notes. The parenthetical terms in column 1 are names used by musicians to describe intervals between the note shown and middle C.

Column 4 of Table 8.1 shows that a few tones have simple ratios of frequency to middle C. For example, the frequency of G is approximately $\frac{3}{2}$ times the frequency of middle C (musicians call this interval a *fifth*), and the frequency of F is approximately $\frac{4}{3}$ times the frequency of middle C (musicians call this interval a *fourth*). According to many musicians, the most pleasing combinations of notes, called **consonant tones,** are those whose frequencies have a simple ratio. Referring to consonant tones, the Chinese philosopher Confucius observed that small numbers are the source of perfection in music.

EXAMPLE 1 *The Dilemma of Temperament*

Because the whole-number ratios in Table 8.1 are not exact, tuners of musical instruments have the problem of *temperament,* which can be demonstrated as follows. Start at middle C with a frequency of 260 cps. Using the whole number ratios, find the frequency if you raise C by a sixth to A, raise A by a fourth to D, lower D by a fifth to G, and lower G by a fifth to C. Having returned to the same note, have you also returned to the same frequency?

Solution According to Table 8.1, raising a note by a sixth increases its frequency by a factor of $\frac{5}{3}$; thus, the frequency of A above middle C is $\frac{5}{3} \times 260$ cps = 433.33 cps. Raising this note by a fourth increases its frequency by $\frac{4}{3}$; thus, D has a frequency of $\frac{4}{3} \times 433.33$ cps = 577.77 cps. Lowering D by a fifth (a factor of $\frac{2}{3}$) to G gives a frequency of $\frac{2}{3} \times 577.77 = 385.18$ cps. Finally, lowering G by another fifth (a factor of $\frac{2}{3}$) puts us back to middle C, with a frequency of $\frac{2}{3} \times 385.18$ cps = 256.79. Note that, by using whole-number ratios, we have not quite returned to the proper frequency of 260 cps for middle C. The problem is that the whole-number ratios are not exact. That is, $\frac{5}{3} \times \frac{4}{3} \times \frac{2}{3} \times \frac{2}{3} = \frac{80}{81}$ is close to, but not *exactly,* 1. ■

MUSICAL SCALES AS EXPONENTIAL GROWTH

The increase in frequencies in a scale is just another example of exponential growth. Each successive frequency is $f \approx 1.05946$ times, or approximately 5.9% more than, the previous frequency. In other words, the frequencies increase at a fixed relative growth rate. Thus, we can use an exponential function (see Unit 7C) to find any frequency on the scale. Suppose we start at a frequency that we call Q_0. Then the frequency Q of the note n half-steps higher is given by

$$Q = Q_0 \times 1.05946^n$$

EXAMPLE 2 *Exponential Growth on Musical Scales*

Use the exponential growth law to find the frequency of the note one fifth above middle C, the note one octave and one fifth above middle C, and the note two octaves and one fifth above middle C.

Solution We let the frequency of middle C be the initial value for the scale; that is, we set $Q_0 = 260$ cps. Table 8.1 shows that the note a fifth above middle C is G, which is seven half-steps above middle C. Thus, we let $n = 7$ in the exponential law and find that the frequency of G is

$$Q = Q_0 \times 1.05946^7 = 390 \text{ cps}$$

Of course, Beethoven's Eroica Symphony or Michelangelo's ceiling of the Sistine Chapel can be reduced to an adequate mathematical formula. There is, however, one little catch. The only person capable of making such an all-inclusive analysis must be able to feel the emotions that Beethoven and Michelangelo felt, to think in sound and paint as they did.

—J. MURRAY BARBOUR, Music Historian

The note one octave and a fifth above middle C is 12 + 7 = 19 half-steps above middle C. Letting $n = 19$, we find that the frequency of this note is

$$Q = Q_0 \times 1.05946^{19} = 779 \text{ cps}$$

The note two octaves and a fifth above middle C is (2 × 12) + 7 = 31 half-steps above middle C. Letting $n = 31$, we find that the frequency of this note is

$$Q = Q_0 \times 1.05946^{31} = 1558 \text{ cps}$$

FROM TONES TO MUSIC

Although the simple frequency ratios of "pure" tones are the building blocks of music, the sounds of music are far richer and more complex. For example, a plucked violin string does much more than produce the simple frequency of its vibration. The string motion is transferred through the bridge of the violin to its top, and the ribs transfer those vibrations to the back of the instrument. With the top and back of the violin in oscillation, the entire instrument acts as a resonating chamber, which excites and amplifies the higher harmonics of the original tone.

Similar principles generate rich and complex sounds in all instruments. Figure 8.3 (top) shows a typical sound wave that might be produced by an instrument. Note that it isn't a simple wave like those pictured in Figure 8.1; instead, it consists of a combination of simple waves that are the harmonics of the fundamental. In fact, the complex wave is the *sum* of the three other simple waves shown below it. The fact that a musical sound can be expressed as a sum of simple harmonics is surely the deepest connection between mathematics and music. The French mathematician Jean Baptiste Joseph Fourier first enunciated this principle in about 1810; it was one of the most profound discoveries in mathematics.

FIGURE 8.3 The complex sound wave on the left is the sum of the three simple waves on the right.

Stradivarius was essentially a craftsman of science, one with considerable, demonstrable knowledge of mathematics and acoustical physics.

—THOMAS LEVENSON,
Measure for Measure: A Musical History of Science, pp. 207–208

Although mathematics helps in understanding music, many mysteries remain. For example, in about 1700, an Italian craftsman known as Stradivarius made what are still considered to be the finest violins and cellos ever produced. Despite years of study by mathematicians and scientists, no one has succeeded in reproducing the unique sounds of a Stradivarius instrument.

THE DIGITAL AGE

When we talk about sound waves and imagine music to consist of waves, we are working with the **analog** picture of music. Until about 25 years ago, all musical recordings (phonograph cylinders, records, and tape recordings) were based on the

analog picture of music. Storing music in the analog mode requires storing analogs to sound waves; for example, on records, the grooves in the vinyl surface are etched with the shape of the original musical sound wave. If you have listened to analog recordings, you know that this shape can easily become distorted or damaged.

Most recent musical recordings use a **digital** picture of music. A computer takes the analog version of a piece of music and chops it into very short time intervals (fractions of a second); the time intervals are used much like bins in a histogram. For each time interval, the computer measures the frequencies of the sound waves and records them numerically. This process, called *digitizing*, transforms the music into a list of numbers. The list can then be saved on a storage device such as a compact disc (CD). Inside a playback device such as a CD player, a computer uses a reverse process to convert the numbers back into sound waves so that they can be heard. As long as the time intervals are short enough, the music sounds smooth and authentic, with no evidence of digitizing.

Digital technology provides easy and endless ways to "process" music. Through techniques of **digital signal processing,** the sounds of a musical recording can be modified with computers. For example, extraneous sounds (background noises or "hiss") can be detected and removed. Changing the music by amplifying certain frequencies or attenuating others also is possible. Moreover, once digital music can be modified, it is a short step to creating music. Instruments called **synthesizers** can create and imitate a tremendous variety of sounds digitally without strings, brass tubes, or reeds. In the digital age, the dividing line between mathematics and music all but vanishes.

Review Questions

1. What is pitch? How is it related to the frequency of a musical note?
2. Define *fundamental frequency, first harmonic,* and *octave.* Why are these concepts important in music?
3. What is a 12-tone scale? How are the frequencies of the notes on a 12-tone scale related to one another?
4. Explain how the notes of the scale are generated by exponential growth.
5. How do the wave forms of real musical sounds differ from the wave forms of simple tones? How are they related?
6. What is the difference between an analog and a digital recording of music? What are the advantages of digital recording?

In the News

1. **Numbers and Music.** Look carefully at cassette or CD liner notes, music catalogs, or other materials related to musical products. Find as many different uses of numbers as possible in these materials, and explain how the numbers are related to the music.
2. **Digital Music.** The ease with which digital music can be copied has increased the importance of the copyright issue, especially with respect to music available on the Web. Find a recent article on issues of copying digital music. Discuss the article and its conclusions.

Exercises

1. **Octaves.** Starting with a tone having a frequency of 110 cycles per second, find the frequencies of the tones that are one, two, three, and four octaves higher.
2. **Octaves.** Starting with a tone having a frequency of 880 cycles per second, find the frequencies of the tones that are one, two, three, and four octaves lower.
3. **Notes of a Scale.** Find the frequencies of the 12 notes of the scale that starts at the A above middle C; this A has a frequency of 437 cycles per second.
4. **Notes of a Scale.** Find the frequencies of the 12 notes of the scale that starts at the G above middle C; this G has a frequency of 390 cycles per second.

5. **Exponential Growth and Scales.** Starting at middle C, with a frequency of 260 cps, find the frequency of the following notes:
 a. five half-steps above middle C
 b. a fifth above middle C
 c. an octave and a fourth above middle C
 d. 36 half-steps above middle C
 e. four octaves and three half-steps above middle C

6. **Exponential Growth and Scales.** Starting at middle A, with a frequency of 437 cps, find the frequency of the following notes:
 a. six half-steps above middle A
 b. a third above middle A
 c. an octave and a fourth above middle A
 d. 25 half-steps above middle A
 e. two octaves and two half-steps above middle A

7. **Exponential Decay and Scales.** What is the frequency of the note five half-steps *below* middle A (which has a frequency of 437 cps)? eight half-steps *below* middle A?

8. **The Dilemma of Temperament.** Start at middle A, with a frequency of 437 cps. Using the whole-number ratios in Table 8.1, find the frequency if you raise A by a fifth to E. What is the frequency if you raise E by a fifth to B? What is the frequency if you lower B by a sixth to D? What is the frequency if you lower D by a fourth to A? Having returned to the same note, have you also returned to the same frequency? Explain.

9. **Circle of Fifths.** The circle of fifths is generated by starting at a particular musical note and stepping upward by intervals of a fifth (seven half-steps). For example, starting at middle C, a circle of fifths includes the notes $C \to G \to D' \to A' \to E'' \to B'' \to \ldots$, where each (') denotes a higher octave. Eventually the circle comes back to C several octaves higher.
 a. Show that the frequency of a tone increases by a factor of $2^{7/12} = 1.498$ if it is raised by a fifth. (Hint: Recall that each half-step corresponds to an increase in frequency by a factor of $f = \sqrt[12]{2}$.)
 b. By what factor does the frequency of a tone increase if it is raised by two fifths?
 c. Starting with middle C, at a frequency of 260 cycles per second, find the frequencies of the other notes in the circle of fifths.
 d. How many fifths are required for the circle of fifths to return to a C? How many octaves are covered by a complete circle of fifths?
 e. What is the ratio of the frequencies of the C at the beginning of the circle and the C at the end of the circle?
 f. A circle of *fourths* is generated by starting at any note and stepping upward by intervals of a fourth (five half-steps). By what factor is the frequency of a tone increased if it is raised by a fourth? How many fourths are required to complete the entire circle of fourths? How many octaves are covered in a complete circle of fourths?

10. **Rhythm and Mathematics.** In this unit, we focused on musical sounds, but rhythm and mathematics are also closely related. For example, in "4/4 time," there are four *quarter notes* in a measure. If two quarter notes have the duration of a *half note*, how many half notes are in one measure? If two *eighth notes* have the duration of a quarter note, how many eighth notes are in one measure? If two *sixteenth notes* have the duration of an eighth note, how many sixteenth notes are in one measure?

Web Projects

Find useful links for Web Projects on the text Web site: www.aw.com/bennett-briggs

1. **Mathematics and Music.** Many Web sites are devoted to connections between music and mathematics. Visit one such site and study a connection in some depth. Write a one- to two-page essay that describes this connection between mathematics and music.

2. **Internet Music.** Many Web sites make music available digitally. Some of these Web sites have come under attack for violating copyright laws. Find one such Web site. Describe how it works both from a user's standpoint and from a commercial standpoint. Do you think the site's use of music is legal? Is it ethical? Defend your opinions.

3. **Mathematics and Composers.** Many musical composers, both classical and modern, have used mathematics in their compositions. Research the life of one such composer. Write a one- to two-page essay discussing the role of mathematics in the composer's life and music.

UNIT 8B

PERSPECTIVE AND SYMMETRY

We now turn our attention to connections between mathematics and the visual arts, such as painting, sculpture, and architecture. At least three aspects of the visual arts relate directly to mathematics: *perspective*, *symmetry*, and *proportion*. In this unit, we will see how Renaissance mathematicians and artists discovered perspective and then we will explore the idea of symmetry. (We'll save *proportion* for Unit 8C.)

PERSPECTIVE

People of all cultures have used geometrical ideas and patterns in their artwork. In the western world, the ancient Greeks developed strong ties between the arts and mathematics, because both endeavors were central to their view of the world. Much of the Greek outlook was lost during the Middle Ages, but the Renaissance brought at least two new developments that made mathematics an essential tool of artists. First, there was a renewed interest in natural scenes, which brought a need to paint with realism. Second, many of the artists of the day also worked as engineers and architects.

The desire to paint landscapes with three-dimensional realism brought Renaissance painters face to face with the matter of perspective. In their attempts to capture depth and volume on a two-dimensional canvas, these artists made a science of painting. The painters Brunelleschi (1377–1446) and Alberti (1404–1472) are generally credited with developing, in about 1430, a system of perspective that involved geometrical thinking. Alberti's principle that a painting "is a section of a projection" lies at the heart of drawing with perspective.

Suppose you want to paint a simple view looking down a hallway with a checkerboard tile floor. Figure 8.4 shows a *side view* of the artist's eye, the canvas, and the hallway. Note the four lines labeled L_1, L_2, L_3, and L_4. The two side walls of the

FIGURE 8.4 Side view of a hallway, showing perspectives.

hallway intersect the floor and the ceiling along these four lines. These lines are important because they are parallel to each other in the scene and perpendicular to the canvas (or the plane containing the canvas).

Let's now look at the scene as the artist sees and paints it. The artist looks down the hallway with the point of view shown in Figure 8.5.

FIGURE 8.5 An artist's view of the hallway from Figure 8.4, showing perspectives. Adapted from M. Kline, *Mathematics in Western Culture*.

The lines L_1, L_2, L_3, and L_4, which are parallel in the actual scene, are not parallel in the painting. In fact, they all meet at a single point, labeled P, which is called the **principal vanishing point.** Thus, we have the first principle of perspective discovered by the Renaissance painters:

All lines that are parallel in the real scene and perpendicular to the canvas must intersect at the principal vanishing point of the painting.

Note that other lines that are parallel to lines L_1, L_2, L_3, and L_4, such as the lines going straight down the hallway along the floor tiles, also meet at the principal vanishing point. For example, the line connecting the points B and b intersects P, as do the line connecting C and c and the line connecting D and d.

What happens to lines that are parallel in the actual scene but not perpendicular to the canvas, such as the diagonal lines along the floor tiles? If you study Figure 8.5 carefully, you'll see that such lines intersect at their own vanishing points, all on the horizontal line passing through the principal vanishing point; this line is called the **horizon line.** For example, the right-slanting diagonals of the floor tiles are parallel in the actual scene. But, in the painting, these lines meet at the vanishing point labeled P_1 on the horizon line. Similarly, the left-slanting diagonals meet at a vanishing point called P_2 on the horizon line. In fact, all sets of lines that are mutually parallel in the real scene must meet at their own vanishing point on the horizon line.

Time out to think

Imagine looking along a set of long parallel lines that stretches far into the distance, such as a set of train tracks or a set of telephone lines. The lines will appear to your eyes to get closer as you look into the distance. If you were painting a picture of the scene, where would you put the principal vanishing point? Why?

FIGURE 8.6 *The Last Supper*, by Leonardo da Vinci, shown with several lines that are parallel in the real scene and therefore converge at the principal vanishing point behind Christ.

Let no one who is not a mathematician read my works.

—LEONARDO DA VINCI

Leonardo da Vinci (1452–1519) contributed greatly to the science of perspective. We can see da Vinci's mastery of perspective in many of his paintings. If you study *The Last Supper* (Figure 8.6), you will notice several parallel lines in the actual scene intersecting at the principal vanishing point of the painting, which is directly behind the central figure of Christ.

The German artist Albrecht Dürer (1471–1528) further developed the science of perspective. Near the end of his life, he wrote a popular book that stressed the use of geometry and encouraged artists to paint according to mathematical principles. Figure 8.7 is one of Dürer's woodcuts, showing an artist using his principles of perspective. A string from a point on the lute is attached to the wall at the point corresponding to the artist's eye. At the point where the string passes through the frame, a point is placed on the canvas. As the string is moved to different points on the lute, a drawing of the lute is created in perfect perspective on the canvas.

The artist Jan Vredeman de Vries (1527–1604) summarized much of the science of perspective in a book he published in 1604. Figure 8.8 shows a sketch from his book that illustrates how thoroughly perspective can be analyzed.

FIGURE 8.7 Woodcut by Albrecht Dürer.

FIGURE 8.8 Sketch by Jan Vredeman de Vries.

Time out to think

Identify some of the vanishing points in the sketch in Figure 8.8. What parallel lines from the real scene converge at each vanishing point?

Perspective drawing is sometimes abused deliberately. The painting *False Perspective* (Figure 8.9) by English artist William Hogarth (1697–1764) reminds us of how essential perspective is in art. Note where the fishing line of the near man lands, and how the woman in the window appears to be lighting the pipe of a man on a distant hill. The familiar work of Maurits C. Escher (1898–1972) also confounds us with its use and abuse of perspective. His drawing *Belvedere* (Figure 8.10) illustrates good use of perspective, yet the pillars of the structure are cleverly drawn to show impossible positions.

> **historical note**
>
> Leonardo da Vinci was not only an artist but also a great scientist and engineer. His notebooks contain many ideas that were far ahead of his time. A century before Copernicus, he wrote that the Earth was not the center of the universe. He correctly recognized fossils as remains from extinct species and suggested the possibility of long-term geological change on the Earth. Unfortunately, he wrote his notebooks in code, so his contemporaries knew nothing of his ideas.

FIGURE 8.9 *False Perspective*, by William Hogarth.

FIGURE 8.10 *Belvedere*, by M. C. Escher.

SYMMETRY

The term *symmetry* has many meanings. Sometimes it refers to a kind of balance. For example, *The Last Supper* (see Figure 8.6) is symmetrical because the disciples are grouped in four groups of three, with two groups on either side of the central figure of Christ. A human body is symmetrical because a vertical line drawn through the head and navel divides the body into two (nearly) identical parts (Figure 8.11).

Symmetry can also refer to repetition of patterns. Native American pottery is often decorated with simple borders that use repeating patterns. Similar symmetries are found in African, Moslem, and Moorish art such as that shown in the Figure 8.12.

FIGURE 8.11 Leonardo da Vinci's sketch showing the symmetry of the human body.

FIGURE 8.12 This Islamic tiling shows many elaborate symmetries. It is located in the Medersa Attarine, a 14th-century school of Muslim theology in Fez, Morocco.

In mathematics, *symmetry* is a property possessed by an object that remains unchanged by some operations. For example, a circle still looks the same if it is rotated about its center. A square still looks the same if it is flipped across one of its diagonals (Figure 8.13). There are many such mathematical symmetries, often quite subtle. However, three symmetries are easy to identify:

- **Reflection symmetry:** An object remains unchanged when reflected across a straight line. For example, the letter A has reflection symmetry about a vertical line, while the letter H has reflection symmetry about a vertical and a horizontal line (Figure 8.14).
- **Rotation symmetry:** An object remains unchanged when rotated through some angle about a point. For example, the letters O and S have rotation symmetry because they are unchanged when rotated 180° (Figure 8.15).
- **Translation symmetry:** A pattern remains the same when shifted, say to the right or left. The pattern . . . XXX . . . (with the X's continuing in both directions) has translation symmetry because it still looks the same if we shift it to the left or to the right (Figure 8.16). (In mathematics and physics, *translating* an object means moving it in a straight line, without rotating it.)

FIGURE 8.13

FIGURE 8.14 Reflection symmetry.

FIGURE 8.15 Rotation symmetry.

XXXXXX

FIGURE 8.16 Translation symmetry.

EXAMPLE 1 *Finding Symmetries*

Identify the types of symmetry in each star in Figure 8.17.

(a) (b)

FIGURE 8.17

Solution

a. The five-pointed star has five lines about which it can be flipped (reflected) without changing its appearance, so it has five reflection symmetries (Figure 8.18a). Because it has five vertices that all look the same, it can be rotated by $\frac{1}{5}$ of a full circle, or $360°/5 = 72°$, and still look the same. Similarly, its appearance remains unchanged if it is rotated by $2 \times 72° = 144°$, $3 \times 72° = 216°$, or $4 \times 72° = 288°$. Thus, this star has four rotational symmetries.

b. The six-pointed star has six reflection lines about which it can be flipped (reflected) without changing its appearance, so it has six reflection symmetries (Figure 8.18b). Because of its six vertices, it has rotation symmetry when rotated by $\frac{1}{6}$ of a full circle, or $360°/6 = 60°$. It also has symmetry if rotated by $2 \times 60° = 120°$, $3 \times 60° = 180°$, $4 \times 60° = 240°$, or $5 \times 60° = 300°$. Thus, this star has five rotational symmetries.

(a) (b)

FIGURE 8.18 Each line represents a reflection symmetry. Rotation symmetries require rotation by multiples of the indicated angles.

Gustave Dore's (1832–1883) engraving *The Vision of the Empyrean* offers a dramatic illustration of rotation symmetry (Figure 8.19). This grand image of the cosmos can be rotated by many different angles and, at least on a large scale, appears much the same.

Sometimes, it is the *departures* from symmetry that make art effective. The 20th-century work *Supernovae* (Figure 8.20) by the Hungarian painter Victor

FIGURE 8.19 *The Vision of the Empyrean,* by Gustave Dore.

FIGURE 8.20 *Supernovae,* by Victor Vasarely.

Vasarely (1908–1997) may have started as a symmetric arrangement of circles and squares. But the gradual deviations from that pattern make a powerful visual effect.

Given the strong ties between mathematics and art, you may not be surprised that mathematical algorithms (recipes) can generate art on computers. Figure 8.21 shows an intricate Persian rug design generated on a computer by Anne Burns. The algorithm can be varied to give an endless array of patterns and symmetries.

FIGURE 8.21 A "Persian rug," generated on a computer by Anne Burns.

> **by the way...**
> The light-sensitive photoreceptors in the human eye "tile" the retina in a hexagonal array, much like that in Figure 8.22c.

(a) (b) (c)

FIGURE 8.22 The three possible tilings made from regular polygons: (a) triangles, (b) squares, and (c) hexagons.

TILINGS

A form of art called *tilings* (or *tessellations*) involves covering a flat area, such as a floor, with geometrical shapes. Tilings usually have regular or symmetric patterns. Tilings are found in ancient Roman mosaics, stained glass windows, and the elaborate courtyards of Arab mosques—as well as in many modern kitchens and bathrooms.

> **by the way...**
> The word *tessellation* comes from a Latin word for the small tiles used in mosaics.

More precisely, a **tiling** is an arrangement of *polygons* (many-sided figures) that interlock perfectly with no overlapping. The simplest tilings use just one type of regular polygon. Figure 8.22 shows three such tilings made with equilateral triangles, squares, and regular hexagons, respectively. Note that there are no gaps or overlaps between the polygons in any of the three cases. In each case, the tiling is made by translating (shifting) the same basic polygon in various directions. Thus, these tilings have translation symmetry.

What happens if you try to make a tiling with, say, regular pentagons? If you try, you'll find that it simply does not work. The interior angles of a regular pentagon measure 108°. Thus, as Figure 8.23 shows, the angle that remains when three regular pentagons are placed next to each other is too small to fit another regular pentagon. In fact, a mathematical theorem states that tilings with a single regular polygon are possible only with equilateral triangles, squares, and hexagons (as in Figure 8.22).

Angle = 360° − (3 × 108°) = 36°. This is too small for another pentagon to fit.

FIGURE 8.23 Regular pentagons cannot make a tiling.

Of course, more tilings are possible if we remove the restriction of using only a single type of regular polygon. For example, if we allow more than one different regular polygon but still require that the arrangement of polygons look the same around each vertex (intersection point), it is possible to make the eight different tiling patterns shown in Figure 8.24.

FIGURE 8.24 Eight tilings, each made by combining two or more types of regular polygons.

Time out to think

Verify that each of the tilings in Figure 8.24 uses only regular polygons. How many different regular polygons are used in each of these tilings? Verify that the same arrangement of polygons appears around each vertex. (Note: Look carefully at the polygons; there are no circles in this figure.)

Tilings that use *irregular* polygons (those with sides of different lengths) are endless in number. As an example, suppose we start with an arbitrary triangle that has no special properties (other than three sides). The easiest way to tile a region with this triangle is by translating it parallel to two of its sides, as shown in Figure 8.25. We shift the original triangle to the right so that the new triangle touches the original triangle at a single point. We also shift the original triangle down so that the new triangle touches the original triangle at a single point. Then we repeat these right/left and up/down translations as many times as we like. The gaps created in this process are themselves triangles that interlock perfectly with the translated triangles to create a tiling.

FIGURE 8.25 A tiling made by translating a triangle in two directions.

Figure 8.26 shows another example. This time, we begin by reflecting an arbitrary triangle to produce a wing-shaped object, and then we translate this object up/down and right/left.

FIGURE 8.26 A tiling made by first reflecting a triangle to make a "wing," then translating the wing in two directions.

All of the tilings discussed so far are called *periodic tilings* because they have a pattern that is repeated throughout the tiling. In recent decades, mathematicians have explored tilings that are *aperiodic,* meaning that they do not have a pattern that repeats throughout the entire tiling. Figure 8.27 shows an aperiodic tiling created by British mathematician Roger Penrose. If you look at the center of the figure, there appears to be a fivefold symmetry (a rotational symmetry that you would find in a pentagon). However, if the figure were extended indefinitely in all directions, the same pattern would never be repeated.

FIGURE 8.27 An aperiodic tiling by Roger Penrose.

Tilings can be beautiful and practical for such things as floors and ceilings. However, recent research also shows that tilings may be very important in nature. In particular, many molecules and crystals apparently have patterns and symmetries that can be understood with the same mathematics used to study tilings in art.

EXAMPLE 2 *Quadrilateral Tiling*

Create a tiling by translating the quadrilateral shown in Figure 8.28. As you translate the quadrilateral, make sure that the gaps left behind have the same quadrilateral shape.

FIGURE 8.28

Solution We can find the solution by trial and error, translating the quadrilateral in different directions until we have correctly shaped gaps. Figure 8.29 shows the solution. Note that the translations are along the directions of the two diagonals of the quadrilateral. The gaps between the translated quadrilaterals are themselves quadrilaterals that interlock perfectly to complete the tiling.

FIGURE 8.29 A quadrilateral tiling.

Review Questions

1. Describe the ideas of perspective and symmetry.
2. How is the principal vanishing point in a picture determined?
3. What is the horizon line? How is it important in a painting showing perspective?
4. Briefly describe and distinguish among *reflection symmetry*, *rotation symmetry*, and *translation symmetry*. Draw a simple picture that shows each type of symmetry.
5. What is a tiling? Draw a simple example.
6. Briefly explain why there are only three possible tiling patterns that consist of a single regular polygon. What are the three patterns?
7. Briefly explain why more tilings are possible if we remove the restriction of using regular polygons.
8. What is the difference between periodic and aperiodic tilings?

In the News

1. **Perspective in Life.** Describe at least three ways in which the use of visual perspective affects your life.
2. **Symmetry in Life.** Find at least three objects from your daily life that exhibit some type of mathematical symmetry. Describe the symmetry in each case.

Exercises

1. **Vanishing Points.** Consider the simple drawing in Figure 8.30 of a road and a telephone pole.
 a. Locate a vanishing point for the drawing. Is it the principal vanishing point?

 FIGURE 8.30

 b. With proper perspective, draw three more telephone poles receding into the distance.

2. **Correct Perspective.** Consider the two boxes shown in Figure 8.31. Which one is drawn with proper perspective relative to a single vanishing point? Explain.

 FIGURE 8.31

3. **Drawing with Perspective.** Make the square, circle, and triangle in Figure 8.32 into three-dimensional solid objects: a box, a cylinder, and a triangular prism, respectively. The given objects should be used as the front faces of the three-dimensional objects and all figures should be drawn with correct perspective relative to the given vanishing point P.

 FIGURE 8.32

4. **Drawing MATH with Perspective.** Make the letters MATH in Figure 8.33 into three-dimensional solid letters. The given letters should be used as the front faces of the three-dimensional letters and all letters should be drawn with correct perspective relative to the given vanishing point P.

 FIGURE 8.33

5. **Proportion and Perspective.** The drawing in Figure 8.34 shows two poles drawn with correct perspective relative to a single vanishing point. As you can check, the first pole is 2 centimeters tall in the drawing and the second pole is 2 centimeters away from the first pole in the drawing.
 a. Draw three more identical poles that are equally spaced in the drawing.

FIGURE 8.34

 b. What are the heights of these three new poles in the drawing?

 c. In the actual scene, would these five poles be equally spaced? Explain.

6. **Two Vanishing Points.** Figure 8.35 shows a road receding into the distance. In the direction of the arrow, draw a second road that intersects the first road. Be sure that the vanishing points of the two roads lie on a horizontal horizon line.

FIGURE 8.35

7. **Desargues' Theorem.** An early theorem of projective geometry, proved by French architect and engineer Girard Desargues (1593–1662), says that if two triangles (*ABC* and *abc* in Figure 8.36) are drawn so that the straight lines joining corresponding vertices (*Aa*, *Bb*, and *Cc*) all meet in a point *P* (corresponding to a vanishing point), then the corresponding sides (*AC* and *ac*, *AB* and *ab*, *BC* and *bc*), if extended, will meet in three points that all lie on the same line *L*. Draw two triangles of your own in such a way that the conditions of Desargues' theorem are satisfied. Verify that the conclusions of the theorem are true.

FIGURE 8.36

8. **Symmetry in Letters.** Find all of the capital letters of the alphabet that have
 a. right/left reflection symmetry (such as A)
 b. top/bottom reflection symmetry (such as H)
 c. both right/left and top/bottom reflection symmetry
 d. a rotational symmetry

9. **Star Symmetries.**
 a. How many reflection symmetries does a four-pointed star have (Figure 8.37a)? How many rotational symmetries does a four-pointed star have?

(a) (b)

FIGURE 8.37

 b. How many reflection symmetries does a seven-pointed star have (Figure 8.37b)? How many rotational symmetries does a seven-pointed star have?

10. **Symmetries of Geometric Figures.**
 a. Draw an equilateral triangle (all three sides have equal length). How many degrees can the triangle be rotated about its center so that it remains unchanged in appearance? (There are several correct answers.)
 b. Draw a square (all four sides have equal length). How many degrees can the square be rotated about its center so that it remains unchanged in appearance? (There are several correct answers.)
 c. Draw a regular pentagon (all five sides have equal length). How many degrees can the pentagon be rotated about its center so that it remains unchanged in appearance? (There are several correct answers.)
 d. Can you see a pattern in parts a, b, and c? How many degrees can a regular n-gon be rotated about its center so that it remains unchanged in appearance? How many different angles answer this question for an n-gon?

Identifying Symmetries. Identify all of the symmetries in the figures of Exercises 11–14.

11.

12.

13.

Pattern continues in both directions.

14.

Tilings from Translating Triangles. In Exercises 15–16, make a tiling from each triangle using translations only, as in Figure 8.25.

15.

16.

Tilings from Translating and Reflecting Triangles. For Exercises 17–18, make a tiling from the given triangle using translations and reflections, as in Figure 8.26.

17. The triangle in Exercise 15
18. The triangle in Exercise 16

Tilings from Quadrilaterals. In Exercises 19–20, make a tiling from the quadrilateral using translations, as in Figure 8.29.

19.

20.

21. **Why Quadrilateral Tilings Work.** Consider the tiling with quadrilaterals shown in Figure 8.29. Look at any of the points at which four quadrilaterals meet in the tiling. Call such a point P. Given that the sum of the inside angles of any quadrilateral is 360°, show that the sum of the angles around the point P is also 360°, thus proving that the quadrilaterals interlock perfectly.

22. **Tiling with a Rhombus.** A rhombus is a quadrilateral in which all four sides have the same length and opposite sides are parallel. Show how a tiling can be made from a rhombus using translations only (as in Figure 8.25) and an initial reflection and then translations (as in Figure 8.26).

23. **Project: Symmetry and Proportion in Art.** Find one piece of pre-20th-century art and one piece of 20th-century art that you like. Use as many ideas from this unit as possible (involving perspective and symmetry) to write a two- to three-page analysis and comparison of these two pieces of art.

Web Projects

Find useful links for Web Projects on the text Web site: www.aw.com/bennett-briggs

1. **Art and Mathematics.** Many Web sites are devoted to connections between art and mathematics. Visit one such site and study a connection in some depth. Write a one- to two-page essay that describes this connection between mathematics and art.

2. **Art Museums.** Many art museums have Web sites where you can explore their collections. Visit an art museum on the Web, and study its online collection. Describe a few

3. **Escher.** Look on the Web for art by M. C. Escher. Choose one of his works and write a short essay about his use of perspective in the piece.

4. **Penrose Tilings.** Many Web sites are devoted to Penrose tilings (aperiodic tilings). Visit one such site and learn more about the nature and uses of these tilings. Write a short essay describing your findings.

UNIT 8C

PROPORTION AND THE GOLDEN RATIO

In Unit 8B, we studied how mathematics enters into art through the ideas of symmetry and perspective. In this unit, we turn our attention to the third major mathematical idea involved with art: proportion.

The importance of proportion was expressed well by astronomer Johannes Kepler (1571–1630):

> Geometry has two great treasures: one is the theorem of Pythagoras; the other, the division of a line into extreme and mean ratio. The first we may compare to gold; the second we may name a precious jewel.

Kepler's statement about *the division of a line into extreme and mean ratio* describes one of the oldest principles of proportion, which dates back to the time of Pythagoras (c. 500 B.C.) when scholars asked the following question: How can a line segment be divided into two pieces that have the most visual appeal and balance?

Surprisingly, although this was a question of beauty, there seemed to be general agreement on the answer. Suppose a line segment is divided into two pieces, as shown in Figure 8.38. We will call the length of the long piece L and the length of the short piece 1.

FIGURE 8.38

The Greeks claimed that the most visually pleasing division of the line had the property that the ratio of the length of the long piece to the length of the short piece is the same as the ratio of the length of the entire line segment to the length of the long piece. That is,

$$\frac{L}{1} = \frac{L+1}{L}$$

This statement of proportion can be solved (see Exercise 8) to find that L has a special value, denoted by the Greek letter ϕ (pronounced "fie" or "fee"), which is

$$\phi = \frac{1 + \sqrt{5}}{2} = 1.61803\ldots$$

The number ϕ is more commonly called the **golden ratio** (or *golden section*); it is also sometimes called the *divine proportion*. Note that the golden ratio is an irrational number often approximated as 1.6, or $\frac{8}{5}$. Figure 8.39 shows that, for any line segment

The senses delight in things duly proportioned.
—St. Thomas Aquinas (1225–1274)

by the way...
The Greek letter ϕ is the first letter in the Greek spelling of Phydias, the name of a Greek sculptor who may have used the golden ratio in his work.

divided in two pieces according to the golden ratio, the ratio of the long piece to the short piece is

$$\frac{x}{y} = \frac{1.61803\ldots}{1} = \phi \approx \frac{8}{5}$$

FIGURE 8.39

EXAMPLE 1 Calculating a Golden Ratio

Suppose the line segment in Figure 8.40 is divided according to the golden ratio. If the length of the longer piece labeled *x* is 5 centimeters, how long is the entire line segment?

FIGURE 8.40

Solution Because the line segment is divided in the golden ratio, we know that $x/y = \phi$. We can solve for *y* by multiplying both sides by *y* and dividing both sides by ϕ:

$$\frac{x}{y} = \phi \quad \rightarrow \quad y = \frac{x}{\phi}$$

Substituting $x = 5$ cm and an approximate value for ϕ, we find

$$y = \frac{5 \text{ cm}}{\phi} \approx \frac{5 \text{ cm}}{1.6} \approx 3.1 \text{ cm}$$

The entire segment is made up of both *x* and *y*, so its total length is approximately 5 cm + 3.1 cm = 8.1 cm. ■

THE GOLDEN RATIO IN ART HISTORY

Although the ancient Greeks struggled with the notion of irrational numbers, they embraced the golden ratio. For example, the *pentagram* (Figure 8.41) was the seal of the mystical Pythagorean Brotherhood. This five-pointed star inscribed in a circle produces a pentagon at the center. The golden ratio occurs in at least ten different ways in the pentagram. For example, if the length of each side of the pentagon is 1, then the length of each arm of the star is ϕ.

FIGURE 8.41 A pentagram.

Time out to think

Using a ruler to measure, find at least one other place in the pentagram of Figure 8.41 where the ratio of the lengths of two line segments is the golden ratio.

From the golden ratio it is a short step to the other famous Greek expression of proportion, the **golden rectangle**—a rectangle whose long side is ϕ times the length of its short side. A golden rectangle can be of any size, but its sides must have a ratio of $\phi \approx \frac{8}{5}$. Figure 8.42 shows a golden rectangle.

The golden rectangle had both practical and mystical importance to the Greeks. It became a cornerstone of their philosophy of *aesthetics*—the study of beauty. There is considerable speculation about the uses of the golden rectangle in art and architecture in ancient times. For example, it is widely claimed that many of the great monuments of antiquity, such as the Pyramids in Egypt, were designed in accordance with the golden rectangle. And, whether by design or by chance, the proportions of the Parthenon (in Athens, Greece) match those of the golden rectangle very closely (Figure 8.43).

FIGURE 8.42 A golden rectangle—the side lengths have a ratio $\phi \approx \frac{8}{5}$.

FIGURE 8.43 The proportions of the Parthenon closely match those of a golden rectangle.

> **by the way...**
> The Parthenon was completed in about 430 B.C. as a temple to Athena Parthenos, the Warrior Maiden. It stands on the Acropolis (which means "the uppermost city"), about 500 feet above Athens.

The golden rectangle also appears in many more recent works of art and architecture. The book *De Divina Proportione*, illustrated by Leonardo da Vinci in 1509, is filled with references to and uses of ϕ. Indeed, da Vinci's unfinished painting *St. Jerome* (Figure 8.44) places the central figure inside an imaginary golden rectangle. More recently, the French impressionist painter Georges Seurat is said to have used the golden ratio on every canvas. The abstract geometric paintings of the 20th-century Dutch painter Piet Mondrian (Figure 8.45) are filled with golden rectangles.

Today, the golden rectangle appears in many everyday items. For example, photographs, note cards, cereal boxes, posters, and windows often have the proportions of the golden rectangle. But the question remains as to whether the golden rectangle

FIGURE 8.44 *St. Jerome*, by Leonardo da Vinci, has the central figure positioned as though he were inside a golden rectangle.

FIGURE 8.45 *Place de la Concorde*, by Piet Mondrian, is filled with golden rectangles.

is really more pleasing. In the late 19th century, the German psychologist Gustav Fechner (1801–1887) studied the question statistically. He showed several rectangles with various length to width ratios to hundreds of people and recorded their choices for the most and least visually pleasing rectangles. The results, given in Table 8.2, show that almost 75% of the participants chose rectangles that were very close to the golden rectangle.

> **by the way ...**
> Despite evidence of the appeal of the golden rectangle and ratio, there are theories to the contrary. A 1992 study by George Markowsky discredits claims that the golden ratio was used in art and architecture, attributing them to coincidence and bad science. He also claims that statistical studies of people's preferences, such as Fechner's research, are not conclusive (see Exercise 15).

TABLE 8.2 Fechner's Data

LENGTH TO WIDTH RATIO	MOST PLEASING RECTANGLE (PERCENTAGE RESPONSE)	LEAST PLEASING RECTANGLE (PERCENTAGE RESPONSE)
1.00	3.0	27.8
1.20	0.2	19.7
1.25	2.0	9.4
1.33	2.5	2.5
1.45	7.7	1.2
1.50	20.6	0.4
$\phi \approx 1.62$	35.0	0.0
1.75	20.0	0.8
2.00	7.5	2.5
2.50	1.5	35.7

Time out to think

Do you think that the golden rectangle is visually more pleasing than other rectangles? Explain.

EXAMPLE 2 Household Golden Ratios

The following household items were found to have the dimensions shown. Which item comes closest to having the proportions of the golden ratio?

- Standard sheet of paper: 8.5 in. × 11 in.
- 8 × 10 picture frame: 8 in. × 10 in.
- 35-mm slide: 35 mm × 23 mm

Solution The ratio of the sides of a standard sheet of paper is $11/8.5 = 1.29$, which is 20% less than the golden ratio. The ratio of the sides of a standard picture frame is $10/8 = 1.25$, which is 22% less than the golden ratio. The ratio of the sides of a 35-mm slide is $35/23 = 1.52$, which is 5% less than the golden ratio. Of the three objects, the 35-mm slide comes closest to being a golden rectangle. ∎

THE GOLDEN RATIO IN NATURE

The golden ratio appears to be common in the "artwork" of nature. One striking example is spirals created from golden rectangles. We begin by dividing a golden rectangle to make a square on its left side, as shown in Figure 8.46a. If you measure the sides of the remaining smaller rectangle to the right of the square, you'll see that it is a smaller golden rectangle.

FIGURE 8.46 (a) A logarithmic spiral begins with a golden rectangle that is divided to make a square on the left. (b) The process is repeated in each successive golden rectangle, and a spiral is created by connecting the corners of all the squares. (c) A chambered nautilus shell looks much like a logarithmic spiral.

We now repeat this splitting process on the second golden rectangle, this time making the square on the top (instead of the left) of the golden rectangle. This split makes a third, even smaller, golden rectangle. Continuing to split each new golden rectangle in this manner generates the result shown in Figure 8.46b. Now, we connect opposite corners of all the squares with a smooth curve. The result is a continuous curve called a **logarithmic spiral** (or *equiangular spiral*). This spiral very closely matches the spiral shape of the beautiful chambered nautilus shell (Figure 8.46c).

Another intriguing connection between the golden ratio and nature comes from a problem in population biology, first posed by a mathematician known as Fibonacci in 1202. Fibonacci's problem essentially asked the following question about the reproduction of rabbits.

> *Suppose that a pair of baby rabbits takes one month to mature into adults, then produces a new pair of baby rabbits the following month and each subsequent month. Further suppose that each newly born pair of rabbits matures and gives birth to additional pairs with the same reproductive pattern. If no rabbits die, how many pairs of rabbits are in the population at the beginning of each month?*

Figure 8.47 shows the solution to this problem for the first six months. Note that the number of pairs of rabbits at the beginning of each month forms a sequence that begins 1, 1, 2, 3, 5, 8. Fibonacci found that the numbers in this sequence continue to grow with the following pattern:

$$1, 1, 2, 3, 5, 8, 13, 21, 34, 55, \ldots$$

This sequence of numbers is known as the **Fibonacci sequence**.

Let's call the nth Fibonacci number F_n, so we have $F_1 = 1, F_2 = 1, F_3 = 2, F_4 = 3$, and so forth. The most basic property of the Fibonacci sequence is that the next number in the sequence is the sum of the previous two numbers. For example, note that

$$F_3 = F_2 + F_1 = 1 + 1 = 2 \quad \text{and} \quad F_4 = F_3 + F_2 = 2 + 1 = 3$$

We can express this rule in general as $\quad F_{n+1} = F_n + F_{n-1}.$

Time out to think

Confirm that the above rule works for Fibonacci numbers F_3 through F_{10}. Use the rule to determine the eleventh Fibonacci number (F_{11}).

414 CHAPTER 8 MATHEMATICS AND THE ARTS

Population (pairs):	1	1	2	3	5	8
Beginning of:	Month 1	Month 2	Month 3	Month 4	Month 5	Month 6

FIGURE 8.47

The connection between the Fibonacci numbers and the golden ratio becomes clear when we compute the ratios of successive Fibonacci numbers, as shown in Table 8.3. Note that, as we go further out in the sequence, the ratios of successive Fibonacci numbers get closer and closer to the golden ratio $\phi = 1.61803\ldots$.

TABLE 8.3 Ratios of Successive Fibonacci Numbers

$F_3/F_2 =$	$2/1 = 2.0$	$F_{11}/F_{10} =$	$89/55 = 1.618182$
$F_4/F_3 =$	$3/2 = 1.5$	$F_{12}/F_{11} =$	$144/89 = 1.617978$
$F_5/F_4 =$	$5/3 = 1.667$	$F_{13}/F_{12} =$	$233/144 = 1.618056$
$F_6/F_5 =$	$8/5 = 1.600$	$F_{14}/F_{13} =$	$377/233 = 1.618026$
$F_7/F_6 =$	$13/8 = 1.625$	$F_{15}/F_{14} =$	$610/377 = 1.618037$
$F_8/F_7 =$	$21/13 = 1.6154$	$F_{16}/F_{15} =$	$987/610 = 1.618033$
$F_9/F_8 =$	$34/21 = 1.61905$	$F_{17}/F_{16} =$	$1597/987 = 1.618034$
$F_{10}/F_9 =$	$55/34 = 1.61765$	$F_{18}/F_{17} =$	$2584/1597 = 1.618034$

by the way...
Fibonacci, also known as Leonardo of Pisa, is credited with popularizing the use of Hindu-Arabic numerals in Europe. His book *Liber Abaci* ("Book of the Abacus"), published in 1202, explained their use and the importance of the number zero.

There are many examples of the Fibonacci sequence in nature. The heads of sunflowers and daisies consist of a clockwise spiral superimposed on a counterclockwise spiral (both of which are logarithmic spirals), as shown in Figure 8.48. The number of individual florets in each of these intertwined spirals is a Fibonacci number—for example, 21 and 34 or 34 and 55. Biologists have also observed that the number of petals on many common flowers is a Fibonacci number (for example, irises have 3 petals, primroses have 5 petals, ragworts have 13 petals, and daisies have 34 petals). The arrangement of leaves on the stem of many plants also exhibits the Fibonacci sequence. And the spiraling Fibonacci numbers can be identified on pine cones and pineapples.

(a) (b)

FIGURE 8.48

Review Questions

1. Explain the golden ratio in terms of proportions of line segments.
2. How is a golden rectangle formed?
3. What evidence suggests that the golden ratio and golden rectangle hold particular beauty?
4. What is a logarithmic spiral? How is it formed from a golden rectangle?
5. What is the Fibonacci sequence?
6. What is the connection between the Fibonacci sequence and the golden ratio? Give some examples of the Fibonacci sequence in nature.

In the News

1. **Proportion in Life.** Describe at least three ways in which the use of visual proportion affects your life.
2. **The Golden Ratio.** Find a recent picture of a new building or architectural design. Study the picture and decide whether the golden ratio is involved in any way.

Exercises

1. **Golden Ratio.** Draw a line segment 4 inches long. Now subdivide it according to the golden ratio. Verify your work by computing the ratio of the whole segment length to the long segment length and the ratio of the long segment length to the short segment length.
2. **Golden Rectangles.** Measure the sides of each rectangle in Figure 8.49, and compute the ratio of the long side to the short side for each rectangle. Which one is a golden rectangle?

FIGURE 8.49

Dimensions of Golden Rectangles. Exercises 3–6 give the length of one side of a golden rectangle. Find the length of the other side. Notice that the other side could be either longer or shorter than the given side. Use the approximation $\phi \approx 1.62$ for your work.

3. 1.5 inches
4. 2.5 meters
5. 6.4 kilometers
6. 0.12 centimeter
7. **Everyday Golden Rectangles.** Find at least three everyday objects with rectangular shapes (for example, billboards, cereal boxes, windows). In each case, measure the side lengths and calculate the ratio. Are any of these objects golden rectangles? Explain.

8. **Finding ϕ.** The property that defines the golden ratio is
$$\frac{L}{1} = \frac{L+1}{L}$$
 a. Show that, if we multiply both sides by L and rearrange, this equation becomes
$$L^2 - L - 1 = 0$$
 Confirm that substituting the value of ϕ for L satisfies this equation.

 b. The quadratic formula states that, for any equation of the form $ax^2 + bx + c = 0$, the solutions are given by
$$x = \frac{-b + \sqrt{b^2 - 4ac}}{2a}$$
 and
$$x = \frac{-b - \sqrt{b^2 - 4ac}}{2a}$$
 Use the quadratic formula to solve for L in the formula for the golden ratio. Show that one of the roots is ϕ.

9. **Properties of ϕ**
 a. Enter $\phi = (1 + \sqrt{5})/2$ into your calculator. Show that $1/\phi = \phi - 1$.
 b. Now compute ϕ^2. How is this number related to ϕ?

10. **Logarithmic Spirals.** Draw a rectangle that is 10 centimeters on a side. Follow the procedure described in the text for subdividing the rectangle until you can draw a logarithmic spiral. Do all work carefully, and show your measurements with your work.

11. **The Lucas Sequence.** A sequence called the *Lucas sequence* is closely related to the Fibonacci sequence. The Lucas sequence begins with the numbers $L_1 = 1$ and $L_2 = 3$ and then uses the same relation $L_{n+1} = L_n + L_{n-1}$ to generate $L_3, L_4, \ldots$.
 a. Generate the first ten Lucas numbers.
 b. Compute the ratio of successive Lucas numbers $L_2/L_1, L_3/L_2, L_4/L_3$, and so on. Can you determine if these ratios approach a single number? What number is it?

12. **Graphing Fechner's Data.** Consider Gustav Fechner's data shown in Table 8.2. Make a histogram that displays the responses for both the most pleasing and the least pleasing rectangle proportions.

13. **Project: The Golden Navel.** An old theory claims that, on average, the ratio of the height of a person to the height of his/her navel is the golden ratio. Collect "navel ratio data" from as many people as possible. Graph the ratios in a histogram, find the average ratio over your entire sample, and discuss the outcome. Do your data support the theory?

14. **Project: Mozart and the Golden Ratio.** In a paper called "The Golden Section and the Piano Sonatas of Mozart" (*Mathematics Magazine*, Vol. 68, No. 4, 1995), John Putz gives the lengths (in measures) of the first part (*a*) and the second part (*b*) of the movements of the 19 Mozart piano sonatas. Some of the data are given below.

a = length of first part	b = length of second part
38	62
28	46
56	102
56	88
24	36
77	113
40	69
46	60
15	18
39	63
53	67

 a. The ratio of the length of the whole movement to the length of the longer segment is $(a + b)/b$. Add a third column to the table and compute this ratio for the given data.
 b. Make a histogram of the ratios that you computed in part a (choosing an appropriate bin size). Comment on how well they are approximated by ϕ.
 c. Read the article by John Putz. Do you believe that Mozart composed with the golden ratio in mind?

15. **Project: Debunking the Golden Ratio.** Find the article "Misconceptions about the Golden Ratio" by George Markowsky (*College Mathematics Journal*, Vol. 23, No. 1, 1992). Choose at least one of the misconceptions that Markowsky discusses, summarize it, and then explain whether you find his argument convincing. Discuss your opinion of whether the golden ratio has been consciously used by artists and architects in their work.

Web Projects

Find useful links for Web Projects on the text Web site: www.aw.com/bennett-briggs

1. **Golden Controversies.** Many Web sites are devoted to the controversy concerning the role of the golden ratio in art. Find a specific argument on one side of this controversy and summarize it in a one- to two-page essay.

2. **Fibonacci Numbers.** Learn more about Fibonacci numbers and possible occurrences of the Fibonacci sequence in nature. Write a short essay about one aspect of Fibonacci numbers that you find interesting.

CHAPTER 8 SUMMARY

UNIT	KEY TERMS	KEY IDEAS AND SKILLS
8A	wave pitch frequency harmonic octave musical scale digital recording	**Understand** how a plucked string produces sound. **Measure** frequency in cycles per second, and find harmonics of the frequency. **Understand** the musical scale and the ratios of frequencies among musical notes. **Explain** the difference between analog and digital representations of music.
8B	perspective vanishing point horizon line symmetry reflection rotation translation	**Understand** the use of perspective in painting. **Find** symmetries in paintings and tilings. **Create** tilings with regular or irregular polygons.
8C	proportion golden ratio golden rectangle Fibonacci sequence	**The golden ratio:** $$\phi = \frac{1 + \sqrt{5}}{2} = 1.61803\ldots$$ **The Fibonacci sequence:** 1, 1, 2, 3, 4, 5, 13, 21, 34, 55, . . . **Understand** claimed uses of the golden ratio in both art and nature.

9

MATHEMATICS AND POLITICS

Mathematics and politics have been part of human culture for millennia, so it's not surprising that the two have become interwoven through history. Indeed, the connections between mathematics and politics go all the way to the heart of the democratic process. In this chapter, we'll discuss the prominent role of mathematics in voting and apportionment.

I must study politics and war that my sons may have liberty to study mathematics and philosophy. My sons ought to study mathematics and philosophy, geography, natural history, naval architecture, navigation, commerce and agriculture in order to give their children a right to study painting, poetry, music, architecture, statuary, tapestry, and porcelain.

—JOHN ADAMS, letter to Abigail Adams, May 12, 1780

UNIT 9A Voting: Does the Majority Always Rule? In elections with more than two candidates, there are several acceptable ways to choose a winner. We will study these methods and see that different methods can lead to different winners.

UNIT 9B Apportionment: The House of Representatives and Beyond: As with voting, there are several acceptable ways to apportion seats in the House of Representatives. We will study these methods and, again, find that no single method is always fair.

UNIT 9A

VOTING: DOES THE MAJORITY ALWAYS RULE?

The 2000 presidential election should have shattered any illusions about the simplicity of voting. The election was not decided for more than a month after election day. In the end, the outcome turned on the results in Florida, where George W. Bush's official margin of victory over Al Gore was fewer than 600 votes out of more than 6 million votes cast in Florida. The difficulties of counting large numbers (see Unit 3C) mean that this result undoubtedly has significant uncertainty associated with it—which is why people will probably never stop arguing about who really received more votes in Florida.

Mathematically, the issue of counting votes and associated uncertainties falls under the science of statistics. However, this is only one of the mathematical issues associated with voting. In this unit, we will investigate various different types of voting systems.

VOTING WITH TWO CHOICES

The simplest type of voting involves only two choices. With two choices, the most common way of deciding the vote is by **majority rule:** The choice receiving more than 50% of the votes wins. While we take this rule for granted, it has three properties worth noting:

- Every vote has the same *weight*. That is, no one person's vote counts more or less than any other person's vote.
- There is *symmetry* between the candidates: If all the votes were reversed, the loser would become the winner.
- If a vote for the loser were changed to a vote for the winner, the outcome of the election would not be changed.

We can illustrate these properties with a simple example. Suppose that Sharon and Robert are running for captain of the debate team. Sharon wins with 10 votes to Robert's 9 votes. The first property, that all votes have the same weight, means we care only about the vote totals—not about the names of the people who cast the individual votes. The second property tells us that if we reverse all the votes, so that Sharon receives 9 votes and Robert receives 10, then Robert becomes the winner. The third property tells us that if a vote for Robert (the loser) is changed to a vote for Sharon (the winner), Sharon still wins: This change raises Sharon's vote total to 11 and reduces Robert's to 8.

by the way ...
In 1952, Kenneth May proved that majority rule is the *only* voting system for two candidates that satisfies all three properties.

> **Majority Rule**
> The **majority rule** says that the choice receiving more than 50% of the vote is the winner.

Time out to think

Suppose that a vote for the *winner* is changed to a vote for the loser. Can the outcome of the election change in that case? Explain.

U.S. Presidential Elections

United States presidential elections involve a twist on majority rule. The **popular vote** in a presidential election reflects the total number of votes received by each candidate. However, presidents are elected by **electoral votes.** As mandated by the Constitution, each state gets as many electoral votes as the state has members of Congress (senators *plus* representatives). When you cast a ballot for president, you actually are casting votes for your state's *electors;* the electors meet to cast their votes a few weeks later.

In most states, the electoral votes are determined by a winner-take-all system: *All the electoral votes of that state go to the candidate with the most popular votes.* (As of 2000, the only exceptions to the winner-take-all system are the states of Nebraska and Maine.) The overall winner of the presidential election is chosen by majority rule among the electoral votes. If no candidate receives a majority of the electoral votes, the House of Representatives chooses the president. This has happened twice in U.S. history: The House elected Thomas Jefferson following the undecided election of 1800 and John Quincy Adams following the undecided election of 1824.

EXAMPLE 1 *2000 Presidential Election*

Table 9.1 shows the 2000 presidential election official results. Discuss the outcomes of the popular and electoral votes.

TABLE 9.1 2000 Presidential Election

	POPULAR VOTE	ELECTORAL VOTE
George W. Bush	50,456,141	271
Al Gore	50,996,039	266
Ralph Nader	2,882,807	0
Others	1,069,559	0
Abstentions	—	1*
Total	105,404,546	538

*An elector from the District of Columbia pledged to Al Gore abstained in protest over the District's lack of representation in Congress.

Solution The difference in popular vote totals between Gore and Bush was

$$50{,}996{,}039 - 50{,}456{,}141 = 539{,}898$$

That is, Gore won the popular vote over Bush by a margin of more than a half million votes. However, because of the small percentages of the total vote that went to Nader and other candidates, neither Gore nor Bush won a majority of the popular vote. Gore's percentage of the popular vote was

$$\frac{50{,}996{,}039}{105{,}404{,}546} = 0.4838 = 48.38\%$$

historical note

The 1800 election pitted the Republican party ticket of Thomas Jefferson and Aaron Burr against the Federalist party ticket of John Adams and Charles Pickney. But electoral ballots at the time did not distinguish between votes for president and vice president. Jefferson and his running mate Burr ended up tied, with 73 electoral votes each, while Adams and Pickney received 65 and 64 electoral votes, respectively. It took 7 days and 36 ballots before the House of Representatives finally decided the election in favor of Jefferson. The 12th Amendment to the U.S. Constitution, passed in 1804, instituted separate balloting for president and vice president.

Bush's percentage of the popular vote was

$$\frac{50{,}456{,}141}{105{,}404{,}546} = 0.4787 = 47.87\%$$

In electoral vote, Bush won 271 out of the total 538 votes available, which gave him a slight majority of

$$\frac{271}{538} = 0.5037 = 50.37\%$$

Because he won a majority of the electoral vote, Bush became the president of the United States. ■

Time out to think

Do you think that it would be fairer if presidential elections were decided by popular vote rather than electoral vote? Why or why not?

Variations on Majority Rule

United States senators are generally allowed to speak about a legislative bill for as long as they wish before it is brought to a vote. If a senator opposes a particular bill but fears that it will gain a majority vote, the senator may choose to speak continuously (at least during the hours that the issue is under consideration)—and thereby prevent the vote from ever taking place. This technique for preventing a vote on a bill is called a **filibuster.** The filibuster can be ended only by a vote of 3/5, or 60%, of the senators. This situation, in which 60% rather than 50% of the votes are required to end a filibuster, illustrates that majority rule is not always followed strictly.

Whenever a candidate or issue must receive *more* than a majority of the vote to win—such as 60% of the vote, 75% of the vote, or a unanimous vote—we say that a **super majority** is required. Criminal trials offer an example: All states require a super majority vote of the jury to reach a verdict, and many states require that the vote be unanimous. A jury that cannot reach this super majority or unanimous agreement is called a *hung jury*. When a jury is hung, the judge generally declares a mistrial, and the case must be tried again or dropped.

The U.S. Constitution requires super majority votes for many specific issues. For example, an international treaty can be ratified only by a 2/3 super majority vote of the Senate. Amending the Constitution first requires a 2/3 super majority vote on the amendment in both the House and the Senate, and then the amendment must also be approved by 3/4 of the states.

Another variation on majority rule occurs with **veto** power. For example, a bill proposed in the U.S. Congress normally becomes law if it receives a majority vote in both the House and the Senate. However, if the president vetoes the bill, it can become law only if it then receives a 2/3 super majority vote in both the House and the Senate to override the veto. The courts can also effectively veto a popular vote. For example, even if a proposition in a state election receives a huge majority of the popular vote, it will not become law if the courts declare that the proposition violates the U.S. Constitution.

historical note

Prior to 2000, the winner of the popular vote had lost the electoral vote just twice in U.S. history. In 1876, Samuel J. Tilden won the popular vote by some 264,000 votes over Rutherford B. Hayes, but Hayes won the electoral vote 185 to 184. In 1888, Benjamin Harrison won the popular vote by about 90,000 votes over Grover Cleveland, but Cleveland won the electoral vote by a margin of 233 to 168.

by the way . . .

The U.S. Constitution also specifies that it may be amended if 2/3 of the states call for "a convention for proposing amendments." However, such a call for a convention has never occurred.

EXAMPLE 2 Majority Rule?

Evaluate the outcome in each of the following cases.

a. Of the 100 senators in the U.S. Senate, 59 favor a new bill on campaign finance reform. The other 41 senators are adamantly opposed and start a filibuster. Will the bill pass?

b. A criminal conviction in a particular state requires a vote by 3/4 of the jury members. On a nine-member jury, seven jurors vote to convict. Is the defendant convicted?

c. A proposed amendment to the U.S. Constitution has passed both the House and the Senate with more than the required 2/3 super majority. Each state holds a vote on the amendment. The amendment garners a majority vote in all but 14 of the 50 states. Is the Constitution amended?

d. A bill limiting the powers of the president has the support of 73 out of 100 senators and 270 out of 435 members of the House of Representatives. But the president promises to veto the bill if it is passed. Will it become law?

by the way...
Senate rules do not limit what a senator can say. During some filibusters, senators have read novels aloud in order to keep speaking. In 1964, the Civil Rights Act was passed by Congress only after 75 days of fulibuster by southern senators.

Solution

a. The filibuster can be ended only by a vote of 3/5 of the Senate, or 60 out of the 100 senators. Thus, the 59 senators in favor of the bill cannot stop the filibuster. The bill will not become law.

b. Seven out of nine jurors represents a super majority of 7/9 = 77.8%. This percentage is more than the required 3/4 (75%), so the defendant is convicted.

c. The amendment must be approved by 3/4, or 75%, of the states. But the 14 states that do not support the amendment represent 14/50 = 28% of the 50 states. Thus, only 100% − 28% = 72% of the states voted to ratify the amendment, which is not enough for it to become part of the Constitution.

d. The bill has the support of 73/100 = 73% of the Senate and 270/435 = 62% of the House. But overriding a presidential veto requires a super majority vote of 2/3 = 66.7% in both the House and the Senate. The 62% support in the House is not enough to override the veto, so the bill will not become law. (Note: The law requires a vote of 2/3 of those present in Congress, so this example assumes that all members are present.) ∎

VOTING WITH THREE OR MORE CHOICES

Consider the following results in a three-candidate election for governor:

Candidate	Percentage of vote
Smith	32%
Jones	33%
Wilson	35%

No candidate has a majority, so who should become governor?

One possibility is to award the governorship to the person who received the most votes, called a **plurality** of the vote. Wilson has the greatest percentage and becomes governor if we decide this election by plurality. An alternative way to decide the election is to hold a **runoff** between the top two vote getters. In that case, another election must be held between Jones and Wilson.

The plurality method for deciding the election is simpler (and cheaper) than the runoff method. However, it can lead to a situation in which the majority of the electorate opposes the new governor. For example, suppose that all of Smith's supporters prefer Jones to Wilson. In that case, the combined 65% of the electorate that voted for Smith and Jones would choose Jones over Wilson. By the plurality method, Wilson would become governor despite being opposed by 65% of the voters. By the runoff method, Jones would easily defeat Wilson to become governor.

EXAMPLE 3 *1992 Presidential Election*

Table 9.2 shows the results for the three major candidates in the 1992 U.S. presidential election. Analyze the outcome. Is it possible that Bush could have won if Perot had not run?

TABLE 9.2 1992 Presidential Election Results

	POPULAR VOTE	POPULAR PERCENTAGE	ELECTORAL VOTE	ELECTORAL PERCENTAGE
Clinton	44,909,889	43.3%	370	68.8%
Bush	39,104,545	37.7%	168	31.2%
Perot	19,742,267	19.0%	0	0%
Total	103,756,701	100%	538	100%

Solution Clinton won a *plurality* of the popular vote, but there was no majority winner of the popular vote. However, Clinton won a large majority of the electoral vote. If Perot had not run, the 19% of the votes that he received would presumably have been divided in some way between Clinton and Bush. Note that Bush needed an additional 12.3% of the vote to reach a 50% total (37.7% + 12.3% = 50%). Thus, if $12.3/19 \approx 65\%$ of the Perot voters preferred Bush to Clinton, Bush could have won the popular vote in Perot's absence; of course, winning the popular vote does not guarantee winning the electoral vote.

Time out to think

Do you think that a runoff is fairer than choosing the winner by plurality in a three-way race? Why or why not?

Preference Schedules

So far, we have examined situations in which each voter is allowed to vote for only a single candidate. However, there are many cases in which voters are asked to rank candidates in order of preference. This type of election is used for sports polls, sports awards (such as the Heisman Trophy), consumer surveys, beauty pageants, and the Academy Awards for motion pictures.

Imagine a club of 55 people that holds an election among five candidates for president. For simplicity, let's call the candidates A, B, C, D, and E. Each ballot asks the voter to rank these candidates in order of preference (Figure 9.1).

Table 9.3 shows a special type of table, called a **preference schedule**, that summarizes the election results. Each column shows a different ranking among the candidates, and the number at the bottom of the column is the number of people who chose that ranking. For example, the first column shows that 18 voters ranked the candidates in the order A, D, E, C, B, and the second column shows that 12 voters ranked the candidates in the order B, E, D, C, A.

Study Table 9.3 carefully. Which candidate should be declared the winner? As we'll soon see, a winner cannot be declared without at least some controversy.

FIGURE 9.1 Sample ballot for the club election.

TABLE 9.3 Preference Schedule for the Club Election

FIRST	A	B	C	D	E	E
SECOND	D	E	B	C	B	C
THIRD	E	D	E	E	D	D
FOURTH	C	C	D	B	C	B
FIFTH	B	A	A	A	A	A
	18	12	10	9	4	2

TECHNICAL NOTE
With five candidates, there are 5! = 120 possible orders in which they can be ranked. This example uses only 6 of these 120 possible orders.

Choosing a Winner by Plurality

Let's begin by counting *first-place* votes in Table 9.3:

- A received 18 first-place votes (column 1).
- B received 12 first-place votes (column 2).
- C received 10 first-place votes (column 3).
- D received 9 first-place votes (column 4).
- E received 4 + 2 = 6 first-place votes (columns 5 and 6).

The supporters of Candidate A can argue that A received a **plurality** *(the most first-place votes)* and should be declared the winner.

Choosing a Winner by a Single Runoff

"Not so fast!" yell the supporters of Candidate B. They suggest a runoff between A and B because they were the top two candidates in first-place votes. Table 9.3 shows that, although 18 people voted for A in first place, the other 37 people all ranked A in

last place. These 37 people will vote for B over A in a runoff. Thus, B wins the runoff easily, by a vote of 37 to 18. Supporters of Candidate B can now proclaim victory for their candidate.

Choosing a Winner by Sequential Runoffs

Now the supporters of Candidate C chime in. They claim that a single runoff is unfair because it ignores rankings below the top two. They suggest a series of runoffs called a **sequential runoff:** The candidate with the fewest first-place votes is eliminated at each stage, and runoffs continue until someone claims a majority. By studying the results in Table 9.3, we can determine the outcomes of these runoffs without actually holding separate runoff elections.

We've already found that Candidate E received the fewest first-place votes, so we eliminate E for the first runoff. With E gone, other candidates move up in each ranking. For example, the first column of Table 9.3 shows that 18 people chose the rank order A, D, E, C, B; without E, this ranking becomes A, D, C, B. Similarly, the ranking without E in the second column of Table 9.3 becomes B, D, C, A. Table 9.4 shows all the rankings after eliminating E—which therefore represent the results of the first runoff. Note that Candidate B gained 4 first-place votes that previously had gone to E (second-to-last column) and Candidate C gained 2 first-place votes that previously had gone to E (last column). Because no one has yet claimed a majority of the first-place votes, we need a second runoff.

TABLE 9.4 Rankings after Eliminating E for the First Runoff

FIRST	A	B	C	D	B	C
SECOND	D	D	B	C	D	D
THIRD	C	C	D	B	C	B
FOURTH	B	A	A	A	A	A
	18	12	10	9	4	2

Candidate D has the fewest first-place votes (9) in Table 9.4, so D is eliminated for the second runoff. Table 9.5 shows the results. Note that the 9 people who previously had D in first place (column 4 of Table 9.4) now have C in first place (column 4 of Table 9.5).

TABLE 9.5 Rankings after Eliminating D for the Second Runoff

FIRST	A	B	C	C	B	C
SECOND	C	C	B	B	C	B
THIRD	B	A	A	A	A	A
	18	12	10	9	4	2

Candidate C is now the leader, with $10 + 9 + 2 = 21$ first-place votes (columns 3, 4, and 6). But this total is still short of a majority, so we need one

more runoff. Because B now has the fewest first-place votes (12 + 4 = 16), we eliminate B. Table 9.6 shows the new results. Note that C gets the first-place votes of all 16 people who previously had ranked B first. Candidate C now has 37 first-place votes (all but the 18 votes in column 1), which is a majority. Thus, Candidate C is the winner by the sequential runoff method.

TABLE 9.6 Rankings after Eliminating B for the Third Runoff

FIRST	A	C	C	C	C	C
SECOND	C	A	A	A	A	A
	18	12	10	9	4	2

Choosing a Winner by a Point System

Now it is the turn of Candidate D's supporters to argue that the winner should be selected by a **point system.** Because there are five candidates, first-place votes are worth 5 points, second-place votes are worth 4 points, and so on, down to 1 point for fifth-place votes. We can calculate the total number of points won by each candidate by multiplying the number of votes in each column of Table 9.3 by the assigned point value and adding these products together.

A gets $(18 \times 5) + (12 \times 1) + (10 \times 1) + (9 \times 1) + (4 \times 1) + (2 \times 1) = 127$ points
B gets $(18 \times 1) + (12 \times 5) + (10 \times 4) + (9 \times 2) + (4 \times 4) + (2 \times 2) = 156$ points
C gets $(18 \times 2) + (12 \times 2) + (10 \times 5) + (9 \times 4) + (4 \times 2) + (2 \times 4) = 162$ points
D gets $(18 \times 4) + (12 \times 3) + (10 \times 2) + (9 \times 5) + (4 \times 3) + (2 \times 3) = 191$ points
E gets $(18 \times 3) + (12 \times 4) + (10 \times 3) + (9 \times 3) + (4 \times 5) + (2 \times 5) = 189$ points

Candidate D is now the winner, by virtue of having the largest number of points.

A system that assigns points for every ranking, as in the above example, is called a **Borda count.** A standard Borda count assigns 1 point for last place and an additional point for each higher place. Many common ranking systems use modified Borda counts. For example, voters in college football polls usually rank their top 25 teams (leaving many other teams unranked). A first-place ranking is worth 25 points, second place is worth 24 points, and so on, to 1 point for twenty-fifth place. At a meet between two college swim teams, the winner may be decided by a point system that awards 4 points for first place, 2 points for second place, and 1 point for third place.

historical note

The Borda count is named for French mathematician and astronomer Jean-Charles de Borda (1733–1799). Borda objected to the belief that "in an election by ballot the plurality of voices indicates the will of the electors." He wrote that this belief is "true in the case where the election is conducted between two candidates only, [but] may lead to error in all other cases."

Time out to think

Look for a situation in which a point system is used to determine a winner, such as a football poll or a sports competition. How are points assigned to the different places? Would the outcome be the same if the winner had instead been chose by a plurality?

EXAMPLE 4 *Majority Loser*

The seven sportswriters of Seldom County rank the county's three women's volleyball teams according to the preference schedule in Table 9.7. Determine the winner by plurality and by a Borda count. Discuss the results.

TABLE 9.7 Volleyball Team Rankings

FIRST	A	C	B
SECOND	B	B	C
THIRD	C	A	A
	4	1	2

Solution By plurality, Team A is the winner, having received 4 out of the 7 first-place votes (which is not just a plurality, but also a majority). For the Borda count, we assign 3 points for a first-place vote, 2 points for a second-place vote, and 1 point for a third-place vote, and find the total number of points.

$$\text{Team A gets } (4 \times 3) + (1 \times 1) + (2 \times 1) = 15 \text{ points}$$
$$\text{Team B gets } (4 \times 2) + (1 \times 2) + (2 \times 3) = 16 \text{ points}$$
$$\text{Team C gets } (4 \times 1) + (1 \times 3) + (2 \times 2) = 11 \text{ points}$$

So Team B is the winner by the Borda count method. Although the majority of the sportswriters chose Team A as the best team, Team B gets ranked first by the Borda count. The fact that the Borda count winner can be different from the majority winner is a well-known shortcoming of the method. ∎

Choosing a Winner by Pairwise Comparisons

So far, we have discussed methods by which each of candidates A, B, C, and D can be declared the winner. Now, candidate E's supporters point out an important fact about the election rankings. Suppose, they say, that the vote had been only between E and A, without the other candidates. Column 1 of Table 9.3 has A ranked higher than E, so these 18 voters would choose A over E. However, E is ranked higher than A in all the other columns, so E would get the remaining 37 votes (out of the 55 total). That is, E beats A by 37 to 18 in a one-on-one contest.

Now suppose that the vote had been only between E and B. Candidate E is ranked higher than B in columns 1, 4, 5, and 6 (of Table 9.3), so E gets 18 + 9 + 4 + 2 = 33 votes. The remaining 22 votes go to B. Thus, E beats B by 33 to 22. A similar analysis of the race between only E and C shows that E is the winner by 36 to 19, and E comes out ahead in a one-on-one contest with D by 28 to 27.

Because Candidate E beats every other candidate in one-on-one contests, E's supporters now declare victory in the election. This method of deciding an election by analyzing the outcomes of one-on-one contests is called the **method of pairwise comparisons.** More generally, the winner by this method is the candidate who wins the most one-on-one contests.

The method of pairwise comparisons is also called the *Condorcet method*, because it was invented by French mathematician and political leader Marie Jean Antoine Nicholas de Caritat, Marquis de Condorcet (1743–1794). Condorcet, as he is

historical note

Condorcet did pioneering work in probability and calculus before becoming a leader of the French Revolution. He argued forcefully for equal rights for women, for universal free education, and against capital punishment. In 1794, as extremists took control of the revolution, he was arrested because of his aristocratic background. He died in prison the next day, in a death labeled a suicide by his captors.

usually called, invented this method because he thought it would lead to fairer elections. However, this method suffers a major drawback, of which Condorcet was aware: It does not always produce a clear winner.

EXAMPLE 5 *Condorcet Paradox*

Table 9.8 shows a three-candidate preference schedule. Can you find a winner through pairwise comparisons? Explain.

TABLE 9.8

FIRST	A	C	B
SECOND	B	A	C
THIRD	C	B	A
	14	12	10

Solution With three candidates, three pairwise comparisons are possible: A versus B, B versus C, and A versus C. Columns 1 and 2 show A ranked ahead of B, but B is ahead of A in column 3. Thus, A beats B in a one-on-one contest by 26 to 10. In the B versus C contest, B is ranked ahead of C in columns 1 and 3, so B beats C by 24 to 12. Similarly, A is ranked ahead of C only in column 1, so C beats A by 22 to 14. Summarizing, we have the following results for the pairwise comparisons:

- A beats B
- B beats C
- C beats A

Because A beats B and B beats C, the first two results suggest that A should also beat C. However, the third result shows that A actually loses to C. Thus, the results are paradoxical, and there is no way to establish a clear winner.

Time out to think

Does Example 5 have a plurality winner? Does it have a winner by a Borda count? Explain.

SUMMARY: CHOOSING A WINNER IS NOT SO EASY

Analyzing the preference schedule in Table 9.3 by five different methods, we found that all five candidates could reasonably claim to be the winner! Other election results may not be so ambiguous, but the point should be clear: Different people can reasonably disagree about who should be declared the winner in an election with more than two candidates. In fact, a mathematical theorem ("Arrow's Impossibility Theorem") *proves* that there is *no absolutely fair* way of deciding elections among more than two candidates.

> **Voting Methods with Three or More Candidates**
>
> **Plurality method:** The candidate with the most first-place votes wins.
>
> **Single (top-two) runoff method:** The two candidates with the highest first-place vote totals have a runoff. The winner of this runoff is the winner of the election.
>
> **Sequential runoff method:** A series of runoffs is held, eliminating the candidate with the *fewest* first-place votes at each stage. Runoffs continue until one candidate has a *majority* of the first-place votes and is declared the winner.
>
> **Point system (Borda count):** Points are awarded according to the rank of each candidate on each ballot (first, second, third, . . .). The candidate with the most points wins.
>
> **Pairwise comparisons (Condorcet) method:** The candidate who wins the most pairwise (one-on-one) contests is the winner of the election.

Review Questions

1. What is majority rule? When can it definitively decide an election?
2. Contrast the popular vote and electoral vote in a presidential election.
3. What is a filibuster? What percentage of the vote is required to end one?
4. What is a super majority? Give several examples in which a super majority is required to decide a vote.
5. What is a veto? How does a veto affect the idea of majority rule?
6. Describe how a three-way election can be decided either by plurality or by runoff. Will both methods necessarily give the same results? Explain.
7. What is a preference schedule? Give an example of how to make one.
8. Using the preference schedule in Table 9.3, explain how the vote can be decided in five different ways.

In the News

1. **Election Aftermath.** Find an article dealing with some aspect of the aftermath of the 2000 presidential election. Summarize the article and discuss how the closeness of the election still affects our lives.
2. **Foreign Elections.** Find a news article that describes the outcome of a national election in a foreign country. Summarize the outcome of the election. Is the method for choosing a winner mentioned?
3. **Non-Political Election.** Find an example in the news of an election that is not for a political office. Summarize the method used to choose a winner and describe the outcome of the election.

Exercises

1. **1876 Presidential Election.** In the 1876 presidential election, a total of 8,430,783 votes were counted. Samuel J. Tilden received 4,300,590 votes to Rutherford B. Hayes's 4,036,298 votes. However, Hayes received 185 electoral votes to Tilden's 184. Contrast the outcomes of the popular and electoral votes.
2. **1888 Presidential Election.** In the 1888 presidential election, a total of 11,379,131 votes were counted. Benjamin Harrison received 5,543,892 votes to Grover Cleveland's 5,434,488 votes. However, Cleveland received 233 electoral votes to Harrison's 168. Contrast the outcomes of the popular and electoral votes.

Close Presidential Elections. The following table shows the results for the top two candidates in three other close presidential elections. Use these data in Exercises 3–5.

Year	Major candidates	Electoral vote	Popular vote
1880	James A. Garfield	214	4,449,053
	Winfield S. Hancock	155	4,442,035
1916	Woodrow Wilson	277	9,129,606
	Charles E. Hughes	254	8,538,221
1976	Jimmy Carter	297	40,830,763
	Gerald R. Ford	240	39,147,973

3. Contrast the outcomes of the popular and electoral votes for the 1880 election.
4. Contrast the outcomes of the popular and electoral votes for the 1916 election.
5. Contrast the outcomes of the popular and electoral votes for the 1976 election.
6. **Super Majorities.**
 a. Of the 100 senators in the U.S. Senate, 57 favor a new bill on health care reform. The opposing senators start a filibuster. Is the bill likely to pass?
 b. A criminal conviction in a particular state requires a vote by 2/3 of the jury members. On a 12-member jury, 7 jurors vote to convict. Will the defendant be convicted?
 c. A proposed amendment to the U.S. Constitution has passed both the House and the Senate with more than the required 2/3 super majority. Each state holds a vote on the amendment, and it receives a majority vote in all but 15 of the 50 states. Is the Constitution amended?
 d. A tax increase bill has the support of 70 out of 100 senators and 260 out of 435 members of the House of Representatives. The president promises to veto the bill if it is passed. Is it likely to become law?
7. **Super Majorities.**
 a. According to the bylaws of a corporation, a 2/3 vote of the shareholders is needed to approve a merger. Of the 15,890 shareholders voting on a certain merger, 10,580 approve of the merger. Will the merger happen?
 b. A criminal conviction in a particular state requires a vote by 3/4 of the jury members. On a 15-member jury, 11 jurors vote to convict. Will the defendant be convicted?
 c. A proposed amendment to the U.S. Constitution has passed both the House and the Senate with more than the required 2/3 super majority. Each state holds a vote on the amendment, and it receives a majority vote in 32 of the 50 states. Is the Constitution amended?
 d. A tax increase bill has the support of 68 out of 100 senators and 292 out of 435 members of the House of Representatives. The president promises to veto the bill if it is passed. Is it likely to become law?
8. **1912 Presidential Election.** The 1912 U.S. presidential election featured three major candidates, with the vote split as follows. For this exercise, assume that all votes were cast for one of these three candidates. (Actually, other candidates split 7.6% of the total vote.)

Candidate	Popular votes	Electoral votes
Woodrow Wilson	6,286,214	435
Theodore Roosevelt	4,126,020	88
William Taft	3,483,922	8

 a. Calculate each candidate's percentage of the popular vote. Who won a plurality? Did any candidate win a majority?
 b. Calculate each candidate's percentage of the electoral vote. Who won a plurality? Did any candidate win a majority?
 c. Suppose that Taft had dropped out of the election. Is it possible that Roosevelt would have won the popular vote? Could Roosevelt have become president in that case? Explain.
 d. Suppose that Roosevelt had dropped out of the election. Is it possible that Taft would have won the popular vote? Could Taft have become president in that case? Explain.
9. **Three-Candidate Elections.** Consider an election in which the votes were cast as follows.

Candidate	Percentage of vote
Able	32%
Best	40%
Crown	28%

 a. Who won a plurality? Does any candidate have a majority? Explain.

b. What percentage of Crown's votes would Able need to win a runoff election?

10. **Three-Candidate Elections.** Consider an election in which the votes were cast as follows.

Candidate	Percentage of vote
Davis	21%
Earnest	38%
Fillipo	41%

a. Who won a plurality? Does any candidate have a majority? Explain.

b. What percentage of Davis's votes would Earnest need to win a runoff election?

11. **Three-Candidate Elections.** Consider an election in which the votes were cast as follows.

Candidate	Number of votes
Grand	130
Height	150
Inviglio	185

a. Who won a plurality? Does any candidate have a majority? Explain.

b. How many of Grand's votes would Height need to win a runoff election?

12. **Three-Candidate Elections.** Consider an election in which the votes were cast as follows.

Candidate	Number of votes
Joker	225
King	360
Lord	285

a. Who won a plurality? Does any candidate have a majority? Explain.

b. How many of Joker's votes would Lord need to win a runoff election?

13. **Ballots to Schedule.** Ten voters are asked to rank three brands of soda: A, B, and C. The ten voters turn in the following ballots, showing their preferences in order:

ABC CBA CAB ACB ABC
CAB ABC ABC CAB ACB

Make a preference schedule for these ballots.

14. **Ballots to Schedule.** Twelve voters are asked to rank four brands of candy bar: A, B, C, and D. The twelve voters turn in the following ballots, showing their preferences in order:

ABCD CBAD CADB ACDB DABC ABCD
CADB ABCD ABCD CADB DABC ACDB

Make a preference schedule for these ballots.

Preference Schedules. For the preference schedules in Exercises 15–19, answer the following questions.

a. How many votes were cast?

b. Find the plurality winner. Did the plurality winner also receive a majority? Explain.

c. Find the winner by a runoff of the top two candidates.

d. Find the winner by a sequential runoff.

e. Find the winner by a Borda count.

f. Find the winner, if any, by the method of pairwise comparisons.

g. Summarize the results of the various methods of determining a winner. Based on these results, is there a clear winner? If so, why? If not, which candidate should be selected as the winner, and why?

15.
First	B	D	C	A	D	C
Second	D	A	D	D	A	A
Third	C	C	A	C	B	B
Fourth	A	B	B	B	C	D
	20	15	10	8	7	6

16.
First	B	D	D	C	E
Second	A	B	B	A	A
Third	C	A	E	B	D
Fourth	D	C	C	D	B
Fifth	E	E	A	E	C
	9	7	6	4	3

17.
First	A	A	B	B	C	C
Second	B	C	A	C	A	B
Third	C	B	C	A	B	A
	30	5	20	5	10	30

18.

First	A	B	D
Second	B	A	C
Third	C	D	B
Fourth	D	C	A
	10	10	10

19.

First	E	B	D
Second	D	C	A
Third	A	E	B
Fourth	B	A	C
Fifth	C	D	E
	40	30	20

20. **Condorcet Winner.** If a candidate wins *all* head-to-head (two-way) races with other candidates, that candidate is called the *Condorcet winner*. A Condorcet winner automatically wins by the pairwise comparisons (Condorcet) method. Consider the following preference schedule for a four-candidate election. Is there a Condorcet winner? Explain.

First	B	B	A	A
Second	A	A	C	D
Third	C	D	D	C
Fourth	D	C	B	B
	30	30	30	20

21. **Pairwise Comparisons Question.** Suppose 15 voters rank four candidates. How many pairs of candidates must be examined to carry out the pairwise comparisons method? How many pairs of candidates must be examined to carry out the pairwise comparisons method with five candidates?

22. **Borda Question.** In a preference schedule with five candidates and 25 voters, what is the total number of points awarded to the candidates using the usual Borda count weights?

23. **Borda Question.** Suppose 25 voters rank four candidates: A, B, C, and D. With the usual Borda count weights, A receives 60 points, B receives 50 points, and C receives 10 points. Is it possible that D has won this election by the Borda count? Explain.

24. **Condorcet Paradox.** Consider the following preference schedule. Can you find a winner through pairwise comparisons? Explain.

First	C	B	A
Second	A	C	B
Third	B	A	C
	8	9	10

Web Projects

Find useful links for Web Projects on the text Web site: www.aw.com/bennett-briggs

1. **Election 2000.** Some people claim that the results of the 2000 presidential election were affected by voting irregularities or vote counting problems. Pick one such controversy and examine the evidence on both sides. Write a short report discussing the controversy and defending your personal opinion as to whether the problem affected the election outcome. Also discuss how similar problems might be prevented in the future.

2. **Academy Awards.** The election process for the Academy Awards (for films) involves several stages and several different voting methods. Use the Web site for the Academy Awards to investigate the full election procedure for the Academy Awards. Describe the procedure and comment on its fairness.

3. **Sports Polls.** Most sports have regular polls while the sport is in season. Choose a particular sport and investigate how teams are ranked. Describe the methods used to rank teams, give some typical results, and discuss the fairness of the method.

4. **Elections Around the World.** Many countries have methods much different than the American system for choosing a president or prime minister. Choose a particular country and use the Web to learn about national elections. Describe in detail the voting procedure used and discuss its strengths and shortcomings.

UNIT 9B

APPORTIONMENT: THE HOUSE OF REPRESENTATIVES AND BEYOND

You are probably aware that each state in the United States has two senators and one or more representatives in Congress. But *why* does the number of representatives vary from state to state, and who decides how many of the 435 seats in the House of Representatives each state gets? These questions involve **apportionment,** because the seats in the House of Representatives must be apportioned (divided) among the states in a fair way.

In this unit, we'll discuss the mathematics that underlies apportionment. As with methods for voting, we'll see that no single method of apportionment is unquestionably fairer than others. Nevertheless, the mathematical ideas are very useful and can be applied to many problems beyond those of government.

> **DEFINITION**
>
> For the United States House of Representatives, **apportionment** is a process used to divide the available seats among the states. More generally, apportionment is a process used to divide any set of people or objects among various individuals or groups.

All legislative Powers herein granted shall be vested in a Congress of the United States, which shall consist of a Senate and House of Representatives.

—Article 1, Section #1, of the United States Constitution

THE CONSTITUTIONAL CONTEXT

Much of the mathematics of apportionment arose from historical attempts to meet the letter and spirit of the United States Constitution. According to the Constitution, the legislative (law-making) branch of the United States government consists of two distinct bodies: the **Senate** and the **House of Representatives** (called "the House" for short).

The Constitution (Article 1, Section 3) specifies a simple method of apportionment for senators: Each state gets two senators. According to the original Constitution, senators were chosen by state legislatures. However, the 17th amendment to the Constitution, adopted in 1913, changed the method of selecting senators to direct election by the people. Senators are elected for six-year terms.

The requirements for the House of Representatives are much more complicated. The Constitution mandates that seats in the House be apportioned to the states according to their populations, subject to a minimum of one seat for every state. It allows Congress to set the total number of representatives, so long as the total number does not exceed one for each 30,000 people.

The Constitution directs Congress to reapportion seats in the House every ten years, following a census. For example, the current apportionment is based on results from the 2000 census. (Because elections in 2000 came before the census was completed, the 2000 apportionment affects representation beginning with the

2002 Congress.) However, the Constitution does not specify a particular procedure for apportionment. As we will soon see, it is not easy to choose an apportionment procedure.

EXAMPLE 1 United States House of Representatives

The House currently has 435 representatives, and the 2000 census found a U.S. population of 281 million. On average, how many people are represented by each representative? Suppose that the total number of representatives were the constitutional limit of one for every 30,000 people. How many representatives would there be in that case?

Solution Dividing the population of 281 million by 435 representatives, we find that the number of people served by each representative is

$$\frac{\text{population}}{\text{number of representatives}} = \frac{281{,}000{,}000}{435} \approx 646{,}000$$

On average, each representative serves approximately 646,000 people. If there were one representative for every 30,000 people, the total number of representatives would be

$$\frac{\text{population}}{30{,}000} = \frac{281{,}000{,}000}{30{,}000} \approx 9370$$

That is, the constitutional limit allows more than 9000 representatives, in contrast to the actual number of 435. ■

THE APPORTIONMENT PROBLEM

Example 1 shows that each House member represents 646,000 people, on average. Apportionment would be easy if every state's population were a simple multiple of 646,000. For example, a state with a population of 1,292,000, which is 2 × 646,000, would get 2 representatives. Similarly, a state with a population of 6,460,000, which is 10 × 646,000, would get 10 representatives.

Of course, states do not have such convenient populations. For example, Rhode Island has a population of about 1,050,000, which is about 1.6 × 646,000. We

historical note

The first Congress, which met on March 4, 1789 in the Federal Hall in New York City, had 59 members in the House of Representatives. The number gradually rose, reaching 435 in 1912. It has stayed at 435 since, except when two representatives were added when Hawaii and Alaska became states in 1959. The number returned to 435 at the next apportionment and is now fixed by a law passed in 1941.

by the way ...
The 2000 apportionment changed the House representation of 18 states. Arizona, Georgia, Florida, and Texas each gained two seats, while Nevada, Colorado, North Carolina, and California each gained one seat. On the losing side, New York and Pennsylvania each lost two seats, while Mississippi, Oklahoma, Indiana, Wisconsin, Illinois, Michigan, Ohio, and Connecticut each lost one seat.

might say that Rhode Island is "entitled" to 1.6 representatives, but representatives are people and cannot be divided fractionally. That is, Rhode Island can have one representative or two representatives, but not 1.6 representatives.

The people of Rhode Island would prefer to have two representatives, because that strengthens their voice in the House. However, people of other states might prefer that Rhode Island have only one representative, thereby leaving one more representative for another state.

In essence, the apportionment problem deals with finding a systematic way of deciding whether Rhode Island gets one or two representatives. The system must be as fair to all states as possible, while also ensuring that precisely 435 House seats are awarded in total.

The Standard Divisor and Quota

Look again at how we determined that Rhode Island was "entitled" to 1.6 representatives. First, we divided the total U.S. population by the total number of representatives to find that each representative serves an average of 646,000 people. We then divided Rhode Island's population by 646,000 to find the 1.6 representatives.

In the terminology of apportionment, the 646,000 average is called the **standard divisor** for this problem. The 1.6 representatives is called the **standard quota** of representatives for Rhode Island; it is the number (quota) that Rhode Island would get if it were possible to have fractional representatives.

Standard Divisor and Quota

The **standard divisor** is the average number of people per seat (in the House of Representatives) for the entire population of the United States:

$$\text{standard divisor} = \frac{\text{total U.S. population}}{\text{number of seats}}$$

The **standard quota** for a state is the number of seats it would be entitled to *if* fractional seats were allowed:

$$\text{standard quota} = \frac{\text{state population}}{\text{standard divisor}}$$

The standard divisor and standard quota also apply to apportionment problems besides that of the House of Representatives. Simply replace the number of seats by the number of items to be apportioned and the state and total populations by the relevant populations in the problem.

Note that there is only *one* standard divisor. In contrast, each state has its own standard quota.

EXAMPLE 2 *Finding Standard Quotas*

The 2000 census found a population of 902,000 for Montana and 494,000 for Wyoming. Find the standard quota for each state. The 2000 apportionment left each of these states with the constitutional minimum of one representative. Compare their standard quotas to their actual representations.

Solution The standard divisor is the same for all states; as we found earlier, it is 646,000 people per representative. For Montana, the standard quota is

$$\text{standard quota} = \frac{\text{state population}}{\text{standard divisor}} = \frac{902,000}{646,000} \approx 1.4$$

For Wyoming, the standard quota is

$$\text{standard quota} = \frac{\text{state population}}{\text{standard divisor}} = \frac{494,000}{646,000} \approx 0.76$$

Montana's standard quota is 1.4, but it has only one representative. Montana's people can rightfully feel underrepresented in Congress. Wyoming also has the constitutional minimum of one representative, even though its standard quota is only 0.76. Thus, Wyoming's people are relatively overrepresented in Congress, at least according to population.

EXAMPLE 3 *School Teacher Apportionment*

A small school district is reapportioning its 14 elementary teachers among its three elementary schools, which have the following enrollments: Washington Elementary, 197 students; Lincoln Elementary, 106 students; and Roosevelt Elementary, 145 students. Find the standard quota of teachers for each school.

Solution This problem is just like the House apportionment problem, except we are apportioning teachers rather than representatives. The standard divisor is the average number of students per teacher. We find it by dividing the total student population by the number of teachers:

$$\text{standard divisor} = \frac{\text{total number of students}}{\text{number of teachers}} = \frac{197 + 106 + 145}{14} = 32$$

We find the standard quota for each school by dividing the school's enrollment by the standard divisor. Table 9.9 shows the calculations and results. Note that the total (sum) of the standard quotas is the number of teachers to be apportioned.

TABLE 9.9 Finding Standard Quotas for a School Teacher Apportionment

SCHOOL	WASHINGTON	LINCOLN	ROOSEVELT	TOTAL
ENROLLMENT	197	106	145	448
STANDARD QUOTA	$\frac{197}{32} = 6.15625$	$\frac{106}{32} = 3.3125$	$\frac{145}{32} = 4.53125$	14

enrollments divided by standard divisor of 32

The Challenge of Apportionment

The 435 representatives and 50 states make apportionment in the United States House of Representatives fairly complex. For the purposes of understanding apportionment, it's easier to work with a smaller set of states. Suppose there were only four states, A, B, C, and D, with populations as shown in Table 9.10. Note that the total population is 10,000.

Suppose these four states decide to create a legislature with 100 seats. Then the standard divisor is

$$\text{standard divisor} = \frac{\text{total population}}{\text{number of seats}} = \frac{10,000}{100} = 100$$

> *by the way...*
> Besides the 100 senators and 435 representatives, the U.S. Congress also includes a resident commissioner from Puerto Rico and delegates from American Samoa, the District of Columbia, Guam, and the Virgin Islands. These five individuals may take part in the floor discussions and vote in committees, but have no vote in the full House.

The second row of Table 9.10 shows the standard quotas calculated with this standard divisor. As it must, the total of the standard quotas equals the 100 available seats.

TABLE 9.10 Four-State Standard Quotas for 100 Seats

STATE	A	B	C	D	TOTAL
POPULATION	936	2726	2603	3735	10,000
STANDARD QUOTA	9.36	27.26	26.03	37.35	100

state populations divided by standard divisor of 100

Note: The total population is 10,000 and the standard divisor is 100.

Time out to think

Confirm the standard quotas in Table 9.10 by dividing each state's population by the standard divisor.

We cannot use the fractional standard quotas for the actual apportionment; instead, we must find a way to make integers from the standard quotas. The most obvious solution would be to round all the standard quotas according to standard rounding rules. In that case, all four standard quotas would round down, because all four have fractional parts less than 0.5. However, rounding in this way leads to a total of 9 + 27 + 26 + 37 = 99 seats, one short of the 100 seats that are supposed to be apportioned. Failure to attain 100 seats with standard rounding rules means we must find a different way to make integers from the standard quotas. And herein lies the challenge of apportionment: There are many reasonable ways to make integers from the standard quotas, but they do not all lead to the same apportionment results.

HAMILTON'S METHOD

The United States conducted its first census in 1790, presenting Congress with its first task of apportionment almost immediately after ratification of the Constitution. Alexander Hamilton, then Secretary of the Treasury, championed a simple apportionment procedure that worked as follows.

> **Hamilton's Method of Apportionment**
> After finding the standard quota for each state:
> - First, give each state the number of seats found by rounding its standard quota *down*; this number is the state's **minimum quota** (or *lower quota*).
> - If there are any extra seats after each state has been given its minimum quota, look at each state's **fractional remainder**—the fraction that remains in the standard quota after subtracting the minimum quota. Give the extra seats one at a time to the states with the highest fractional remainders, until the seats are all gone.

Let's apply Hamilton's method to the four states in Table 9.10. The easiest way is by extending the table, as shown in Table 9.11. The first two rows (population and standard quota) are unchanged. The third row shows the minimum quotas found by rounding down the standard quotas. Note that the total of the minimum quotas is 99, leaving 1 extra seat out of the 100 to be allotted. The fourth row shows the fractional remainders after rounding down; the extra seat goes to the state with the largest fractional remainder, which is State A. The last row shows the final apportionment. Note that State A gets one seat more than its minimum quota, while the other states get precisely their minimum quotas.

TABLE 9.11 Applying Hamilton's Method to the Four-State Data from Table 9.10

STATE	A	B	C	D	TOTAL
POPULATION	936	2726	2603	3735	10,000
STANDARD QUOTA	9.36	27.26	26.03	37.35	100
MINIMUM QUOTA	9	27	26	37	99
FRACTIONAL REMAINDER	0.36 (largest)	0.26	0.03	0.35	1
FINAL APPORTIONMENT	10	27	26	37	100

- state populations divided by standard divisor of 100 → STANDARD QUOTA
- standard quotas rounded down → MINIMUM QUOTA
- remainders after rounding → FRACTIONAL REMAINDER
- the extra seat goes to the state with the largest fractional remainder (State A) → FINAL APPORTIONMENT

EXAMPLE 4 *Applying Hamilton's Method*

Apply Hamilton's method to determine the apportionment of teachers among the schools in Example 3.

Solution Table 9.12 repeats the enrollments and standard quotas found in Example 3. Hamilton's method tells us first to round down all the standard quotas; we get the minimum quotas in the third row. The total of the minimum quotas is 13, one less than the 14 teachers to be apportioned. The school with the largest fractional remainder (fourth row), which is Roosevelt, gets the extra teacher. Thus, in the final apportionment, Washington and Lincoln get their minimum quotas of 6 and 3 teachers, respectively; Roosevelt gets 5 teachers, rather than its minimum quota of 4.

TABLE 9.12 Applying Hamilton's Method to the School Teacher Apportionment from Table 9.9

SCHOOL	WASHINGTON	LINCOLN	ROOSEVELT	TOTAL
ENROLLMENT	197	106	145	448
STANDARD QUOTA	$\frac{197}{32} = 6.15625$	$\frac{106}{32} = 3.3125$	$\frac{145}{32} = 4.53125$	14
MINIMUM QUOTA	6	3	4	13
FRACTIONAL REMAINDER	0.15625	0.3125	0.53125 (largest)	1
FINAL APPORTIONMENT	6	3	5	14

- enrollments divided by standard divisor of 32 → STANDARD QUOTA
- standard quotas rounded down → MINIMUM QUOTA
- remainders after rounding → FRACTIONAL REMAINDER
- the extra teacher goes to the school with the largest fractional remainder (Roosevelt) → FINAL APPORTIONMENT

> **historical note**
>
> President Washington's veto of Hamilton's method and the subsequent adoption of Jefferson's method affected the representation of only two states. Delaware would have had 2 seats by Hamilton's method, but ended up with only 1 by Jefferson's method. Virginia would have had 18 seats by Hamilton's method, but ended up with 19 by Jefferson's. It's worth noting that Virginia was Jefferson's home state.

The First Presidential Veto

Hamilton's method is simple and appears reasonably fair. Perhaps as a result, both the Senate and the House voted in 1791 to adopt Hamilton's method for apportionment. However, recall that for a bill to become law, it must not only be passed by both the Senate and the House, but also be signed by the president. If the president vetoes a bill, super majority votes of 2/3 in the House and Senate are needed to override the veto (see Unit 9A). In 1792, the bill authorizing Hamilton's method received the first presidential veto in the history of the United States. In his veto message of April 5, 1792, President George Washington wrote:

> *I have maturely considered the act passed by the two Houses entitled "An act for an apportionment of representatives among the several States according to the first enumeration," and I return it to your House, wherein it originated, with the following objections:*
>
> *First. The Constitution has prescribed that representatives shall be apportioned among the several States according to their respective numbers, and there is no one proportion or divisor which, applied to the respective numbers of the States, will yield the number and allotment of representatives proposed by the bill.*
>
> *Second. The Constitution has also provided that the number of representatives shall not exceed 1 for every 30,000, which restriction is by the context and by fair and obvious construction to be applied to the separate and respective numbers of the States; and the bill has allotted to eight of the States more than 1 for every 30,000.*

Congress was unable to override the veto, and instead passed a bill authorizing a different apportionment method proposed by Thomas Jefferson, then Secretary of State. We'll discuss Jefferson's method shortly.

Time out to think

The Constitution actually says, "The number of representatives shall not exceed one for every thirty thousand, but each state shall have at least one representative . . ." Read Washington's veto statement carefully. Why did he have to add the words "which restriction is by the context and by fair and obvious construction . . ." in order to justify his veto?

Fairness Issues with Hamilton's Method

Although Hamilton's method did not become law at the time it was proposed, it was reintroduced and adopted as the apportionment procedure in 1850. It remained in use until 1900. During this time, several problems with Hamilton's method emerged; the most famous is called the **Alabama paradox.**

After the 1880 census, the Chief Clerk of the Census Office, C. W. Seaton, used Hamilton's method to compute apportionments for various possible House sizes. (Unlike today, in those days the House size often changed with each apportionment.) He discovered that Alabama would get 8 seats in a House of 299 representatives, but only 7 seats in a House of 300 representatives. This curious result seems unfair, because Alabama loses a seat when the total number of seats increases.

> **historical note**
>
> Hamilton's method was reintroduced to Congress by Representative Samuel Vinton of Ohio. However, Vinton was apparently unaware of Hamilton's proposal more than 50 years earlier. As a result, the method was at least temporarily called Vinton's method.

> **The Alabama Paradox**
> In a fair apportionment system, adding extra seats must not result in fewer seats for any state. The **Alabama paradox** occurs when the total number of available seats increases, yet one state (or more) loses seats as a result. It can occur with Hamilton's method.

Alabama did not actually lose a seat in the 1882 apportionment, because Congress chose a larger house size (325) for which the Alabama paradox did not present itself. But knowing that the paradox was possible reduced support for Hamilton's method, and it was abandoned by Congress in 1900.

Hamilton's method has not been used for House apportionment since 1900, but it has been studied, leading to the discovery of two other paradoxes. The **population paradox** was discovered around 1900, when it was found that Hamilton's method would award a seat to Maine at Virginia's expense, even though Virginia was growing much faster than Maine.

> **The Population Paradox**
> When apportionment changes because of population growth, we would expect faster-growing states to gain seats at the expense of slow-growing states. When the opposite occurs—a slow-growing state gains a seat at the expense of a faster-growing state—we have what we call the **population paradox.**

The **new states paradox** was discovered in 1907, when Oklahoma became the 46th state in the United States. Because it was not yet time for reapportionment, Congress decided to add *new* seats for Oklahoma. Based on its population, Oklahoma was clearly entitled to 5 seats, so Congress increased the number of seats in the House from 386 to 391. Surprisingly, calculations with Hamilton's method showed that the addition of Oklahoma's 5 seats would have caused New York to lose a seat while Maine gained one. (See the Exercises for examples of the population and new states paradoxes.)

> **The New States Paradox**
> When additional seats are added to accommodate a new state, we do not expect this addition to change the apportionment for existing states. If it does, we have an example of the **new states paradox.**

historical note

In the 1876 election, Democrat Samuel J. Tilden won the popular vote but lost the presidency to Republican Rutherford B. Hayes by one electoral vote. The electoral vote was severely disputed, with some states appointing rival slates of electors. Tilden eventually stepped aside in the "Compromise of 1877," which led to Hayes's inauguration on March 4, 1877. Interestingly, if electors had been apportioned by Hamilton's method, as the law required, Tilden would have won in the first place. Instead, electors had been apportioned according to a compromise from a few years earlier, even though the compromise violated the legal requirement.

EXAMPLE 5 *Four-State Alabama Paradox*

Using Hamilton's method, recompute the apportionment in Table 9.11 if there are 101 seats instead of 100. Does the Alabama paradox occur? Explain.

Solution The total population is 10,000, so with 101 seats the standard divisor is

$$\text{standard divisor} = \frac{\text{total population}}{\text{number of seats}} = \frac{10,000}{101} \approx 99.0099$$

We find the standard quota for each state by dividing its population by this standard divisor. You should confirm that the standard quotas come out as shown in Table 9.13; note that they are slightly different from the standard quotas shown in Table 9.11 for 100 seats. The third row of Table 9.13 shows the minimum quotas, which are unchanged from Table 9.11. However, the fractional remainders (fourth row) have changed, which changes the final apportionment.

This time, there are two extra seats after the minimum quotas have been filled. By Hamilton's method, these two seats go to States D and B, because they have the two largest fractional remainders. Meanwhile, State A ends up with its minimum quota of 9 seats, even though it had 10 seats in Table 9.11. In other words, increasing the total number of seats from 100 to 101 has cost State A a seat—an illustration of the Alabama paradox.

TABLE 9.13 Recomputed Apportionment for Table 9.11 with 101 Seats

STATE	A	B	C	D	TOTAL
POPULATION	936	2726	2603	3735	10,000
STANDARD QUOTA	9.4536	27.5326	26.2903	37.7235	101
MINIMUM QUOTA	9	27	26	37	99
FRACTIONAL REMAINDER	0.4536	0.5326 (2nd largest)	0.2903	0.7235 (largest)	2
FINAL APPORTIONMENT	9	28	26	38	101

- state populations divided by standard divisor of 99.0099
- standard quotas rounded down
- remainders after rounding
- the extra seats go to the two states with the largest fractional remainders (D and B)

JEFFERSON'S METHOD

Thomas Jefferson's rival apportionment method became law after President Washington vetoed Hamilton's method. Jefferson's method essentially tries to avoid decisions about fractional remainders by seeking minimum quotas that use up all available seats. As we've seen, the minimum quotas used in Hamilton's method—which are simply the standard quotas rounded down—often leave extra seats. Jefferson realized that he could achieve his goal by changing the standard quotas to new values, called **modified quotas,** so that the new set of minimum quotas would use up all the seats.

Finding modified quotas means dividing the state populations by a **modified divisor** that is different from (lower than) than the standard divisor. The trick in Jefferson's method is choosing a modified divisor that leads to the desired result, which is a set of minimum quotas that uses all the seats. Choosing a modified divisor generally requires trial and error: If your first choice doesn't work, try again until you find one that does.

Jefferson's Method of Apportionment

Begin by finding the standard divisor, standard quotas, and minimum quotas (as in Hamilton's method). If there are no extra seats after each state has been given its minimum quota, then the apportionment is complete. If there *are* extra seats, *start the computations over* as follows.

- Choose a **modified divisor** that is *lower* than the standard divisor. Use it to compute **modified quotas** by dividing the state populations by the modified divisor. Round these modified quotas down to find a new set of minimum quotas.
- If there are just enough seats to fill all the minimum quotas, the apportionment is complete. Otherwise, do one of the following:

 1. If there are still extra seats with the new minimum quotas, start again with a lower modified divisor.
 2. If there are not enough total seats to fill all the minimum quotas, start again with a higher modified divisor (but still smaller than the standard divisor).

Let's apply Jefferson's method to the same four states to which we applied Hamilton's method. The first three rows of Table 9.14 are unchanged from Table 9.11, showing the minimum quotas (third row) that result from the standard divisor of 100. Because the minimum quotas leave an extra seat, Jefferson's method tells us to try a new (lower) divisor. Study Table 9.14 carefully to see how we implement Jefferson's method:

TABLE 9.14 Applying Jefferson's Method to the Four-State Data from Table 9.11

STATE	A	B	C	D	TOTAL
POPULATION	936	2726	2603	3735	10,000
STANDARD QUOTA	9.36	27.26	26.03	37.35	100
MINIMUM QUOTA	9	27	26	37	99
MODIFIED QUOTA (WITH DIVISOR 99)	9.45	27.54	26.29	37.73	101.01
MINIMUM QUOTA (WITH DIVISOR 99)	9	27	26	37	99
MODIFIED QUOTA (WITH DIVISOR 98)	9.55	27.82	26.56	38.11	102.04
MINIMUM QUOTA (WITH DIVISOR 98)	9	27	26	38	100

Annotations:
- state populations divided by standard divisor of 100 → STANDARD QUOTA
- standard quotas rounded down → MINIMUM QUOTA
- state populations divided by *modified divisor* of 99 → MODIFIED QUOTA (WITH DIVISOR 99)
- modified quotas rounded down → MINIMUM QUOTA (WITH DIVISOR 99)
- state populations divided by *modified divisor* of 98 → MODIFIED QUOTA (WITH DIVISOR 98)
- modified quotas rounded down; all 100 seats are used, so this is final apportionment → MINIMUM QUOTA (WITH DIVISOR 98)

- The standard divisor is 100, so we try a lower *modified divisor* of 99. We divide the state populations by this divisor to find the modified quotas in the fourth row. The fifth row shows the new set of minimum quotas, found by rounding down the modified quotas.

- Because the new set of modified quotas still leaves an extra seat, we try an even lower modified divisor of 98. The last two rows show the modified quotas and minimum quotas found with this divisor. This time they add up to 100, so the apportionment is complete.

Note that State A gets 9 seats by Jefferson's method, as opposed to 10 seats by Hamilton's method (see Table 9.11). Meanwhile, State D gets 38 seats by Jefferson's method, one more than it gets by Hamilton's method.

Is Jefferson's Method Fair?

We've already seen that Hamilton's method is susceptible to several paradoxes that make it appear unfair in some cases. Is Jefferson's method better?

Mathematicians have studied these methods carefully. Jefferson's method avoids all three of the paradoxes found with Hamilton's method (the Alabama, population, and new states paradoxes). However, Jefferson's method sometimes leads to a different problem. Let's investigate.

Remember that standard quotas represent the number of seats that would be apportioned *if* fractional apportionment were allowed. For example, we earlier found a standard quota of 1.6 for Rhode Island, suggesting that it is "entitled" to 1.6 seats. We might therefore expect that any "fair" apportionment method would either round this standard quota up to give Rhode Island two seats or round it down to give Rhode Island one seat. If instead Rhode Island ended up with zero seats or three or more seats, we would probably conclude that the apportionment was unfair. This idea, which holds that the actual apportionment for any state should be its standard quota rounded either up or down, is often called the **quota criterion.** An apportionment that fails the quota criterion is generally considered unfair. As the following example shows, Jefferson's method sometimes fails the quota criterion.

> **The Quota Criterion**
>
> For a fair apportionment, the number of seats assigned to each state should be its standard quota rounded either up or down to the nearest integer.

by the way...
Hamilton's method always satisfies the quota criterion because it starts with the standard quota rounded down, then at most adds one—which gives the standard quota rounded up. Hamilton's method and other apportionment methods that always satisfy the quota criterion are generically called *quota methods*. In contrast, methods that change the divisor, like Jefferson's method, are called *divisor methods* and sometimes violate the quota criterion.

EXAMPLE 6 *Jefferson's Method and the Quota Criterion*

Consider a four-state legislature with 100 seats, in which the states have the following populations: State A, 680; State B, 1626; State C, 1095; and State D, 6599. Use Jefferson's method to apportion the 100 seats. Is the quota criterion satisfied?

Solution We begin by finding the standard divisor, which is

$$\text{standard divisor} = \frac{\text{total population}}{\text{number of seats}} = \frac{10{,}000}{100} = 100$$

Table 9.15 shows the computations by Jefferson's method. The second row gives the standard quotas and the third row gives the resulting minimum quotas. There are three extra seats, so Jefferson's method tells us to try a lower modified divisor. Row 4 shows the modified quotas with a divisor of 98 and the last row shows the new minimum quotas. Because this new set of minimum quotas uses all 100 seats, it represents the final apportionment.

Note that State D has a standard quota of 65.99. By the quota criterion, State D should get either 65 or 66 seats. However, State D ends up with 67 seats, so this apportionment violates the quota criterion.

TABLE 9.15 A Case Where Jefferson's Method Violates the Quota Criterion

STATE	A	B	C	D	TOTAL
POPULATION	680	1626	1095	6599	10,000
STANDARD QUOTA	6.80	16.26	10.95	65.99	100
MINIMUM QUOTA	6	16	10	65	97
MODIFIED QUOTA (WITH DIVISOR 98)	6.94	16.59	11.17	67.34	102.04
MINIMUM QUOTA (WITH DIVISOR 98)	6	16	11	67	100

- state populations divided by standard divisor of 100
- standard quotas rounded down
- state populations divided by *modified divisor* of 98
- modified quotas rounded down

What Became of Jefferson's Method

After being signed into law by President Washington in 1792, Jefferson's method was used through the 1830s. However, violations of the quota criterion occurred in the apportionments following both the 1820 and the 1830 census. Moreover, Congress learned that the violations tend to give extra seats to larger states at the expense of smaller states. As a result, Congress abandoned Jefferson's method for the apportionment following the 1840 census.

OTHER APPORTIONMENT METHODS

As we've discussed, Jefferson's method was abandoned by 1840 and Hamilton's method, adopted in 1850, was abandoned in 1900. Many other apportionment methods have been proposed, but only two others have been used for the United States House of Representatives.

Webster's Method

In 1832, the famous senator and orator Daniel Webster, of Massachusetts, proposed a variation on Jefferson's method. Webster's method was adopted for the apportionment following the 1840 census. It was then abandoned in favor of Hamilton's method in 1850, but resurrected in 1900. Webster's method remained in use until 1940.

Webster's method is very similar to Jefferson's method with one exception: Instead of looking for a set of modified quotas that can all be rounded *down* to give the correct total number of seats, Webster's method seeks modified quotas that give the correct total number of seats by using standard rounding rules. That is, we round *up* when the fractional part of a modified quota is 0.5 or more, and down when the fractional part is less than 0.5. Note that, whereas the modified divisor in Jefferson's method is always less than the standard divisor, the modified divisor in Webster's method may be either greater or less than the standard divisor.

EXAMPLE 7 Applying Webster's Method

Consider a four-state legislature with 100 seats, in which the states have the following populations: State A, 948; State B, 749; State C, 649; and State D, 7654. Use Webster's method to apportion the 100 seats.

Solution The total population is 10,000 and the number of seats is 100, so the standard divisor is 100 (as in Example 6). Table 9.16 shows the computations by Webster's method. The second row gives the standard quotas and the third row gives the resulting minimum quotas; there are two extra seats. Webster's method tells us to try a modified divisor and then round the modified quotas to the nearest integer. Row 3 shows the modified quotas with a divisor of 99.85; finding a divisor that works generally requires some trial and error (not shown here). The last row shows the rounded quotas, which represent the final apportionment because they use all 100 seats.

TABLE 9.16 A Four-State Apportionment with Webster's Method

STATE	A	B	C	D	TOTAL
POPULATION	948	749	649	7654	10,000
STANDARD QUOTA	9.48	7.49	6.49	76.54	100
MINIMUM QUOTA	9	7	6	76	98
MODIFIED QUOTA (WITH DIVISOR 99.85)	9.4942	7.5013	6.4997	76.6550	100.1502
ROUNDED QUOTA	9	8	6	77	100

- state populations divided by standard divisor of 100 → STANDARD QUOTA
- standard quotas rounded down → MINIMUM QUOTA
- state populations divided by *modified divisor* of 99.85 → MODIFIED QUOTA
- modified quotas rounded to nearest integer → ROUNDED QUOTA

Time out to think

Is the quota criterion satisfied in Example 7? Explain.

FIGURE 9.2 A scatter plot of number of representatives versus population for the 50 states (2000 data). A perfect straight-line correlation would mean all states were equally represented.

The Hill-Huntington Method

In 1941, Congress replaced Webster's method with an alternative called the Hill-Huntington method. It remains in use today (Figure 9.2).

The Hill-Huntington method gets its name from Joseph Hill, who served as Chief Statistician of the Census Bureau, and Harvard mathematician Edward Huntington. After Hill proposed the method in 1911, mathematicians debated whether it was fairer than Webster's method, while Congress debated both the fairness and the political aspects of its use. For the 1920 census, Congress continued to use Webster's method. For the 1930 census, Webster's method and the Hill-Huntington method led to the same apportionments, so controversy was temporarily averted. But in the apportionment following the 1940 census, Webster's method gave one extra seat to Michigan while the

Hill-Huntington method gave the seat to Arkansas. Michigan tended to vote Republican while Arkansas tended to vote Democratic. Because Democrats had a majority in Congress, they made a political decision: They voted to use the Hill-Huntington method, thereby giving the extra seat to Democratic-leaning Arkansas. President Roosevelt (also a Democrat) signed the bill into law.

The Hill-Huntington method is almost identical to Webster's method, except it uses a different rule to decide whether modified quotas should be rounded up or down. In Webster's method, rounding follows the usual rule in which we round up if the fractional part is 0.5 or more, and round down if it is less than 0.5. In the Hill-Huntington method, rounding is based instead on the **geometric mean** of the integers on either side of the modified quota.

> **DEFINITIONS**
>
> The **geometric mean** of any two numbers x and y is $\sqrt{x \times y}$. The more familiar mean, $(x + y)/2$, is called the **arithmetic mean.** In general, we assume a "mean" is an arithmetic mean unless told otherwise. (Thus, the mean we have used elsewhere in this book has always been the arithmetic mean.)

If the modified quota is less than the geometric mean of the two nearest integers, it gets rounded down; if it is more than the geometric mean, it gets rounded up. As an example, suppose a state has a modified quota of 2.47, which we could potentially round to either 2 or 3. Under Webster's method, we would round it down to 2, because the fractional part of 2.47 is less than 0.5. Under the Hill-Huntington method, we first find the geometric mean of 2 and 3, which is

$$\sqrt{2 \times 3} = \sqrt{6} \approx 2.45$$

Because the modified quota of 2.47 is *greater* than this geometric mean, it gets rounded up to 3 in the Hill-Huntington method.

Note that the geometric mean of two consecutive integers is always less than their arithmetic mean. For example, the geometric mean of 2 and 3 is 2.45, which is less than their arithmetic mean of 2.5 by 0.05. However, the two means tend to be closer for larger consecutive integers. For example, the geometric mean of 10 and 11 is $\sqrt{10 \times 11} = \sqrt{110} \approx 10.488$, which is only 0.012 less than their arithmetic mean of 10.5. In practical terms, use of the geometric mean in the Hill-Huntington method therefore increases the chance that extra seats will go to smaller states rather than larger states.

EXAMPLE 8 *Applying the Hill-Huntington Method*

Apply the Hill-Huntington method to the case described in Example 7. Is the resulting apportionment the same as by Webster's method?

Solution Table 9.17 shows the calculations; the first three rows are the same as in Table 9.16, because they are computed the same way by all methods. The fourth row shows modified quotas with a modified divisor of 100.06; again, finding a modified divisor that works generally requires some trial and error which is not shown here. To

determine the rounded quotas, we first find the geometric means (row 5). The Hill-Huntington method then tells us to round down if the modified quota is less than the geometric mean and to round up otherwise. The last row shows the rounded quotas; because they use all 100 seats, the apportionment is complete. Note that States C and D end up with different numbers of seats in this apportionment than in Webster's apportionment in Example 7.

TABLE 9.17 Applying the Hill-Huntington Method to the Four-State Data from Table 9.16

STATE	A	B	C	D	TOTAL
POPULATION	948	749	649	7654	10,000
STANDARD QUOTA	9.48	7.49	6.49	76.54	100
MINIMUM QUOTA	9	7	6	76	98
MODIFIED QUOTA (WITH DIVISOR 100.06)	9.47	7.49	6.49	76.49	99.94
GEOMETRIC MEAN	$\sqrt{9 \times 10}$ ≈ 9.487	$\sqrt{7 \times 8}$ ≈ 7.483	$\sqrt{6 \times 7}$ ≈ 6.481	$\sqrt{76 \times 77}$ ≈ 76.498	
ROUNDED QUOTA	9	8	7	76	100

- state populations divided by standard divisor of 100
- standard quotas rounded down
- state populations divided by modified divisor of 100.06
- geometric mean of integers on either side of modified quota
- modified quotas rounded down if less than geometric mean, up otherwise

Time out to think

Consider the results in Examples 7 and 8. Which apportionment method would State C prefer? Which would State D prefer? Explain.

IS THERE A BEST METHOD FOR APPORTIONMENT?

The history of apportionment is a quest for fairness. Hamilton's method proved to be unfair because of the three paradoxes (Alabama, population, and new states). Jefferson's method proved to be unfair because it can violate the quota criterion. How do Webster's method and the Hill-Huntington method compare?

Mathematicians have studied the methods by running simulations of many possible apportionments. Results show that, like Jefferson's method, Webster's method and the Hill-Huntington method can also violate the quota criterion. However, the violations occur much less frequently with Webster's or Hill-Huntington than with Jefferson's method, making them arguably fairer.

For example, if Jefferson's method had remained in use, nearly every apportionment since 1850 would have violated the quota criterion. In contrast, no violations have occurred with Webster's method or the Hill-Huntington method. Simulations show that, by chance alone, Webster's and Hill-Huntington are expected to violate the quota criterion only once in several hundred apportionments. Interestingly, Webster's appears slightly less prone to violations of the quota criterion than Hill-Huntington, suggesting that it is slightly fairer.

practical matters

REDISTRICTING

Apportionment determines the number of representatives for each state. Politically, however, this number is only the beginning of the practical issues associated with electing representatives. In states with more than one seat in the House, the seats are divided among congressional districts within the state. For example, a state with six seats has six districts, with the people in each district electing one representative. Just as apportionment is recomputed every ten years following a census, so are districts redrawn with new boundaries at these times. Although few people follow the details of this redistricting, it can have significant political impact.

The Constitution leaves it to the states to decide how to draw the boundaries of their districts. In most states, redistricting is up to the state legislature, which can use the process for political advantage. To see how, consider a state that has six House seats and in which half the people are Democrats and half are Republicans. If you ignored redistricting, you might guess that this state would have three Democratic representatives and three Republican representatives.

However, suppose the state elects its six representatives in six districts (District 1 through District 6) with 1 million people each. The even division of voter preferences among the two parties means that there are 3 million Democrats and 3 million Republicans in the state. Now, to take an extreme case, suppose the legislature draws the boundaries of District 1 so that all of its 1 million people are Democrats. Clearly, District 1 will elect a Democratic representative. But the remaining five districts will have 3 million Republicans and only 2 million Democrats. If they are drawn so that their populations reflect this general distribution, every one of these five districts will have Republicans outnumbering Democrats by a 3 to 2 margin. The end result: Despite having equal numbers of Democratic and Republican voters, the state ends up with five Republican representatives and only one Democratic representative. Of course, the same idea could occur in reverse, giving Democrats the advantage rather than Republicans.

The practice of forming voting districts to give an unfair advantage to a political party is so common that it has its own name—*gerrymandering*. (The term originated in 1812 when Massachusetts, under Governor Elbridge *Gerry*, created a district with roughly the shape of a sala*mander*.) Since the 1960s, the Supreme Court has found gerrymandering illegal, ruling that it violates the Constitution's intention to provide equal representation to all citizens. Nevertheless, the issue of fairness is sufficiently subjective to leave the party that controls redistricting in any state with broad influence over the makeup of the state's House delegation. Similar redistricting considerations apply to state and local representation. Today, political parties use sophisticated mathematical models to help draw districts in a way most favorable to their interests. (But if they go too far, opposing parties may claim illegal gerrymandering and challenge the district boundaries in court.)

Redistricting is more complex mathematically than apportionment, but it offers the same basic lesson: Democracy is much more difficult in practice than it is in principle, which is why it is so important to be politically and mathematically aware if you hope to influence the future of your town, state, country, or world.

Is there any apportionment system that is unquestionably fair in all circumstances? Such an apportionment system would have to satisfy the quota criterion in all cases, while also being immune to the three paradoxes that affect Hamilton's method. Unfortunately, a theorem proved by mathematicians M. L. Balinsky and H. P. Young states that such a system is impossible. It tells us that, in the end, we cannot

choose between apportionment procedures on the basis of fairness alone. As a result, apportionment will always involve political decisions, and we can expect it to be a subject of continued debate.

Review Questions

1. What is apportionment? What does it mean for the U.S. House of Representatives?
2. Briefly describe the nature of the apportionment problem. How does this problem arise from the requirements of the Constitution?
3. Describe how to find the standard divisor and standard quotas for an apportionment problem. Give an example.
4. Explain how Hamilton's method apportions seats. Briefly describe the history of Hamilton's method.
5. What is the Alabama paradox? What other paradoxes affect Hamilton's method? Why do these paradoxes make the method seem unfair in the cases where they arise?
6. Explain how Jefferson's method apportions seats. Briefly describe the history of Jefferson's method.
7. What is the quota criterion? Why are violations of this criterion considered unfair?
8. Briefly describe how Webster's method and the Hill-Huntington method differ from Jefferson's method.
9. Explain why Webster's method and the Hill-Huntington method are considered fairer than Jefferson's method, but still not fair in all cases. What is the significance of the Balinsky and Young theorem?

In the News

1. **Apportionment News.** Find a news story that concerns the most recent apportionment for the U.S. House of Representatives. Summarize the story and the role of apportionment methods in any controversies.
2. **Local Apportionment.** Find a recent news story concerning an apportionment problem that applies locally or at the state level. Discuss the apportionment procedure used. Does the apportionment seem fair?
3. **Your State's Representatives.** How many seats does your state have in the House of Representatives? How does this number compare to your state's standard quota? Based on your findings, is your state over- or underrepresented in the House? Explain.

Exercises

State Representation. The following table shows populations and numbers of seats in the House for four states, based on the 2000 census (taking effect with the 2002 Congress). For the states indicated in Exercises 1–4, find the standard quota and compare it to the actual number of seats for the state. Then explain whether the state is relatively under- or overrepresented in the House. Use the census 2000 value for the total U.S. population, which was 281 million.

State	Population	Seats in House
Connecticut	3,406,000	5
Georgia	8,186,000	13
Florida	15,982,000	25
Ohio	11,353,000	18

1. Connecticut
2. Georgia
3. Florida
4. Ohio

Alabama Paradox. In Exercises 5–8, assume that 100 delegates must be apportioned to states with the given populations. Determine the number of representatives for each state using Hamilton's method. Then assume that the number of delegates is increased to 101. Determine the new number of representatives for each state using Hamilton's method. State whether the change in total delegates results in the Alabama paradox.

5. A: 950; B: 670; C: 246
6. A: 2540; B: 1140; C: 6330
7. A: 770; B: 155; C: 70; D: 673
8. A: 562; B: 88; C: 108; D: 242

The New States Paradox. In Exercises 9–10, the populations of three states are given. Determine how 100 delegates should be apportioned among these states using Hamilton's method. Then suppose that a new state with a population of 500 is added to the system and 5 new seats are added to the system. Determine the apportionment for the four states (assuming the populations of the first three states remain the same). Does the addition of the new state decrease the representation of any of the original states?

9. A: 1140; B: 6320; C: 250
10. A: 5310; B: 1330; C: 3310

Jefferson's Method. In Exercises 11–14, apply Jefferson's method to states with the given populations. Assume that 100 delegates are to be apportioned. In each case, state whether the quota criterion is satisfied.

11. A: 98; B: 689; C: 212 (modified divisor 9.83)
12. A: 1280; B: 631; C: 2320 (modified divisor 42.00)
13. A: 69; B: 680; C: 155; D: 75 (modified divisor 9.60)
14. A: 1220; B: 5030; C: 2460; D: 690 (modified divisor 92.00)

Comparing Methods. Exercises 15–18 list populations for three or four states. In each case, assume 100 delegates are to be apportioned and do the following:

a. Find the apportionment by Hamilton's method.
b. Find the apportionment by Jefferson's method.
c. Find the apportionment by Webster's method.
d. Find the apportionment by the Hill-Huntington method.
e. Compare the results of the various methods. Which methods give the same results? Do any of the results violate the quota criterion? Overall, which method do you think is best for this apportionment? Why?

15. A: 535; B: 344; C: 120
16. A: 144; B: 443; C: 389
17. A: 836; B: 2703; C: 2626; D: 3835
18. A: 1234; B: 3498; C: 2267; D: 5558

Non-House Apportionments. Exercises 19–22 describe apportionment problems for situations besides the House of Representatives. In each case, do the following:

a. Find the apportionment by Hamilton's method.
b. Find the apportionment by Jefferson's method.
c. Find the apportionment by Webster's method.
d. Find the apportionment by the Hill-Huntington method.
e. Compare the results of the various methods.

19. A high school is creating a student committee to allocate use of classrooms after hours. The committee is to consist of 10 student members, chosen from three interest groups: social groups, which have 48 members; political groups, which have 97 members; and athletic groups, which have 245 members.

20. A city plans to purchase 16 new emergency vehicles. They are to be apportioned among 485 members of the police department, 213 members of the fire department, and 306 members of the paramedic squad.

21. A chain of hardware stores is reapportioning its 25 managers among stores in four locations according to their monthly gross sales. The sales at the four stores are as follows: Boulder, $2.5 million; Denver, $7.6 million; Broomfield, $3.9 million; Ft. Collins, $5.5 million.

22. The city parks commission plans to build nine new parks, to be apportioned among three neighborhoods by population. The neighborhoods are Greenwood, population 4300; Willowbrook, population 3040; and Cherryville, population 2950.

Web Projects

Find useful links for Web Projects on the text Web site: www.aw.com/bennett-briggs

1. **2000 Census Apportionment.** How did your home state fare in the apportionment based on the 2000 census? How have changes in population affected your state's standard quota since the 2000 apportionment? Overall, do you think the apportionment was fair to your state?

2. **House Districts.** Once a state knows its number of seats in the U.S. House, it is still up to the state to draw boundaries for the districts of various representatives. This process can be very contentious. Find the district boundaries for your state. Do the boundaries make sense geographically, or do they appear to be politically motivated? Explain.

3. **Redistricting Roulette.** The Center for Voting and Democracy offers a game that you can play to study the process of redistricting. Try the game; then write a short report on what you learned.

CHAPTER 9 SUMMARY

UNIT	KEY TERMS	KEY IDEAS AND SKILLS
9A	majority rule popular vote electoral vote filibuster super majority veto preference schedule plurality	**Understand** U.S. presidential elections. **Understand** variations on majority rule. **Apply** five methods for deciding an election with three or more candidates: plurality method single (top-two) runoff sequential runoff point system (Borda count) pairwise comparisons (Condorcet method)
9B	apportionment standard divisor standard quota minimum quota modified divisor modified quota Alabama paradox population paradox new states paradox quota criterion	**Know** the history of apportionment mathematics. **Apply** four apportionment methods: Hamilton's method Jefferson's method Webster's method Hill-Huntington method **Know** the potential flaws of each method. **Understand** the significance of the Balinsky and Young theorem.

SUGGESTED READINGS

A Mathematician Reads the Newspaper, J. A. Paulos, Basic Books, New York, 1995.

All the Math That's Fit to Print: Articles from the Manchester Guardian, Keith Devlin, Mathematical Association of America, Washington, DC, 1994.

Arithmetic of Life and Death, The, G. Shaffner, Ballantine Books, New York, 1999.

Bell Curve Debate, The, R. Jacoby and N. Glauberman, eds., Times Books, New York, 1995.

Beyond the Limits: Confronting Global Collapse, Envisioning a Sustainable Future, D. H. Meadows, D. L. Meadows, and J. Randers, Chelsea Green Publishing Co., White River Junction, VT, 1992.

Billions and Billions, C. Sagan, Random House, New York, 1997.

Broken Dice and Other Mathematical Tales of Chance, The, I. Ekeland, University of Chicago Press, Chicago, 1993.

Can You Win?, M. Orkin, W. H. Freeman, New York, 1991.

Cartoon Guide to Statistics, The, L. Gonick and W. Smith, HarperCollins, New York, 1993.

Challenging Lateral Thinking Puzzles, P. Sloane and D. MacHale, Sterling Publishing Co., New York, 1993.

"Choice of Voting Systems, The," R. Niemi and W. Riker, *Scientific American*, Vol. 234, June 1976, pp. 21–27.

Concise Introduction to Logic, P. J. Hurley, Wadsworth Publishing, Belmont, CA, 1988.

Damned Lies and Statistics, J. Best, University of California Press, Berkeley, 2001.

Divine Proportion, The: A Study in Mathematical Beauty, H. E. Huntley, Dover Books, New York, 1970.

Envisioning Information, E. Tufte, Connecticut Graphics Press, Cheshire, CT, 1983.

"Forgotten Fundamentals of the Energy Crisis," A. A. Bartlett, *American Journal of Physics*, September 1978.

Fractals: Form, Chance, and Dimension, Benoit Mandelbrot, W. H. Freeman, San Francisco, 1977.

Gödel, Escher, Bach: An Eternal Golden Braid, D. R. Hofstadter, Basic Books, New York, 1979.

Golden Mean, The, C. F. Linn, Doubleday, New York, 1974.

Goodbye, Descartes: The End of Logic and the Search for a New Cosmology of the Mind, K. Devlin, Wiley, New York, 1997.

Graphs, Models, and Finite Mathematics, J. Malkevitch and W. Meyer, Prentice Hall, Englewood Cliffs, NJ, 1974.

Great Lateral Thinking Puzzles, P. Sloane and D. MacHale, Sterling Publishing Co., New York, 1994.

How Many People Can the Earth Support, J. E. Cohen, W. W. Norton, New York, 1995.

How to Lie with Statistics, D. Huff, W. W. Norton, New York, 1952.

How to Solve It, G. Polya, Princeton University Press, Princeton, NJ, 1945/1985.

How to Tell the Liars from the Statisticians, R. Hooke, Marcel Dekker, New York, 1983.

Innumeracy, J. A. Paulos, Hill and Wang, New York, 1988.

Introduction to Logic, I. Copi, Macmillan, New York, 1968.

Introduction to Management Science, B. Taylor, W. C. Brown, Dubuque, IA, 1986.

"Is Using a Car Phone Like Driving Drunk?" *Chance*, Vol. 10, No. 2, 1997.

Lady Luck: The Theory of Probability, W. Weaver, Anchor Books, Garden City, NJ, 1963.

Logic and Knowledge: Essays, 1901–1950, Bertrand Russell, edited by Robert Charles Marsh, G. Allen & Unwin, 1956.

Mathematical Puzzles of Sam Loyd, S. Loyd, Dover Books, New York, 1959.

Mathematical Tourist, The, I. Peterson, W. H. Freeman, New York, 1988.

Mathematics, David Bergamai, ed., Time Life Books, Alexandria, VA, 1980.

Mathematics: Problem Solving Through Recreational Mathematics, B. Averbach and O. Chein, W. H. Freeman, New York, 1980.

Mathematics, the Science of Patterns: The Search for Order in Life, Mind, and the Universe, K. Devlin, Scientific American Library, New York, 1994.

Measure for Measure: A Musical History of Science, T. Levenson, Simon and Schuster, New York, 1994.

Mismeasure of Man, The, S. J. Gould, W. W. Norton, New York, 1981.

Mismeasure of Woman, The, C. Tavris, Touchstone Books, New York, 1993.

Misused Statistics: Straight Talk for Twisted Numbers, A. Jaffe and H. Spirer, Marcel Dekker, New York, 1987.

More Mathematical Puzzles of Sam Loyd, S. Loyd, Dover Books, New York, 1960.

Nature's Numbers, I. Stewart, Basic Books, New York, 1995.

Number: The Language of Science, T. Dantzig, Macmillan, New York, 1930.

Once Upon a Number, J. Paulos, Basic Books, New York, 1998.

Pi in the Sky: Counting, Thinking, and Being, J. D. Barrow, Little Brown, Boston, 1992.

Population Explosion, The, P. R. Ehrlich and A. H. Ehrlich, Simon and Schuster, New York, 1990.

Principles of Operations Research, F. Budnick, Irwin Press, Burr Ridge, IL, 1988.

Probabilities in Everyday Life, J. McGervey, Ivy Books, New York, 1989.

Psychology of Judgment and Decision Making, The, Scott Plous, McGraw Hill, New York, 1993.

Randomness, Deborah Bennett, Harvard University Press, Cambridge, MA, 1998.

Science of Fractal Images, The, H. O. Peitgen and D. Saupe, eds., Springer-Verlag, New York, 1988.

Scientific American Book of Mathematical Puzzles and Diversions, The, Martin Gardner, Simon and Schuster, New York, 1959.

Second Scientific American Book of Mathematical Puzzles and Diversions, The, Martin Gardner, Simon and Schuster, New York, 1961.

Seeing Through Statistics, J. M. Utts, Duxbury Press, Belmont, CA, 1999.

Social Choice and Individual Values, K. Arrow, Wiley, New York, 1963.

Statistical Methods for Psychology, D. C. Howell, Duxbury Press, Belmont, CA, 1992.

Statistical Reasoning for Everyday Life, J. Bennett, W. Briggs, and M. Triola, Addison Wesley Longman, Reading, 2001.

Statistical Reasoning in Law and Public Policy, J. L. Gastwirth, Academic Press, New York, 1988.

Statistics: Concepts and Controversies, D. Moore, W. H. Freeman, New York, 1991.

Statistics by Example, T. Sincich, Dellen Publishing Co., San Francisco, 1990.

Statistics for Business and Economics, H. Kohler, HarperCollins, New York, 1994.

Statistics with a Sense of Humor, F. Pyrczak, Fred Pyrczak Publisher, 1989.

"Symmetry and Complexity of Elections, The," Donald Saari, *Complexity*, Vol. 2, January/February 1997.

Tainted Truth: The Manipulation of Fact in America, C. Crossen, Simon and Schuster, New York, 1994.

Theory of Voting, R. Farquharson, Yale University Press, New Haven, 1969.

Topics in the Theory of Voting, P. Straffin, UMAP Expository Monograph, Birkhauser, Boston, 1980.

200% of Nothing, A. K. Dewdney, Wiley, New York, 1993.

Universe and the Teacup, The: The Mathematics of Truth and Beauty, K. C. Cole, Harcourt Brace, Orlando, FL, 1998.

Visual Display of Quantitative Information, The, E. Tufte, Connecticut Graphics Press, Cheshire, CT, 1983.

Vital Signs 1998, compiled by the Worldwatch Institute, W. W. Norton, New York, 1998 (updated annually).

What the Odds Are, L. Krantz, Harper Perennial, New York, 1992.

CREDITS

TEXT AND FIGURE CREDITS

Prologue
pp. 1, 5, 7: Excerpts from *Everybody Counts* reprinted with permission from EVERYBODY COUNTS: A REPORT TO THE NATION ON THE FUTURE OF MATHEMATICS EDUCATION. Copyright 1989 by the National Academy of Sciences. Courtesy of the National Academy Press, Washington, DC, p. 76. *p. 2:* Excerpt from "On the Pulse of the Morning" from ON THE PULSE OF THE MORNING by Maya Angelou. Copyright © 1993 by Maya Angelou. Reprinted by permission of Random House, Inc.

Chapter 1
p. 32: © Joel Pett, *Lexington Herald-Leader*, Universal Press Syndicate. All rights reserved. *p. 37:* Drawing of the planets from THE COSMIC PERSPECTIVE by Jeff Bennett, Megan Donahue, Nicholas Schneider, and Mark Voit, Addison-Wesley, 1999.

Chapter 2
p. 103: Sidney Harris cartoon—No reproduction other than for personal enjoyment without permission.

Chapter 3
Fig. 3.5: Copyright © 1995 by THE NEW YORK TIMES. Reprinted with permission.

Chapter 5
p. 241: Sidney Harris cartoon—No reproduction other than for personal enjoyment without permission. *Fig. 5.12:* Copyright © 1995 by THE NEW YORK TIMES. Reprinted by permission. *Fig. 5.13:* From THE WALL STREET JOURNAL ALMANAC by Editors of The Wall Street Journal. Copyright © 1997 by Dow Jones & Company, Inc. Reprinted by permission of Ballantine Books, a Division of Random House, Inc. *Fig. 5.19:* Copyright © 1995 by THE NEW YORK TIMES. Reprinted by permission. *Fig. 5.20:* Copyright © 2000 by THE NEW YORK TIMES. Reprinted by permission. *Fig. 5.24:* Copyright © 1999 by THE NEW YORK TIMES. Reprinted by permission. *Fig. 5.25:* Copyright 1999, USA TODAY. Reprinted with permission. *Fig. 5.27:* From TIME ALMANAC 1997, p. 886. Reprinted with permission of Time-Life Books. *Fig. 5.30:* Copyright © 2000 by THE NEW YORK TIMES. Reprinted by permission. *Fig. 5.43:* NASA.

Chapter 8
Fig. 8.6: Scala/Art Resource, NY. *Fig. 8.7:* The Granger Collection, New York. *Fig. 8.8:* From "Perspective, Jan Vredeman de Vries." Reprinted by permission of Dover Publications, Inc., NY. *Fig. 8.9:* © CORBIS/Bettmann. *Fig. 8.10:* M. C. Escher's "Belvedere" © 2001 Cordon Art B.V.-Baarn-Holland. All rights reserved. *Fig. 8.11:* The Granger Collection, New York. *Fig. 8.12:* © John R. Jones; Papilio/CORBIS. *Fig. 8.19:* The Granger Collection, New York. *Fig. 8.20:* Tate Gallery, London/Art Resource, NY. *Fig. 8.21:* Courtesy of Anne Burns. *Fig. 8.24:* © James Egleson, from *Scientific American*, April, 1961. *Fig. 8.27:* Courtesy of Paul J. Steinhardt, Princeton University. *Fig. 8.43a:* Art Resource/ NY. *Fig. 8.43b:* From Huntley, THE DIVINE PROPORTION, © 1970 Dover Publications. *Fig. 8.44:* Scala/Art Resource, NY. *Fig. 8.45:* Dallas Museum of Fine Art, Foundation for the Arts Collection, Gift of the James H. and Lillian Clark Foundation. *Fig. 8.46c:* © Grant Heilman from Grant Heilman Photography, Inc. *Fig. 8.48a:* From Huntley, THE DIVINE PROPORTION, © 1970 Dover Publications. *Fig. 8.48b:* © Runk/Schoenberger from Grant Heilman Photography, Inc.

PHOTO CREDITS

Prologue
p. 1: AP/Wide World Photos.

Chapter 1
p. 13: © Ken Fisher/Stone; *p. 14:* © Hugh Sitton/Stone; *p. 20:* Jeffrey Bennett; *p. 21:* © Bettmann/CORBIS; *p. 23:* © PhotoDisc; *p. 40:* AP/Wide World Photos; *p. 48:* © Vincent Oliver/Stone; *p. 63:* © PhotoDisc; *p. 65:* © PhotoDisc.

Chapter 2

p. 69: © Michael Simpson/FPG International; *p. 77:* © Ron Chapple/FPG International; *p. 79:* © PhotoDisc; *p. 80:* © PhotoDisc; *p. 85 (left and right):* © PhotoDisc; *p. 86:* AP/Wide World Photos; *p. 89:* © Jerry Cooke/CORBIS; *p. 92:* © PhotoDisc; *p. 99:* AP/Wide World Photos; *p. 111:* © Patrick Field; Eye Ubiquitous/CORBIS.

Chapter 3

p. 117: © Stuart Dee/The Image Bank; *p. 133:* © Herbert Hartmann/The Image Bank; *p. 135:* © Vic Bider/PhotoEdit; *p. 138:* © PhotoDisc; *p. 158:* NASA; *p. 157:* Jet Propulsion Lab/NASA; *p. 160* © PhotoDisc; *p. 161:* AP/Wide World Photos; *p. 162:* AP/Wide World Photos; *p. 168:* © Mark Burnett/Stock Boston, Inc./PictureQuest.

Chapter 4

p. 177: AP/Wide World Photos; *p. 192:* © PhotoDisc; *p. 203:* © PhotoDisc; *p. 204:* AP/Wide World Photos; *p. 213:* © Ken Fisher/Stone; *p. 221* © Mark Richards/PhotoEdit/PictureQuest.

Chapter 5

p. 223: © Steve Niedorf/The Image Bank; *p. 225:* © Spencer Grant/PhotoEdit; *p. 236:* © PhotoDisc; *p. 238:* © PhotoDisc; *p. 240:* © PhotoEdit; *p. 242:* © James P. Blair/CORBIS; *p. 246:* Photo by Michael Okoniewski, Copyright © 1994; *p. 248:* © PhotoDisc; *p. 276:* © Michael Newman/PhotoEdit; *p. 286:* AP/Wide World Photos; *p. 289:* © PhotoDisc; *292:* AP/Wide World Photos; *p. 298:* AP/Wide World Photos.

Chapter 6

p. 303: © Steve Niedorf/The Image Bank; *p. 306:* © PhotoDisc; *p. 308:* AP/Wide World Photos; *p. 313:* © PhotoDisc; *p. 318:* AP/Wide World Photos; *p. 330:* © Bettmann/CORBIS; *p. 331:* © PhotoDisc; *p. 343:* © PhotoDisc; *p. 344:* AP/Wide World Photos; *p. 347:* © Louis B. Wallach Inc./The Image Bank; *p. 348:* AP/Wide World Photos.

Chapter 7

p. 351: AP/Wide World Photos; *p. 355:* Courtesy of Professor Albert Allen Bartlett; *p. 365:* AP/Wide World Photos; *p. 369:* © PhotoDisc; *p. 370:* © PhotoDisc; *p. 374:* © KerenSu/CORBIS; *p. 381:* © American Museum of Natural History; *p. 382:* © PhotoDisc; *p. 383:* AP/Wide World Photos.

Chapter 8

p. 387: © Shaun Egan/Stone.

Chapter 9

p. 419: © PhotoDisc; *p. 420:* AP/Wide World Photos; *p. 421:* © PhotoDisc; *p. 423:* © Bettmann/CORBIS; *p. 434:* © Joseph Sohm; Visions of American/CORBIS; *p. 435:* (*top*) AP/Wide World Photos, (*bottom*) © Bettmann/CORBIS; *p. 438:* © Lee Snider/CORBIS: *p. 442:* © PhotoDisc; *p. 445:* © Lee Snider/CORBIS.

ANSWERS

CHAPTER 1

Unit 1A

1. *Premise:* Following President Reagan's defense buildup, the Soviet Union began the process of democratization that ultimately led to its breakup. *Conclusion:* Reagan is responsible for the changes that led to the demise of the Soviet Union. False cause.
2. *Premise:* The Golden Rule is basic to every system of ethics in every culture. *Conclusion:* The Golden Rule is a sound ethical principle. Circular reasoning or diversion.
3. *Premise:* The mother has used drugs in the past. *Conclusion:* The decision in the custody case depends on the mother's present drug use. Limited choice.
4. *Premise:* Everyone I know is voting for the Governor. *Conclusion:* You should vote for the Governor. Appeal to popularity.
5. *Premise:* Mom smoked when she was my age. *Conclusion:* I won't heed her advice to stop smoking. Personal attack.
6. *Premise:* Most of the great mathematicians in history have been men. *Conclusion:* Men must be better at mathematics than women. Hasty generalization.
7. *Premise:* No one has ever proven that telepathy doesn't exist. *Conclusion:* I believe in telepathy. Appeal to ignorance.
8. *Premise:* My two best teachers were both women. *Conclusion:* Women make better teachers than men. Hasty generalization.
9. *Premise:* Senator Smith is one of the biggest recipients of campaign contributions from the National Rifle Association. *Conclusion:* Senator Smith's bill cannot help the cause of gun control. Personal attack.
10. *Premise:* Violent crime by youth has risen in virtual lockstep with increased violence on television. *Conclusion:* Television violence leads to real violence. False cause or limited choice.
11. *Premises:* The percentage of the population over 18 who smoke has decreased from 40% to about 20%. The percentage of overweight people has increased from 25% to 35%. *Conclusion:* Quitting smoking leads to overeating. False cause or hasty generalization.
12. *Premises:* Boys score higher than girls on standardized mathematics tests. Girls score higher than boys on verbal tests. *Conclusions:* Boys have a greater aptitude for mathematics than girls. Girls are better at verbal skills. Hasty generalization or limited choice.
13. *Premise:* My little boy loves dolls and my little girl loves trucks. *Conclusion:* There's no truth to the claim that boys are more interested in mechanical toys while girls prefer maternal toys. Appeal to ignorance or hasty generalization.
14. *Premise:* Goldwater's policies will lead to the annihilation of children and you love children. *Conclusion:* Vote for Johnson. Appeal to emotion.
15. *Premise:* Bush favors repealing the estate tax, which falls most heavily on the rich. *Conclusion:* Bush favors the rich; vote for Gore. Straw man.
16. *Premise:* Gore claims his regulatory changes will reduce the threat of global warming. *Conclusion:* Gore believes that government is the solution to all our problems; vote for Bush. Straw man.

Unit 1B

1. Proposition. 2. Not a proposition. 3. Proposition.
4. Proposition. 5. Not a proposition. 6. Proposition.
7. London is not the capital of England. Original is true. Negation is false.
8. The Gettysburg Address was not given by George Washington. Original is false. Negation is true.
9. Caesar was not a Roman. Original is true. Negation is false.
10. Mark Twain did not write *Tom Sawyer*. Original is true. Negation is false.
11. Some cowboys do not wear boots. Original is false. Negation is true.
12. Some people are above the law. Original is true. Negation is false.
13. No dogs are obedient. Original is true. Negation is false.
14. All folks know how to listen. Original is true. Negation is false.
15. The city council supported the police chief.
16. The Senator supports demonstrations.
17. Congress upheld the veto (opposed the bill).
18. The Senate supported the bill.
19. The anti-discrimination policy is supported by the Constitution.
20. Demonstrations are now allowed.
21. Quebec is the capital of Canada. Moscow is the capital of Russia. The first proposition is false, and the conjunction is false.
22. Emily Dickinson was an American poet. Boris Yeltsin was a Russian president. Both propositions are true, and the conjunction is true.
23. Ben is married. Ben is a bachelor. Both propositions cannot be true, and the conjunction is false.
24. $3 \times 7 = 21$. $6 + 3 = 9$. Both propositions are true, and the conjunction is true.
25. Grass is green. Skies are blue. Both propositions are true, and the conjunction is true.
26. Some people are nice. Some people are mean. Both propositions are true, and the conjunction is true.
27. Exclusive. 28. Exclusive, whichever comes first.
29. Probably exclusive. 30. Exclusive.
31. Exclusive. 32. Inclusive.

33. Seattle is the capital of the United States. Paris is the capital of France. The first proposition is false, and the disjunction is true.
34. Julia Roberts was an 18th-century monk. Mark Twain was a writer. The first proposition is false, and the disjunction is true.
35. Bill Clinton was a U.S. president or Martin Luther King was a U.S. president. The second proposition is false, and the disjunction is true.
36. Ben is married. Ben is a bachelor. One proposition is true, and the disjunction is true.
37. $8 \times 4 = 32$. $7 \times 3 = 22$. The second proposition is false, and the disjunction is true.
38. $3 \times 7 = 21$. $6 + 3 = 9$. Both propositions are true, and the disjunction is true.
39. Search on earthquakes and California.
40. Search on (earthquakes or volcanic eruptions) and Japan.
41. Search on Martin Luther King and assassination.
42. Search on United States and Argentina and cold war.
43. *Hypothesis:* San Francisco is in California. *Conclusion:* San Francisco is in the Pacific time zone. Both propositions are true, and the implication is true.
44. *Hypothesis:* London is the capital of England. *Conclusion:* Paris is the capital of China. The consequent is false, and the implication is false.
45. *Hypothesis:* Paris is the capital of China. *Conclusion:* London is the capital of England. The antecedent is false, and the implication is true.
46. *Hypothesis:* The Pope is unmarried. *Conclusion:* The Pope is a bachelor. Both propositions are true, and the implication is true.
47. *Hypothesis:* Sparrows can fly. *Conclusion:* Sparrows are birds. Both propositions are true, and the implication is true.
48. *Hypothesis:* Butterflies can fly. *Conclusion:* Butterflies are birds. The consequent is false, and the implication is false.
49. *Hypothesis:* $2 + 2 = 5$. *Conclusion:* Primates are not mammals. Both propositions are false, and the implication is true.
50. *Hypothesis:* $5 \times 7 = 0$. *Conclusion:* Fish can fly. Both propositions are false, and the implication is true.
51. If a person is a resident of Miami, then that person is a resident of the United States. Proposition is true.
52. If she's a member of Congress, then she's a lawyer. Proposition is false.
53. If you are a musician, then you play the saxophone. Proposition is false.
54. If you take a biology course, then you satisfy a science requirement. Proposition could be true.
55. If you have prostate cancer, then you are a male. Proposition is true.
56. If you live in Santa Fe, then you live in New Mexico. Proposition is true.
57. *Converse:* If Marco lives in the United States, then he lives in Chicago; false. *Inverse:* If Marco does not live in Chicago, then he does not live in the United States; false. *Contrapositive:* If Marco does not live in the United States, then he does not live in Chicago; true.
58. *Converse:* If the patient is sleeping, then the patient is alive; true. *Inverse:* If the patient is not alive, then the patient is not sleeping; true. *Contrapositive:* If the patient is not sleeping, then the patient is not alive; false.
59. *Converse:* If I get wet, then I went swimming; false. *Inverse:* If I don't go swimming, then I won't get wet; false. *Contrapositive:* If I don't get wet, then I didn't go swimming; true.
60. *Converse:* If it can fly, then it's a penguin; false. *Inverse:* If it's not a penguin, then it cannot fly; false. *Contrapositive:* If it cannot fly, then it's not a penguin; false.
61. *Converse:* If there is gas in the tank, then the engine is running; false. *Inverse:* If the engine is not running, then there is no gas in the tank; false. *Contrapositive:* If there is no gas in the tank, then the engine is not running; true.
62. *Converse:* If you live in Hawaii, then you live near the ocean; true. *Inverse:* If you don't live near the ocean, then you don't live in Hawaii; true. *Contrapositive:* If you don't live in Hawaii, then you don't live near the ocean; false.
63. If a young man has not wept, then he is a savage. If an old man will not laugh, then he is a fool.
64. If you laugh, then the world will laugh with you.
65. If a person has no vices, then that person has few virtues.
66. If a man reads nothing at all, then he is better educated than the man who reads only newspapers.
67. If Sue lives in Cleveland, then she lives in Ohio.
68. If a number is even, then it is divisible by 2.
69. If Sue lives in Ohio, then she lives in Cleveland.
70. If an animal is a fish, then it has hair.
71. **a.** Yes. **b.** Yes. **c.** Yes. **d.** Yes. **e.** No. **f.** Yes.
72. (1) If the payer did not know you had remarried, then alimony you receive after you remarry is taxable. (2) If the payer did know you had remarried, then alimony you receive after you remarry is not taxable. (3) If you pay alimony, then it is never deductible.
73. If a computer exhibits human-level intelligence, then it must have the requisite processing power.
74.

p	q	p and q	not (p and q)	not p or not q
T	T	T	F	F
T	F	F	T	T
F	T	F	T	T
F	F	F	T	T

The propositions are equivalent.

75.

p	q	p or q	not (p or q)	not p and not q
T	T	T	F	F
T	F	T	F	F
F	T	T	F	F
F	F	F	T	T

The propositions are equivalent.

76.

p	q	p and q	not (p and q)	not p and not q
T	T	T	F	F
T	F	F	T	F
F	T	F	T	F
F	F	F	T	T

The propositions are not equivalent.

77.

p	q	p or q	not (p or q)	not p or not q
T	T	T	F	F
T	F	T	F	T
F	T	T	F	T
F	F	F	T	T

The propositions are not equivalent.

78.

p	q	r	p and q	(p and q) or r	p or r	p or q	(p or r) and (p or q)
T	T	T	T	T	T	T	T
T	T	F	T	T	T	T	T
T	F	T	F	T	T	T	T
T	F	F	F	F	T	T	T
F	T	T	F	T	T	T	T
F	T	F	F	F	F	T	F
F	F	T	F	T	T	F	F
F	F	F	F	F	F	F	F

The propositions are not equivalent.

79.

p	q	r	p or q	(p or q) and r	(p and r)	(q and r)	(p and r) or (q and r)
T	T	T	T	T	T	T	T
T	T	F	T	F	F	F	F
T	F	T	T	T	T	F	T
T	F	F	T	F	F	F	F
F	T	T	T	T	F	T	T
F	T	F	T	F	F	F	F
F	F	T	F	F	F	F	F
F	F	F	F	F	F	F	F

The propositions are equivalent.

Unit 1C

1. **a.** Rational. **b.** Rational. **c.** Natural.
 d. Integers. **e.** Rational. **f.** Rational.
2. **a.** Rational. **b.** Real. **c.** Natural. **d.** Integers.
 e. Rational. **f.** Rational.
3. **a.** {January, February, ..., November, December}.
 b. {1, 2, 3, ..., 98, 99, 100}.
 c. {Washington, Adams, Jefferson}.
 d. {Nebraska, Nevada, New Hampshire, New Jersey, New Mexico, New York, North Carolina, North Dakota}.
4. **a.** {California, Colorado, Connecticut}.
 b. {n, o, p, ..., x, y, z}.
 c. {100, 101, ..., 199, 200}.
 d. {Bush, Clinton, Bush}.

5. (Venn diagram: overlapping circles "doctors" and "women")

6. (Venn diagram: "poetry" inside "literature")

7. (Venn diagram: disjoint circles "negative integers" and "natural numbers")

8. (Venn diagram: "positive integers" inside "whole numbers")

9. (Venn diagram: "verbs" inside "words")

10. (Venn diagram: "pianists" inside "musicians")

11. (Venn diagram: "painters" inside "artists")

12. (Venn diagram: "genetics" inside "science")

13. *Subject:* bachelors. *Predicate:* men.

(Venn diagram: "bachelors" inside "men")

14. *Subject:* pilots. *Predicate:* women.

15. *Subject:* U.S. presidents. *Predicate:* men.

16. Some mammals are swimming animals. *Subject:* mammals. *Predicate:* swimming animals.

17. No fish are flying animals. *Subject:* fish. *Predicate:* flying animals.

18. No dogs are swimming animals. *Subject:* dogs. *Predicate:* swimming animals.

19. All nurses are people who know CPR. *Subject:* nurses. *Predicate:* people who know CPR.

20. All lawyers are highly educated people. *Subject:* lawyers. *Predicate:* highly educated people.

21. *Subject:* U.S. Senators. *Predicate:* Republicans. *Negation:* No U.S. Senators are Republicans. Original statement is true. Negation is false.
22. *Subject:* snakes. *Predicate:* mammals. *Negation:* Some snakes are not mammals. Original statement is false. Negation is true.
23. *Subject:* doctors. *Predicate:* left-handed people. *Negation:* Some doctors are not left-handed. Original statement is false. Negation is true.
24. *Subject:* countries. *Predicate:* countries with coastlines. *Negation:* All countries are countries with coastlines. Original statement is true. Negation is false.
25. *Subject:* people. *Predicate:* sociable people. *Negation:* All people are sociable. Original statement is true. Negation is false.
26. *Subject:* days. *Predicate:* rainy days. *Negation:* No days are rainy days. Original statement is true. Negation is false.
27. The diagram assumes that no pilots are dentists.

28.

29. The diagram assumes that no songs are poetry and that some poems and songs are unpublished.

30.

[Venn diagram: "oceans" circle overlapping "salt water" circle; separate "fresh water" circle]

31.

[Diagram: large "clothes" oval containing two separate circles "hats" and "shirts"]

32.

[Nested ovals: outer "places in North America", containing "states of U.S.", containing "coastline on Pacific"]

33. a. 28. b. 15. c. 13. d. 13.

34. a. 7. b. 6. c. 6. d. 9.

35.

[Venn diagram: soccer 12, overlap 2, softball 6; outside 4]

36.

[Venn diagram: feature 8, overlap 7, thumbs up 9; outside 1]

37.

[Venn diagram: drug 3, overlap 12, improvement 6; outside 9]

38.

[Venn diagram: Toyota 30, overlap 30, SUV 10; outside 30]

39.

[Venn diagram: math 9, overlap 7, biology 16; outside 11]

16 students are taking only biology; 11 students are taking neither biology nor math.

40.

[Venn diagram: road 9, overlap 3, mountain 5; outside 0]

41.

[Venn diagram: women 110, overlap 32, biology 21; outside 87]

3 students own both types of bike.

42.

[Venn diagram: tomato per day 191, overlap 9, oral cancer 16; outside 164]

43.

[Venn diagram: coffee 4806, overlap 91, gall stones 385; outside 14,068]

44. a.

[Three-circle Venn diagram: NYT 24, WSJ 27, WP 26; NYT∩WSJ 14, NYT∩WP 16, WSJ∩WP 13, center 8; outside 15]

b. 21 (includes those who read *NYT* also). **c.** 8.
d. 128. **e.** 38.

45. a.

[Three-circle Venn diagram: AC 45, 4WD 8, CP 12; AC∩4WD 21, AC∩CP 16, 4WD∩CP 8, center 10; outside 3]

b. 31 (includes those who chose CP also). **c.** 61.
d. 120. **e.** 45.

46.

[Venn diagram: rectangle containing overlapping circles "protein", "fat", "carbohydrates"; cheese, eggs in fat; meats, nuts in protein∩fat∩carbohydrates area; rice in protein∩carbohydrates; potatoes, apples in carbohydrates; lettuce, coffee, water outside]

47.

[Diagram: "animals" contains "house pets" which contains "dogs", "cats", "canaries"]

48.

[Diagram: "things that fly" contains "birds" (containing "eagles"), "airplanes" (containing "jets"), and "hang gliders"]

49.

[Diagram: "food" contains "dessert ice cream", "snack", "fruit" (containing "citrus fruit" and "strawberries")]

50.

[Diagram: "people" contains "men" (containing "NFL" containing "quarterbacks"), "lawyers", "surgeons", "women"]

51.

[Diagram with overlapping circles: "mammals" (containing "hairy animals"), "swimmers", "fish", "land walkers"]

There could not be hairy fish. There could be hairy animals that swim. There could be walking mammals. There could be hairy animals that walk on land.

52.

[Diagram: "contains protein" contains "meat", "dairy products", and overlaps with "plants" which contains "beans"]

There could not be beans that are dairy products. There could not be meat that is a dairy product. No dairy products are plants. There could be plants with protein.

53.

[Diagram: "conservatives" containing "Republicans"; overlapping with "Democrats", "liberals", "Greens"]

There could be conservative Democrats. There could be liberal Green Party members. There could not be liberal Republicans.

54. a.

women, Republicans, Democrats (Venn diagram with women and Democrats overlapping; Republicans separate)

 b. Yes. c. Yes.
55. No, it cannot be done with two circles.
56. a. 16 different sets of options. b. Not possible.
 c. 32 different sets of options.
 d. 2^n different sets of options.

Unit 1D
1. Were there more than 350 people on each jet or both jets combined?
2. In which year were there 2.9 million incidents?
3. Is $83 billion the gross domestic product or 6% of the gross domestic product?
4. The word "regularly" does not describe the seriousness of the charge; it could mean once every day or once every month.
5. a. The state constitution is less strict than the U.S. Constitution.
 b. A *yes* vote would tighten (strengthen) laws against obscenity.
 c. False.
6. a. No, term limits in the state are not mandatory.
 b. No, declarations would be voluntary.
 c. No, there would be no penalty.
7. Plan A: If you go, it costs $1600; if you cancel, it costs $400. Plan B: If you go, it costs $2900; if you cancel, it costs nothing. If you go, Plan B costs $1300 more. If you cancel, Plan A costs $400 more.
8. You need estimates of the number of minutes you spend calling during the weekday daytime and during the weekend. If the number of weekday daytime minutes is less than one-third the number of weekend minutes, then the new plan is better.
9. a. No. b. No. c. Yes. d. No. e. Yes.
10. a. No. b. Yes. c. No.
 d. No (note that 110% of last year's tax < withholdings < 90% of next year's tax).
 e. No (note that 90% of next year's tax < withholdings < last year's tax).
11. b. Jack does not appear to qualify because he works 150 hours over a period of less than five weeks; the regulation may be stated incorrectly.
 c. Jill does appear to qualify because she works 150 hours over a period of more than five weeks; the regulation may be stated incorrectly.
 d. No, Ivan must also satisfy either (c) or (d) under provision (1).
 e. Yes, Joan satisfies three of the conditions under provision (1).
12. a. Yes. b. Yes. c. Yes.
13. a. They could be consistent because Alice does not specify the time period for her 253 cases.
 b. They could be consistent if Alice prosecuted 127 cases more than five years ago, most of which resulted in conviction, and 126 cases in the last five years, only four of which resulted in conviction (as Zack claimed). This seems an unlikely explanation.
14. The official is affirming the conclusion, which is a fallacy. Premise: If athletes use drugs, then their performance will improve. Premise: Athletes' performances have improved. Conclusion: Athletes are using drugs. There could be other explanations.
15. If the estimate of the computer's resale value is accurate, the purchase option costs $1005, while the lease option costs $1050. All facts are "hard" except for the resale value.
16. a. Total bill = service charge + (Block 1 rate × 100% of AWC) + (Block 2 rate × 150% of AWC).
 b. Total bill = service charge + (Block 1 rate × 100% of AWC) + (Block 2 rate × 250% of AWC) + (Block 3 rate × 50% of AWC).
 c. Total bill = service charge + (Block 2 rate × 350% of AWC) + (Block 3 rate × 100% of AWC).
 d. Irrigation-only account.
17. The package is the better choice.
18. The current policy costs $5035 for nine months. The upgrade costs $2475 for nine months.
19. If you go, Plan 2 costs $250 more. If you cancel, Plan 1 costs $75 more.
20. The one-day trip costs $750. The weekend trip costs $655 (two nights and two days).
21. If you plan to fly 10 times (or a multiple of 10 times), then Plan A ($3150) is better than Plan B ($3250).
22. One interpretation: The poet was 20 years old when he wrote the poem and will live to age 70.

CHAPTER 2

Unit 2A
1. a. 2/3. b. 11/6. c. 8/3. d. 5/6. e. 19/20.
 f. 35/52. g. 2/17. h. 8/15.

2. a. 5/2. b. 5/4. c. 11/12. d. 13/12. e. 1/11.
 f. 1/16. g. 9/25. h. 1/24.
3. a. 3/10. b. 31/250. c. 39/50. d. 1/200.
 e. 46/25. f. 3009/1000. g. 1/10,000.
 h. 1001/00,000.
4. a. 7/10. b. 7/5. c. 9/20. d. 247/250.
 e. 401/1000. f. 1001/1000. g. 307/100.
 h. 1/50,000.
5. a. 1.5. b. 0.8. c. 0.167. d. 0.143. e. 0.222.
 f. 0.636. g. 0.967. h. 4.067.
6. a. 0.583. b. 0.386. c. 0.533. d. 0.72.
 e. 0.589. f. 3.0. g. 0.833. h. 5.333.
7. a. Dollars per pound, $/lb.
 b. Kilometers per second, km/s.
 c. Dollars per square foot, $/ft^2.
 d. Cubic feet per second, ft^3/s.
 e. Pounds per square inch, lb/in^2.
8. a. Dollars per square foot, $/ft^2.
 b. Miles per gallon, mi/gal.
 c. Dollars per square yard, $/yd^2.
 d. Grams per cubic centimeter, gm/cm^3.
 e. Foot-pounds, ft-lb.
9. a. 50 m^2; 150 m^3. b. 330 in^2; 3960 in^3.
 c. 25,000,000 ft^3.
10. a. 1000 yd^2; 3000 yd^3. b. 60 ft^2; 210 ft^3. c. 24 in^3.
11. a. 54 ft. b. 648 in. c. 8.875 lb. d. 8712 ft^2.
 e. 768 oz.
12. a. 12,320 yd. b. 15,840 ft. c. 40,320 min.
 d. 2240 acres. e. 0.028 km/s.
13. a. 1 ft^2 = 144 in^2; 1 ft^2/144 in^2; 144 in^2/1 ft^2.
 b. 720 in^2. c. 6000 yd^2; 54,000 ft^2.
14. a. 1 ft^3 = 1728 in^3; 1 ft^3/1728 in^3; 1728 in^3/1 ft^3.
 b. 139,968 in^3. c. 4000 ft^3; 148.15 yd^3.
15. a. 1 British pound. b. 10,753 yen. c. $268.25.
 d. $5.25.
16. a. 1 peso. b. $147.84 Canadian. c. $113.58.
 d. $9.30.
17. $6.06. 18. $9.80. 19. 56 mi/hr. 20. 570 mi/hr.
21. $5000/acre. 22. 300 apples/acre. 23. $20.
24. $3750. 25. $3040. 26. $12,500. 27. 60 mi/hr.
28. $28,080. 29. 2737.5 hr. 30. 2,365,200,000 beats.
31. Wrong; $0.85. 32. Wrong; 15 mi.
33. Wrong; $0.22/lb vs. $0.39/lb. 34. Wrong; 10.7 cans.
35. 2025 ft^3. 36. 2722.5 ft^2. 37. 4320 liters.
38. 250,000 pages.
39. a. 105 years. b. 15 to 22 weeks.
40. 15,603,840,000 ft^3; 0.013 = 1.3%.

Unit 2B
1. a. 10^{11}. b. 10. c. 10. d. 10^7. e. 10^{-6}.
 f. 10^2. g. 1.1×10^6. h. 9.9999×10^6.
2. a. 10^{21}. b. 10^{-2}. c. 10^3. d. 10^{-11}. e. 10^{-10}.
 f. 10^4. g. 100,100,000. h. 9.9999×10^3.
3. a. For example, 5 ft 10 in. = 70 in. b. 8 lb.
 c. 4 liquid pt; 3.44 dry pt.
 d. 322,560 million in^3; 187 million ft^3.
4. a. For example, 150 lb = 2400 oz av = 0.075 ton.
 b. 34.5 mi/hr. c. 21.66 in^3. d. 9.26 ft.
5. a. 1000 times smaller. b. 1000 times smaller.
 c. 1,000,000,000,000 times smaller.
6. a. 100 times smaller. b. 1,000,000,000 times larger.
 c. 1,000,000 times larger.
7. a. 32.8 ft. b. 0.80 km. c. 75.7 L. d. 0.31 in^3.
 e. 111.5 m^2.
8. a. 1.15 yd. b. 68.04 kg. c. 62.14 mi/hr.
 d. 343.35 lb/ft^3. e. 40.23 km/hr.
9. a. 7.2°C. b. 68°F. c. 5.0°F. d. −22.0°F.
 e. 21.1°C.
10. a. 17.6°F. b. −9.4°C. c. 59°F. d. 23.9°C.
 e. −6.7°C.
11. a. −223.15°C. b. −33.15°C. c. 283.15 K.
12. a. 233.15 K. b. 126.85°C. c. 398.15 K.
13. Sensible. 14. Ridiculous. 15. Sensible.
16. Ridiculous. 17. Ridiculous. 18. Sensible.
19. Ridiculous. 20. Ridiculous.
21. a. Metric mile is 6.8% shorter than a USCS mile.
22. 1/16 in. = 1.59 mm; the tools are not interchangeable.
23. 31,996 ft; 6.06 mi; 9.75 km.
24. 621,116 mg, 1.37 lb; 106,026 mg, 0.23 lb.
25. 6 karats. 26. 8.75 g. 27. 0.5 g.
28. No, it would be 125% pure!
29. a. 4.5 billion joules. b. 1.7 kilowatts.
 c. 375 L, 99 gal.
30. a. 3.5 billion joules. b. 1.3 kilowatts. c. 291 L, 77 gal.
31. 930 watts. Yes. 32. 775 watts. Yes.
33. $245.
34. $0.027 per day; $9.86 per year.
35. 750 kilowatt-hr; $60 saved.
36. a. 121 watts.
 b. 3.8 billion joules per year, about 1% of total energy used.
37. 720 million kilowatt-hr per month; 1.6 million kg of coal;
 720,000 homes.
38. 238 kilowatt-hr per month; 15 kg of uranium; 240,000
 homes.
39. 4.3 million joules; 50 watts.
40. 20 m^2.

Unit 2C
1. a. (Truck, car) = (11, 1), (9, 4), (7, 7), (5, 10), (3, 13),
 (1, 16).
 b. (Truck, car) = (13, 1), (11, 4), (9, 7), (7, 10), (5, 13),
 (3, 16), (1, 19).

2. **a.** (4-axle, 3-axle, 2-axle) = (0, 0, 5), (1, 0, 3), (2, 0, 1), (0, 2, 2), (1, 2, 0).
 b. (4-axle, 3-axle, 2-axle) = (0, 0, 10), (0, 2, 7), (0, 4, 4), (0, 6, 1), (1, 0, 8), (1, 2, 5), (1, 4, 2), (2, 0, 6), (2, 2, 3), (2, 4, 0), (3, 0, 4), (3, 2, 1), (4, 0, 2), (5, 0, 0).
3. 0.03 second. **4.** Hack wins. **5.** 100 km.
6. 50 km. **10.** 33 million men. **11.** 75 seconds.
12. Reuben spoke on January 1, and he was born on December 31.
13. 4 apples. **14.** That man is my son.
15. The blind fiddler is a woman. **16.** $200 gain.
17. Select a fruit from the box labeled *Apples and Oranges*.
18. Select one ball from the first barrel, two balls from the second barrel, and so on, up to ten balls from the tenth barrel.
19. 7 crossings.
21. The flagpoles are next to each other (no distance between them).
22. Start by dividing the coins into three groups of four coins.
23. 3 socks.
24. Alma visited at 8:00 p.m., Bess at 9:00 a.m., Cleo at 10:00 p.m., and Dina at 11:00 a.m.
25. Prisoner 1 does not see red hats on both Prisoners 2 and 3; if he did, he would know that he had a white hat (and would be freed). Therefore, the hats on Prisoners 2 and 3 must be WW, WR, or RW, and Prisoner 2 knows this! So if Prisoner 2 saw a red hat on Prisoner 3, Prisoner 2 would know that he had a white hat (and would be freed). Therefore, Prisoner 3 knows that he must have a white hat, and he is freed.

CHAPTER 3

Unit 3A

1. **a.** 1/4 = 0.25 = 25%. **b.** 0.45 = 45% = 9/20.
 c. 0.33333... = 33.333...% = 1/3.
 d. 23% = 0.23 = 23/100. **e.** 1/8 = 0.125 = 12.5%.
 f. 1.34 = 134% = 67/50. **g.** 0.65 = 65% = 13/20.
 h. 99% = 0.99 = 99/100.
2. **a.** 120% = 1.2 = 6/5. **b.** 0.652 = 65.2% = 163/250.
 c. 2.34 = 234% = 117/50. **d.** 0.98 = 98% = 49/50.
 e. 45% = 0.45 = 9/20. **f.** 1.01 = 101% = 101/100.
 g. 3/8 = 0.375 = 37.5%. **h.** 125% = 1.25 = 5/4.
3. **a.** 5; 0.2; 500%. **b.** 6; 1/6 = 0.167; 600%.
 c. 11; 1/11 = 0.091; 1100%. **d.** 0.082; 12.2; 8.2%.
 e. 0.156; 6.4; 15.6%. **f.** 0.268; 3.73; 26.8%.
4. **a.** 0.455; 2.2; 45.5%. **b.** 3.73; 0.268; 373%.
 c. 1.24; 0.809; 124%. **d.** 1.13; 0.888; 113%.
 e. 0.176; 5.70; 17.6%.
5. **a.** 22.1%. **b.** 86.7%. **c.** 41.0%.
6. **a.** 16.1%. **b.** 18.7%. **c.** 58.3%.
7. Absolute change = −717 newspapers; percentage change = −32.2%.
8. Absolute change = −63,000 deaths; percentage change = −6.3%.
9. Absolute change = $690 billion; percentage change = 431.3%.
10. Absolute change = 10 percentage points; percentage change = 22.7%.
11. Absolute change = 18 percentage points; percentage change = 150%.
12. Absolute change = 0.53 gallon; percentage change = 72.6%.
13. Absolute change = 4366 libraries; percentage change = 15.3%.
14. John's salary increased $8000, or 40%. Mary's salary increased $10,000, or 40%.
15. Absolute difference = 0.7 million newspapers; percentage difference = 65.4%.
16. Absolute difference = 18,751 cars; percentage difference = 4.9%.
17. Absolute difference = 18 million arrivals; percentage difference = 37.5%.
18. Absolute difference = 1.8 million barrels; percentage difference = 28.1%.
19. Absolute difference = 295,000 operations; percentage difference = 141.1%.
20. Absolute difference = 3.1 years; percentage difference = 4.1%.
21. 140%. **22.** 124%. **23.** 80%. **24.** 55%.
25. The retail cost is 2 times the wholesale cost.
26. A sale price is 0.85 times an item's original price.
27. The retail cost is 1.4 times the wholesale cost.
28. $48.
29. −15.5 percentage points; −22.7%.
30. −7 percentage points; −25.9%.
31. 21 percentage points; 53.8%.
32. 17 percentage points; 63.0%.
33. 33%; 40%.
34. 180 inches; 270 inches.
35. $752.12. **36.** $101.40. **37.** $18,744.
38. $3428.58. **39.** 17.3%. **40.** 17.6%. **41.** 12.6%.
42. 18.3%. **43.** About $75 billion. **44.** 66.7%.
45. $3.6 billion. **46.** 533 man-days. **47.** $0.72 trillion.
48. 69%.
49. False. Her final weight was 1% less than her original weight.
50. False. Your final salary is 0.25% less than your original salary.
51. False. Your taxes increased 11.3% over two years.
52. False. Overall, scores increased by 4%.
53. Not sensible. **54.** Not sensible. **55.** Sensible.
56. Sensible. **57.** Sensible. **58.** Not sensible.
59. No, you cannot average averages.
60. No, you cannot average averages.

61. True.
62. False; 40% × 20% = 8%.
63. False, some of the hotels with restaurants may also have pools.
64. True, because those who drive do not take a bus.

Unit 3B

1. a. 5,000,000; 5 million. **b.** 7,000,000,000; 7 billion.
 c. 0.02; 2 hundredths. **d.** 800,000,000,000; 800 billion.
 e. 0.000 000 1; 1 ten-millionth.
 f. 0.0009; 9 ten-thousandths.
2. a. 7000; 7 thousand. **b.** 1,500,000,000; 1.5 billion.
 c. 3090.6; 3 thousand ninety and six tenths.
 d. 2,300,000; 2.3 million.
 e. 0.00022; 22 hundred-thousandths.
 f. 0.706; 706 thousandths.
3. a. 6×10^2. **b.** 9×10^{-1}. **c.** 5×10^4.
 d. 3×10^{-3}. **e.** 5×10^{-4}. **f.** 7×10^{10}.
4. a. 6.4598×10^4. **b.** 9.2×10^{-1}. **c.** 5.00098×10^2.
 d. 2×10^3. **e.** 2.5×10^8. **f.** -2.3800×10^4.
5. a. 2.4×10^{10}. **b.** 6.45×10^2. **c.** 4.5×10^{-3}.
 d. 2.2×10^{88}.
6. a. 2×10^3. **b.** 5.0×10^8. **c.** 2.0×10^{36}.
 d. 5.5×10^{10}.
7. a. 10^{35} is 10^9, or 1 billion, times larger than 10^{26}.
 b. 10^{27} is 10^{10}, or 10 billion, times larger than 10^{17}.
 c. 1 billion is 10^3, or 1000, times larger than 1 million.
 d. 7 trillion is 10^9, or 1 billion, times larger than 7 thousand.
 e. 2×10^{-6} is 10^3, or 1000, times larger than 2×10^{-9}.
 f. 6.1×10^{29} is 10^2, or 100, times larger than 6.1×10^{27}.
8. a. 5 billion is 20 times larger than 250 million.
 b. 9.3×10^2 is 30,000 times larger than 3.1×10^{-2}.
 c. 10^{-8} is 50,000 times larger than 2×10^{-13}.
 d. 3.5×10^{-2} is 500,000 times larger than 7×10^{-8}.
 e. 1000 is 10^6 (1 million) times larger than 0.001.
 f. 10^{12} is 10^{21} times larger than 10^{-9}.
9. $\$6.32 \times 10^{11}$. **10.** $\$8 \times 10^{11}$. **11.** 5.4×10^9 bytes.
12. 7.39860×10^5 deaths. **13.** 5.096×10^8 km^2.
14. 2.626165×10^{15} watt-hours.
15. a. 2×10^6; exact value is 2×10^6.
 b. $3 \times 10^2 = 300$; exact value is 311.0.
 c. -1.44×10^5; exact value is -1.49×10^5.
 d. 10^{10}; exact value is 10^{10}.
 e. 2100; exact value is 2169.
 f. 2×10^5; exact value is 2.00948×10^5.
16. a. 3×10^7; exact value is 2.8×10^7.
 b. 20,000; exact value is 15,385.
 c. 5×10^8; exact value is 4.95×10^8.
 d. 10^5; exact value is 1.2×10^5.
 e. 2.1×10^6; exact value is 2.04×10^6.
 f. 0.05; exact value is 0.056.

17. A 10-story building is between 100 and 120 feet high. A football field is 100 yards, or 300 feet, long—almost three times longer than the building is high.
18. It's possible at 2 mi/hr and 8 hours per day.
19. It is not unusual for Americans to put 15,000 to 20,000 miles on a single car in a year. Assume that an average American (excluding traveling business people) takes two 5000-mile airplane trips per year. This amounts to 10,000 miles flown.
20. A stadium holds many more people.
23. 100,800 breaths (assuming 10 breaths per minute).
24. Approximately 90 gallons per year (assuming 2 pints per day).
25. Approximately 420,000 words.
26. 1000 gallons.
28. 2 trillion kilograms (assuming 1000 gal/driver/yr and 200 million drivers).
29. 7.6 births per minute.
30. 6027 marriages per day.
31. 251 births per minute.
32. 82 suicides per day.
33. Approximately 2.2×10^4 km (assuming 10 bills per mm).
34. Approximately 5.8×10^5 km (assuming 10 bills per mm).
35. 1973 heart attacks per day. **36.** About 25 stadiums.
37. 16 candy bars. **38.** 125 liters.
39. Uranium releases 35,000 times more energy.
40. Fusion of hydrogen releases about 6 million times more energy.
41. About 7×10^{-7} liter. **42.** About 7 million liters.
43. About 1.8 million kilograms. **44.** Yes.
45. a. 100,000 to 1. **b.** 15,840 to 1. **c.** 2,000,000 to 1.
 d. 328 to 1.
46. a. 63,360 to 1. **b.** 42,240 to 1. **c.** 100,000 to 1.
 d. 3280 to 1.
47.

Planet	Model diameter	Model distance from sun
Mercury	0.5 mm	6 m
Venus	1.2 mm	11 m
Earth	1.3 mm	15 m
Mars	0.7 mm	23 m
Jupiter	14.3 mm	78 m
Saturn	12.0 mm	143 m
Uranus	5.2 mm	287 m
Neptune	4.8 mm	450 m
Pluto	0.2 mm	590 m

48. 6.7 m; 0.067 mm.
49. About 9000 gallons per month (assuming 100 million households).
50. a. 10^{12} watts.
 b. 1 terawatt, which is 0.1 of the world's electricity demand.

c. To avoid depleting plant life, it's necessary to use fossil fuels, which provide the energy stored by plants over thousands of years.
51. a. About 1821 kg/cm^3.
 b. 7600 kg, or the mass of a large car.
 c. 7×10^{11} kg/cm^3.
52. 7 grams, or 0.016 pound, or about a quarter of an ounce.
53. 1.5 mm; 1.8 mm; 1.3 mm; 1.7 mm; take your height in quarters.

Unit 3C

1. a. 2. b. 15. c. 779. d. 4. e. 14. f. 235. g. 2000. h. 89. i. -14.
2. a. 2365.985; 2366.0; 2370; 2400.
 b. 322,354.090; 322,354.1; 322,350; 322,400.
 c. 6000.000; 6000.0; 6000; 6000.
 d. 11.333; 11.3; 10; 0.
 e. 78.555; 78.6; 80; 100.
 f. 0.452; 0.5; 0; 0.
 g. -12.100; -12.1; -10; 0.
 h. -850.765; -850.8; -850; -900.
 i. -12.001; -12.0; -10; 0.
3. a. Three significant digits; it is precise to the nearest tenth (0.1) of a km/hr.
 b. Six significant digits; it is precise to the nearest thousandth (0.001) of a second.
 c. Two significant digits; it is precise to the nearest 0.00001 mm.
 d. Four significant digits; it is precise to the nearest 0.000 000 1 m.
 e. One significant digit; it is precise to the nearest 100,000.
 f. Five significant digits; it is precise to the nearest 0.0001.
4. a. Three significant digits; it is precise to the nearest person.
 b. Four significant digits; it is precise to the nearest tenth (0.1) of a liter.
 c. Six significant digits; it is precise to the nearest 0.00001 mm.
 d. Three significant digits; it is precise to the nearest 0.000 001 m.
 e. Two significant digits; it is precise to the nearest 10 million.
 f. Four significant digits; it is precise to the nearest 10 million.
5. Random error due to miscounting; systematic error due to interpretation of what counts as a sport utility vehicle.
6. Random error due to miscounting; systematic error due to undercounting homeless people or overcounting college students.
7. Random error due to miscalculation of average; systematic error due to underreporting income on tax returns.
8. Random error due to miscalculation of average; systematic error due to overreporting income during interviews.
9. Random error due to errors in reading scale; no systematic error if the scale is well calibrated.
10. Random error due to errors in reading scale; systematic error due to clothes and calibration of the scale.
11. Random error due to miscounting.
12. Random error due to errors in reading the watch; systematic errors due to calibration.
13. (1) is a random error; (2) is a systematic error (most likely due to underreporting).
14. (1) is a random error; (2) is a systematic error (most likely due to underreporting).
15. All altitude readings will be about 2780 feet too low; this is a systematic error.
16. Random error.
17. Absolute error = $0.50; relative error = 2.7%.
18. Absolute error = 2 lb; relative error = 11.1%.
19. Absolute error = $-$$17; relative error = $-$26.2%.
20. Absolute error = 0.5 in.; relative error = 1.0%.
21. Absolute error = $-$0.25 cup; relative error = $-$4.5%.
22. Absolute error = 5 min; relative error = 8.3%.
23. Absolute error = $-$0.3 mi; relative error = $-$2.5%.
24. Absolute error = 0.1 cm; relative error = 100%.
25. Tape measure is more accurate; laser is more precise.
26. Laser is more precise and more accurate.
27. Digital scale is more precise and more accurate.
28. Clinic scale is more accurate; digital scale is more precise.
29. Random and systematic errors could be present; the claim is not believable with the given precision.
30. Random and systematic errors could be present; the claim is believable with the given precision.
31. Random and systematic errors could be present; the claim is believable with the given precision.
32. Random errors could be present; the claim is believable with the given precision.
33. Random and systematic errors could be present; the claim is believable with the given precision.
34. Random and systematic errors could be present; the claim is not believable with the given precision.
35. Random and systematic errors could be present; the claim is not believable with the given precision.
36. Because officially declared species are designated by humans, the claim is believable with the given precision.
37. 140 liters. 38. 3 hours, 59 minutes. 39. 1600 kg^2.
40. 2.0 km/min. 41. 38 mi. 42. 51 kg. 43. $230.
44. $2.7 million.

Unit 3D

1. Josh; Josh; Jude. 2. Allan, Allan, Abner.
3. a. New Jersey.

b. The percentage of nonwhites is significantly less in Nebraska than in New Jersey.

4. **a.** The SAT scores decreased between 1988 and 1998 in all grade categories.
 b. Over all grade categories, the average SAT score increased between 1988 and 1998.

5. **a.** Whites: 0.18%; nonwhites: 0.54%; total: 0.19%.
 b. Whites: 0.16%; nonwhites: 0.34%; total: 0.23%.
 c. The rate for both whites and nonwhites was higher in New York than in Richmond, yet the overall rate was higher in Richmond than in New York. The percentage of nonwhites was significantly less in New York than in Richmond.

6. The Gazelles showed more improvement in both categories than the Cheetahs, yet the overall improvement was greater for the Cheetahs than for the Gazelles. The weight trainers comprised 50% of the Gazelles and 65% of the Cheetahs.

7. Within each category, the percentage of women hired was greater than the percentage of men hired. Overall, the hiring rate was higher for men (55%) than for women (41.7%). Note that the number of blue-collar jobs was over five times greater than the number of white-collar jobs.

8. Within each trial, treatment A had the higher cure rate. Overall, treatment B had a higher cure rate (55%) than A (41.7%). Note that the largest number of people had treatment B in the second trial.

9. **a.** Spelman had a better record for home games (34.5% vs. 32.1%) and away games (75.0% vs. 73.7%) individually.
 b. Morehouse has the better overall average (62.5% vs. 48.9%).
 c. Teams are generally rated on their overall record.

10. **a.** Drug B had the better cure rate for men (98.0% vs. 50.0%) and for women (11.2% vs. 5.0%).
 b. Drug A had the better cure rate overall (45.0% vs. 27.0%).
 c. Because the drugs had such different effects on men and women, the individual cure rates should be cited and used for evaluating the drugs.

11. **b.** 216 people were accused of lying; of these, 18 were actually lying and 198 were telling the truth; 91.7% were falsely accused.
 c. 1784 people were telling the truth according to the polygraph; of these, 1782 were actually telling the truth and 2 were lying; 99.9% of those found to be telling the truth were actually telling the truth.

12. **b.** 48/60 = 80%.

c. 5.7%; in part b the comparison is to all those with the disease, and in part c the comparison is to all those who test positive.
d. The chance of the patient having the disease increases to 5.7%, given that the test was positive, compared to the overall chance of 1.5% in the absence of a test.

13. **b.** 95% of those at risk for HIV test positive; 67.9% of those at risk who test positive have HIV.
 c. The chance of the patient having HIV is 67.9%, which is greater than the overall incidence rate of 10%.
 d. 95% of those in the general population with HIV test positive; 5.4% of those in the general population who test positive have HIV.
 e. The chance of the patient having HIV is 5.4%, compared to the overall incidence rate of 0.3%.

14. The housing program will see an increase of 1%, which represents an increase in actual dollars. However, the increase is less than the increase in the CPI and thus effectively amounts to a cut.

CHAPTER 4

Unit 4A

1. **a.** $x = 10$. **b.** $y = 5$. **c.** $z = 6$. **d.** $x = 5$.
 e. $x = 9$. **f.** $n = 3$. **g.** $x = 4$. **h.** $w = -2$.
2. **a.** $x = 10$. **b.** $x = -3$. **c.** $y = 7$. **d.** $n = 20$.
 e. $z = 8$. **f.** $x = 1$. **g.** $s = 4$. **h.** $y = -3$.

3.

	Yancy		Samantha	
Year	Interest	Balance	Interest	Balance
0	———	$500	———	$500
1	$25	$525	$25	$525
2	$25	$550	$26.25	$551.25
3	$25	$575	$27.56	$578.81
4	$25	$600	$28.94	$607.75
5	$25	$625	$30.39	$638.14

Yancy's balance increases by $125 or 25%. Samantha's balance increases by $138.14 or 27.6%.

4.

	Trevor		Kendra	
Year	Interest	Balance	Interest	Balance
0	———	$1000	———	$1000
1	$60	$1060	$60	$1060
2	$60	$1120	$63.60	$1123.60
3	$60	$1180	$67.42	$1191.02
4	$60	$1240	$71.46	$1262.48
5	$60	$1300	$75.75	$1338.23

Trevor's balance increases by $300 or 30%. Kendra's balance increases by $338.23 or 33.8%.

5. $2687.83. 6. $26,532.98. 7. $162,822.98.
8. $4803.10. 9. $42,918.71. 10. $462,330.07.
11. After 10 years, Chang has $705.30; after 30 years, he has $1403.40. After 10 years, Kio has $722.52; after 30 years, she has $1508.74. Kio has $17.22 or 2.4% more than Chang after 10 years. Kio has $105.34 or 7.5% more than Chang after 30 years.
12. After 10 years, José has $2586.61; after 30 years, he has $7691.46. After 10 years, Marta has $2611.21; after 30 years, she has $7912.99. Marta has $24.60 or 0.95% more than José after 10 years. Marta has $221.53 or 2.9% more than José after 30 years.
13. $1731.08. 14. $2323.65. 15. $21,248.27.
16. $13,623.37. 17. $2848.95. 18. $4945.99.
19. $31,663.68. 20. $48,147.25. 21. 6.72%.
22. 4.59%. 23. 3.30%. 24. 7.45%.
25. 6.77%; 6.80%; 6.82%. 26. 8.24%; 8.30%; 8.33%.
27.

| | Account 1 | | Account 2 | |
Year	Interest	Balance	Interest	Balance
0	—	$1000	—	$1000
1	$55	$1055	$57	$1057
2	$58	$1113	$59	$1116
3	$61	$1174	$63	$1179
4	$65	$1239	$67	$1246
5	$68	$1307	$70	$1317
6	$72	$1379	$74	$1391
7	$76	$1455	$79	$1470
8	$80	$1535	$83	$1553
9	$84	$1619	$87	$1640
10	$89	$1708	$93	$1733

Account 1 has increased in value by $708 or 70.8%. Account 2 has increased in value by $733 or 73.3%.

28. a. If there is only one compounding during the year, the balance will increase by the APR after one year. Thus, the APY is the APR.
 b. With more than one compounding per year, the APY is always greater than the APR. The APY reflects the cumulative effects of several compoundings during the year. At the end of the year, the balance will be greater than if there had been only one compounding.
 c. APY does *not* depend on the starting principal. It gives the percentage increase in the balance.
 d. APY does depend on the number of compoundings per year. The more compoundings during the year, the greater the APY.

29. $1040.81; $1221.40; $2225.54; APY = 4.08%.
30. $2102.54; $2568.05; $5436.56; APY = 5.13%.
31. $10,618.37; $13,498.59; $33,201.17; APY = 6.18%.
32. $3122.43; $3664.21; $6676.62; APY = 4.08%.
33. $5335.80; $6920.15; $18,346.48; APY = 6.72%.
34. $536.25; $709.53; $2027.60; APY = 7.25%.
35. After 5 years, Bernard has $1946.64; after 20 years, he has $3505.80. After 5 years, Carla has $1797.60; after 20 years, she has $3805.33.
36. After 5 years, Brian has $2091.14; after 20 years, he has $4668.41. After 5 years, Celeste has $1815.70; after 20 years, she has $3960.90.
37. $4224.
38. $4106.
39. $4079.
40. $4066.
41. a. $33,963. b. $25,924. c. $13,932.
42. Plan A: $13,597; Plan B: $15,526.
44. a.

n	1	4	12	365	500	1000
APY	12.00	12.55	12.68	12.75	12.75	12.75

b. 12.75%.

c. **Annual Percentage Yield (APY) with APR = 12%**

d. The APY is higher with continuous compounding than with any other periodic compounding.
e. $563.75; $911.06.
45. a. 14.3 years. b. 6.0 years. c. 68.1 years.

Unit 4B
1. a. 64. b. 256. c. 81. d. 1/16. e. 64.
 f. 78,125.
2. a. 5. b. 3. c. 81. d. 1/4. e. 1/4. f. 3.
3. a. $x = \pm 10$. b. $x = 3$. c. $x = 8$. d. $x = 32$.
4. a. $x = 8$ or $x = -4$. b. $x = 3$. c. $x = \pm 10$.
 d. $x = \pm 6$.
5. $174,550.39; $24,000.
6. $187,696.03; $24,000.

7. $86,144.21; $43,200.
8. $90,920.11; $43,200.
9. Yolanda: balance = $15,528.23; deposits = $12,000. Zach: balance = $15,093.47; deposits = $12,000.
10. Polly: balance = $8193.97; deposits = $6000. Quint: balance = $6736.12; deposits = $4800.
11. Juan: balance = $32,775.87; deposits = $24,000. Maria: balance = $33,736.06; deposits = $25,000.
12. George: balance = $6923.39; deposits = $4800. Harvey: balance = $8818.79; deposits = $6000.
13. $329.96. 14. $1092.45. 15. $256.13.
16. $947.95.
17. Balance of $15,848.11 is not enough.
18. Balance of $23,772.17 is not enough.
19. Balance of $29,081.87 is not enough.
20. Balance of $31,085.28 is not enough.
21. Monthly deposit = $419.36.
22. Monthly deposit = $716.13.
23. 87.3%; 13.4%. 24. 56.3%; 2.3%. 25. 105.5%; 7.5%.
26. −44.4%; −17.8%. 27. −40.0%; −9.7%.
28. 70.0%; 19.3%. 29. 68.0%; 6.7%.
30. 439,900%; 18.3%.

Unit 4C
1. a. Starting principal = $40,000; APR = 7%; number of payments per year = 12; loan term = 20 years; payment amount = $310.
 b. Number of payments = 240; amount paid = $74,400.
 c. Interest payments = $34,400; payments toward principal = $40,000.
2. a. Starting principal = $7500; APR = 9%; number of payments per year = 12; loan term = 10 years; payment amount = $95.
 b. Number of payments = 120; amount paid = $11,400.
 c. Interest payments = $3900; payments toward principal = $7500.
3. a. $241.26. b. $57,902.40.
 c. Interest payments = $32,902.40; payments toward principal = $25,000.
4. a. $145.59. b. $17,471.17.
 c. Interest payments = $5471.17; payments toward principal = $12,000.
5. a. $1048.82. b. $377,575.20.
 c. Interest payments = $227,575.20; payments toward principal = $150,000.
6. a. $1390.52. b. $250,293.60.
 c. Interest payments = $100,293.60; payments toward principal = $150,000.
7. a. $1014.27. b. $182,567.99.
 c. Interest payments = $82,567.99; payments toward principal = $100,000.
8. a. $984.74. b. $177,253.20.

c. Interest payments = $77,253.20; payments toward principal = $100,000.
9. a. $166.07. b. $5978.52.
 c. Interest payments = $978.52; payments toward principal = $5000.
10. a. $212.47. b. $12,748.20.
 c. Interest payments = $2748.20; payments toward principal = $10,000.
11. a. $477.83. b. $86,009.40.
 c. Interest payments = $36,009.40; payments toward principal = $50,000.
12. a. $665.30. b. $239,508.00.
 c. Interest payments = $139,508.00; payments toward principal = $100,000.
13. Monthly payments = $768.91.

End of month	Interest	Payment toward principal	New principal
1	$708.33	$60.58	$99,939.42
2	$707.90	$61.01	$99,878.41
3	$707.47	$61.44	$99,816.97

14. Monthly payments = $114.68.

End of month	Interest	Payment toward principal	New principal
1	$80.00	$34.68	$11,965.32
2	$79.77	$34.91	$11,930.41
3	$79.54	$35.14	$11,895.27

15. Monthly payments: $308.77; $241.79; $202.76. You can afford only option 3.
16. Monthly payments: $180.91; $127.20; $101.45. You can afford options 2 and 3.
17. $229.20; $2750.40. 18. $127.24; $3053.76.
19. $94.19; $3390.84. 20. $233.99; $2807.88.
21.

Month	Payment	Expenses	Interest	New balance
0				$1200.00
1	$200	$75	$18.00	$1093.00
2	$200	$75	$16.40	$984.40
3	$200	$75	$14.77	$874.17
4	$200	$75	$13.11	$762.28
5	$200	$75	$11.43	$648.71
6	$200	$75	$9.73	$533.44
7	$200	$75	$8.00	$416.44
8	$200	$75	$6.25	$297.69
9	$200	$75	$4.47	$177.16
10	$200	$75	$2.66	$54.82

A partial 11th payment will pay off the loan.

22.

Month	Payment	Expenses	Interest	New balance
0				$1200.00
1	$300	$75	$18.00	$993.00
2	$300	$75	$14.90	$782.90
3	$300	$75	$11.74	$569.64
4	$300	$75	$8.54	$353.18
5	$300	$75	$5.30	$133.48

A partial 6th payment will pay off the loan.

23.

Month	Payment	Expenses	Interest	Balance
0	—	—	—	$300
1	$300	$175	$4.50	$179.50
2	$150	$150	$2.69	$182.19
3	$400	$350	$2.73	$134.92
4	$500	$450	$2.02	$86.94
5	0	$100	$1.30	$188.24
6	$100	$100	$2.82	$191.06
7	$200	$150	$2.87	$143.93
8	$100	$80	$2.16	$126.09

24. a. Interest is $20 per month.
 b. Interest is $80 per month.
25. Option 1: monthly payments = $733.76; total payments = $264,153.60. Option 2: monthly payments = $927.01; total payments = $166,861.80.
26. Option 1: monthly payments = $550.32; total payments = $198,115.20. Option 2: monthly payments = $674.12; total payments = $121,341.60.
27. Option 1: monthly payments = $810.49; total payments = $291,776.40. Option 2: monthly payments = $1061.89; total payments = $191,140.20.
28. Option 1: monthly payments = $1227.92; total payments = $442,051.20. Option 2: monthly payments = $1597.83; total payments = $287,609.40.
29. Option 1: monthly payments = $587.01; closing costs = $1200. Option 2: monthly payments = $559.37; closing costs = $2800.
30. Option 1: monthly payments = $615.13; closing costs = $0. Option 2: monthly payments = $559.37; closing costs = $4400.
31. Option 1: monthly payments = $545.74; closing costs = $2000. Option 2 monthly payments = $518.88; closing costs = $3600.
32. Option 1: monthly payments = $559.37; closing costs = $1000. Option 2: monthly payments = $505.65; closing costs = $4700.
33. a. $224.93. b. $316.69.
 c. Total payments for 20-year loan = $53,983.20; total payments for 10-year loan = $38,002.80.
34. a. $463.09. b. $573.39.
 c. Total payments for 25-year loan = $138,927.00; total payments for 15-year loan = $103,210.20.
35. Savings of about $250 per month with the ARM; about $188 per month more with the ARM.
36. Savings of about $312 per month with the ARM; about $156 per month more with the ARM.
37. $62,141; $77,676.
38. $171,621; $214,526.
39. a. $95.57; $130.17; $158.34. b. $67,444.20.
 c. $325.43; $78,103.
40. a. $440. b. $5280. c. $64.53.

CHAPTER 5

Unit 5A

1. *Population*: registered voters in California; *sample:* 1026 people selected for interviews; *population parameters:* approval ratings of candidates among all registered voters; *sample statistics:* approval ratings of candidates among those in the sample.
2. *Population:* all stars in the galaxy; *sample:* the few stars selected for measurements; *population parameter:* mean of distances between all stars and the Earth; *sample statistics:* mean of distances between the stars in the sample and the Earth.
3. *Population:* all new computers; *sample:* one computer selected for measurements; *population parameter:* speed of all new computers on specific tasks; *sample statistic:* speed of the selected computer on specific tasks.
4. *Population:* all adult Americans; *sample:* 1200 selected adults; *population parameters:* percentages of all adult Americans who respond yes and no to the question; *sample statistics:* percentages of those in sample who respond yes and no to the question.
5. *Population:* all adult Americans; *sample:* 1010 selected adults; *population parameters:* rankings of prestigious professions among all adult Americans; *sample statistics:* rankings of prestigious professions by those in the sample.
6. *Population:* 1.6 million first-year college students; *sample:* 261,000 selected first-year college students; *population parameters:* attitudes of all first-year college students; *sample statistics:* attitudes of those in sample.
7. *Step 1:* Population is all students at the school; goal is to determine annual pizza consumption. *Step 2:* Choose a representative sample. *Step 3:* Determine annual pizza

consumption for those in the sample. *Step 4:* Infer annual pizza consumption for all students. *Step 5:* Assess results and formulate conclusion.

8. *Step 1:* Population is all business travelers; goal is to determine level of frustration among business travelers. *Step 2:* Choose a representative sample. *Step 3:* Determine level of frustration for those in the sample. *Step 4:* Infer level of frustration for all business travelers. *Step 5:* Assess results and formulate conclusion.
9. *Step 1:* Population is adult American women; goal is to determine mean height of adult American women. *Step 2:* Choose a representative sample. *Step 3:* Determine mean height for those in the sample. *Step 4:* Infer mean height for all adult American women. *Step 5:* Assess results and formulate conclusion.
10. *Step 1:* Population is all people who go to restaurants; goal is to determine average tip (as a percentage) left in restaurants. *Step 2:* Choose a representative sample. *Step 3:* Determine average tip for those in the sample. *Step 4:* Infer average tip for all people who go to restaurants. *Step 5:* Assess results and formulate conclusion.
11. *Step 1:* Population is all alkaline batteries; goal is to determine mean lifetime of all alkaline batteries. *Step 2:* Choose a representative sample. *Step 3:* Determine mean lifetime of alkaline batteries in the sample. *Step 4:* Infer mean lifetime of all alkaline batteries. *Step 5:* Assess results and formulate conclusion.
12. *Step 1:* Population is all high school students at the school; goal is to determine percentage of high school students who are vegetarians. *Step 2:* Choose a representative sample. *Step 3:* Determine percentage of vegetarians among students in the sample. *Step 4:* Infer percentage of vegetarians among all students. *Step 5:* Assess results and formulate conclusion.
13. Not representative (age could be an important factor); not representative; not representative; representative.
14. Not representative; not representative; not representative; most representative.
15. Stratified.
16. Convenience 17. Stratified. 18. Systematic.
19. Simple random. 20. Simple random.
21. Observational, not case-control.
22. Observational, case-control; cases are patients with melanoma; controls are cancer-free patients.
23. Observational, not case-control.
24. Observational, case-control; cases are participants who exercise regularly; controls are participants who do not exercise regularly.
25. Experiment; treatment group consists of plants treated with fertilizer; control group consists of plants treated with no fertilizer; placebo and blinding are not needed.
26. Observational, not case-control. 27 Observational.
28. Experiment. 29. Observational, case-control.
30. Experiment. 31. Experiment. 32. Experiment.
33. Good evidence. 34. No evidence.
35. Good evidence. 36. No evidence.
37. 50.5% to 55.5%; yes. 38. 46.5% to 50.5%; no.
39. 4.0% to 4.4%; 4.1% to 4.5%.
40. 67% to 73%; no, respondents may not have reported accurately.
41. a. *Goal:* To determine public opinion on standard of living. Population is all parents; population parameter is percentage of parents who respond yes to the question.
 b. Sample is 748 parents surveyed; sample statistic is percentage of parents who respond yes to the question.
 c. 59.4% to 66.6%.
42. a. *Goal:* To determine public opinion on self-reliance. Population is all people (perhaps all Americans); population parameter is percentage of people who agree with the statement.
 b. Sample is 4000 people surveyed; sample statistic is percentage of people who agree with the statement.
 c. 68.4% to 71.6%.
43. a. *Goal:* To determine the U.S. unemployment rate. Population is all people of working age; population parameter is percentage of population that is unemployed.
 b. Sample is people in 60,000 households; sample statistic is percentage of those in sample who are unemployed.
 c. 4.0% to 4.4%.
44. a. *Goal:* To determine the opinion of Americans about the likelihood of landing on Mars in the next 50 years. Population is all Americans; population parameter is percentage of Americans who respond yes to the question.
 b. Sample is 1546 people surveyed; sample statistic is percentage of people in the sample who respond yes to the question.
 c. 73% to 79%.
45. a. *Goal:* To determine the opinion of Americans about the likelihood of human cloning in the next 50 years. Population is all Americans; population parameter is percentage of Americans who respond yes to the question.
 b. Sample is 1546 people surveyed; sample statistic is percentage of people in the sample who respond yes to the question.
 c. 48% to 54%.
46. a. *Goal:* To determine public opinion on the use of savings accounts. Population is all (American) adults; population parameter is percentage of adults who use savings accounts.
 b. Sample is 2000 adults surveyed; sample statistic is percentage of people in sample who use savings accounts.
 c. 62% to 66%.

Unit 5B

1. Possible conflict of interest.
2. Possible selection and participation bias.
3. Possible selection bias.
4. Possible selection bias.
5. Possible selection bias.
6. Possible participation bias.
7. Possible conflict of interest.
8. Possible conflict of interest.
11. What are the sample and sample size? How is "great deal of confidence" measured? Which military leaders are considered?
12. What are the sample and sample size? What criteria are used to judge restaurants?
13. What are the sample and sample size? Is self-reporting accurate?
14. What are the sample and sample size? How were responses obtained?
15. What are the sample and sample size?
16. How were the data obtained?
17. 98% of all movies is different from 98% of top rental movies.
18. The difference between the responses is not great, perhaps not statistically significant.
19. It may be true, but the sample is very small and there could be bias in the study.

Unit 5C

1. Qualitative.
2. Quantitative.
3. Qualitative.
4. Quantitative.
5. Qualitative.
6. Quantitative.
7. Qualitative.
8. Quantitative.

9.

Grade	Freq.	Rel. freq.	Cum. freq.
A	4	0.17	4
B	7	0.29	11
C	8	0.33	19
D	3	0.13	22
F	2	0.08	24
Total	24	1.00	24

10.

Rating	Freq.	Rel. freq.	Cum. freq.
5-star	5	0.09	5
4-star	10	0.18	15
3-star	20	0.37	35
2-star	15	0.27	50
1-star	5	0.09	55
Total	55	1.00	55

11.

Bin	Freq.	Rel. freq.	Cum. freq.
95–99	2	0.08	2
90–94	2	0.08	4
85–89	6	0.24	10
80–84	3	0.12	13
75–79	4	0.16	17
70–74	1	0.04	18
65–69	4	0.16	22
60–64	0	0.00	22
55–59	1	0.04	23
50–54	1	0.04	24
< 50	1	0.04	25
Total	25	1.00	25

12.

Bin	Freq.	Rel. freq.	Cum. freq.
90–99	4	0.16	4
80–89	9	0.36	13
70–79	5	0.20	18
60–69	4	0.16	22
50–59	2	0.08	24
< 50	1	0.04	25
Total	25	1.00	25

13. a. The teams are qualitative categories.
 b.

14. a. The companies are qualitative categories.
 b.

15. a. The ethnic groups are qualitative categories.

b.

16. a. The continents are qualitative categories.

b.

17. a. The land masses are qualitative categories.

b.

18. a. The age groups are quantitative categories.

b.

19. a. The religions are qualitative categories.

b.

20. a. The numbers of colleges are quantitative categories.

b.

21. a. The years are quantitative categories.

b.

[Graph: Average family size vs year, 1940–1995, declining from ~3.7 to ~3.0]

22.

Bin	Frequency
30–39	12
40–49	13
50–59	5
60–69	3
70–79	1
Total	34

[Bar chart: Number of actors by age bin, matching table above]

Unit 5D

1. **a.** China, India, Russia. **b.** China, India, Russia.
2. **a.** Each set of three bars corresponds to three years: 1975, 1985, and 1995.
 b. The 20-year change was much greater for those with a college degree. **c.** Two dimensions would suffice.
3. **a.** Cancer increased; tuberculosis decreased; cardiovascular disease increased and then decreased, for an overall decrease; pneumonia decreased.
 b. The death rate for cardiovascular disease reached a maximum of about 500 (deaths per 100,000) in 1950.
 c. The death rate for cancer in 2000 was about 210 (deaths per 100,000).
 d. If the lines of the graph were extended, cancer would overtake cardiovascular disease as the leading cause of death.
4. **a.** 1930 men: 75,000; 1930 women: 50,000; 2000 men: 500,000; 2000 women: 600,000.
 b. In 1980, slightly more men than women earned degrees; in 2000, more women than men earned degrees. **c.** During the 1960s.
 d. In 1950, about 450,000 degrees were awarded; in 2000, about 1.15 million degrees were awarded.
5. **a.** 1980: about 8%; 1990: about 15%; 1995: about 15%; 2000: about 12%.
 b. 1960: about 52%; 1980: about 23%; 2000: about 17%.
 c. 1960: about 27%; 1980: about 47%; 2000: about 57%.
6. Defense spending decreased; payments for individuals increased; net interest increased and then decreased.
7. There are significant regional differences; probabilities tend to be much higher in the northern states.
8. The land is much flatter on the right (east) half of the map; the land on the left (west) half of the map is steep and has two peaks.
9. **a.** Each bar represents the percentage of the population for a particular age group in a particular year.
 b. There has been a general decrease in the percentage of people in the youngest age category.
 c. The data are truly three-dimensional.
10. **a.** In 1860, about 35% of the population was in the 5- to 19-year-old category. In 1990, about 15% of the population was in this category.
 b. In 1860, about 35% of the population was in the 20- to 44-year-old category. In 1990, about 40% of the population was in this category.
 c. The percentage of under-5-year-olds increased because of the baby boom.
11. Two graphs work best:

Percentage of Women Never Married

[3D bar chart by age group (20–24, 25–29, 30–34, 35–39, 40–44) comparing 1970 and 1996]

14. It is difficult to include U.S. data on the same graph.

Percentage of Men Never Married [bar chart: 1970 and 1996, by age 20–24 through 40–44]

Annual Deaths Due to Firearms (excluding U.S., which has 30 times more deaths than Germany and Canada) [stacked bar chart: Accidents, Suicides, Homicides for Germany, Canada, Australia, Spain, UK, Sweden, Vietnam, Japan]

12.

Amount of Alcohol and Auto Fatalities [bar chart: None, Moderate, High from 1982 to 1996]

15. The area of the 1997 TV is about 16 times greater than the area of the 1980 TV. The volume of the 1997 TV is about 64 times greater than the volume of the 1980 TV.

16. The actual values for the 1990 data (87.5%, 11.3%, and 1.2%) and the 2050 data (80.0%, 15.4%, and 4.6%) are not reflected accurately in the tilted pie charts.

17. Note that the zero point for braking distance is not on the graph. The braking distance of Lincolns is actually only 11% greater than the braking distance of Oldsmobiles.

18. The zero point for earnings is not on the graph.

[bar chart: Average weekly earnings, Men ~750, Women ~520]

13. Circulation is in millions.

Number of Newspapers and Circulation [bar chart: Number and Circulation, 1920–2000]

19. *Personal Computer Sales* [line graph: Sales (millions), 1981–1999]

ANSWERS A-21

Personal Computer Sales

20. Most of the growth occurred after 1950.

Unit 5E

1. Pounds, Calories; weak positive correlation.
2. Inches, shoe size; weak positive correlation.
3. Revolutions per minute, miles per hour; strong positive correlation.
4. Feet, degrees; strong negative correlation.
5. % alcohol, seconds; strong positive correlation.
6. Pounds, miles per gallon; weak negative correlation.
7. Pounds, dollars; weak positive correlation.
8. Dollars, number of items; strong negative correlation.
9. a.

 b. Weak positive correlation.

10. a.

 b. No correlation.

11. a.

 b. Weak positive correlation.

12. a.

 b. Strong negative correlation.

13. a.

[scatter plot: Earnings (billions of dollars) vs Total sales (billions of dollars)]

b. Strong positive correlation.

14. a.

[scatter plot: Infant mortality rate (per 1000 births) vs Mean daily Calories]

b. Strong negative correlation.
15. There is a positive correlation between the crime rate and the number of people in prison; common underlying cause.
16. There is a positive correlation between the miles of freeway and the amount of traffic congestion; common underlying cause.
17. There is a positive correlation between gasoline prices and the number of airline passengers; possible direct cause.
18. There is a positive correlation between ice tea sales and ticket sales; common underlying cause.
19. The cost of deer hunting permits is positively correlated with deer sightings; direct cause.
20. There is a positive correlation between the number of priests and attendance at movies; common underlying cause.
21. a. Guideline 1.
 b. Guideline 2.
 c. Guideline 3.
 d. The above observations suggest that perhaps your headaches are caused by bad air in the building in which you work (there is a "disease" called sick building syndrome).
22. The causes of cancer are often random. A cancer cell is produced when the growth control mechanisms of a normal cell are altered by a random mutation. Smoking may increase the chance of such a mutation occurring, but the mutation will not occur in all individuals. Therefore, by chance, some smokers escape cancer.
23. People do not decide to become conductors until relatively late in life—say, 30 years of age or over. Having already survived this many years, a person has a longer-than-average life expectancy.
24. Cause cannot be established until a mechanism is confirmed. There could be an underlying cause.

Unit 5F

1. Mean = 671.37; median = 672.2; mode = 672.2.
2. Mean = 98.44; median = 98.4; mode = 98.4.
3. Mean = 0.188; median = 0.165; mode = 0.16.
4. Mean = 78.4; median = 79; modes = {70, 73, 76, 81}.
5. Mean = 9.5; median = 10; modes = {5, 10, 15}.
6. Mean = 0.9194; median = 0.920; no mode.
7. Median. 8. Mean. 9. Median. 10. Mean.
11. Mean or median. 12. Mean or median.
13. Mean = 0.8124 pound; median = 0.8161 pound; 0.7901 is an outlier; mean = 0.8161 pound; median = 0.8163 pound.
14. Mean = 157,586 acres; median = 104,100 acres; Alaska is a high outlier; mean = 81,317 acres; median = 78,650 acres; Connecticut is a low outlier; mean = 182,933 acres; median = 109,050 acres.
15. Distribution has two peaks (bimodal), is not symmetric, and has high variation.
16. Distribution has one peak, is right-skewed, and has low variation.
17. Distribution has one peak, is symmetric, and has moderate variation.
18. a. Because the mean is greater than the median, the distribution is right-skewed.
 b. About 50% of the families, or 150 families, earn less than $35,000.
 c. No.
19. One peak; left-skewed; small variation.
20. One peak; symmetric; moderate variation.
21. Two peaks; symmetric; large variation.
22. No peak; symmetric; large variation.
23. One peak; symmetric; moderate variation.
24. No peak; symmetric; large variation.
25. One peak; right-skewed; large variation.
26. One peak; right-skewed; large variation.

CHAPTER 6

Unit 6A

1. a. 24 choices. b. 10 stations. c. 32 styles.
 d. 30 systems.

2. a. 336 packages. b. 24 Commissions. c. 360 sets.
 d. 64 versions.
3. 0.1. 4. 1/24 = 0.042. 5. 1/4 = 0.25.
6. 3/8 = 0.375. 7. 4/36 = 0.11. 8. 15/45 = 0.33.
9. 6/22 = 0.27. 10. 1/5 = 0.2. 11. 12/30 = 0.4.
12. 1/100 = 0.01. 13. 0.86. 14. 34/100 = 0.34.
15. 34/100 = 0.34. 16. 1/200 = 0.005. 17. 5/6 = 0.83.
18. 3/4 = 0.75. 19. 0.24. 20. 9/12 = 0.75.
21. 5/6 = 0.83. 22. 0.8. 23. 3/6 = 0.5.
24. 35/36 = 0.97. 25. 52/100 = 0.52; 51/100 = 0.51.
26. 2000 probability = 0.13; 2050 probability = 0.20.
27. P(0 heads) = 1/16 = 0.0625; P(1 head) = 4/16 = 0.25;
 P(2 heads) = 6/16 = 0.375; P(3 heads) = 4/16 = 0.25;
 P(4 heads) = 1/16 = 0.0625.
28. P(2) = 1/16 = 0.0625; P(3) = 2/16 = 0.125;
 P(4) = 3/16 = 0.1875; P(5) = 4/16 = 0.25;
 P(6) = 3/16 = 0.1875; P(7) = 2/16 = 0.125;
 P(8) = 1/16 = 0.0625.
29. The coins appear to be unfair. There should be about 125 occurrences of 0 heads and 3 heads.
30. 0.045.

	Women	Men
Under 20	0.166	0.066
20–24	0.408	0.360
25–29	0.272	0.343
30–34	0.101	0.148
35–44	0.045	0.071
45–64	0.007	0.011
Over 65	0.001	0.001

31. 1/2. 32. 1/3. 33. 1/2. 34. 1/1.
35. Gain of $15. 36. Gain of $25.
37. Odds for A = 99 to 1; odds for B = 24 to 1.

Unit 6B

1. Independent; 0.001.
2. Independent; 0.000 772.
3. Dependent; 0.00686.
4. Independent; 0.00001.
5. Independent; 0.000 225.
6. Independent; 0.0313.
7. Independent; 0.00391.
8. Dependent; 0.0632.
9. Non-overlapping; 0.333.
10. Non-overlapping; 0.5.
11. Non-overlapping; 0.306.
12. Non-overlapping; 0.167.
13. Overlapping; 0.75.
14. Overlapping; 0.722.
15. Overlapping; 0.308.
16. Non-overlapping; 0.154.
17. 0.598. 18. 0.518. 19. 0.875. 20. 0.999.
21. 0.672. 22. 0.651. 23. 0.798. 24. 0.984.
25. a. 0.986. b. 0.619.
26. a. 0.733. b. 1.000. c. 0.643. d. 0.217.
27. a. Dependent. b. 0.303. c. 0.309.
 d. Nearly equal.
28. P(**AA**) = 0.25; P(**Aa** or **aA**) = 0.5; P(**aa**) = 0.25;
 P(dominant) = 0.75; P(recessive) = 0.25.
29. 0.506.
30. a. 0.122. b. 0.636. c. 0.632.
31. a. 1/40 = 0.025. b. 0.000 625. c. 0.224.
32. a. 0.183. b. 0.332.

Unit 6C

1. You shouldn't expect to get exactly 5000 heads. The proportion of heads should approach 0.5 as the number of tosses increases.
2. It means the person is "due" for an accident or citation. If accidents happen randomly, it is not true.
3. Expected value = −$0.25; outcome of one game cannot be predicted; over 100 games, you should expect to lose.
4. Expected value = $0.375; outcome of one game cannot be predicted; over 100 games, you should expect to win.
5. Expected value = $0.50; outcome of one game cannot be predicted; over 100 games, you should expect to win.
6. Expected value = $0.83; outcome of one game cannot be predicted; over 100 games, you should expect to win.
7. Expected value = $350; expected profit = $35,000,000.
8. Expected value = $150; expected profit = $15,000,000.
9. 15 minutes.
10. a. P = 1/3 = 0.333; mean wait time = 10 minutes.
 b. P = 2/3 = 0.667; mean wait time = 20 minutes.
 c. 16.67 minutes.
 d. Shorter.
11. Expected value = −$0.43; expected loss = $157.
12. Expected value = $0.44; expected gain = $161.
13. Expected value = −$0.78; expected loss = $285.
14. Expected value = −$0.49; expected loss = $179.
15. 1-point expected value = 0.94; 2-point expected value = 0.74.
16. 36.2 years.
17. 2.64 people.
18. a. 47% heads; lost $6. b. Lost $12.
 c. You need 56 heads, which is possible.
 d. The probability of a head on any toss is always 0.5.
19. a. Lost $10. b. Lost $16. c. Lost $20.
 d. 60 even numbers.
 e. 45%, 46%, 48%; the proportion of even numbers approaches 0.5, but the net loss increases.
20. a. If you toss a head, the difference becomes 15; if you toss a tail, the difference becomes 17.

b. On each toss, the difference in heads and tails is equally likely to increase or decrease. After 1000 tosses, the difference is equally likely to be greater than 16 or less than 16.

c. Once you have fewer heads than tails, the deficit of heads is likely to remain.

21. $7 million.
22. **a.** −$0.014. **b.** Lose $1.40. **c.** Lose $7.
 d. $14,000.
23. **a.** −$0.24. **b.** −$7.24.
24. **a.** Decision 1: Option A expected value = $1,000,000; Option B expected value = $1,140,000. Decision 2: Option A expected value = $110,000; Option B expected value = $250,000.
 b. Consistent for Decision 2; not consistent for Decision 1.

Unit 6D

1. **a.** 5040. **b.** 840. **c.** 35. **d.** 3,991,680.
 e. 495. **f.** 21.
2. **a.** 720. **b.** 120. **c.** 5. **d.** 332,640. **e.** 330.
 f. 16.
3. $10^4 = 10,000$ addresses.
4. $7^{10} = 282,475,249$ tunes.
5. $26^5 = 11,881,376$ words.
6. $30^3 = 27,000$ combinations.
7. 720 passwords. **8.** 479,001,600 ways.
9. $4^3 = 64$ words. **10.** 30,240 ways.
11. 252 subcommittees. **12.** 270,725 hands.
13. 210 sets. **14.** 311,875,200 sequences.
15. 495 subcommittees. **16.** 4368 hands.
17. 17,576,000 plates. **18.** 125,970 teams.
19. **a.** 72 sundaes. **b.** 1728 cones. **c.** 1320 cones.
 d. 220 cones.
20. **a.** 8,000,000 numbers.
 b. Two exchanges, assuming four people per telephone number.
21. 9 toppings; 10 toppings.
22. **a.** 100,000 ZIP codes. **b.** 2700 people per ZIP code.
 c. 10^9 ZIP codes; yes.
23. 1/201,376. **24.** 1/3,838,380. **25.** 1/649,740.
26. 1/649,740. **27.** 1/720. **28.** 1/560.
29. 0.024; 0.117. **30.** 0.0377; 0.253.
31. **a.** 0.025. **b.** 0.00065. **c.** 50 people.
 d. 1.3 people.
33. **a.** 0.870. **b.** 0.0004. **c.** 2×10^{-7}.

CHAPTER 7

Unit 7A

1. Linear; 16,515 people.
2. Exponential; 106,121 people.
3. Exponential; $169.
4. Linear; $1.40.
5. Exponential; $61.95.
6. Exponential; $8100.
7. Linear; $130,000.
8. Exponential; $133,100.
9. 524,288 grains; 1,048,575 grains; about 150 pounds.
10. 536,870,912 grains; 1,073,741,824 grains; about 153,392 pounds.
11. 1.3×10^{12} tons.
12. Wheat on the chessboard weighs 650 times more.
13. $327.68.
14. About 33 meters or 36 yards (assuming 1.5 mm per penny).
15. About 37 days.
16. About 65 days (assuming 1.5 mm per penny).
17. 2^{50} bacteria; 1/1024 full.
18. $2^{10} = 1024$ bacteria; $1/2^{50}$ full.
19. 2.6 meters.
20. 333 minutes.
21. **a.**

Year	Population
2000	6.0×10^9
2100	2.4×10^{10}
2200	9.6×10^{10}
2300	3.8×10^{11}
2400	1.5×10^{12}
2500	6.1×10^{12}
2600	2.5×10^{13}
2700	9.8×10^{13}
2800	3.9×10^{14}
2900	1.6×10^{15}
3000	6.3×10^{15}

 b. Between 2800 and 2850.
 c. About 2150.

Unit 7B

1. **a.** False. **b.** True. **c.** False. **d.** True.
 e. True.
2. **a.** False. **b.** True. **c.** False. **d.** False.
 e. False.
3. **a.** 0.602. **b.** 4.301. **c.** −0.301. **d.** 1.505.
 e. −0.602. **f.** −0.699.
4. **a.** 1.699. **b.** 3.699. **c.** −1.301. **d.** 1.398.
 e. −0.699. **f.** −1.398.
5. $2^8 = 256$; 2^{56}. **6.** $2^3 = 8$; $2^5 = 32$. **7.** 44 years.
8. $2^{13} = 8192$.
9. 23,645 people; 35,839 people.
10. 44,123 people; 124,800 people.

11. 16,777,216 cells; 4,294,967,296 cells.
12. 64 cells; 4096 cells.
13. 7.0 billion; 15.1 billion; 28.0 billion.
14. 6.7 billion; 12.0 billion; 19.0 billion.

15.

Month	Beginning-of-month population	End-of-month population
1	100	107
2	107	114
3	114	123
4	123	131
5	131	140
6	140	150
7	150	161
8	161	172
9	172	184
10	184	197
11	197	210
12	210	225
13	225	241
14	241	258
15	258	276

Doubling time is approximately 10 months.

16.

Month	Beginning-of-month population	End-of-month population
1	1,000	1,200
2	1,200	1,440
3	1,440	1,728
4	1,728	2,074
5	2,074	2,488
6	2,488	2,986
7	2,986	3,583
8	3,583	4,300
9	4,300	5,160
10	5,160	6,192
11	6,192	7,430
12	7,430	8,916
13	8,916	10,699
14	10,699	12,839
15	12,839	15,407

Doubling time is approximately 4 months.

17. 10 years; 1.23. **18.** 35 years; 4.42.
19. 100 months; 1.09; 1.95. **20.** 36.8 years; 1.21.
21. 0.25; 0.063. **22.** 0.82; 0.016. **23.** 0.25; 0.13.
24. 0.016; 0.002.
25. 62,500 animals; 7813 animals.
26. 329,877 animals; 143,587 animals.
27. 0.41 kg; 0.037 kg. **28.** 0.65 kg; 0.013 kg.
29. Essentially zero. **30.** 0.056 kg.
31. 7 years; 0.00005. **32.** 8.75 weeks; 0.016.
33. 10 years; 312 elephants.
34. 1219 kg.
35. Approximate: 7 years; exact: 7.27 years; price = $1331.
36. Approximate: 0.9 month; exact: 1.18 months; price = $1,156,831.
37. Approximate: 23 years; exact: 23.4 years; population = 438 million.
38. Approximate: 3.5 weeks; exact: 3.8 weeks; population = 1,310,463 termites.

Unit 7C

1. a. 4.52. **b.** 1.63. **c.** 1.13. **d.** 1.19.
2. a. 2.46. **b.** 2.37. **c.** 12.97. **d.** 1.48.
3. a. 31.62. **b.** 0.01. **c.** 0.032. **d.** 5000.
4. a. 10,000. **b.** 50. **c.** 7. **d.** 1.16.
5. a. $Q = 85,000 \times (1.024)^t$.

b.

Year	Population
0	85,000
1	87,040
2	89,129
3	91,268
4	93,458
5	95,701
6	97,998
7	100,350
8	102,759
9	105,225
10	107,750

c.

6. a. $Q = 800 \times (1.03)^t$.

b.

Year	Homicides
0	800
1	824
2	849
3	874
4	900
5	927
6	955
7	984
8	1,013
9	1,043
10	1,075

c.

7. a. $Q = 10^6 \times (0.93)^t$.

b.

Year	Millions of acres
0	1
1	0.93
2	0.86
3	0.80
4	0.75
5	0.70
6	0.65
7	0.60
8	0.56
9	0.52
10	0.48

c.

8. a. $Q = 10,000 \times (0.997)^t$.

b.

Month	Population
0	10,000
1	9,970
2	9,940
3	9,910
4	9,881
5	9,851
6	9,821
7	9,792
8	9,763
9	9,733
10	9,704

c.

9. a. $Q = \$125{,}000 \times (1.07)^t$.

b.

Year	Average price
0	$125,000
1	$133,750
2	$143,113
3	$153,130
4	$163,850
5	$175,319
6	$187,591
7	$200,723
8	$214,773
9	$229,807
10	$245,894

c.

10. a. $Q = 8 \times (0.85)^t$.

b.

Hour	Concentration
0	8.00
1	6.80
2	5.78
3	4.91
4	4.18
5	3.55
6	3.02
7	2.56
8	2.18
9	1.85
10	1.57

c.

11. a. $Q = 2000 \times (1.05)^t$.

b.

Year	Monthly salary
0	$2,000.00
1	$2,100.00
2	$2,205.00
3	$2,315.25
4	$2,431.01
5	$2,552.56
6	$2,680.19
7	$2,814.20
8	$2,954.91
9	$3,102.66
10	$3,257.79

c.

12. a. $Q = \$10{,}000 \times (0.5)^t$.

b.

Year	Value of rubles
0	$10,000
1	$5,000
2	$2,500
3	$1,250
4	$625
5	$313
6	$156
7	$78
8	$39
9	$20
10	$10

c.

13. 114,233; 1,633,584.
14. $115.93.
15. 7.4%.
16. 9.2%.
17. Increased by a factor of 1157 per year; 1.98% per month.
18. Increased by a factor of 4096 per year; 2.34% per month.
19. 33.28 years.
20. **a.** 2.3 billion tons. **b.** 13.3 billion tons.
 c. 4.9 billion tons. **d.** About 1.4% per year.
21. **a.** 15.9 mg. **b.** 120 hours.
22. **a.** 141.4 mg; 100.0 mg; 50 mg. **b.** After about 52 days.
23. **a.** 1.06 billion years old. **b.** 3.88 billion years old.
24. **a.** 2149 years ago. **b.** 22,866 years ago.
 c. No, the half-life is too small.
25. 9.5 mg/cm^2.

CHAPTER 8

Unit 8A

1. 220 cps; 440 cps; 880 cps; 1760 cps.
2. 440 cps; 220 cps; 110 cps; 55 cps.

3.

Note	Frequency (cps)
A	437
A#	463
B	491
C	520
C#	551
D	583
D#	618
E	655
F	694
F#	735
G	779
G#	825
A	874

5. **a.** 347 cps. **b.** 390 cps. **c.** 694 cps.
 d. 2080 cps.
 e. 4947 cps.
6. **a.** 618 cps. **b.** 551 cps. **c.** 1167 cps.
 d. 1852 cps.
 e. 1962 cps.
7. 327 cps; 275 cps.
8. 656 cps; 983 cps; 590 cps; 442 cps; frequency is not the same.
9. **b.** 2.24.

c.

Note	Frequency (cps)
C	260
G	390
D	584
A	875
E	1,311
B	1,964
F#	2,943
C#	4,410
G#	6,608
D#	9,901
A#	14,835
F	22,227
C	33,303

 d. 12 fifths; 7 octaves.
 e. 128 times the frequency.
 f. By a factor of 1.335; 12 fourths; 5 octaves.
10. In 4/4 time, there are four quarter notes in a measure. This means that there are two half notes in a measure, eight eighth notes in a measure, and 16 sixteenth notes in a measure.

4.

Note	Frequency (cps)
G	390
G#	413
A	438
A#	464
B	491
C	521
C#	551
D	584
D#	619
E	656
F	695
F#	736
G	780

Unit 8B

1. **a.** A vanishing point of the picture is that point at which the edges of the road appear to meet. According to the

definition given in the book, this vanishing point is not the principal vanishing point. The principal vanishing point corresponds to all lines that are parallel in the real scene and perpendicular to the canvas. The road in this scene is not perpendicular to the canvas.

b.

2. The left box.

3.

4.

5. a.

b. 1.28 cm; 1.02 cm; 0.82 cm.
c. No.

6.

8. a. A, H, I, M, O, T, U, V, W, X, Y. **b.** B, C, D, E, H, I, K, O, X.
c. H, I, O, X. **d.** H, I, O, X, N, S, Z.

9. a. 4 reflection; 3 rotation. **b.** 7 reflection; 6 rotation.

10. a. 120°; 240°. **b.** 90°; 180°; 270°.
c. 72°; 144°; 216°; 288°.
d. A regular polygon with n sides can be rotated through 360° ÷ n, and multiples of this angle, and its appearance remains the same.

11. It has reflection symmetries; it can be reflected across a vertical line through its center, a horizontal line through its center, or either of its diagonals, and its appearance remains the same. It has rotation symmetries; it can be rotated though 90°, 180°, or 270°, and its appearance remains the same.

12. It has reflection symmetries; it can be reflected across a vertical line through its center or a horizontal line through its center, and its appearance remains the same. It has rotation symmetry; it can be rotated though 180°, and its appearance remains the same.

13. It has translation symmetry (if imagined to extend in both directions). It also has reflection symmetry (across a horizontal line between the waves).

14. It has rotation symmetry with angles of 360° ÷ 6 = 60° and multiples of 60°. It can also be reflected across many lines through its center and retain its appearance.

15.

16. *(figure)*

17. *(figure)*

18. *(figure)*

19. *(figure)*

20. *(figure)*

21. The angles around a point *P* are precisely the angles that appear inside of a single quadrilateral. Thus, the angles around *P* have a sum of 360°, and the quadrilaterals around *P* fit together perfectly.

22. *(figure)*

(figure)

Unit 8C

1. The longer segment should have a length of 2.47 inches, and the shorter segment should have a length of 1.53 inches.
2. Third and fifth rectangles.
3. 2.43 in. or 0.93 in.
4. 4.05 m or 1.54 m.
5. 10.37 km or 3.95 km.
6. 0.19 cm or 0.074 cm.
9. **a.** $1/\phi = 1/1.62 = 0.62 = \phi - 1$.
 b. $\phi^2 = (1.62)^2 = 2.62 = \phi + 1$.

11.

n	L_n	L_n/L_{n-1}
1	1	—
2	3	3
3	4	1.33
4	7	1.75
5	11	1.57
6	18	1.64
7	29	1.61
8	47	1.62
9	76	1.617
10	123	1.618

CHAPTER 9

Unit 9A

1. Although Tilden narrowly won the popular vote, Hayes narrowly won the electoral vote and became president.
2. Although Harrison narrowly won the popular vote, Cleveland easily won the electoral vote and became president.
3. Garfield narrowly won the popular vote and easily won the electoral vote to become president.
4. Wilson narrowly won the popular vote and narrowly won the electoral vote to become president.
5. Carter narrowly won the popular vote and comfortably won the electoral vote to become president.
6. **a.** Only $57/100 = 57\%$ of the senators would vote to end the filibuster, so it is unlikely that the filibuster will end; the bill is at risk.
 b. A 2/3 vote of a 12-member jury requires at least 8 votes. Thus, there will be no conviction.

c. Only 70% of the states support the amendment, so it fails to pass.
d. The override will get a 70/100 = 70% vote in the Senate and a 260/435 = 60% vote in the House, so the veto will not be overturned.

7. a. The percentage of shareholders favoring the merger is 10,580/15,890 = 66.58%, which is just shy of the required 2/3 = 66.67% vote needed.
b. A 3/4 vote of a 15-member jury requires at least 12 votes; thus, there will be no conviction.
c. Only 64% of the states support the amendment, so it fails to pass.
d. The override will get a 68/100 = 68% vote in the Senate and a 292/435 = 67.1% vote in the House, so the veto can be overturned.

8. a. Wilson won the popular vote by a plurality but not by a majority.
b. Wilson won the electoral vote by a plurality and also by a majority.
c. If Taft had dropped out of the election and Roosevelt had won most of Taft's popular votes, then Roosevelt could have won the popular vote. However, even if Roosevelt had won *all* of Taft's electoral votes, he could not have been elected president.
d. If Roosevelt had dropped out of the election and Taft had won most of Roosevelt's popular votes, then Taft could have won the popular vote. However, even if Taft had won *all* of Roosevelt's electoral votes, he could not have been elected president.

9. a. Best won a plurality but not a majority.
b. Imagine that there were 100 votes cast in the entire election. Then Able would need 18 votes to win 50% of the votes. Eighteen votes represents 18/28 = 64.3% of Crown's votes.

10. a. Fillipo won a plurality but not a majority.
b. Imagine that there were 100 votes cast in the entire election. Then Earnest would need 12 votes to win (a majority). Twelve votes represents 12/21 = 57.1% of Davis's votes.

11. a. Inviglio has a plurality but not a majority.
b. To overtake Inviglio, Height would need 83 votes, which is 83/130 = 63.8% of Grand's votes.

12. a. King has a plurality but not a majority.
b. To overtake King, Lord would need 150 votes, which is 150/225 = 66.7% of Joker's votes.

13.
First	A	A	C	C
Second	B	C	A	B
Third	C	B	B	A
	4	2	3	1

14.
First	A	A	C	C	D
Second	B	C	B	A	A
Third	C	D	A	D	B
Fourth	D	B	D	B	C
	4	2	1	3	2

15. a. A total of 66 votes were cast.
b. D is the plurality winner (but not by a majority).
c. D is the winner of the top-two runoff.
d. The winner by sequential runoff is D.
e. D is the winner by the Borda count.
f. D wins by the pairwise comparisons method.
g. As the winner by all five methods, candidate D is clearly the winner of the election.

16. a. A total of 29 votes were cast.
b. C is the plurality winner (but not by a majority).
c. A is the winner of the top-two runoff.
d. The winner by sequential runoff is A.
e. B is the winner by the Borda count.
f. B wins by the pairwise comparisons method.
g. Candidates A and B both win by two methods, so the outcome is inconclusive.

17. a. A total of 100 votes were cast.
b. C is the plurality winner (but not by a majority).
c. A is the winner of the top-two runoff.
d. The winner by sequential runoff is A.
e. B is the winner by the Borda count.
f. B wins by the pairwise comparisons method.
g. There is no clear winner.

18. a. A total of 30 votes were cast.
b. A, B, and D tie by the plurality method.
c. The top-two runoff method is inconclusive because only one candidate can be eliminated based on fewest first-place votes. The remaining three candidates tie.
d. The sequential runoff method gives a tie among A, B, and D.
e. B is the winner by the Borda count.
f. B wins by the pairwise comparisons method.
g. As the winner by the only two conclusive methods, candidate B is the winner of the election.

19. a. A total of 90 votes were cast.
b. E is the plurality winner (but not by a majority).
c. B is the winner of the top-two runoff.
d. Since only three candidates received first-place votes, the sequential runoff and the top-two runoff methods are the same.
e. E is the winner by the Borda count.
f. By the pairwise comparison method, D is the winner.
g. Candidates B and E each win two of the five methods, so the outcome is debatable.

20. B is the Condorcet winner.

21. 6 pairwise races; 10 pairwise races.
22. 375 points.
23. D must have 130 points, which is enough to win the election.
24. A wins over B, B wins over C, and C wins over A. This curious situation, in which A is preferred to B, who is preferred to C, who is preferred to A, is called the *Condorcet paradox* or *voting paradox*.

Unit 9B
1. 5.27; underrepresented.
2. 12.67; overrepresented.
3. 24.74; overrepresented.
4. 17.57; overrepresented.
5. 51, 36, 13; 52, 36, 13; no paradox.
6. 25, 12, 63; 26, 11, 64; paradox occurs.
7. 46, 9, 4, 41; 47, 9, 4, 41; no paradox.
8. 56, 9, 11, 24; 57, 9, 11, 14; no paradox.
9. 15, 82, 3; 15, 81, 3, 6; state B's representation decreases.
10. 54, 13, 33; 53, 14, 33, 5; state A's representation decreases.
11. 9, 70, 21; quota criterion is violated because state B's standard quota is 68.97.
12. 30, 15, 55; quota criterion is satisfied.
13. 7, 70, 16, 7; quota criterion is satisfied.
14. 13, 54, 26, 7; quota criterion is satisfied.
15. a. 54, 34, 12.
 b. 54, 34, 12; modified divisor 9.90.
 c. 54, 34, 12; modified divisor 9.99.
 d. 54, 34, 12; modified divisor 9.99.
16. a. 15, 45, 40.
 b. 14, 46, 40; modified divisor 9.63.
 c. 15, 45, 40; modified divisor 9.8.
 d. 15, 45, 40; modified divisor 9.8.
17. a. 9, 27, 26, 38.
 b. 8, 27, 26, 39; modified divisor 98.3.
 c. 8, 27, 26, 39; modified divisor 99.5.
 d. 8, 27, 26, 39; modified divisor 99.5.
18. a. 10, 28, 18, 44.
 b. 9, 28, 18, 45; modified divisor 123.5.
 c. 10, 28, 18, 44; modified divisor 125.0.
 d. 10, 28, 18, 44; modified divisor 125.0.
19. a. 1, 3, 6.
 b. 1, 2, 7; modified divisor 35.0.
 c. 1, 3, 6; modified divisor 38.0.
 d. 1, 3, 6; modified divisor 38.0.
20. a. 8, 3, 5.
 b. 8, 3, 5; modified divisor 60.0.
 c. 8, 3, 5; modified divisor 62.0.
 d. 8, 3, 5; modified divisor 62.0.
21. a. 3, 10, 5, 7.
 b. 3, 10, 5, 7; modified divisor 0.76.
 c. 3, 10, 5, 7; modified divisor 0.77.
 d. 3, 10, 5, 7; modified divisor 0.78.
22. a. 4, 3, 2.
 b. 4, 3, 2; modified divisor 1000.
 c. 4, 3, 2; modified divisor 1200.
 d. 4, 3, 2; modified divisor 1220.

INDEX

A

A priori methods, 305
Absolute change, 120, 156, 176
Absolute difference, 122, 156, 176
Absolute error, 156, 176
Absolute zero, 93
Accuracy, 157–159
Acre, 80
Ad hominem, 17
Adams, John, 419
Adams, John Quincy, 421
Adjustable rate mortgage (ARM), 216, 217
Advertising, 14
Alabama paradox, in apportionment, 440–442
Alberti, Leone Battista, 395
Alcoholism, 123
Algebra, 8
 with logarithms, 373
 review of, 189, 200–201
Allais, Maurice, 336
Allen, Fred, 342
Allende meteorite, 381
Analog mode, in music recording, 393
And probability, 316–320, 350. *See also* Probability
And statements, 24–26, 27
Angelou, Maya, 2
Annual percentage rate (APR), 181–182, 185
Annual percentage yield (APY), 185–186, 222
Annual return on investment, 199, 201–202, 222
Annuity, 194
 due, 195
 ordinary, 195
Antarctica, 82, 284–285
Antecedent, 28, 29
Apothecary measures of weight, 87, 88
Appeal to emotion, 17
Appeal to force, 17
Appeal to ignorance, 16
Appeal to majority, 15
Appeal to popularity, 15
Appleton, D. R., 166
Apportionment. *See also* Census; Voting
 best method for, 448–450
 Hamilton's method of, 438–442
 Hill-Huntington method of, 446–448
 Jefferson's method of, 442–445
 problem of, 435–438
 and U.S. Constitution, 434–435
 Webster's method of, 445–446
"Approximately equal to," symbol for, 75
Aquinas, St. Thomas, 409
Aquino, Corazon, 5
Archimedes, 96
Argument, 14
Aristotle, 15, 35
Arithmetic, 8
Arithmetic mean, 131, 290–294, 302, 447
Arrangements with repetition, 337–338, 350
Art, and mathematics, 9, 395–404
Assumptions, 14, 67
Astronomical unit (AU), 143
At least once rule, 332–324, 350
Atomic weight, 364
Atoms, 146
Auto accidents, 297
Availability error, in surveys, 245
Average, 290, 293. *See also* Mean; Median; Mode
Avoirdupois measures of weight, 87, 88

B

Babylonians, 92
Bach, Johann Sebastian, 390
Balinsky, M. L., 449
Balinsky and Young theorem, for apportionment, 450
Bar graphs, 252–254
Barbour, J. Murray, 391
Bartlett, Albert A., 351, 355
Bayes, Thomas, 167
Bayes rule, 167
Beethoven, Ludwig von, 391
Bellow, Saul, 40
Bernoulli, Jacob, 330
Berra, Yogi, 24
Bias, 229–230, 240, 241–243
Bickel, P., 166
Blackmun, Harry, 5
Blinding, 231–233
Blood groups, 49
Blue moon, 313
Borda, Jean-Charles de, 427
Borda count, in voting, 427, 430
British thermal unit (BTU), 95
Brunelleschi, Filippo, 395
Buckley, William F., 18
Budget. *See* Federal budget
Burgess, Anthony, 34
Burns, Anne, 401
Burr, Aaron, 421
Bush, George W., 20, 420, 421, 424
Buyske, Steven G., 5

C

Calculus, 8, 382
Calorie, 95
Caminiti, Ken, 288
Carat, 89
Carbon dioxide emissions, 378–379
Carbon-14, 364
Carlin, George, 157
Carroll, Lewis, 5, 21
Carrying capacity, 357. *See also* Population, world
Case-control studies, 233–234
Cases, 233
Categorical propositions, 67
 negations of, 44–45
 standard form for, 42
 Venn diagrams for, 41–44

I-1

Categories, of data in frequency table, 249, 250
Causality, 281–286
Celsius, Anders, 93
Celsius scale, 93–94
Census, 434, 436, 438, 440. *See also* Apportionment
Center for Voting and Democracy, 451
Chesterton, G. K., 14
China
 carbon dioxide emissions in, 254, 378–379
 population in, 373–374
 population policy in, 110–111
Chlorofluorocarbons (CFCs), 284–285
Circular reasoning (fallacy), 18
Cleveland, Grover, 422
Clinton, Bill, 424
Closing costs, on mortgage, 212
Cole, K. C., 9
Colorado amendment, 54
Combinations, 342–344, 350
Common logs, 367
Compared value, 122
Comparisons, 140–141
Compound interest, 178–179, 361. *See also* Annual percentage yield (APY)
 compounding periods for, 181–188
 continuous, formula for, 186–188, 222
 as exponential growth, 180–181, 372
 general formula for, 180, 222
 paid n times per year, formula for, 183–184, 185, 222
 paid once a year, formula for, 181, 222
Compound proposition, 24
Conclusion (of argument), 14
Conditional propositions, 28–30
Condorcet, Marquis de (Marie Jean Antoine Nicholas de Caritat), 428, 429
Condorcet method, in voting, 428–430
Confidence interval, 234
Confucius, 391
Conjunctions, 24–26, 27, 67
Consequent, 28, 29
Consonant tones, in musical scale, 391
Constitution. *See* U.S. Constitution
Continuous compounding, 186–188, 222
Contour map, 263
Contrapositive, 30, 31
Control group, 230, 231, 233
Convenience sampling, 228
Converse, 30
Conversion
 of currency, 76–79
 metric-USCS, 92–93
 of temperature, 94
 of units, 73–76
Conversion factor, 74
Correlation, 276–281
Correlation coefficient, 279
Counting, with repetition, 337–338
Counting systems, bases for, 92
Credit cards, 210–212
Critical thinking
 defined, 13
 guidelines for, 54–61, 67
Crutzen, Paul, 285
Cullinan diamond, 99
Cumulative frequency, 249, 250
Curie, Marie, 69
Currency, conversion of, 76–79
Cycles per second (cps), 388, 390

D

da Vinci, Leonardo, 397, 398, 399, 411
Data, 224. *See also* Statistics; Variation
 binning of, 251, 252
 distribution of, 290
 geographical patterns in, 262–265
 types of, 250–251
de Maistre, Joseph, 118
de Vries, Jan Vredeman, 397
Debt, 245
Denominator, 72
Density, units of, 96
Dependent events, probability of, 318–320, 350
Desargues, Girard, 407
Dewdney, A. K., 4, 130
Dickerson v. United States, 23
Digital mode, in music recording, 393
Digital signal processing, 393
DiMaggio, Joe, 348
Dinkins, David, 5
Dirksen, Everett, 145
Discounting, 188
Disjoint sets, 39
Disjunctions, 26–27, 67
Distribution, of data set, 290. *See also* Probability distribution
 shape of, 294–297
 variation of, 297–298
Diversion, 18
Divine proportion. *See* Golden ratio
Dodgson, Charles, 5
Doré, Gustave, 400
Dostoyevsky, Fyodor, 331
Double-blind experiment, 231, 232
Doubling, in exponential growth, 353–356, 358, 360
Doubling time, 360–363, 374–375
 approximate formula for, 361–362, 385
 exact formula for, 365–366, 385
Down payment, on mortgage, 212
Drug testing, 170–171
Dürer, Albrecht, 397

E

e (irrational number), 188
Earth, approximate age of, 144
Edward IV (King of England), 178, 179
Effective yield, 185
Either/or probability, 320–322, 350. *See also* Probability
Electoral votes, 421
Ellipsis, 35, 37
Empire State Building, 138
Endowment, 199
Energy, 94, 95, 140
 units of, 94–96, 115
Error
 absolute, 156, 176
 size of, 156–157, 159
 types of, 154–156, 159
Escher, Maurits C., 398
Estimation, 138–140
Euclid, 15
Euler, Leonhard, 38, 188
Euro, 77
European Union, 77
Events, 304. *See also* Probability
 non-overlapping, 320, 350
 overlapping, 321–322
Exams, preparing for, 11
Exclusive *or*, 26
Expected value, 327–330, 350
Experiment, 230, 231, 232
Exponential astonishment, 351
Exponential decay, 363–365
Exponential functions, 371–377, 385
 applications of, 377–381
Exponential growth, 352–358, 385. *See also* Exponential scale; Population, world
 compound interest as, 180–181, 372
 doubling in, 353–356, 358
 musical scales as, 391–392
Exponential scale, 268

F

Factorials, 340
Fahrenheit, Gabriel, 93
Fahrenheit scale, 93–94

Fallacies, 14–18
False cause, 15
False dilemma, 17
False negative, 167–168
False positive, 167–168
Faulkner, William, 40
Fechner, Gustav, 411–412
Federal budget, 151, 245
Fermat, Pierre de, 316, 317
Fermat's Last Theorem, 317
Fibonacci, Leonardo (Leonardo of Pisa), 413, 414
Fibonacci sequence, 413–414, 418
Filibuster, 422, 423
Financial statements, 18
First harmonic, 388
Fixed rate mortgage, 213–215
Flaubert, Gustave, 390
Flooding, 306, 308
Flow chart, 56–58
Fluorescent bulbs, 96
Fourier, Jean Baptiste Joseph, 392
Fractional growth rate, 365
Fractional remainder, for apportionment, 438
Fractions
 common, 72
 decimal form of, 73
French, J. M., 166
Frequency, 249, 250
 of sound, 388, 390
Frequency tables, 249–250
Fujimori, Alberto, 5
Functions, exponential, 371–381, 385
Fundamental frequency, 388
Future value (FV), 180, 195
Fuzzy logic, 29, 66

G

Galileo, 388
Gallup, George, 242
Gambler's fallacy, 330–333
Games of chance, 309
Gardner, Martin, 110
Garfunkel, Art, 5
Gates, Bill, 140
Geometric mean, 447
Geometry, 8, 15
Gerry, Elbridge, 449
Gerrymandering, 449
Ginsburg, Ruth Bader, 286
Global warming, 283–284, 286
Godel, Kurt, 56
Gold, 89, 257

Golden ratio, 409–414, 418
Golden rectangle, 410–412
Golden section. *See* Golden ratio
Goldwater, Barry, 20
Gombaud, Antoine, Chevalier de Mère, 316, 323, 324
Gonzalez, Juan, 288
Gore, Al, 20, 420, 421
Grain (unit of weight), 88
Graphics, 261–270. *See also* Bar graphs; Histograms; Line charts; Statistics, tables and graphs of
Graunt, John, 328
Great Pyramids of Egypt, 411
Greeks, 29, 395, 410
Greenhouse effect, 283
Griffey, Ken, Jr., 288
Grosvenor, Charles H., 293
Gun control, 66

H

Half-life, 363–365, 374–375, 385
 approximate formula for, 365, 385
 exact formula for, 365–366, 385
Half-step, in musical scale, 390
Halifax, George Savile, 306
Halley, Edmund, 328
Hamilton, Alexander, 438
Harrison, Benjamin, 422
Harte, Bret, 332
Hasty generalization, 16
Hayes, Rutherford B., 422, 441
Hemingway, Ernest, 40
Henry I (King of England), 87
Hill, Joseph, 446
Hiroshima, Japan, 365
Histograms, 255–257
Hogarth, William, 398
Hopper, Grace, 5
Horizon line, 396
House edge (in casinos), 332–333
House of Representatives, 434–435, 437, 440. *See also* Apportionment; U.S. Constitution; Voting
Houseman, A. E., 65
Huntington, Edward, 446
Huxley, Aldous, 355

I

If . . . then statements, 28–30
Implications, 28, 29
Inclusive *or*, 26

Independent events, probability of, 317–318, 350
Inference, 226
Infinite series, 106
Inflation, modeling effects of, 377
Installment loan, 206–208
Integers, 36
Internal Revenue Service (IRS), guidelines of, 56–58
Inverse, 30–31
Investments, portfolio of, 257
Irrational numbers, 36

J

Jefferson, Thomas, 34, 90, 421, 440, 442
Jemison, Mae, 5
Johnson, Lyndon, 20
Jones, Chipper, 288
Jordan, Michael, 292
Joule, 95
Joyce, James, 34, 159

K

Karat, 89
Keeling, Dave, 284
Kelvin scale, 93–94
Kepler, Johannes, 409
Keynes, John Maynard, 5
Kilogram, 90
Kilowatt-hour, 95
Knebel, Fletcher, 276
Kurzweil, Ray, 34

L

Landon, Alf, 241–242
Landsteiner, Karl, 49
Language, skills in, 2–3
Larkin, Barry, 288
Law of large numbers, 326–327
Lee, Hsien Loong, 5
Left-skewed distribution, 295–297
Leibniz, Gottfried Wilhelm von, 56
Levenson, Thomas, 392
Libby, Willard, 381
Life expectancy, 123–124
Life insurance, 328
Light-year, 144
Limited choice, 16–17
Lincoln, Abraham, 34
Line charts, 255–257
Linear growth, 352–356, 358, 385

Liter, 90
Literacy, 2
Literary Digest, 241–242
Lloyd, Edward, 328
Loan payment formula, 206–208, 222
Loans, 205–208, 209–210
Logarithmic scales, 268. *See also* Exponential growth
Logarithmic spiral, 413
Logarithms, 367, 373
Logic, 27. *See also* Symbolic logic
 binary, 29
 defined, 8, 14
 fuzzy, 29, 66
Logical connectors, 24, 27
Logically equivalent statements, 31, 67
Long distance calling, 60
Longfellow, Henry Wadsworth, 388
Lotteries, 344
Low birth weight babies, 47–48
Lugar, Richard, 336

M

Majority rule, in voting, 420
 variations on, 422–423
Mammogram, 167–168
Margin of error, 234
Marianas Trench, 90
Markowsky, George, 412, 416
Mars Climate Orbiter, 157
Marx, Groucho, 22
Mathematical model, 9. *See also* Modeling
Mathematics
 branches of, 8
 defined, 7–9
 as a language, 9
 learning, 9
 misconceptions about, 4–7
 people who studied, 5
 skills in, 2–3
Mauna Kea, 99
May, Kenneth, 420
Mean
 arithmetic, 131, 290–294, 302, 447
 geometric, 447
Median, 290–294
Melanoma, 263, 264, 265
Members of a set, 35, 37
Mersenne, Marin, 388
Meter, 90
Metric system, 90–92
 and U.S. customary system, 92–93
Michelangelo, 391
Military services, 35

Mill, John Stuart, 282
Mill's methods, 282
Milner, Florence, 387
Minimum quota, for apportionment, 438
Miranda, Ernesto, 23
Miranda v. State of Arizona, 23
Mode, 290–292
Modeling, 8–9
 exponential, 371–381
Models, 8. *See also* Modeling
Modified divisor, for apportionment, 442–443
Modified quota, for apportionment, 442–443
Molina, Mario, 285
Mondrian, Piet, 411
Monte Carlo method, 308
Moore, Gordon E., 269
Moore's law, 269
Morrison, Toni, 40
Mortgage interest tax deduction, 217
Mortgages, 212–217
Moses, Edwin, 5
Mozart, Wolfgang Amadeus, 416
Mt. Everest, 99
Multiple bar graph, 261
Multiplication principle, 307
Music, and mathematics, 2, 388–393
Mutually exclusive events, 320

N

Nader, Ralph, 421
Nagasaki, Japan, 365
Nash, Ogden, 303
National debt, 245
National Foundation for Credit Counseling, 212
National Institute of Standards and Technology, 90
Natural numbers, 36
Negation, 21–22, 67
 with *all*, *some*, or *none*, 23–24
 of categorical propositions, 44–45
 double, 22–23
Negative correlation, 279
New College of Oxford, 178, 179–181, 182
New states paradox, in apportionment, 441
New value, 120
Newborns, genders of, 110
Nielsen, Arthur C., 225
Nielsen ratings, 224–225, 226, 230, 235
Nightingale, Florence, 5
Nobel Prize
 in chemistry, 285, 381
 in literature, 40
 in medicine, 49
Non-overlapping events, 320, 350

North Pole, 284
"Not equal to," symbol for, 21, 36
Nuclear fission, 141
Nuclear fusion, 141
Number line, 36
Numbers
 measured, combining, 159–160
 sets of, 36
Numerator, 72

O

Observational study, 230, 231
O'Connell, J. W., 166
Octave, 389
Odds, 313–314
Oil, world consumption of, 378
Opinion polls, 234–235
Or statements, 26–27
Order of magnitude estimates, 138–139
Outcome, 304
Outlier, 292–293
Overlapping events, 321–322, 350
Overlapping sets, 39
Overtone, 388
Ozone, 284–285

P

Pairwise comparisons method, 428–430, 432
Pareto, Vilfredo, 253
Pareto chart, 253
Parthenon, 411
Participation bias, 241, 242
Pascal, Blaise, 316, 317, 329
Paulos, John Allen, 7
Penrose, Roger, 404
Percent change, 120
Percent difference, 123
Percentage point, 126
Percentages, 118–124
 abuses of, 129–131
 of vs. *more than* with, 124–125, 126–127
 solving problems involving, 127–129
Perceptual distortion, in graphics, 267
Permutations, 338–341, 350
Perot, Ross, 424
Perry, William, 5
Personal attack, 17, 18
Perspective, in visual arts, 395–398
Philip Morris Company, 240
Phydias, 409
Physiological processes, exponential modeling of, 379–380
Pi (π), 36

Picasso, Pablo, 330
Pickney, Charles, 421
Pictographs, 270
Pie charts, 252–254
Pilon, Juliana Geran, 13
Pitch, of sound, 388
Placebo, 231
Placebo effect, 231–233
Plurality, of votes, 424, 425, 430
Plutonium-239, 364, 365, 369
Point system, in voting, 427–428, 430
Points, on mortgage, 212–213
Polar ice caps, 81–82
Polya, George, 101
Polygraphs, 164, 168–170
Popular vote, 421
Population
 statistical, 224, 225, 226
 world, 121, 351, 357, 361
Population paradox, in apportionment, 441
Population parameter, 226, 227
Positive correlation, 279
Possible cause, 286
Pound, abbreviation for, 80
Power, units of, 94, 95
Powers, 91, 200–201
 in scientific notation, 136
Precision, 157–159
Predicate set, 41
Preference schedule, in voting, 425–429
Premature babies, 48
Premises, 14, 67
 hidden, 55–56
Present value (PV), 180, 197
Principal
 of an investment, 178–179
 of a loan, 205, 222
Principal vanishing point, 396
Probability, 8, 304–305. *See also And* probability; Combinations; Dependent events, probability of; *Either/or* probability; Independent events, probability of; Permutations
 and coincidence, 344–346
 conditional, 319
 and counting, 337–346
 empirical, 306–308, 309
 expressing, 305, 312–313
 of nonoccurrence of event, 309–310
 subjective, 309
 theoretical, 305–306, 309, 350
Probability distribution, 310–312
Probable cause, 286
Problem solving, guidelines and hints for, 101–111

Proctor, Richard, 304
Proportion, in visual art, 409–414
Propositions, 21
Pulitzer Prize, 40
Push polls, 234
Putz, John, 416
Pythagoras, 388, 409

Q

Qualitative data, 250
Quantitative data, 250
Quantitative information, 2
Quantitative literacy, 2
Quantitative reasoning, 2–4
Quota criterion, in apportionment, 444–445

R

Radiation, 22–23
Radioactive dating, 380–381
Radon gas, 244
Random error, 154
Rational numbers, 36
Ratios, 125
Reagan, Ronald, 19
Real numbers, 36
Reasonable doubt, 286
Reciprocal, 72–73
Red herring, 18
Redistricting, 449. *See also* Apportionment; Voting
Reference value, 120, 122
 shifting, 129–130
Reflection symmetry, 399
Rehnquist, William, 23
Relative change, 120, 156, 176
Relative difference, 122, 156, 176
Relative error, 156, 176
Relative frequency, 249, 250
Relative frequency method, of calculating probability, 308
Renaissance, painting during, 395, 396
Repeating digits, 73
Representative sample, 227
Resource depletion, exponential model of, 378–379
Ride, Sally, 5
Right-skewed distribution, 295–297
Robinson, David, 5
Rodriguez, Ivan, 288
Roosevelt, Franklin, 241–242, 447
Roots, 200–201
Rotation symmetry, 399
Rounding numbers, 153

Rowland, F. Sherwood, 285
Rule of 70, 362, 365
Runoff election, 424, 425–427, 430

S

Sagan, Carl, 246
Sample, 225, 226
 bias in, 241–243
 choosing, 227–230
Sample statistics, 225, 226
Santayana, George, 34
Savings plan, 194, 197–199
Savings plan formula, 194–197, 222
Scale, musical, 390–391
Scaling, 142–144
Scatter diagram, 277–280
Scheiber, Anne, 204
Scientific notation, 136–137, 140, 152
Seaton, C. W., 440
Second, 90
Selection bias, 241–243
Self-selection, by survey participants, 243
Senate, 434, 437, 440. *See also* Apportionment; U.S. Constitution; Voting
Sequential runoff, 426–427, 430
Sets, 35, 37
 of numbers, 36, 41
 predicate, 41
 relationships among, 35–38
 subject, 41
Seurat, Georges, 411
Shelley, Percy Bysshe, 147
Significant digits, 151–153
Simon, Pierre, Marquis de Laplace, 308
Simple interest, 178–179, 222
Simple random sample, 228
Simpson, Edward, 165
Simpson's paradox, 165, 166
Single-blind experiment, 231, 232
Single runoff method, 425–426, 430
Skills, language and mathematics, 2–3
Smoking, 240, 241, 277
Social Security, 245
Solzhenitsyn, Alexander, 5
Sosa, Sammy, 288
Sound, and music, 388–389
Spock, Benjamin, 319
Stack plot, 262
Standard divisor, for apportionment, 436
Standard quota, for apportionment, 436
Star of Africa, 99
Statistical studies
 basic steps in, 226–230
 guidelines for evaluating, 239–247

Statistical studies
 types of, 230–235
 variables of interest in, 243–244
Statistics, 8
 defined, 224
 descriptive, 226
 inferential, 226
 sample, 225, 226
 tables and graphs of, 249–257
Steinbeck, John, 40
Stockman, David, 151
Stoker, Bram, 5
Stradivarius, Antonius, 392
Stratified sampling, 228
Straw man, 18
Studying, strategies for, 11
Subject set, 41
Subjective estimate, of probability, 309
Subset, 38, 39
Super majority, in voting, 422
Surveys, 234–235
 problems with, 244–245
Symbolic logic, 21, 24, 26, 28, 56
Symmetric distribution, 295–297
Symmetry, in visual arts, 398–401
Synthesizer, musical, 393
Systematic error, 154
Systematic sampling, 228

T

Tables
 frequency, 249–250
 truth, 21–22
Tax cuts, 170–171
Teenagers' allowances, 291
Temperature, units of, 93–94
Tesselations. *See* Tilings
Thomson, William, Lord Kelvin, 93
Tick marks, 253
Tilden, Samuel J., 422, 441
Tilings, 402–405
Time, budgeting, 10, 11
Time-series diagrams, 256
Tobacco Research Institute, 241
Top-two runoff method, 425–426, 430

Total return on investment, 199–200, 202, 222
Translation symmetry, 399
Treatment group, 230, 231
Treaty of the Meter, 90
Tribe, Laurence, 5
Trigonometry, 8
Troy measures of weight, 87, 88
True negative, 167–168
True positive, 167–168
Truman, Harry S, 127
Truth table, 21, 22
Truth value, 21, 22
Tuberculosis, 262
Tukey, John, 263

U

United Nations, 242
United States
 carbon dioxide emissions in, 378–379
 presidential elections in, 421–422
Units, 70–72
 conversion of, 73–76, 92–93
 of energy and power, 94–96, 115
 problem solving with, 79–82
 raised to powers, 75–76
 standardized, 87–93
 of temperature, 93–94
University of California, Berkeley, 165–166
Urban heat island effect, 155
U.S. Census Bureau, 155, 158, 370. *See also* Census
U.S. Coast Guard, 35
U.S. Constitution, 66, 155, 421, 422, 440, 449
 and apportionment, 434–435
 2nd Amendment to, 66
 12th Amendment to, 421
U.S. customary system (USCS), 87–90
 and metric system, 92–93
U.S. Department of Defense, 35
U.S. Supreme Court, 23, 54

V

Vanderpump, M. P., 166
Variables, 243–244

explanatory vs. response, 278
Variation, of a distribution, 297–298
Vasarely, Victor, 401
Vaughn, Mo, 288
Venn, John, 38
Venn diagrams, 38–41, 47–50, 56
 for categorical propositions, 41–45
 with three sets, 45–47
Veto power, 422
Vinton, Samuel, 440
Visual aids, 56–58
Visual arts, mathematics and, 9, 395–404
Voting. *See also* Apportionment; Redistricting
 with three or more choices, 423–430
 with two choices, 420–423

W

Wade, Virginia, 5
Walker, Larry, 288
Washington, George, 440, 445
Watt, 95
Wave, sound, 388
Webster, Daniel, 445
Weight, measurement systems for, 87, 88
Wells, H. G., 223
White dwarf, 150
Whole numbers, 36
Whorf, Tim, 284
Wilde, Oscar, 281
Wing, Steve, 22, 23
World population, 121, 351, 357, 361

Y

Young, H. P., 449
Yule, George, 165

Z

Zeno's paradox, 106